ADAPTIVE SIGNAL PROCESSING

PRENTICE-HALL SIGNAL PROCESSING SERIES

Alan V. Oppenheim, Editor

Andrews and Hunt *Digital Image Restoration*
Brigham *The Fast Fourier Transform*
Burdic *Underwater Acoustic System Analysis*
Castleman *Digital Image Processing*
Cowan and Grant *Adaptive Filters*
Crochiere and Rabiner *Multirate Digital Signal Processing*
Dudgeon and Mersereau *Multi-Dimensional Digital Signal Processing*
Hamming *Digital Filters, 2e*
Haykin, et al. *Array Signal Processing*
Jayant and Noll *Digital Coding of Waveforms*
Lea, ed. *Trends in Speech Recognition*
Lim *Speech Enhancement*
McClellan and Rader *Number Theory in Digital Signal Processing*
Oppenheim, ed. *Applications of Digital Signal Processing*
Oppenheim and Schafer *Digital Signal Processing*
Oppenheim, Willsky, with Young *Signals and Systems*
Rabiner and Gold *Theory and Application of Digital Signal Processing*
Rabiner and Schafer *Digital Processing of Speech Signals*
Robinson and Treitel *Geophysical Signal Analysis*
Tribolet *Seismic Applications of Homomorphic Signal Processing*
Widrow and Stearns *Adaptive Signal Processing*

ADAPTIVE SIGNAL PROCESSING

Bernard Widrow
Stanford University

Samuel D. Stearns
Sandia National Laboratories

Prentice-Hall, Inc.
Englewood Cliffs, N.J. 07632

Library of Congress Cataloging in Publication Data

Widrow, Bernard, (date)
Adaptive signal processing.

(Prentice-Hall signal processing series)
Includes index.
1. Adaptive signal processing. I. Stearns, Samuel D. II. Title. III. Series.
TK5102.5.W537 1985 621.38′043 84-18057
ISBN 0-13-004029-0

Editorial/production supervision and interior design: *Tracey L. Orbine*
Cover design: *Sue Behnke*
Manufacturing buyer: *Anthony Caruso*

Printed in the United States of America

10 9 8 7

ISBN 0-13-004029 01

Prentice-Hall International, Inc., *London*
Prentice-Hall of Australia Pty. Limited, *Sydney*
Editora Prentice-Hall do Brasil, Ltda., *Rio de Janeiro*
Prentice-Hall Canada Inc., *Toronto*
Prentice-Hall Hispanoamericana, S. A., *Mexico*
Prentice-Hall of India Private Limited, *New Delhi*
Prentice-Hall of Japan, Inc., *Tokyo*
Prentice-Hall of Southeast Asia Pte. Ltd., *Singapore*
Whitehall Books Limited, *Wellington*, *New Zealand*

This book is dedicated to the memory of Moses Widrow, William K. Linvill, and Thomas J. Flanagan. It is also dedicated to the cause of peace on earth. We hope and trust that its contents will be used to improve the lot of mankind everywhere.

Contents

Preface

This book has grown out of nearly three decades of research and teaching in the field of adaptive signal processing. It is designed primarily to be a basic text on adaptive signal processing and, at the time of its publication, it is believed to be the only basic text on the subject, or at least the only textbook covering the breadth of subject matter shown in the table of contents.

The book is based on class notes for a one– or two–semester senior or graduate level course in adaptive signal processing taught at Stanford University, the University of New Mexico, and Sandia National Laboratories. Every chapter except Chapter 1 has exercises at the end, and these are considered to be an essential part of any course using the text. The exercises are often used to complete the reader's understanding of a concept or to present different applications of ideas in the text.

Referring to the table of contents, the reader can see that the book is divided into four main parts. The first three parts—General Introduction, Theory of Adaptation with Stationary Signals, and Adaptive Algorithms and Structures—make up a little less than half of the text. The material in these parts is considered basic theory and would normally be included in any first course on adaptive signal processing. The fourth part—Applications—consists of six chapters on various engineering applications of adaptive signal processing. In this part the instructor may wish to concentrate on subjects of special interest. However, even in a one–semester course, the instructor will probably wish to include at least the first portion of each chapter.

For prerequisites, we assume that the student has at least senior-level academic experience in engineering and mathematics, and has the ability to write and run computer programs. The latter is essential for doing many of the exercises. A course in linear systems analysis, particularly in discrete systems with the use of the z-transform, would provide a very useful (if not essential) background. Also, a course in engineering statistics or probability, or the equivalent, provides a helpful background.

In the first part of the text, Chapter 1 introduces the concept of adaptation as a property or characteristic of certain systems in engineering. Chapter 2 introduces the adaptive linear combiner, which is the simplest and most widely used adaptive structure. Chapter 2 also describes a geometric "performance surface" which is useful in the analysis of all adaptive systems.

Part II, Theory of Adaptation with Stationary Signals, contains an analysis of the performance surface and its properties. The analysis begins in Chapter 3, and in Chapter 4 adaptation is viewed as the process of searching the performance surface for its minimum. Chapter 5 contains a statistical analysis of gradient estimation on the performance surface and a comparison of search methods.

In Part III, Adaptive Algorithms and Structures, the least mean squares (LMS) algorithm is introduced and discussed in Chapter 6. In Chapter 7 basic signal processing concepts that are required for the rest of the book are introduced. These include primarily the z-transform relationships linking the time and frequency domains. To conclude Part III, Chapter 8 introduces adaptive algorithms other than the LMS algorithm and adaptive structures other than the adaptive linear combiner, including the adaptive lattice structure. The latter is considered, at the time of this writing, to be a rapidly developing area, and our introduction to it is therefore less comprehensive than we would wish it to be.

Finally, Part IV covers the major application areas of adaptive signal processing. Once the basics in Chapters 1–8 have been learned, subjects can be chosen selectively from Part IV. In Chapters 9 and 10, forward and inverse adaptive modeling are introduced and applied to areas such as multipath communication, geophysical exploration, digital filter design, and telephone channel equalization. Adaptive control systems are introduced in Chapter 11, and Chapter 12 introduces adaptive interference canceling, with several examples of application. Chapters 13 and 14 cover adaptive arrays and beamformers.

While writing this text, the authors have had the benefit of critiques, comments, and suggestions from many talented colleagues. We are very grateful for the reviews and ideas we have received, and thankful for the friendships engendered and increased through this work. We especially wish to acknowledge the help of Robert D. Fraser, Dennis R. Morgan, Dae H. Youn, Eugene Walach, Richard Gooch, Ruth A. David, Sharon K. Fletcher, Claude S. Lindquist, Daksheesh Parikh, Delores M. Etter, Edward S. Angel, Lloyd J. Griffiths, Nasir Ahmed, John R. Treichler, C. Richard Johnson, Jr., Michael G. Larimore, Glenn R. Elliott, John M. McCool, John M. Cioffi, and T. C. Hsia. The book would not be in its present form without the contributions of these special friends.

We also wish to thank all of the students who took the adaptive signal processing courses mentioned above. In effect, they have edited and corrected the text far beyond our ability to do so. We thank all of these students for their patience, interest, and enthusiasm.

The only ones with more patience and perseverance than our students have been the talented ladies who have typed and retyped this text, Debra Shepperd at Sandia and Mieko Parker at Stanford. We also acknowledge their help with gratitude.

Bernard Widrow
Samuel D. Stearns

List of Symbols

SYMBOL	USE(S) IN THIS BOOK
a	(1) forward weight in a linear filter (2) bit in genetic algorithm
b	(1) recursive weight in a linear filter (2) bit in genetic algorithm
c	(1) plant output signal (2) signal propagation velocity
d	(1) desired response (2) antenna element spacing
e	natural logarithmic base, 2.71828...
$f(\ )$	continuous function of
g	plant output signal
h	impulse response
j	$\sqrt{-1}$
k	sample number
l	(1) weight number (2) element spacing
n	(1) general index (2) noise sample value
p	total white input noise power
r	(1) convergence ratio in gradient search algorithms (2) uniform random number in $(0, 1)$ (3) reference input signal
s	(1) signal in lattice filter (2) input signal
t	continuous time

u	(1) inverse of z (2) plant input signal
v	translated weight, $w - w^*$
v'	weight value in principal axis coordinate system
w	weight value
x	input signal
y	output signal
z	variable in the z-transform
z^{-1}	inverse of z (unit delay)
A	(1) z-transform of a (2) amplitude gain
B	z-transform of b
C	(1) function used in lattice conversion (2) constant signal amplitude
D	signal distortion
$E[\ ]$	expected (mean, average) value of
F	transfer function
G	transfer function
H	transfer function
$\mathbf{I}$	the identity matrix, diag [1 1 1 . . . 1]
J	(1) transfer function (2) jamming signal
K	number of beamforming elements
L	index of the last filter weight, w_L
M	(1) misadjustment (2) number of feedback weights
N	(1) number of samples per cycle (2) number of error samples taken with perturbed weights (3) number of discrete frequencies
$\mathbf{N}$	gradient noise, $\hat{\nabla} - \nabla$
$\mathbf{N}'$	$\mathbf{N}$ in principal axis coordinate system
P	(1) perturbation due to derivative measurement (2) estimated signal power (3) plant transfer function
$\mathbf{P}$	correlation vector of input and desired signals
PS	transfer function of pseudofilter
Q	filter quality factor
$\mathbf{Q}$	(1) eigenvector matrix of $\mathbf{R}$ (2) scaled estimated $\mathbf{R}$ matrix

$\mathbf{R}$	correlation matrix of input signal, x
$\mathbf{S}$	(1)matrix used in SER algorithm (2) signal vector in adaptive arrays
T	(1) transpose of a vector or matrix (2) time step between samples in seconds
T	time constant of adaptation
$\mathbf{U}$	augmented signal vector
$\mathbf{V}$	translated weight vector, $\mathbf{W} - \mathbf{W}^*$
$\mathbf{V}'$	weight vector in principal axis system
$\mathbf{W}$	weight vector
X	z-transform of x
$\mathbf{X}$	input signal vector
Y	z-transform of y
$Z^{-1}[\ ]$	inverse z-transform of
α	(1) exponential decay constant (2) output signal derivative (3) forgetting factor in SER and lattice algorithms
α_r	rth moment of ε_k
$\hat{\alpha}_r$	estimate of α_r
β	(1) output signal derivative (2) adjustable gain factor
γ	(1) performance penalty (2) leakage factor
δ	(1) small perturbation in a weight value (2) translated lattice filter weight (3) beam-steering delay
ε, ϵ	error signal
κ	lattice filter weight
λ	(1) eigenvalue (2) wavelength
μ	convergence parameter in gradient search algorithms
ν	(1) convergence parameter (2) lattice filter weight
ν^2	input noise power
ξ	mean-square error (MSE) performance function
$\hat{\xi}$	estimate of ξ
π	3.14159265...
ρ	signal-to-noise ratio
σ^2	signal variance or power

τ	time constant of weight convergence
ϕ	(1) average random signal power (2) correlation function
ω	angular frequency in rad (sampling freq. = 2π)
Δ	delay value
Θ	phase angle (rad)
θ	signal arrival angle
Λ	eigenvalue matrix, diag $[\lambda_0 \lambda_1 \ldots \lambda_L]$
Φ	power density (z-transform of ϕ)
ψ	(1) input signal (2) signal arrival angle
Ω	angular frequency in rad/s (sampling freq. = $2\pi/T$)
*	denotes an optimal value, as in $\mathbf{W}^*$
∇	gradient vector of the performance function
$\hat{\nabla}$	estimate of ∇

part I

GENERAL INTRODUCTION

(Chapters 1 and 2)

OBJECTIVES OF PART I

In the first two chapters of this book we have three major objectives. The first is to introduce the basic meaning of "adaptation" (or "adaption") in the engineering sense, and to set adaptive signal processing into the general signal processing context.

The second objective is to describe the adaptive linear combiner, which is the simplest and most widely applicable adaptive processor. It is the basic adaptive device that will be used exclusively through Chapter 6, as well as in much of the rest of the text.

The third objective is to persuade the reader to think of the overall process of adaptation in geometrical terms. We wish to think of adaptation as a procedure for moving generally downhill on a "performance surface" like the one shown on page 2, which is the L-dimensional surface in $(L + 1)$-dimensional space formed by plotting the mean-square error versus the adaptive parameters. These geometrical concepts and terms are described in Chapter 2.

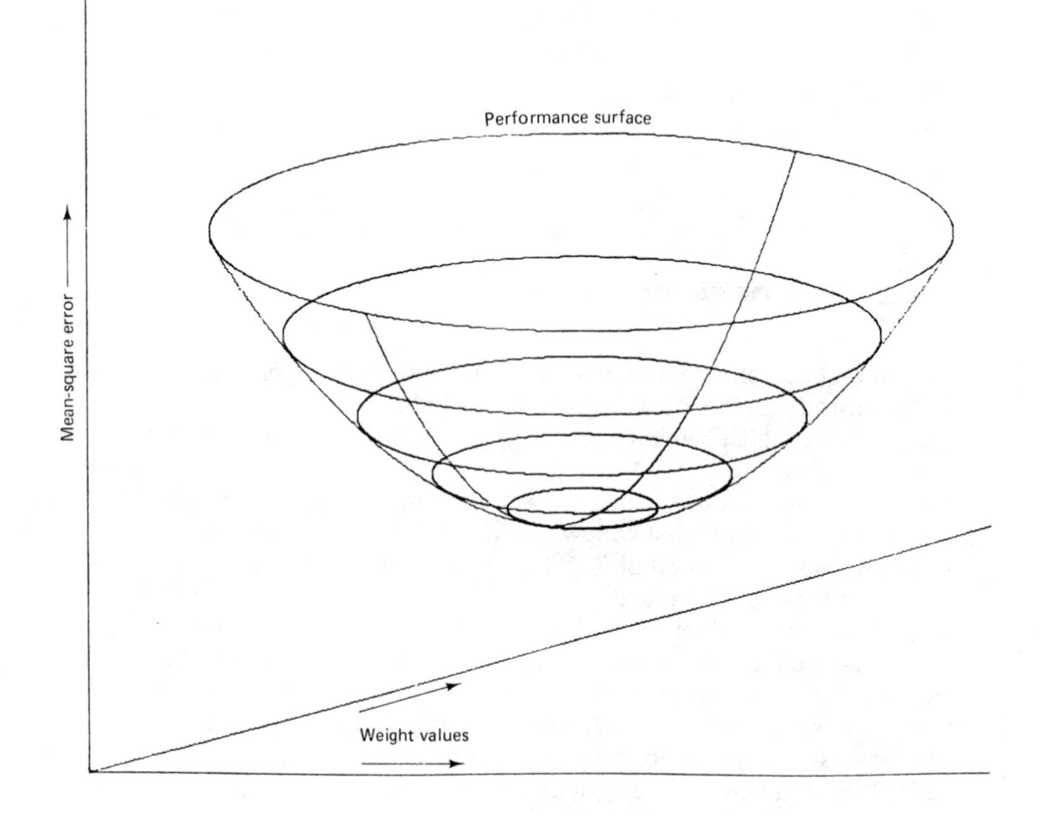
Performance surface
Mean-square error
Weight values

chapter 1

Adaptive Systems

DEFINITION AND CHARACTERISTICS

> adapt, v.t., 1. to make suitable to requirements or conditions; adjust or modify fittingly; ... v.i., 2. to adjust oneself to different conditions, environments, etc. (*Random House Dictionary*, 1971)

In recent years, a growing field of research in "adaptive systems" has resulted in a variety of adaptive automatons whose characteristics in limited ways resemble certain characteristics of living systems and biological adaptive processes. According to the *Random House Dictionary*, some of the meanings of "adaptation" are:

> 1. the act of adapting. 2. the state of being adapted; adjustment. 3. *Biol.* a. any alteration in the structure or function of an organism or any of its parts that results from natural selection and by which the organism becomes better fitted to survive and multiply in its environment. b. a form or structure modified to fit changed environment. 5. *Physiol.* the decrease in response of sensory receptor organs, as those of vision, touch, temperature, olfaction, audition, and pain, to changed, constantly applied, environmental conditions. 6. *Ophthalm.* the regulating by the pupil of the quantity of light entering the eye. 7. *Sociol.* a slow, usually unconscious modification of individual and social activity in adjustment to cultural surroundings.

It will be noted that the definition above is expressed primarily in terms of biological adaptation to environment. The same definitions serve at least to some extent for "artificial" or human-made adaptive systems, which are the central concern of this book.

An adaptive automaton is a system whose structure is alterable or adjustable in such a way that its behavior or performance (according to some desired criterion) improves through contact with its environment. A simple example of an automaton or automatic adaptive system is the automatic gain control (AGC) used in radio and television receivers. The function of this circuit is to adjust the sensitivity of the receiver inversely as the average incoming signal strength. The receiver is thus able to adapt to a wide range of input levels and to produce a much narrower range of output intensities.

The purpose of this book is to present certain basic principles of adaptation; to explain the design, operating characteristics, and applications of the simpler forms of adaptive systems; and to describe means for their physical realization. The types of systems discussed include those designed primarily for the purposes of adaptive control and adaptive signal processing. Such systems usually have some or all of the following characteristics:

1. They can automatically adapt (self-optimize) in the face of changing (nonstationary) environments and changing system requirements.
2. They can be trained to perform specific filtering and decision-making tasks. Synthesis of systems having these capabilities can be accomplished automatically through training. In a sense, adaptive systems can be "programmed" by a training process.
3. Because of the above, adaptive systems do not require the elaborate synthesis procedures usually needed for nonadaptive systems. Instead, they tend to be "self-designing."
4. They can extrapolate a model of behavior to deal with new situations after having been trained on a finite and often small number of training signals or patterns.
5. To a limited extent, they can repair themselves; that is, they can adapt around certain kinds of internal defects.
6. They can usually be described as nonlinear systems with time-varying parameters.
7. Usually, they are more complex and difficult to analyze than nonadaptive systems, but they offer the possibility of substantially increased system performance when input signal characteristics are unknown or time varying.

AREAS OF APPLICATION

Recent progress in microcircuit design and production has resulted in very compact, economical, and reliable signal processors that rival biological nervous systems in size and are clearly superior to biological systems in speed. The result has been a very fast-growing field of applications for all types of digital signal processing, including adaptive processing. Current applications for adaptive systems are in such fields as communications, radar, sonar, seismology, mechanical design, navigation systems, and biomedical electronics.

Part IV of this book concerns certain classes of applications, and the chapter headings provide a rough picture of the application areas. Chapter 9 covers adaptive modeling and system identification, in which an adaptive system models an unknown system or "plant," which may be slowly varying with time. For a given input signal, the adaptive system tries to match the plant output. Adaptive modeling is used in electrical and mechanical design and testing.

Chapter 10 is on inverse modeling, deconvolution, and equalization. All of these terms, as well as the term "inverse filtering," refer to the process of removing the effect of some device or medium on a signal. For example, we may wish to make an audio system respond with equal gain to all speech frequencies, or to remove the effects of a transmission line on a radar pulse.

Chapter 11 is on adaptive control systems, that is, control systems whose characteristics change with time and adapt to the environment. An example is the flight control system whose gain and response times change with air density. Adaptive noise canceling, discussed in Chapter 12, has been applied to areas such as speech communications, electrocardiography, and seismic signal processing. Adaptive noise canceling is in fact applicable to a wide variety of signal-enhancement situations, because noise characteristics are not often stationary in real-world situations. Chapters 13 and 14, on adaptive arrays, describe an area where adaptive signal processing concepts have proved to be especially useful.

GENERAL PROPERTIES

The essential and principal property of the adaptive system is its time-varying, self-adjusting performance. The need for such performance may readily be seen by realizing that if a designer develops a system of *fixed design* which he or she considers optimal, the implications are that the designer has foreseen all possible input conditions, at least statistically, and knows what he or she would like the system to do under each of these conditions. The designer has then chosen a specific criterion whereby performance is to be judged, such as the amount of error between the output of the actual system and that of some selected model or "ideal" system. Finally, the designer has chosen the system that appears best according to the performance criterion selected, generally choosing this system from an a priori restricted class of designs (such as linear systems).

In many instances, however, the complete range of input conditions may not be known exactly, or even statistically; or the conditions may change from time to time. In such circumstances, an adaptive system that continually seeks the optimum within an allowed class of possibilities, using an orderly search process, would give superior performance compared with a system of fixed design.

By their very nature, adaptive systems must be time varying and nonlinear. Their characteristics depend, among other things, on their input signals. If an input signal x_1 is applied, an adaptive system will adapt to it and produce an output—let us call it y_1. If another input signal, x_2, is applied, the system will adapt to this second signal and will again produce an output—let us call it, this time, y_2. Generally, the form or the structure or the adjustments of the adaptive system will be different for the two different inputs. If the sum of the two inputs is applied to

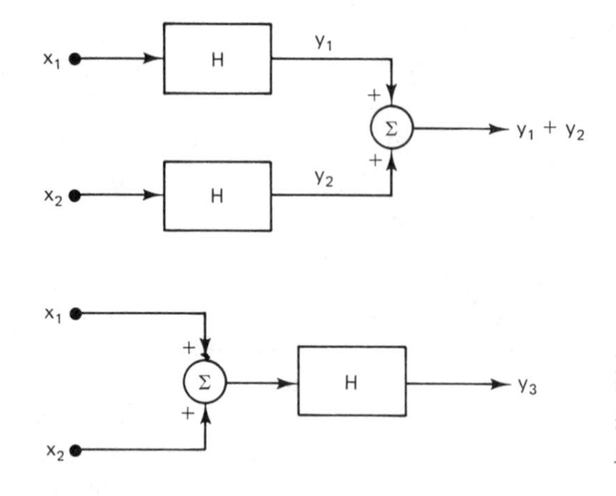

Figure 1.1 The lower output, y_3, is equal to $y_1 + y_2$ if H is a linear system. If H is adaptive, y_3 is generally different from $y_1 + y_2$.

the adaptive system, the latter will adapt to this new input—but it will produce an output that will generally not be the same as $y_1 + y_2$, the sum of the outputs that would have corresponded to inputs x_1 and x_2. In such a case, as illustrated in Figure 1.1, the principle of superposition does not work as it does with linear systems. If a signal is applied to the input of an adaptive system to test its response characteristics, the system adapts to this specific input and thereby changes its own form. Thus the adaptive system is inherently difficult to characterize in conventional terms.

Within the realm of nonlinear systems, adaptive systems cannot be distinguished as belonging to an absolutely clear subset. However, they have two features that generally distinguish them from other forms of nonlinear systems. First, adaptive systems are adjustable, and their adjustments usually depend on finite-time average signal characteristics rather than on instantaneous values of signals or instantaneous values of the internal system states. Second, the adjustments of adaptive systems are changed purposefully in order to optimize specified performance measures.

Certain forms of adaptive systems become linear systems when their adjustments are held constant after adaptation. These may be called "linear adaptive systems." They are very useful; they tend to be mathematically tractable; and they are generally easier to design than other forms of adaptive systems.

OPEN- AND CLOSED-LOOP ADAPTATION

Several ways to classify adaptive schemes have been proposed in the literature. It is most convenient here to begin by thinking in terms of *open-loop* and *closed-loop* adaptation. The open-loop adaptive process involves making measurements of input or environmental characteristics, applying this information to a formula or to a computational algorithm, and using the results to set the adjustments of the adaptive system. Closed-loop adaptation, on the other hand, involves automatic experimentation with these adjustments and knowledge of their outcome in order to optimize a measured system performance. The latter process may be called adaptation by "performance feedback."

The principles of open- and closed-loop adaptation are illustrated in Figures 1.2 and 1.3. In both cases it is helpful to envisage the adaptive process as it might be performed manually by a human operator or "supervisor." In Figures 1.2(a) and 1.3(a) the supervisor is shown adjusting the controls of the processor by reading a display that registers measurements of the preselected performance criterion. In the open-loop system this criterion is a characteristic of the input signal and perhaps other data, and in the closed-loop system it is also a function of the output signal. The adjustments in Figure 1.3 are made even though the operator may have no knowledge of what is inside the processor or of the functions performed by the controls. The operator does not process the input signal; he or she only controls the adjustments of the processor to keep its performance optimized according to the preselected criterion. Thus the operator's function is purely supervisory, and in real automatic adaptive systems the operator is replaced by computational or "adaptive" algorithms, as suggested by Figures 1.2(b) and 1.3(b). The "other data" in these figures may be data about the environment of the adaptive system, or in the closed-loop case, it may be a desired version of the output signal.

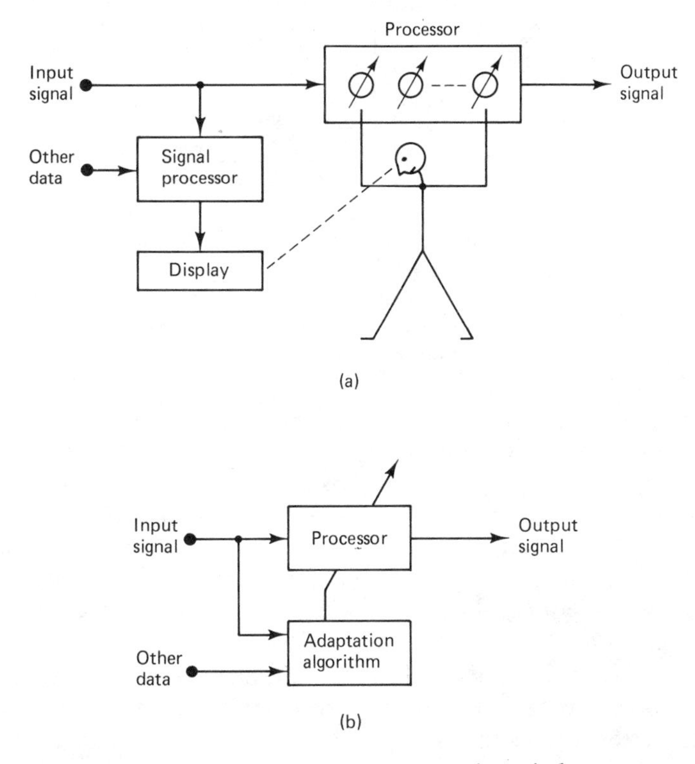

Figure 1.2 Open-loop adaptation: (a) concept; (b) equivalent system.

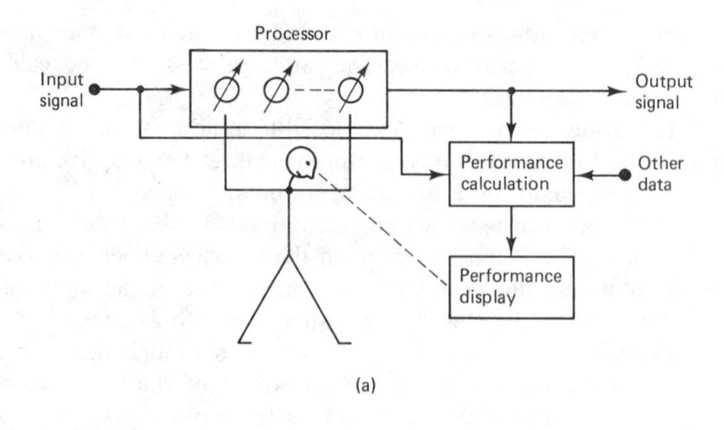

(a)

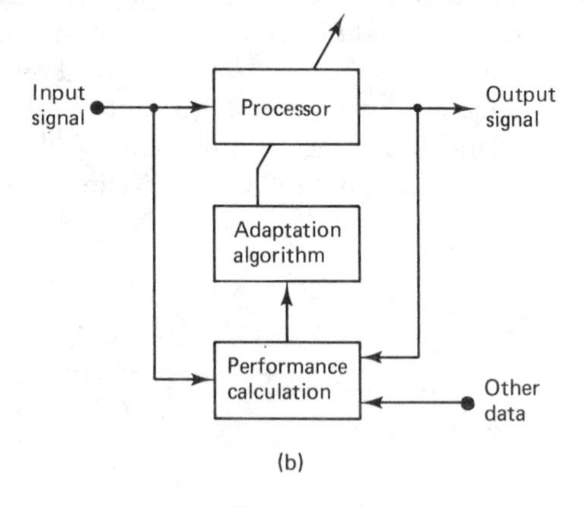

(b)

Figure 1.3 Closed-loop adaptation.

When designing an adaptive process, many factors determine the choice of closed-loop versus open-loop adaptation. The availability of input signals and performance-indicating signals is a major consideration. Also, the amount of computing capacity and the type of computer required to implement the open-loop and closed-loop adaptation algorithms will generally differ. Certain algorithms require the use of a general-purpose digital computer, whereas other algorithms could be implemented far more economically with special-purpose chips or other apparatus. Some of these structural considerations are discussed in later chapters. It is difficult to develop general principles to guide all choices, but several advantages and a few disadvantages of closed-loop adaptation, which is the main subject of this book, can be pointed out here.

Closed-loop adaptation has the advantage of being workable in many applications where no analytic synthesis procedure either exists or is known; for example, where error criteria other than mean-square are used, where systems are nonlinear or time variable, where signals are nonstationary, and so on. Closed-loop adaptation can also be used effectively in situations where physical system component values are variable or inaccurately known. Closed-loop adaptation will find the best choice of component values. In the event of partial system failure, an adaptation mechanism that continually monitors performance will optimize this performance by adjusting and reoptimizing the intact parts. As a result, system reliability can often be improved by the use of performance feedback.

The closed-loop adaptation process is not always free of difficulties, however. In certain situations, performance functions do not have unique optima. Automatic optimization is an uncertain process in such situations. In other situations, the closed-loop adaptation process, like a closed-loop control system, could be unstable. The adaptation process could diverge rather than converge. In spite of these possibilities, performance feedback is a powerful, widely applicable technique for implementing adaptation. Most of the adaptive processes described in this book will be closed-loop processes utilizing performance feedback.

APPLICATIONS OF CLOSED-LOOP ADAPTATION

Let us now consider briefly some applications of the closed-loop, performance feedback concept. These will anticipate the applications chapters in Part IV, beginning with Chapter 9.

We begin by representing the performance feedback process [Figure 1.3(b)] more specifically in Figure 1.4. We call the input signal x and define a "desired response" signal d, which is assumed to represent the desired output of the adaptive system. The signal d is, for our purpose here, the "other data" in Figure 1.3(b).

The error signal, ε, is the difference between the desired output signal and the actual output signal, y, of the adaptive system. Using the error signal, an adaptive algorithm adjusts the structure of the adaptive system, thus altering its response characteristics by minimizing some measure of the error, thereby closing the performance loop.

Different structures are possible for the adaptive system, and these are discussed later, beginning in Chapter 2. Adaptive algorithms for adjusting these structures are also discussed later, beginning in Chapter 4. Here we wish only to show in a general way how the scheme in Figure 1.4 is applied in practical situations.

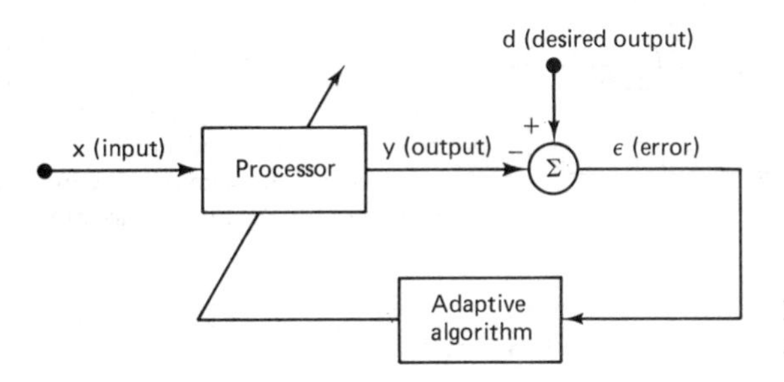

Figure 1.4 Signals in closed-loop adaptation.

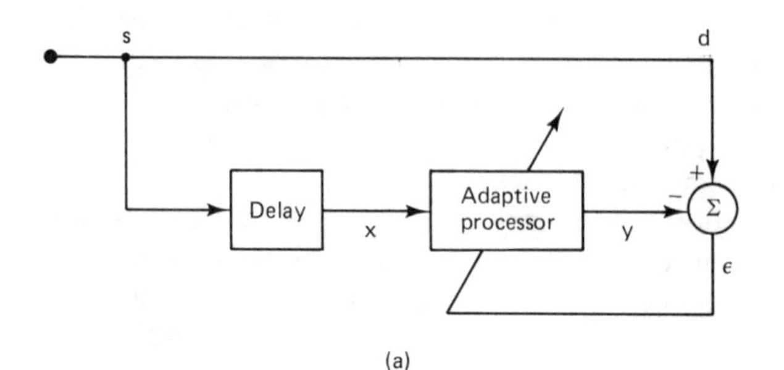

(a)

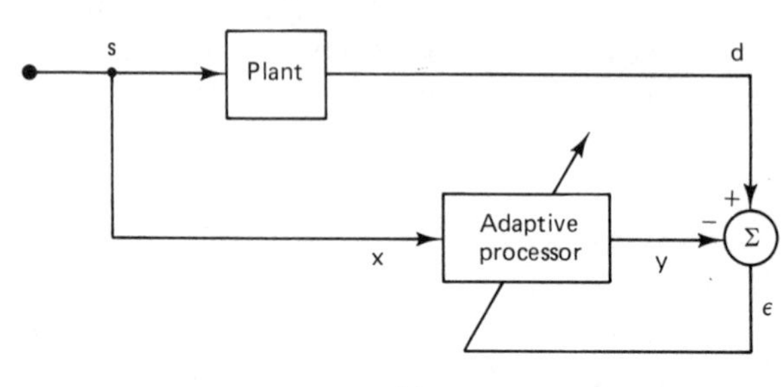

(b)

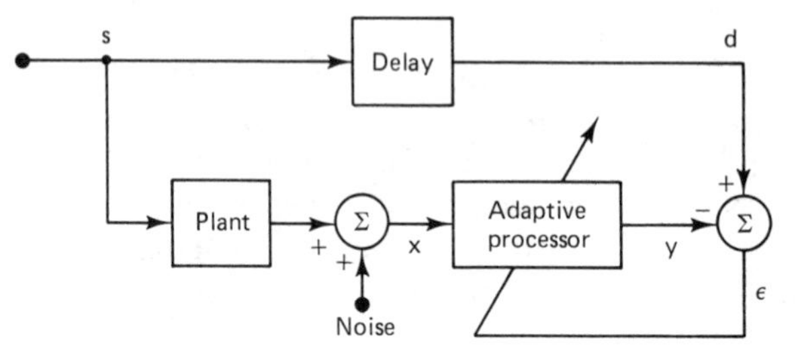

(c)

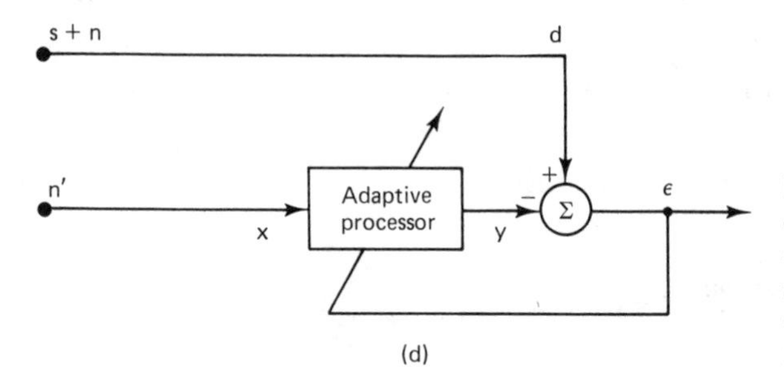

(d)

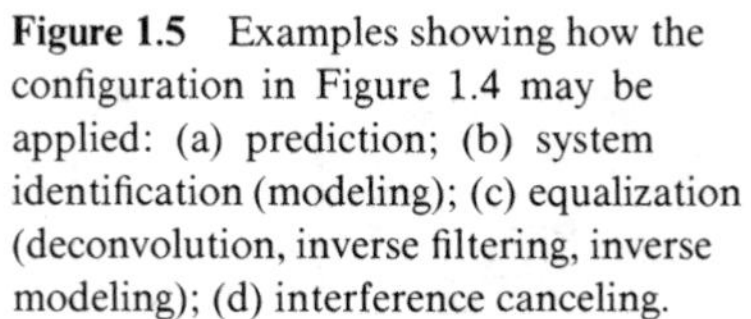

Figure 1.5 Examples showing how the configuration in Figure 1.4 may be applied: (a) prediction; (b) system identification (modeling); (c) equalization (deconvolution, inverse filtering, inverse modeling); (d) interference canceling.

Some examples of applications are given in Figure 1.5. Notice that Figure 1.4, which illustrates the basic closed-loop adaptive process, is simplified slightly and embedded in each part of this figure, and that the application determines how the desired signal, d, is obtained.

The prediction application in Figure 1.5(a) is perhaps the simplest of the four. The desired signal is the input signal, s, and a delayed version of the latter is sent to the adaptive processor, which must therefore try to "predict" the current input signal in order to have y cancel d and drive ε toward zero. Prediction is used in signal encoding and noise reduction, and is discussed in Chapters 8, 9, and 12.

The system identification application in Figure 1.5(b) is also easy to understand. Here a broadband signal, s, is the input to the adaptive processor as well as to an unknown "plant" (a term originating in the control literature). To reduce ε, the adaptive processor tries to emulate the plant's transfer characteristic. After adaptation the plant is "identified" in the sense that its transfer function can be specified as essentially the same as that of the adaptive processor. Adaptive system identification or modeling can be used as such, to model a slowly varying plant whose input and output signals are available, for example, in vibration studies of mechanical systems. It can also be used in many other ways, some of which are described in Chapters 9 and 11.

The inverse modeling application [Figure 1.5(c)] is discussed primarily in Chapter 10, and its uses in control are discussed in Chapter 11. In this application the adaptive processor attempts to recover a delayed version of the signal, s, which is assumed to have been altered by the slowly varying plant and to contain additive noise. The delay in the figure is to allow for the delay, or propagation time, through the plant and the adaptive processor. Adaptive equalization could be used to undo (deconvolve) the effects of a transducer, a communication channel, or some other system, or to produce an inverse model of an unknown plant. It is also applicable in the design of digital filters, as well as in adaptive control problems, and so on.

Finally, Figure 1.5(d) shows the adaptive processor in an interference-canceling configuration. Here the signal, s, is corrupted by additive noise, n, and a distorted but correlated version of the noise, n', is also available. The goal of the adaptive processor in this case is to produce an output, y, that closely resembles n, so that the overall output, ε, will closely resemble s. We will show, under certain very broad conditions, that the optimal adaptive processor is that which minimizes the mean-square value of ε. The subject of adaptive interference canceling is discussed in Chapters 12 through 14.

EXAMPLE OF AN ADAPTIVE SYSTEM

By now we have seen that adaptive signal processing is really a very general and basic term which implies the use of time-varying, self-adjusting signal processors. We think of the performance of these systems as being purposeful, useful, and sometimes even "intelligent" in a limited sense. Before concluding this introductory chapter we introduce just one specific example of such a system, to provide a more concrete illustration of adaptation in general and closed-loop adaptation in particular. We will use the "prediction" application in Figure 1.5(a), keeping in mind that in doing so we have a very restricted and specific example of a very broad and general class of systems.

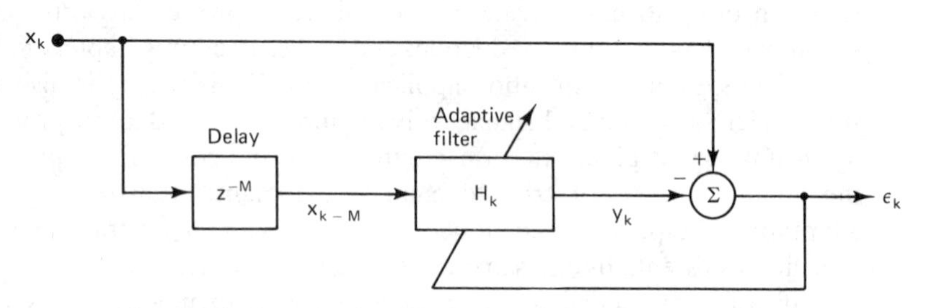

Figure 1.6 Example of an adaptive system: an adaptive predictor.

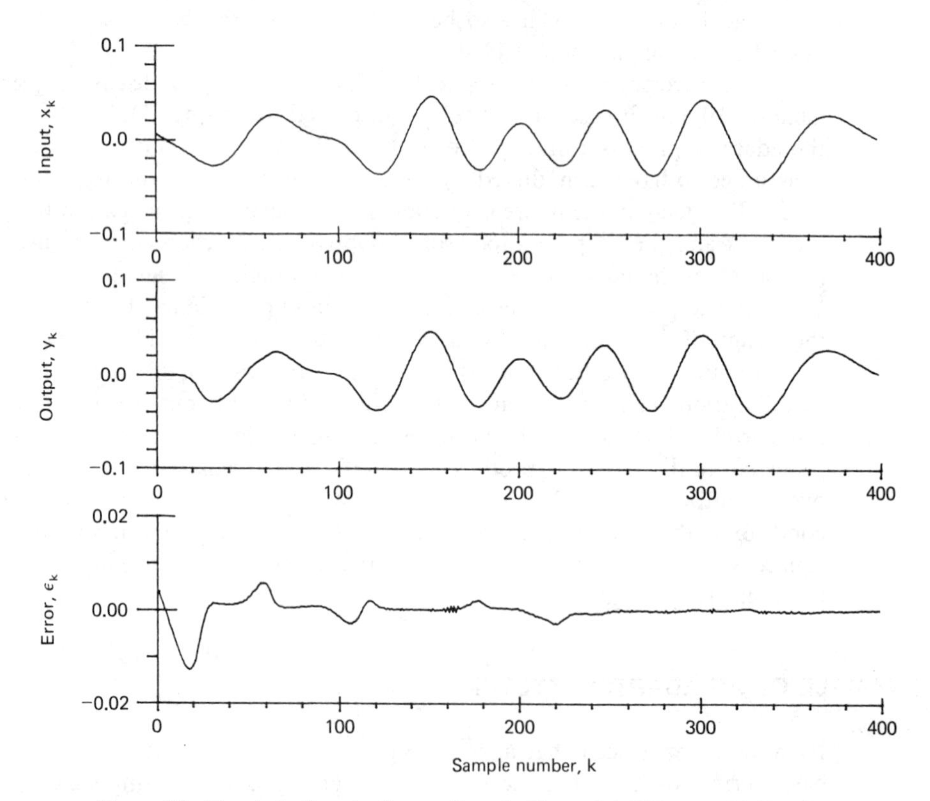

Figure 1.7 Signals in the adaptive predictor in Figure 1.6. The error (ε) becomes nearly zero as the system learns to predict the input (x) and the output (y) approaches x.

The adaptive predictor diagram in Figure 1.5(a) is redrawn in Figure 1.6, with examples of several conventions of notation that are used throughout this text. The symbols x_k, y_k, and ε_k represent the kth elements in the time series represented by x, y, and ε. Usually, the time series may be assumed to be obtained by sampling continuous signals. Thus $x_k = x(kT)$, and so on, where T is the time step or interval between samples. The symbol z^{-M} stands for a fixed delay of M time steps, so that in Figure 1.6 the output of the z^{-M} block is labeled x_{k-M}. The symbol H_k represents the transfer characteristic of the adaptive processor or "adaptive filter." We leave a detailed description of H_k to subsequent chapters, except to note here that the subscript k indicates that the transfer characteristic possibly changes at each sample point.

Thus in Figure 1.6 we see that the input signal, x_k, is delayed and then filtered to produce y_k, and that y_k is subtracted from x_k to obtain the error, ε_k. The transfer function, H_k, is adjusted to keep the average squared value of this error as small as possible. In this way, the processor always uses past values of x to predict the present value of x while using ε to adjust H, and is thus involved in a performance-feedback process.

An example of adaptive predictor performance is shown in Figure 1.7. The waveforms are constructed from digital data (here as well as in examples ahead) by drawing straight lines between the sample points. The input, x, is a frequency-limited random signal, and $M = 1$ in this case. Note that y_k becomes a better prediction of x_k, and that the magnitude of ε_k decreases, as k increases and the adaptive system learns progressively to predict x_k. More detailed examples of adaptive prediction are covered in subsequent chapters.

THE CHAPTERS AHEAD

In Chapters 2 through 8 we present some basic discussion and mathematical description of the adaptive process. The performance of adaptive systems can be analyzed either in the time domain or in the frequency domain; in general, the frequency-domain analysis tends to be more elegant and difficult. Therefore, in Chapters 2 through 6 we stay in the time domain and avoid using transforms, transfer functions, and so on, as much as possible. In Chapter 2 we introduce the adaptive linear combiner, which is the basic nonrecursive form of the adaptive filter. Then in Chapters 3 through 5 we introduce the geometry of the "performance surface" and consider various methods of adaptively seeking the minimum point on this surface.

In Chapter 6 we introduce the well-known least-mean-square (LMS) algorithm, which is the simplest, most important, and most widely used algorithm for adjusting the weights in a linear adaptive system. After reviewing Chapter 2, the more advanced reader or the reader wishing simply to apply the LMS algorithm without covering all of the theory may wish to skip to Chapter 6 and beyond, without covering all of the material in Chapters 3 through 5.

Some areas of adaptive signal processing, including recursive adaptive filters, cannot be analyzed conveniently without reference to the frequency domain, so Chapter 7 is concerned mainly with developing foundations for this type of analysis. Then in Chapter 8 we introduce other types of algorithms that require frequency-domain analysis. In Chapter 8 we also introduce the adaptive lattice, a structure different from the adaptive linear combiner of Chapter 2. Finally, Chapters 9 through 14 deal with some of the important applications of adaptive signal processing, as described above.

chapter 2

The Adaptive Linear Combiner

GENERAL DESCRIPTION

The adaptive linear combiner, or nonrecursive adaptive filter, is fundamental to adaptive signal processing. It appears, in one form or another, in most adaptive filters and systems, and it is the single most important element in "learning" systems and adaptive processes in general. Thus a primary purpose of this book is to develop a thorough exposition of the properties, behavior, means of adaptation, rate of adaptation, and useful applications of the adaptive linear combiner.

Because of its nonrecursive structure, the adaptive linear combiner is relatively easy to understand and analyze. In essence it is a time-varying, nonrecursive digital filter, and as such its performance is quite simple. Its behavior and the means of adapting it, as well as its implementation in different forms, are well understood. Furthermore, we know of specific applications where its performance is "best" in some sense.

A diagram of the general form of the adaptive linear combiner is shown in Figure 2.1. There is an input signal vector with elements $x_0, x_1, \dots, x_L$, a corresponding set of adjustable weights, $w_0, w_1, \dots, w_L$, a summing unit, and a single output signal, y. A procedure for adjusting or adapting the weights is called a "weight adjustment," "gain adjustment," or "adaptation" procedure. The combiner is called "linear" because for a fixed setting of the weights its output is a linear combination of the input components. However, when the weights are in the process of being adjusted, they, too, are a function of the input components, and the output of the combiner is no longer a linear function of the input. That is, its operation becomes nonlinear in accordance with the general principle set forth in Chapter 1.

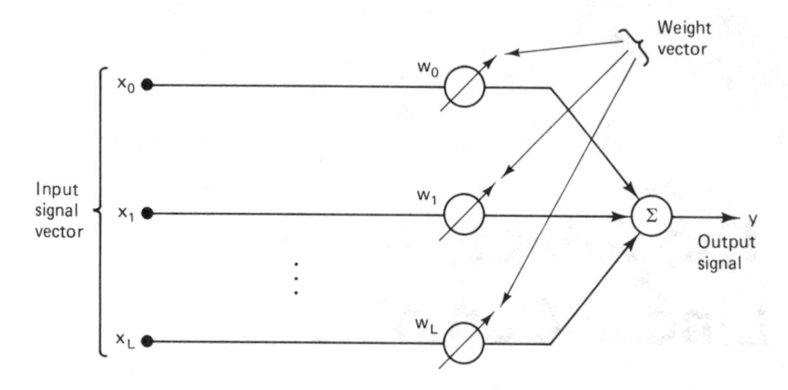

Figure 2.1 General form of adaptive linear combiner.

INPUT SIGNAL AND WEIGHT VECTORS

There are two important ways to interpret physically the elements of the input vector in Figure 2.1. First, they may be considered to be *simultaneous* inputs from $L + 1$ *different* signal sources. An example of this interpretation would be an adaptive antenna or an adaptive acoustic detection system, in which each input line is connected to a separate sensor.

Alternatively, the elements $x_0 - x_L$ may be considered to be $L + 1$ *sequential* samples from the *same* signal source. One example of this interpretation is the adaptive processor described in Chapter 1 and illustrated in Figure 1.6.

We refer to these two interpretations as the *multiple-input* and *single-input* cases, and it is convenient to label the input vector differently for the two cases, as follows:

$$\textit{Multiple inputs:} \quad \mathbf{X}_k = [x_{0k} \quad x_{1k} \quad \cdots \quad x_{Lk}]^{\mathrm{T}} \tag{2.1}$$

$$\textit{Single input:} \quad \mathbf{X}_k = [x_k \quad x_{k-1} \quad \cdots \quad x_{k-L}]^{\mathrm{T}} \tag{2.2}$$

In this notation T stands for transpose, so $\mathbf{X}_k$ is actually a column vector in both cases. The subscript k is used as a time index. Thus in the multiple-input case, all of the elements are taken at the kth sampling time, whereas in the single-input case, the elements are sequential samples taken at points $k, k - 1, \ldots,$ going back in time through the sequence of data samples.

In the single-input case, the adaptive processor can be implemented with an adaptive linear combiner and unit delay elements, as shown in Figure 2.2. The structure in Figure 2.2 is called an adaptive transversal filter. Note that a second subscript, k, has been added to the weights, to make them explicitly time varying. The adaptive transversal filter is the temporal (as opposed to spatial) form of the nonrecursive adaptive filter. It has a wide range of applications in the fields of adaptive modeling and adaptive signal processing. Most of the adaptive systems

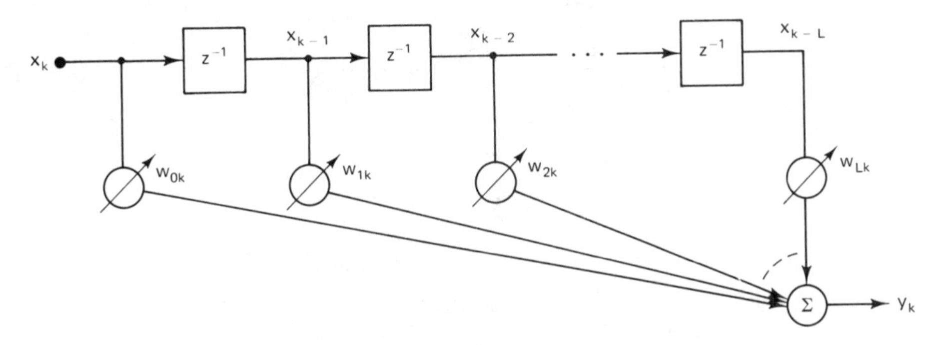

Figure 2.2 Adaptive linear combiner in the form of single-input adaptive transversal filter.

described in later chapters of this book are based on the use of an adaptive transversal filter.

In some multiple-input systems, we need a *bias weight* that simply adds a variable bias into the sum, y_k. We obtain this conveniently where it is needed by setting the first input element, x_{0k} in Figure 2.1 and (2.1), permanently equal to one (or some other constant value) as illustrated in Figure 2.3. The bias weight is not normally required in single-input systems.

From the input signal notation in (2.1) and (2.2), we obtain the input–output relationships for Figures 2.2 and 2.3 as follows:

$$\textit{Single input:} \qquad y_k = \sum_{l=0}^{L} w_{lk} x_{k-l} \tag{2.3}$$

$$\textit{Multiple inputs:} \qquad y_k = \sum_{l=0}^{L} w_{lk} x_{lk} \tag{2.4}$$

When x_{0k} is set identically equal to 1 in (2.4) as just described, w_{0k} becomes a bias weight.

Corresponding with (2.1) and (2.2), we have a weight vector,

$$\mathbf{W}_k = [w_{0k} \quad w_{1k} \quad \cdots \quad w_{Lk}]^{\mathrm{T}} \tag{2.5}$$

With this definition we can express (2.3) and (2.4) in a single relationship using vector notation:†

$$y_k = \mathbf{X}_k^{\mathrm{T}} \mathbf{W}_k = \mathbf{W}_k^{\mathrm{T}} \mathbf{X}_k \tag{2.6}$$

With this description of the operation of the adaptive linear combiner, we can

† We assume that the reader understands vector and matrix multiplication. Any elementary text on matrix algebra should serve as a review.

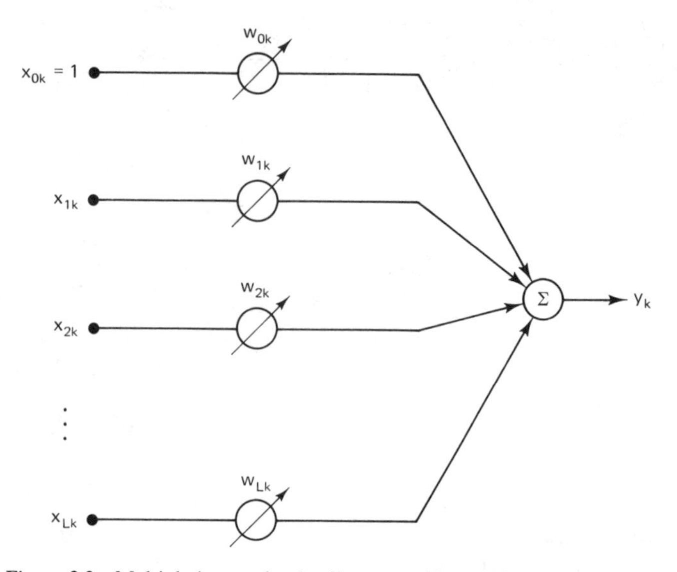

Figure 2.3 Multiple-input adaptive linear combiner with bias weight w_{0k}.

proceed to a discussion of how it adapts, that is, of the effect of changing the vector $\mathbf{W}_k$ as the time index, k, changes.

DESIRED RESPONSE AND ERROR

The adaptive linear combiner can be used in both open- and closed-loop adaptive systems. As discussed in connection with Figure 1.2, the adjustment of the weight vector in open-loop systems does not depend explicitly on properties of the output, but only on the input and on properties of the environment.

With closed-loop systems, however, as in Figure 1.3, the weight vector depends on the output signal as well as other data. Generally, for the adaptive linear combiner the other data include a "desired response" or "training signal." Our discussions in this book will be involved primarily with closed-loop, performance-feedback systems, so we need a thorough understanding of these signals.

In the adaptation process with performance feedback, the weight vector of the linear combiner is adjusted to cause the output, y_k, to agree as closely as possible with the desired response signal. This is accomplished by comparing the output with the desired response to obtain an "error" signal and then adjusting or optimizing the weight vector to minimize this signal. In most practical instances the adaptive process is oriented toward minimizing the mean-square value, or average power of the error signal. Optimization by this criterion, in both adaptive and nonadaptive systems, has been widely practiced in the past and has many advantages [1, 8–16].

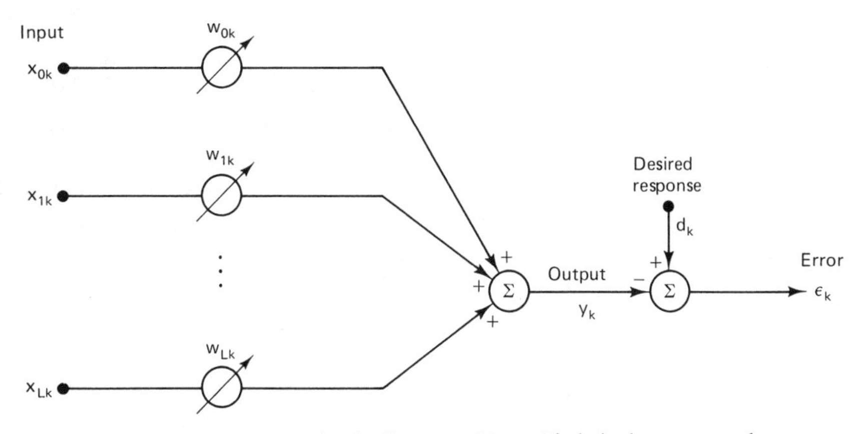

Figure 2.4 Multiple-input adaptive linear combiner with desired response and error signals.

The method of deriving the error signal by means of the desired response input is shown in the multiple-input diagram of Figure 2.4. The output signal, y_k, is simply subtracted from the desired signal, d_k, to produce the error signal, ε_k.

The source of the desired response signal, d_k, depends on the application of the adaptive combiner. For the present we will assume the availability of such a signal. In later chapters we discuss its derivation in more detail. We note, however, that considerable ingenuity is often required to find a suitable signal, since if the actual desired response were available one would generally not need the adaptive system.

We now proceed to a discussion of the performance function, which is a function of the error signal just described.

THE PERFORMANCE FUNCTION

As in Figure 2.4, the error signal with time index k is

$$\varepsilon_k = d_k - y_k \tag{2.7}$$

Substituting (2.6) into this expression yields

$$\varepsilon_k = d_k - \mathbf{X}_k^{\mathrm{T}}\mathbf{W} = d_k - \mathbf{W}^{\mathrm{T}}\mathbf{X}_k \tag{2.8}$$

Here we have dropped the subscript k from the weight vector $\mathbf{W}$ for convenience, because in this discussion we do not wish to adjust the weights. We now square (2.8) to obtain the instantaneous squared error,

$$\varepsilon_k^2 = d_k^2 + \mathbf{W}^{\mathrm{T}}\mathbf{X}_k\mathbf{X}_k^{\mathrm{T}}\mathbf{W} - 2d_k\mathbf{X}_k^{\mathrm{T}}\mathbf{W} \tag{2.9}$$

We assume that ε_k, d_k, and $\mathbf{X}_k$ are statistically stationary and take the expected value† of (2.9) over k:

$$E[\varepsilon_k^2] = E[d_k^2] + \mathbf{W}^{\mathrm{T}}E[\mathbf{X}_k\mathbf{X}_k^{\mathrm{T}}]\mathbf{W} - 2E[d_k\mathbf{X}_k^{\mathrm{T}}]\mathbf{W} \tag{2.10}$$

Note that the expected value of any sum is the sum of expected values, but that the expected value of a product is the product of expected values when the variables are statistically independent. The signals x_k and d_k are not generally independent.

The mean-square-error function can be more conveniently expressed as follows. Let $\mathbf{R}$ be defined as the square matrix

$$\mathbf{R} = E[\mathbf{X}_k\mathbf{X}_k^{\mathrm{T}}] = E\begin{bmatrix} x_{0k}^2 & x_{0k}x_{1k} & x_{0k}x_{2k} & \cdots & x_{0k}x_{Lk} \\ x_{1k}x_{0k} & x_{1k}^2 & x_{1k}x_{2k} & \cdots & x_{1k}x_{Lk} \\ \vdots & \vdots & \vdots & & \vdots \\ x_{Lk}x_{0k} & x_{Lk}x_{1k} & x_{Lk}x_{2k} & \cdots & x_{Lk}^2 \end{bmatrix} \tag{2.11}$$

This matrix is designated the "input correlation matrix." The main diagonal terms are the mean squares of the input components, and the cross terms are the cross correlations among the input components. Let $\mathbf{P}$ be similarly defined as the column vector

$$\mathbf{P} = E[d_k\mathbf{X}_k] = E[d_kx_{0k} \quad d_kx_{1k} \quad \cdots \quad d_kx_{Lk}]^{\mathrm{T}} \tag{2.12}$$

This vector is the set of cross correlations between the desired response and the input components. The elements of both $\mathbf{R}$ and $\mathbf{P}$ are all constant second-order statistics when $\mathbf{X}_k$ and d_k are stationary. Note that the multiple-input form of $\mathbf{X}_k$ was used in (2.11) and (2.12), but the single-input form could just as easily have been used.

We now let the mean-square error in (2.10) be designated by ξ and reexpress it in terms of (2.11) and (2.12) as

$$\text{MSE} \triangleq \xi = E[\varepsilon_k^2] = E[d_k^2] + \mathbf{W}^{\mathrm{T}}\mathbf{R}\mathbf{W} - 2\mathbf{P}^{\mathrm{T}}\mathbf{W} \tag{2.13}$$

It is clear from this expression that the mean-square error ξ is precisely a quadratic function of the components of the weight vector $\mathbf{W}$ when the input components and desired response input are stationary stochastic variables. That is, when (2.13) is expanded, the elements of $\mathbf{W}$ will appear in first and second degree only.

A portion of a typical two-dimensional mean-square-error function is illustrated in Figure 2.5. The vertical axis represents the mean-square error and the horizontal axes the values of the two weights. The bowl-shaped quadratic error function, or *performance surface*, formed in this manner is a paraboloid (a hyperparaboloid if there are more than two weights). It must be concave upward; otherwise, there would be weight settings that would result in a negative mean-square error, an impossible result with real, physical signals. Contours of constant mean-

† For discussions of statistical concepts such as stationarity, expected value, and so on, we refer to the texts on signal processing and time-series analysis [5, 6].

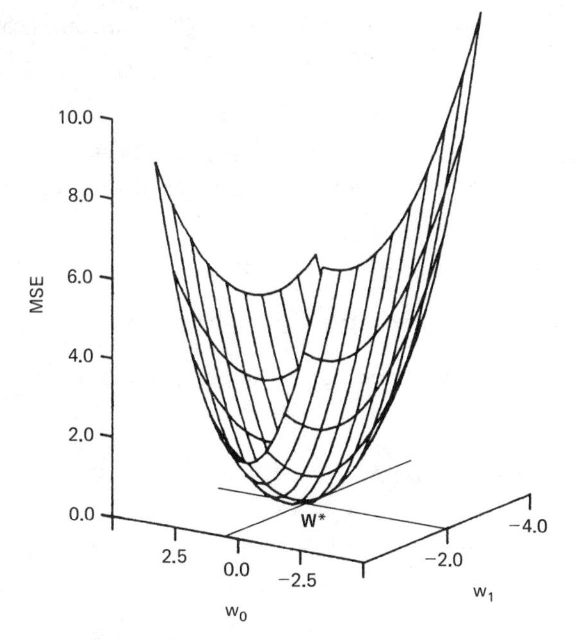

Figure 2.5 Portion of a two-dimensional quadratic performance surface. The mean-square error is plotted vertically, w_0 ranges from -3 to 4, w_1 ranges from -4 to 0, and the optimum weight vector is $\mathbf{W}^* = (0.65, -2.10)$. The minimum MSE is 0.0 in this example.

square error are elliptical, as can be seen by setting ξ constant in (2.13). The point at the "bottom of the bowl" is projected onto the weight-vector plane as $\mathbf{W}^*$, the optimal weight vector or point of minimum mean-square error. With a quadratic performance function there is only a single global optimum; no local minima exist.

GRADIENT AND MINIMUM MEAN-SQUARE ERROR

Many useful adaptive processes that cause the weight vector to seek the minimum of the performance surface do so by gradient methods. The gradient of the mean-square-error performance surface, designated $\nabla(\xi)$ or simply ∇, can be obtained by differentiating (2.13) to obtain the column vector

$$\nabla \triangleq \frac{\partial \xi}{\partial \mathbf{W}} = \left[\frac{\partial \xi}{\partial w_0} \quad \frac{\partial \xi}{\partial w_1} \quad \cdots \quad \frac{\partial \xi}{\partial w_L} \right]^{\mathrm{T}} \tag{2.14}$$

$$= 2\mathbf{RW} - 2\mathbf{P} \tag{2.15}$$

where $\mathbf{R}$ and $\mathbf{P}$ are given by (2.11) and (2.12), respectively. This expression is obtained by expanding (2.13) and differentiating with respect to each component of the weight vector. Differentiation of the term $\mathbf{W}^{\mathrm{T}}\mathbf{RW}$ can be treated as differentiation of the product $(\mathbf{W}^{\mathrm{T}})(\mathbf{RW})$.

To obtain the minimum mean-square error the weight vector $\mathbf{W}$ is set at its optimal value $\mathbf{W}^*$, where the gradient is zero:

$$\nabla = \mathbf{0} = 2\mathbf{RW}^* - 2\mathbf{P} \tag{2.16}$$

Assuming that **R** is nonsingular, the optimal weight vector **W** *, sometimes called the Wiener weight vector, is found from (2.16) to be

$$\mathbf{W}^* = \mathbf{R}^{-1}\mathbf{P} \tag{2.17}$$

This equation is an expression of the Wiener–Hopf equation [8, 9, 12] in matrix form. The minimum mean-square error is now obtained by substituting **W** * from (2.17) for **W** in (2.13):

$$\begin{aligned}\xi_{\min} &= E\left[d_k^2\right] + \mathbf{W}^{*\mathrm{T}}\mathbf{R}\mathbf{W}^* - 2\mathbf{P}^{\mathrm{T}}\mathbf{W}^* \\ &= E\left[d_k^2\right] + [\mathbf{R}^{-1}\mathbf{P}]^{\mathrm{T}}\mathbf{R}\mathbf{R}^{-1}\mathbf{P} - 2\mathbf{P}^{\mathrm{T}}\mathbf{R}^{-1}\mathbf{P}\end{aligned} \tag{2.18}$$

We now simplify this result using three rules that are of general utility in discussions of the performance surface:

1. Identity rule for any square matrix: $\mathbf{A}\mathbf{A}^{-1} = \mathbf{I}$
2. Transpose of a matrix product: $[\mathbf{A}\mathbf{B}]^{\mathrm{T}} = \mathbf{B}^{\mathrm{T}}\mathbf{A}^{\mathrm{T}}$
3. Symmetry of the input correlation matrix: $\mathbf{R}^{\mathrm{T}} = \mathbf{R}$; $[\mathbf{R}^{-1}]^{\mathrm{T}} = \mathbf{R}^{-1}$ [see (2.11)].

Using these rules, (2.18) becomes

$$\xi_{\min} = E\left[d_k^2\right] - \mathbf{P}^{\mathrm{T}}\mathbf{R}^{-1}\mathbf{P} = E\left[d_k^2\right] - \mathbf{P}^{\mathrm{T}}\mathbf{W}^* \tag{2.19}$$

We now introduce an example to help clarify the concepts of quadratic performance surface, gradient, and mean-square error set forth so far.

EXAMPLE OF A PERFORMANCE SURFACE

A simple example of a single-input adaptive linear combiner with two weights is shown in Figure 2.6. The input and desired signals are sampled sinusoids at the same frequency, with N samples per cycle. We assume that $N > 2$ so that the input

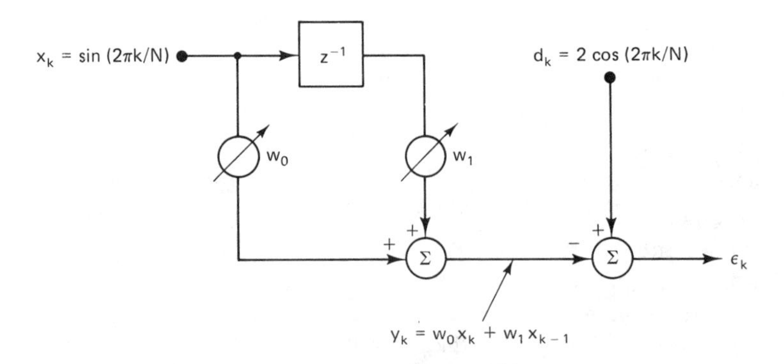

Figure 2.6 Example of an adaptive linear combiner with two weights.

samples are not all zero. We are not concerned here with the origin of these signals, only with the resulting performance surface and its properties.

To obtain the performance function [i.e., ξ in (2.13)], we need the expected signal products in (2.11) and (2.12). Note that we must change the subscripts of x for the single-input case, as in (2.2). The expected products may be found for any product of sinusoidal functions by averaging over one or more periods of the product. Thus,

$$E[x_k x_{k-n}] = \frac{1}{N}\sum_{k=1}^{N} \sin\frac{2\pi k}{N}\sin\frac{2\pi(k-n)}{N}$$

$$= 0.5\cos\frac{2\pi n}{N}; \qquad n = 0, 1 \tag{2.20}$$

$$E[d_k x_{k-n}] = \frac{2}{N}\sum_{k=1}^{N} \cos\frac{2\pi k}{N}\sin\frac{2\pi(k-n)}{N}$$

$$= -\sin\frac{2\pi n}{N}; \qquad n = 0, 1 \tag{2.21}$$

We note further that obviously $E[x_{k-1}^2] = E[x_k^2]$, because the average is over k.

With these results, the input correlation matrix $\mathbf{R}$ and the correlation vector $\mathbf{P}$ can be obtained from (2.11) and (2.12) for this two-dimensional, single-input example:

$$\mathbf{R} = E\begin{bmatrix} x_k^2 & x_k x_{k-1} \\ x_{k-1}x_k & x_{k-1}^2 \end{bmatrix} = \begin{bmatrix} 0.5 & 0.5\cos\frac{2\pi}{N} \\ 0.5\cos\frac{2\pi}{N} & 0.5 \end{bmatrix} \tag{2.22}$$

$$\mathbf{P} = E[d_k x_k \quad d_k x_{k-1}]^{\mathrm{T}} = \begin{bmatrix} 0 & -\sin\frac{2\pi}{N} \end{bmatrix}^{\mathrm{T}} \tag{2.23}$$

As in (2.20) and (2.21) we also obtain $E[d_k^2] = 2$. Using these results in (2.13), we obtain the performance function for this example:

$$\xi = E\left[d_k^2\right] + \mathbf{W}^{\mathrm{T}}\mathbf{R}\mathbf{W} - 2\mathbf{P}^{\mathrm{T}}\mathbf{W}$$

$$= 2 + 0.5[w_0 \quad w_1]\begin{bmatrix} 1 & \cos\frac{2\pi}{N} \\ \cos\frac{2\pi}{N} & 1 \end{bmatrix}\begin{bmatrix} w_0 \\ w_1 \end{bmatrix} - 2\begin{bmatrix} 0 & -\sin\frac{2\pi}{N} \end{bmatrix}\begin{bmatrix} w_0 \\ w_1 \end{bmatrix}$$

$$= 0.5\left(w_0^2 + w_1^2\right) + w_0 w_1 \cos\frac{2\pi}{N} + 2w_1 \sin\frac{2\pi}{N} + 2 \tag{2.24}$$

This performance surface is plotted in Figure 2.5 for $N = 5$ samples per cycle. Note that it is quadratic in w_0 and w_1 and has a single global minimum. The gradient vector at any point w_0, w_1 can be found by substituting (2.22) and (2.23) into (2.15) and is

$$\nabla = 2\mathbf{RW} - 2\mathbf{P}$$

$$= \begin{bmatrix} 1 & \cos\frac{2\pi}{N} \\ \cos\frac{2\pi}{N} & 1 \end{bmatrix} \begin{bmatrix} w_0 \\ w_1 \end{bmatrix} - 2 \begin{bmatrix} 0 \\ -\sin\frac{2\pi}{N} \end{bmatrix}$$

$$= \begin{bmatrix} w_0 + w_1\cos\frac{2\pi}{N} \\ w_0\cos\frac{2\pi}{N} + w_1 + 2\sin\frac{2\pi}{N} \end{bmatrix} \tag{2.25}$$

The Wiener weight vector for this example, $\mathbf{W}^*$, may be found formally from (2.17) by inverting $\mathbf{R}$, or it may be found by setting ∇ equal to zero in (2.25). These are, of course, equivalent operations, and in either case the result is

$$\mathbf{W}^* = \left[2\cot\frac{2\pi}{N} \quad -2\csc\frac{2\pi}{N}\right]^{\mathrm{T}} \tag{2.26}$$

Again, note that $N > 2$ was specified earlier, so that w_0^* and w_1^* in this result are always finite.

Finally, the minimum mean-square error for this example is obtained by substituting (2.23) and (2.26) into (2.19):

$$\xi_{\min} = E\left[d_k^2\right] - \mathbf{P}^{\mathrm{T}}\mathbf{W}^*$$

$$= 2 - \left[0 \quad -\sin\frac{2\pi}{N}\right] \begin{bmatrix} 2\cot\frac{2\pi}{N} \\ -2\csc\frac{2\pi}{N} \end{bmatrix} = 0 \tag{2.27}$$

This result, which says in effect that the weights in Figure 2.6 can be adjusted to reduce ε_k to zero for any value of N, may seem surprising at first. The unit delay by itself can change x_k from a sine into a cosine function only when $N = 4$, that is, only when the unit delay is one-fourth of a cycle. Note that in this case (2.26) gives $w_0^* = 0$ and $w_1^* = -2$. However, with two weights in addition to the delay, the adaptive linear combiner can always shift x_k so that it becomes the proper cosine function, for any N greater than 2 (see also Exercise 3).

ALTERNATIVE EXPRESSION OF THE GRADIENT

Since the mean-square error is a quadratic form in $\mathbf{W}$ which reaches its minimum value when $\mathbf{W}$ equals $\mathbf{W}^*$, one might expect that it could be expressed as

$$\xi = \xi_{\min} + (\mathbf{W} - \mathbf{W}^*)^{\mathrm{T}}\mathbf{R}(\mathbf{W} - \mathbf{W}^*) \tag{2.28}$$

We demonstrate that this expression is valid in the following manner. Noting that in general $(\mathbf{A} - \mathbf{B})^{\mathrm{T}}$ is just $\mathbf{A}^{\mathrm{T}} - \mathbf{B}^{\mathrm{T}}$, we expand (2.28) to obtain

$$\xi = \xi_{\min} + \mathbf{W}^{*\mathrm{T}}\mathbf{R}\mathbf{W}^* + \mathbf{W}^{\mathrm{T}}\mathbf{R}\mathbf{W} - \mathbf{W}^{\mathrm{T}}\mathbf{R}\mathbf{W}^* - \mathbf{W}^{*\mathrm{T}}\mathbf{R}\mathbf{W} \tag{2.29}$$

Each term in (2.29) is a scalar and therefore equal to its own transpose. Thus the last two terms are equal. We now combine these and also substitute (2.19) for $\xi_{\min}$, and get

$$\xi = E[d_k^2] - \mathbf{P}^{\mathrm{T}}\mathbf{W}^* + \mathbf{W}^{*\mathrm{T}}\mathbf{R}\mathbf{W}^* + \mathbf{W}^{\mathrm{T}}\mathbf{R}\mathbf{W} - 2\mathbf{W}^{\mathrm{T}}\mathbf{R}\mathbf{W}^* \tag{2.30}$$

Next, substituting (2.17) for $\mathbf{W}^*$ and recalling again that $\mathbf{R}$ is symmetric, we get

$$\begin{aligned}\xi &= E[d_k^2] - \mathbf{P}^{\mathrm{T}}\mathbf{R}^{-1}\mathbf{P} + \mathbf{P}^{\mathrm{T}}\mathbf{R}^{-1}\mathbf{R}\mathbf{R}^{-1}\mathbf{P} + \mathbf{W}^{\mathrm{T}}\mathbf{R}\mathbf{W} - 2\mathbf{W}^{\mathrm{T}}\mathbf{R}\mathbf{R}^{-1}\mathbf{P} \\ &= E[d_k^2] + \mathbf{W}^{\mathrm{T}}\mathbf{R}\mathbf{W} - 2\mathbf{W}^{\mathrm{T}}\mathbf{P} \\ &= E[d_k^2] + \mathbf{W}^{\mathrm{T}}\mathbf{R}\mathbf{W} - 2\mathbf{P}^{\mathrm{T}}\mathbf{W}\end{aligned} \tag{2.31}$$

This result corresponds with (2.13) and thus validates (2.28).

The quadratic form in (2.28) can be expressed more conveniently when we define a weight deviation vector, as follows:

$$\mathbf{V} = \mathbf{W} - \mathbf{W}^* = [v_0 \quad v_1 \quad \cdots \quad v_L]^{\mathrm{T}} \tag{2.32}$$

Accordingly, (2.28) becomes

$$\xi = \xi_{\min} + \mathbf{V}^{\mathrm{T}}\mathbf{R}\mathbf{V} \tag{2.33}$$

The quantity $\mathbf{V}$ is the deviation of the weight vector from the Wiener optimal weight vector. Any departure of $\mathbf{W}$ from $\mathbf{W}^*$ would cause an excess mean-square error according to the quadratic form $\mathbf{V}^{\mathrm{T}}\mathbf{R}\mathbf{V}$.

In order that ξ be nonnegative for all possible $\mathbf{V}$, it is necessary that $\mathbf{V}^{\mathrm{T}}\mathbf{R}\mathbf{V} \geqq 0$ for all $\mathbf{V}$. When $\mathbf{V}^{\mathrm{T}}\mathbf{R}\mathbf{V} > 0$ for all $\mathbf{V} \neq 0$, the matrix $\mathbf{R}$ is said to be "positive definite" [7]. When $\mathbf{V}^{\mathrm{T}}\mathbf{R}\mathbf{V} = 0$ for certain finite values of $\mathbf{V}$ or for all $\mathbf{V}$, the matrix $\mathbf{R}$ is said to be "positive semidefinite." In physical situations, $\mathbf{R}$ will almost always be positive definite, but a positive semidefinite $\mathbf{R}$ could occur. Texts on matrix theory generally discuss conditions of positive definiteness and positive semidefiniteness.

The gradient of the mean-square error with respect to $\mathbf{V}$ is obtained by differentiating (2.33):

$$\frac{\partial \xi}{\partial \mathbf{V}} = \left[\frac{\partial \xi}{\partial v_0} \quad \frac{\partial \xi}{\partial v_1} \quad \cdots \quad \frac{\partial \xi}{\partial v_L}\right] = 2\mathbf{R}\mathbf{V} \tag{2.34}$$

This gradient is the same as that given by (2.15), because $\mathbf{W}$ and $\mathbf{V}$ differ only by a constant. Thus

$$\nabla = \frac{\partial \xi}{\partial \mathbf{W}} = \frac{\partial \xi}{\partial \mathbf{V}} = 2\mathbf{R}\mathbf{V} = 2(\mathbf{R}\mathbf{W} - \mathbf{P}) \tag{2.35}$$

Expression (2.35) will be used in developing and analyzing a variety of adaptive algorithms.

DECORRELATION OF ERROR AND INPUT COMPONENTS

A useful and important statistical condition exists between the error signal and the components of the input signal vector when $\mathbf{W} = \mathbf{W}^*$. Recall from (2.8) that

$$\varepsilon_k = d_k - \mathbf{X}_k^{\mathrm{T}}\mathbf{W} \tag{2.36}$$

We multiply both sides of this equation by $\mathbf{X}_k$. Since each term is a scalar, we can put $\mathbf{X}_k$ on either side of each term. Thus

$$\varepsilon_k \mathbf{X}_k = d_k \mathbf{X}_k - \mathbf{X}_k \mathbf{X}_k^{\mathrm{T}}\mathbf{W} \tag{2.37}$$

Next, we take the expected value of (2.37) and obtain

$$E[\varepsilon_k \mathbf{X}_k] = \mathbf{P} - \mathbf{R}\mathbf{W} \tag{2.38}$$

Finally, we let $\mathbf{W}$ take its optimum value, $\mathbf{W}^* = \mathbf{R}^{-1}\mathbf{P}$ in (2.17), and get

$$E[\varepsilon_k \mathbf{X}_k]_{\mathbf{W}=\mathbf{W}^*} = \mathbf{P} - \mathbf{P} = \mathbf{0} \tag{2.39}$$

This result is the same as the well-known result of Wiener filter theory: that when the impulse response of a filter is optimized, the error signal is uncorrelated with (orthogonal to) the input signals to the weights.

EXERCISES

1. Much of the discussion in this chapter and in subsequent chapters uses matrix algebra. Prove that the following simple relationships are true for any matrices **A**, **B**, and **C**:
 (a) $\mathbf{AB} \neq \mathbf{BA}$ in general.
 (b) $\mathbf{A}(\mathbf{B} + \mathbf{C}) = \mathbf{AB} + \mathbf{AC}$.
 (c) $(\mathbf{AB})^{\mathrm{T}} = \mathbf{B}^{\mathrm{T}}\mathbf{A}^{\mathrm{T}}$ and $(\mathbf{AB})^{-1} = \mathbf{B}^{-1}\mathbf{A}^{-1}$.
 (d) If $\mathbf{A}$ is symmetric, then $\mathbf{A}^{-1}$ is symmetric.

2. Starting with Equation (2.13), give a detailed derivation of Equation (2.15).

3. In the adaptive linear combiner example shown in Figure 2.6, let $N = 10$. Then:
 (a) Find the optimum weight vector.
 (b) Write an expression for y_k using the answer to part (a).
 (c) Using the result from part (b) and x_k from Figure 2.6, prove that $y_k = d_k$.

4. In Figure 2.6, what weight values will produce a root-mean-square (rms) value of ε_k equal to 2:
 (a) With $N = 5$?
 (b) With $N = 10$?

5. On the performance surface of Figure 2.6, with $N = 8$, what is the gradient vector when $w_1 = 0$ and the mean-squared value of ε_k is:
 (a) 2.0?
 (b) 4.0?
 Why is the gradient steeper in the latter case?

6. Consider the system shown here, which is an adaptive linear combiner with a single weight. Suppose that switch S is open and $E[x_k^2] = 1$, $E[x_k x_{k-1}] = 0.5$, $E[d_k^2] = 4$, $E[d_k x_k] = -1$, and $E[d_k x_{k-1}] = 1$. Derive an expression for the performance function. Make a plot of the performance function.

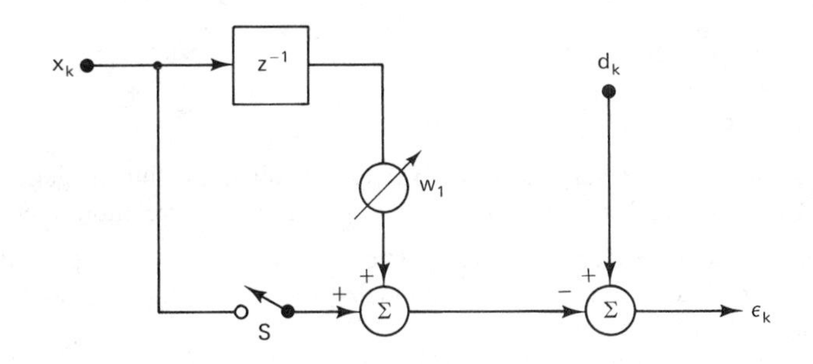

7. Do Exercise 6 with switch S closed.
8. What is the optimum value of w_1 for Exercise 6? What is the corresponding minimum mean-square error?
9. In the adaptive linear combiner diagram of Exercise 6, let x_k and d_k be as given in Figure 2.6, and let $N = 5$. With switch S open:
   (a) Find and plot an expression for ξ.
   (b) Find the optimum value of w_1.
   (c) Find the minimum value of ξ.
10. Do Exercise 9 with switch S closed.
11. The student should have some experience with correlation functions such as those derived in Equations (2.20) and (2.21). Consider the following continuous, periodic waveforms:

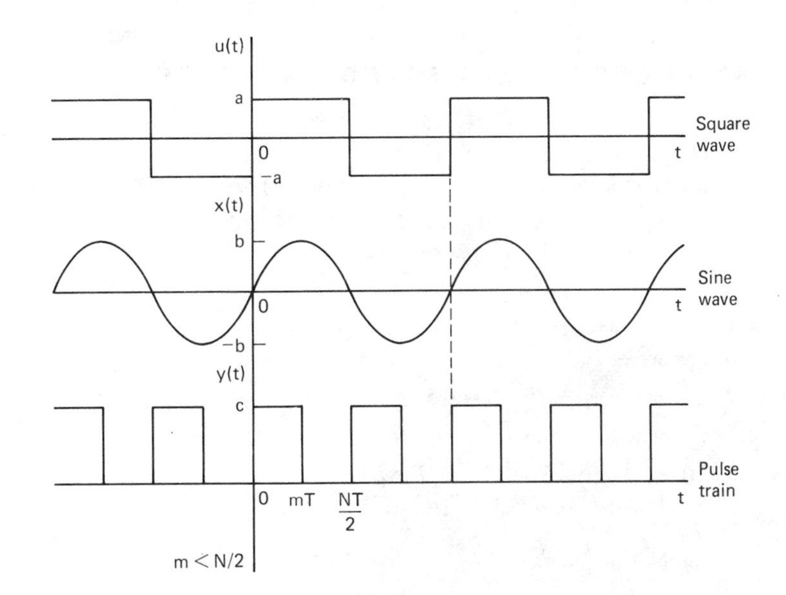

Suppose that each waveform is sampled at $t = 0, T, 2T, \ldots$, such that there are exactly N samples per cycle in the first two cases, and $N/2$ samples per cycle in the third case. If a sample is taken where $f(t)$ is discontinuous, give it the value just to the right of the discontinuity.

(a) Find $E[u_k u_{k-n}]$.
(b) Find $E[y_k y_{k-n}]$.
(c) Find $E[u_k x_{k-n}]$.
(d) Find $E[u_k y_{k-n}]$.
(e) Find $E[x_k y_{k-n}]$.

12. By using an example from Exercise 11, illustrate that in general, autocorrelation is an even function of the shift variable n, but cross correlation is not.

13. Consider the adaptive linear combiner shown below. Suppose that we wish to minimize $E[\varepsilon_k^4]$ rather than $E[\varepsilon_k^2]$. Assume that the signals x_{0k}, x_{1k}, and d_k are all stationary.
(a) Derive an expression for $E[\varepsilon_k^4]$.
(b) Is $E[\varepsilon_k^4]$ quadratic in w_0 and w_1?
(c) Is $E[\varepsilon_k^4]$ a unimodal function of w_0 and w_1?

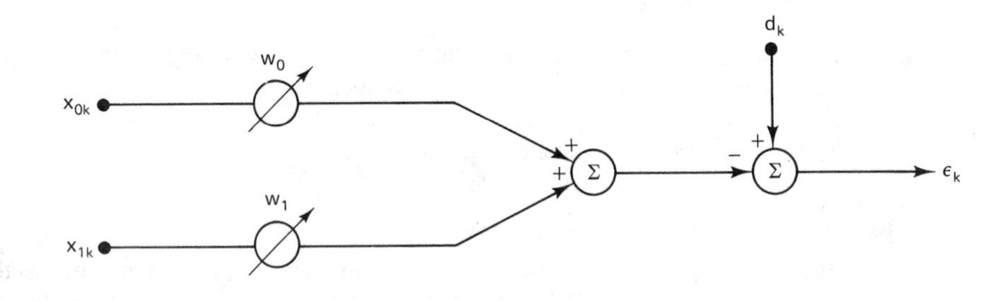

ANSWERS TO SELECTED EXERCISES

3. **(a)** $\mathbf{W}^* = [2.753 \quad -3.403]^{\mathrm{T}}$
(b) $y_k = 2.753 \sin(k\pi/5) - 3.403 \sin[(k-1)\pi/5]$

4. **(a)** $w_0^2 + w_1^2 + 0.6180 w_0 w_1 + 1.9021 w_1 - 4 = 0$
(b) $w_0^2 + w_1^2 + 1.6180 w_0 w_1 + 1.1756 w_1 - 4 = 0$

5. **(a)** $\nabla = [0 \quad 1.4142]^{\mathrm{T}}$
(b) $\nabla = [2 \quad 2.8284]^{\mathrm{T}}, [-2 \quad 0]^{\mathrm{T}}$

6. $\xi = w_1^2 - 2w_1 + 4;\ w_1^* = 1$

7. $\xi = w_1^2 - w_1 + 7;\ w_1^* = 0.5$

8. $w_1^* = 1;\ \xi_{\min} = 3$

13. **(a)** $E[\varepsilon_k^4] = w_0^4 E[x_{0k}^4] + w_1^4 E[x_{1k}^4] - 4w_0^3 E[d_k x_{0k}^3] + \cdots + 12 w_0 w_1 E[d_k x_{0k} x_{1k}] - 4w_0 E[d_k^3 x_{0k}] - 4w_1 E[d_k^3 x_{1k}] + E[d_k^4]$
(b) No
(c) Not in general

REFERENCES AND ADDITIONAL READINGS

1. B. Widrow, "Adaptive Filters," in *Aspects of Network and System Theory*, R. E. Kalman and N. De Claris (Eds.). New York: Holt, Rinehart and Winston, 1970, p. 563.
2. A. V. Oppenheim and R. W. Schafer, *Digital Signal Processing*. Englewood Cliffs, N.J.: Prentice-Hall, 1975, Chap. 8.
3. L. R. Rabiner and B. Gold, *Theory and Application of Digital Signal Processing*. Englewood Cliffs, N.J.: Prentice-Hall, 1975.
4. N. Ahmed and T. Natarajan, *Discrete-Time Signals and Systems*. Reston, Va.: Reston, 1983.
5. L. H. Koopmans, *The Spectral Analysis of Time Series*. New York: Academic Press, 1974.
6. S. D. Stearns, *Digital Signal Analysis*. Rochelle Park, N.J.: Hayden, 1975, Chap. 13.
7. L. G. Kelly, *Handbook of Numerical Methods and Applications*. Reading, Mass.: Addison-Wesley, 1967, Sec. 7.8.
8. N. Wiener, *Extrapolation, Interpolation and Smoothing of Stationary Time Series, with Engineering Applications*. New York: Wiley, 1949.
9. H. W. Bode and C. E. Shannon, "A simplified derivation of linear least squares smoothing and prediction theory," *Proc. IRE*, vol. 38, pp. 417–425, Apr., 1950.
10. R. E. Kalman, "On the general theory of control," in *Proc. First IFAC Congress*. London: Butterworth, 1960.
11. R. E. Kalman and R. S. Bucy, "New results in linear filtering and prediction theory," *Trans. ASME, Ser. D, J. Basic Eng.*, vol. 83, pp. 95–107, Dec. 1961.
12. T. Kailath, "A view of three decades of linear filtering theory," *IEEE Trans. Inf. Theory*, vol. IT-20, pp. 145–181, Mar. 1974.
13. B. Widrow and M. E. Hoff, Jr., "Adaptive switching circuits," *IRE WESCON Conv. Rec.*, pt. 4, pp. 96–104, 1960.
14. J. S. Koford and G. F. Groner, "The use of an adaptive threshold element to design a linear optimal pattern classifier," *IEEE Trans. Inf. Theory*, vol. IT-12, pp. 42–50, Jan. 1966.
15. D. Gabor, W. P. L. Wilby, and R. Woodcock, "A universal nonlinear filter predictor and simulator which optimizes itself by a learning process," *Proc. Inst. Electr. Eng.*, vol. 108B, July 1960.
16. R. W. Lucky, J. Salz, and E. J. Weldon, Jr., *Principles of Data Communication*. New York: McGraw-Hill, 1968.

part II

THEORY OF ADAPTATION WITH STATIONARY SIGNALS

(Chapters 3 – 5)

OBJECTIVES OF PART II

If we have satisfied the objectives of Part I, the reader now understands the idea of the performance or error surface and how it represents the adaptive environment. An important property of the performance surface in adaptive signal processing is the following: If the signals are stationary and have invariant statistical properties, the performance surface remains fixed and rigid in its coordinate system. The adaptive process then consists of starting at some point on the surface, proceeding downhill to the neighborhood of the minimum, and staying there.

On the other hand, if the signals are nonstationary and have statistical properties that slowly change, we can view the performance surface as being "fuzzy," or undulating, or moving in its coordinate system. The adaptive process then consists not only of moving downhill to the minimum, but also of *tracking* the minimum as it moves about in the coordinate system.

Chapters 3 through 5 concern the stationary situation where the performance surface is rigid, which is of course the simplest situation to discuss. In Chapter 3 our objective is to introduce certain mathematical properties of the fixed performance surface. These properties will be useful in later chapters, in comparing the performance of adaptive systems.

The primary objective in Chapter 4 is to understand two basic methods for searching the performance surface — Newton's method and the method of steepest descent — and to compare these two methods. Then the primary objective in Chapter 5 is to study the effect on these two methods of a noisy gradient estimate, which one generally must use in practical situations.

These chapters are all "theoretical background" for Part III, in which we begin to discuss practical adaptive algorithms. Thus the reader may wish to omit sections of Part II, or skip to Part III and then come back to Part II.

chapter 3

Properties of the Quadratic Performance Surface

Having defined the performance surface for a class of adaptive systems, our goal is now to proceed to a discussion of algorithms for adjusting the weights and descending on the performance surface to the minimum mean-square error. We do this in the next three chapters, but first we need to discuss some important properties of the quadratic performance surface. The reader interested in simply having a description of the best way to search the performance surface may proceed to Chapter 6, but the intervening discussion is necessary for an understanding of the search process and how to evaluate and control it.

The properties of the performance surface that we wish to introduce in this chapter are, in turn, due primarily to properties of the input correlation matrix, **R**. In (2.33) we showed that when an adaptive linear combiner is used with stationary inputs, the mean-square error can be expressed in terms of the input signal correlation matrix **R** as

$$\begin{aligned} \xi &= \xi_{\min} + (\mathbf{W} - \mathbf{W}^*)^{\mathrm{T}}\mathbf{R}(\mathbf{W} - \mathbf{W}^*) \\ &= \xi_{\min} + \mathbf{V}^{\mathrm{T}}\mathbf{R}\mathbf{V} \end{aligned} \qquad (2.33)$$

Note that since there are $L + 1$ weights (components of **W**), the matrix **R** has $L + 1$ columns and $L + 1$ rows.

It is clear from (2.33) that the orientation and shape of the quadratic mean-square-error performance surface is a function of **R**. Much can be learned about the performance surface by expressing **R** in normal form, in terms of its eigenvalues and eigenvectors. Any text on linear algebra will provide useful reference material. An excellent reference is Anton [5]. The next two sections contain a brief review of the basic concepts.

NORMAL FORM OF THE INPUT CORRELATION MATRIX

The characteristic values (eigenvalues) of the matrix $\mathbf{R}$ are developed from the homogeneous equation

$$[\mathbf{R} - \lambda\mathbf{I}]\mathbf{Q}_n = \mathbf{0} \tag{3.1}$$

where λ is a scalar variable, $\mathbf{Q}_n$ is a column vector, $\mathbf{I}$ is the identity matrix, and $\mathbf{0}$ is a vector with all elements equal to zero. This homogeneous equation has nontrivial solutions for λ and $\mathbf{Q}_n$ if and only if the following determinant vanishes:

$$\det[\mathbf{R} - \lambda\mathbf{I}] = 0 \tag{3.2}$$

Equation (3.2), called the "characteristic equation" of $\mathbf{R}$, is an algebraic equation in λ of degree $L + 1$. Its $L + 1$ solutions are designated $\lambda_0, \lambda_1, \ldots, \lambda_L$. They are the *eigenvalues* of $\mathbf{R}$, all of which may not be distinct from one another.

Corresponding to each eigenvalue, λ_n, there exists at least one vector solution of (3.1), $\mathbf{Q}_n$, which is determined as follows:

$$\mathbf{R}\mathbf{Q}_n = \lambda_n\mathbf{Q}_n \tag{3.3}$$

The vector $\mathbf{Q}_n$ is the nth eigenvector of $\mathbf{R}$ and is associated with λ_n.

Extending (3.3), we obtain

$$\mathbf{R}[\mathbf{Q}_0 \quad \mathbf{Q}_1 \quad \cdots \quad \mathbf{Q}_L] = [\mathbf{Q}_0 \quad \mathbf{Q}_1 \quad \cdots \quad \mathbf{Q}_L]\begin{bmatrix} \lambda_0 & \cdots & 0 \\ & \lambda_1 & \\ \vdots & \ddots & \vdots \\ 0 & \cdots & \lambda_L \end{bmatrix} \tag{3.4}$$

which can be rewritten as

$$\mathbf{R}\mathbf{Q} = \mathbf{Q}\Lambda \qquad \text{or} \qquad \mathbf{R} = \mathbf{Q}\Lambda\mathbf{Q}^{-1} \tag{3.5}$$

Equation (3.5) gives the *normal form* of $\mathbf{R}$, in which the eigenvalues appear explicitly in Λ. The "eigenvalue matrix" Λ is diagonal, as indicated by (3.4). All its elements are zero except for the main diagonal, whose elements are the set of eigenvalues of $\mathbf{R}$. The modal matrix $\mathbf{Q}$ is called the "eigenvector matrix" of $\mathbf{R}$, because its columns are the eigenvectors of $\mathbf{R}$. Both Λ and $\mathbf{Q}$ are square, with dimensions $(L + 1) \times (L + 1)$, like $\mathbf{R}$.

EIGENVALUES AND EIGENVECTORS OF THE INPUT CORRELATION MATRIX

From the definition (2.11), it is apparent that $\mathbf{R}$ is a symmetric matrix, with $\mathbf{R} = \mathbf{R}^T$. The eigenvectors corresponding to distinct eigenvalues must therefore be orthogonal; that is, $\mathbf{Q}_m^T\mathbf{Q}_n = 0$ for any vector pair. This fact can be easily demonstrated as follows. Let λ_1 and λ_2 be two distinct eigenvalues. Then

$$\mathbf{R}\mathbf{Q}_1 = \lambda_1 \mathbf{Q}_1 \tag{3.6}$$

and

$$\mathbf{R}\mathbf{Q}_2 = \lambda_2 \mathbf{Q}_2 \tag{3.7}$$

Let (3.6) be transposed and then postmultiplied by $\mathbf{Q}_2$:

$$\mathbf{Q}_1^{\mathrm{T}} \mathbf{R}^{\mathrm{T}} \mathbf{Q}_2 = \lambda_1 \mathbf{Q}_1^{\mathrm{T}} \mathbf{Q}_2 \tag{3.8}$$

Next let (3.7) be premultiplied by $\mathbf{Q}_1^{\mathrm{T}}$:

$$\mathbf{Q}_1^{\mathrm{T}} \mathbf{R} \mathbf{Q}_2 = \lambda_2 \mathbf{Q}_1^{\mathrm{T}} \mathbf{Q}_2 \tag{3.9}$$

Recalling that $\mathbf{R} = \mathbf{R}^{\mathrm{T}}$ and combining (3.8) and (3.9), we then obtain

$$\lambda_1 \mathbf{Q}_1^{\mathrm{T}} \mathbf{Q}_2 = \lambda_2 \mathbf{Q}_1^{\mathrm{T}} \mathbf{Q}_2 \tag{3.10}$$

Since $\lambda_1 \neq \lambda_2$ by hypothesis, it follows that

$$\mathbf{Q}_1^{\mathrm{T}} \mathbf{Q}_2 = 0 \tag{3.11}$$

so the eigenvectors of λ_1 and λ_2 are orthogonal.

Since $\mathbf{R}$ is real (all its elements are real numbers) in addition to being symmetric, all its eigenvalues must be real. This fact can be demonstrated by contradiction. Assume that λ_1 is a complex eigenvalue of $\mathbf{R}$. The characteristic equation of $\mathbf{R}$ (3.2) is an $(L + 1)$-degree polynomial in λ set equal to zero. Since complex roots of such a polynomial would occur in conjugate pairs, if λ_1 is a complex eigenvalue, its conjugate $\bar{\lambda}_1$ must also be an eigenvalue. Furthermore, if λ_1 is complex, its eigenvector $\mathbf{Q}_1$ must also be complex, since $\mathbf{R}$ is real [Equation (3.6)]. Moreover, the eigenvalue associated with $\bar{\lambda}_1$ must be the conjugate of $\mathbf{Q}_1$, which is $\bar{\mathbf{Q}}_1$. Since λ_1 is assumed to be complex, it cannot be equal to its conjugate (i.e., $\lambda_1 \neq \bar{\lambda}_1$). Since λ_1 and $\bar{\lambda}_1$ are distinct, the corresponding eigenvectors must be orthogonal (i.e., $\mathbf{Q}^{\mathrm{T}}\bar{\mathbf{Q}} = 0$). But this result is impossible because the inner product of a complex vector and its conjugate is equal to the sum of squares of its components, which must be a positive quantity. Therefore, the assumption that λ_1 is complex leads to a contradiction, and all eigenvalues of the input correlation matrix $\mathbf{R}$ must be real.

Another important result from matrix theory is that if an eigenvalue λ_k is repeated with multiplicity m, there are m corresponding linearly independent eigenvectors. These can always be constructed to be mutually orthogonal and orthogonal to all of the other eigenvectors, if so desired.

In forming the modal matrix $\mathbf{Q}$, it is convenient to scale the eigenvectors so that they are normalized to have unit magnitudes. We have just shown that eigenvectors of distinct eigenvalues are orthogonal. If repeated eigenvalues occur, let the corresponding eigenvectors be chosen so that they too are all orthogonal. Then all $L + 1$ eigenvectors that make up the columns of $\mathbf{Q}$ are mutually orthogonal and normalized, and $\mathbf{Q}$ is said to be "orthonormal." Henceforth we will assume that the modal matrix as represented by the symbol $\mathbf{Q}$ is orthonormal. We can thus write

$$\mathbf{Q}\mathbf{Q}^{\mathrm{T}} = \mathbf{I} \tag{3.12}$$

and

$$\mathbf{Q}^{-1} = \mathbf{Q}^{\mathrm{T}} \tag{3.13}$$

Thus the inverse of $\mathbf{Q}$ always exists.

Finally, we can show that the eigenvalues of the input correlation matrix are always greater than or equal to zero. As discussed in Chapter 2 in connection with (2.33), $\mathbf{R}$ is in general positive semidefinite, and thus

$$\mathbf{V}^{\mathrm{T}}\mathbf{R}\mathbf{V} \geq 0 \tag{3.14}$$

Using the result (3.13) in (3.5), we also have

$$\mathbf{R} = \mathbf{Q}\Lambda\mathbf{Q}^{-1} = \mathbf{Q}\Lambda\mathbf{Q}^{\mathrm{T}} \tag{3.15}$$

Now $\mathbf{V}$ is the deviation of the weight vector from its optimum value, and can be chosen to be any vector in (3.14). Let us therefore set $\mathbf{V}$ successively equal to each column of $\mathbf{Q}$, that is, equal to $\mathbf{Q}_0, \mathbf{Q}_1, \ldots, \mathbf{Q}_L$ in order. Then (3.14) holds for each case, and the $L + 1$ cases can be expressed as

$$\mathbf{Q}^{\mathrm{T}}\mathbf{R}\mathbf{Q} \geq \mathbf{0} \tag{3.16}$$

Substituting (3.15) for $\mathbf{R}$ in (3.16), we have

$$\mathbf{Q}^{\mathrm{T}}\mathbf{Q}\Lambda\mathbf{Q}^{\mathrm{T}}\mathbf{Q} \geq \mathbf{0} \tag{3.17}$$

and if we now use the result in (3.13), we have simply

$$\Lambda \geq \mathbf{0} \tag{3.18}$$

To summarize this section:

1. Eigenvectors corresponding to distinct eigenvalues of $\mathbf{R}$ are mutually orthogonal.
2. The eigenvalues of $\mathbf{R}$ are all real and greater than or equal to zero.
3. The eigenvector matrix $\mathbf{Q}$ can be normalized (made orthonormal) such that $\mathbf{Q}\mathbf{Q}^{\mathrm{T}} = \mathbf{I}$.

AN EXAMPLE WITH TWO WEIGHTS

Solving the characteristic equation (3.2) requires in general that the roots of a polynomial of degree $L + 1$ be found. A computer is usually needed for this type of problem, and various well-known algorithms are available for finding the eigenvalues of a matrix [3, 6]. In fact, there are special algorithms for real, symmetric matrices like $\mathbf{R}$ [4].

In this section we examine a case with $L + 1 = 2$ weights, where the characteristic equation is quadratic and therefore easy to solve. Consider the input correlation matrix used as an example in Chapter 2, and take the specific case where $N = 6$ in (2.22). We have

$$\mathbf{R} = \begin{bmatrix} 0.5 & 0.25 \\ 0.25 & 0.5 \end{bmatrix} \tag{3.19}$$

Solving for the eigenvalues using (3.2), we obtain

$$\begin{aligned} \det[\mathbf{R} - \lambda\mathbf{I}] &= \det\begin{bmatrix} 0.5-\lambda & 0.25 \\ 0.25 & 0.5-\lambda \end{bmatrix} \\ &= \lambda^2 - \lambda + 0.1875 = 0 \end{aligned}$$

Therefore,

$$\lambda_0, \lambda_1 = 0.25, 0.75 \tag{3.20}$$

The eigenvectors are also easy to obtain in this case using (3.3) or, equivalently, (3.1):

$$\begin{bmatrix} 0.25 & 0.25 \\ 0.25 & 0.25 \end{bmatrix}\begin{bmatrix} q_{00} \\ q_{10} \end{bmatrix} = \mathbf{0} \qquad \begin{bmatrix} q_{00} \\ q_{10} \end{bmatrix} = \begin{bmatrix} c_1 \\ -c_1 \end{bmatrix} \tag{3.21}$$

$$\begin{bmatrix} -0.25 & 0.25 \\ 0.25 & -0.25 \end{bmatrix}\begin{bmatrix} q_{01} \\ q_{11} \end{bmatrix} = \mathbf{0} \qquad \begin{bmatrix} q_{01} \\ q_{11} \end{bmatrix} = \begin{bmatrix} c_2 \\ c_2 \end{bmatrix} \tag{3.22}$$

Note that since the determinant of $\mathbf{R} - \lambda\mathbf{I}$ vanishes as in (3.1), the coefficient matrices in (3.21) and (3.22) are singular, and thus the q-values can be determined only with the arbitrary constants, c_1 and c_2. These constants are now chosen to *normalize* $\mathbf{Q}$ as in (3.12) and (3.13):

$$\mathbf{Q}\mathbf{Q}^{\mathrm{T}} = \begin{bmatrix} c_1 & c_2 \\ -c_1 & c_2 \end{bmatrix}\begin{bmatrix} c_1 & -c_1 \\ c_2 & c_2 \end{bmatrix} = \begin{bmatrix} c_1^2 + c_2^2 & -c_1^2 + c_2^2 \\ -c_1^2 + c_2^2 & c_1^2 + c_2^2 \end{bmatrix} = \mathbf{I} \tag{3.23}$$

Therefore,

$$c_1 = \pm c_2 = \frac{1}{\sqrt{2}} \tag{3.24}$$

and

$$\mathbf{Q} = \frac{1}{\sqrt{2}}\begin{bmatrix} 1 & 1 \\ -1 & 1 \end{bmatrix} \tag{3.25}$$

Having the eigenvalues and eigenvectors, we can now write $\mathbf{R}$ in normal form, that is, in the form of (3.5):

$$\mathbf{R} = \mathbf{Q}\mathbf{\Lambda}\mathbf{Q}^{-1} = \mathbf{Q}\mathbf{\Lambda}\mathbf{Q}^{\mathrm{T}}$$

Therefore,

$$\begin{bmatrix} 0.5 & 0.25 \\ 0.25 & 0.5 \end{bmatrix} = \frac{1}{2}\begin{bmatrix} 1 & 1 \\ -1 & 1 \end{bmatrix}\begin{bmatrix} 0.25 & 0 \\ 0 & 0.75 \end{bmatrix}\begin{bmatrix} 1 & -1 \\ 1 & 1 \end{bmatrix} \tag{3.26}$$

Notice that the elements of $\mathbf{\Lambda}$ and $\mathbf{Q}$ all have the properties described in the preceding section. The eigenvectors are orthogonal; that is, their inner product vanishes:

$$q_{00}q_{01} + q_{10}q_{11} = 1 - 1 = 0 \tag{3.27}$$

The two eigenvalues are real and positive, and $\mathbf{Q}$ is orthonormal with $\mathbf{QQ}^{\mathrm{T}} = \mathbf{I}$, as in (3.23).

GEOMETRICAL SIGNIFICANCE OF EIGENVECTORS AND EIGENVALUES

The eigenvectors and eigenvalues are related directly to certain properties of the error surface, ξ. Recall from Chapter 2 that ξ, defined in (2.31), describes a hyperparabolic error surface in a space of $L + 2$ dimensions with coordinate axes corresponding to $\xi, w_0, w_1, \ldots, w_L$. Let us for the moment discuss systems with just two weights, that is, a three-dimensional error space. We can then generalize to higher-dimensional spaces, or to the two-dimensional space of a one-weight system.

In the two-weight case, ξ is a paraboloid as in the example in Figure 2.5. If we cut the paraboloid with planes parallel to the w_0w_1–plane, we obtain concentric ellipses of constant mean-square error, as illustrated in Figure 3.1. From (2.13), the general form of any of these curves projected onto the w_0w_1–plane is

$$\mathbf{W}^{\mathrm{T}}\mathbf{RW} - 2\mathbf{P}^{\mathrm{T}}\mathbf{W} = \text{constant} \tag{3.28}$$

As shown in Figure 3.1, we can translate from $\mathbf{W}$ to new coordinates $\mathbf{V}$, with origin at the centers of the ellipses. The origin is of course at the coordinates of the minimum mean-square error and, as in (2.17),

$$\mathbf{V} = \mathbf{W} - \mathbf{R}^{-1}\mathbf{P} = \mathbf{W} - \mathbf{W}^* \tag{3.29}$$

Then (3.28) becomes

$$\mathbf{V}^{\mathrm{T}}\mathbf{RV} = \text{another constant} \tag{3.30}$$

which represents an ellipse (or, in general, a hyperellipse) centered at the origin of the v_0v_1–plane. In this new coordinate system there are two (or in general $L + 1$) lines normal to the ellipses. These are known as the *principal axes* of all ellipses [1, 2] and of the error surface, and are the lines labeled v_0' and v_1' in Figure 3.1.

Any vector normal to the ellipses can be expressed by thinking of the ellipses as contours of $F(\mathbf{V}) = \mathbf{V}^{\mathrm{T}}\mathbf{RV}$ and taking the gradient of F, which is also the gradient of ξ, since ξ and F differ only by a constant. The gradient is

$$\begin{aligned} \nabla &= \begin{bmatrix} \dfrac{\partial F}{\partial v_0} & \dfrac{\partial F}{\partial v_1} & \cdots & \dfrac{\partial F}{\partial v_L} \end{bmatrix}^{\mathrm{T}} \\ &= 2\mathbf{RV} \end{aligned} \tag{3.31}$$

(One can arrive at this result by writing $\mathbf{V}^{\mathrm{T}}\mathbf{RV}$ in the form of a double sum and taking each derivative in turn.)

On the other hand, any vector passing through the origin at $\mathbf{V} = \mathbf{0}$ must be of the form $\mu\mathbf{V}$. But a principal axis passes through the origin and is normal to $F(\mathbf{V})$.

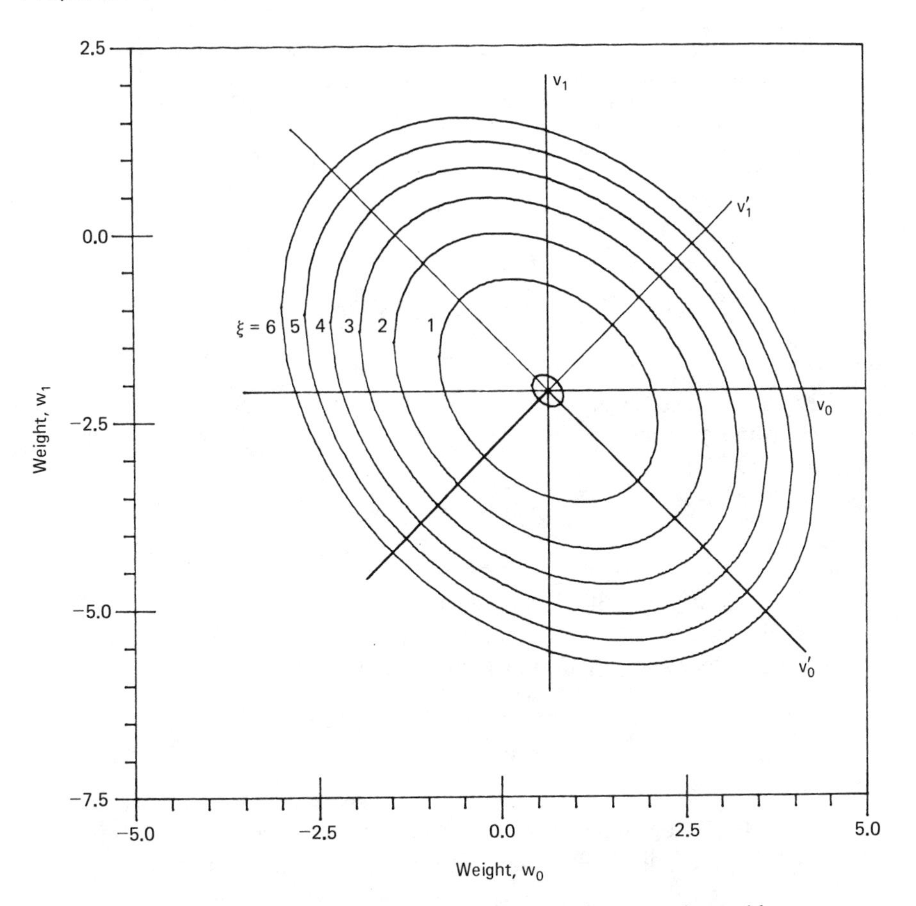

Figure 3.1 Ellipses of constant mean-square error in the w_0w_1–plane, with translated axes v_0 and v_1, and principal axes v_0' and v_1'. The ellipses are contours of the surface in Figure 2.5.

Thus we must have

$$2\mathbf{RV}' = \mu\mathbf{V}'$$

or

$$\left[\mathbf{R} - \frac{\mu}{2}\mathbf{I}\right]\mathbf{V}' = \mathbf{0} \tag{3.32}$$

where $\mathbf{V}'$ now represents a principal axis. This result is in the form of (3.1), so $\mathbf{V}'$ must be an eigenvector of the matrix $\mathbf{R}$. Thus

> The eigenvectors of the input correlation matrix define the principal axes of the error surface.

As a summary of these geometrical transformations, consider the following expressions of the error surface in the three coordinate systems. From (2.28), (2.33), and (3.5), we have

$$\xi = \xi_{\min} + (\mathbf{W} - \mathbf{W}^*)^{\mathrm{T}}\mathbf{R}(\mathbf{W} - \mathbf{W}^*) \tag{3.33}$$

$$= \xi_{\min} + \mathbf{V}^{\mathrm{T}}\mathbf{R}\mathbf{V} \tag{3.34}$$

$$= \xi_{\min} + \mathbf{V}^{\mathrm{T}}(\mathbf{Q}\boldsymbol{\Lambda}\mathbf{Q}^{\mathrm{T}})\mathbf{V}$$

$$= \xi_{\min} + (\mathbf{Q}^{\mathrm{T}}\mathbf{V})^{\mathrm{T}}\boldsymbol{\Lambda}(\mathbf{Q}^{\mathrm{T}}\mathbf{V})$$

$$= \xi_{\min} + \mathbf{V}'^{\mathrm{T}}\boldsymbol{\Lambda}\mathbf{V}' \tag{3.35}$$

Equations (3.33), (3.34), and (3.35) give ξ in the natural, translated, and principal-coordinate systems, respectively. We can verify (3.35) by again taking the gradient as in (3.31):

$$\nabla = 2\boldsymbol{\Lambda}\mathbf{V}'$$

$$= 2\begin{bmatrix}\lambda_0 v_0' & \lambda_1 v_1' & \cdots & \lambda_L v_L'\end{bmatrix}^{\mathrm{T}} \tag{3.36}$$

In contrast with (3.31), we see that if only one component, v_n', is nonzero, the gradient vector lies along that axis; therefore, $\mathbf{V}'$ in (3.35) represents the principal-coordinate system. The transformations corresponding to (3.34) and (3.35) are

$$\textit{Translation}: \quad \mathbf{V} = \mathbf{W} - \mathbf{W}^* \tag{3.37}$$

$$\textit{Rotation}: \quad \mathbf{V}' = \mathbf{Q}^{\mathrm{T}}\mathbf{V} = \mathbf{Q}^{-1}\mathbf{V} \tag{3.38}$$

We can picture these transformations in the example of Figure 3.1.

The eigenvalues of **R** also have important geometrical significance. As seen in (3.36), the gradient of ξ along any principal axis, v_n', could be written

$$\frac{\partial \xi}{\partial v_n'} = 2\lambda_n v_n' \tag{3.39}$$

and also

$$\frac{\partial^2 \xi}{\partial v_n'^2} = 2\lambda_n; \qquad n = 0, 1, \ldots, L \tag{3.40}$$

Thus the second derivative of ξ is twice the eigenvalue along any principal axis, so

> The eigenvalues of the input correlation matrix, **R**, give the second derivatives of the error surface, ξ, with respect to the principal axes of ξ.

We can see a simple example of this result by considering the one-weight case, where ξ becomes a parabola. Let r_{mn} represent the (m, n)th element of **R**. Then, from (2.33), we have for this case

$$\xi = \xi_{\min} + r_{00}(w - w^*)^2 \tag{3.41}$$

There is only one dimension to **W** here, so the w-axis is also the principal axis, and the eigenvalue is $\lambda = r_{00}$. Differentiating twice with respect to w_1, we obtain

$$\frac{\partial^2 \xi}{\partial w^2} = 2r_{00} = 2\lambda \tag{3.42}$$

as in (2.34). Thus, with one weight, the second derivative of the parabola at any point equals $2r_{00}$. A numerical example with two weights is considered in the next section.

A SECOND EXAMPLE

Let us now consider another example with two weights, similar to the first example above but more complete. Let the signal characteristics needed to specify the error surface be the following:

$$\mathbf{R} = \begin{bmatrix} 2 & 1 \\ 1 & 2 \end{bmatrix} \qquad \mathbf{P} = \begin{bmatrix} 7 \\ 8 \end{bmatrix} \qquad E[d_k^2] = 42 \tag{3.43}$$

Substituting these values into (2.13), we have

$$\begin{aligned} \xi &= 42 + [w_0 \quad w_1]\begin{bmatrix} 2 & 1 \\ 1 & 2 \end{bmatrix}\begin{bmatrix} w_0 \\ w_1 \end{bmatrix} - 2[7 \quad 8]\begin{bmatrix} w_0 \\ w_1 \end{bmatrix} \\ &= 2w_0^2 + 2w_1^2 + 2w_0 w_1 - 14w_0 - 16w_1 + 42 \end{aligned} \tag{3.44}$$

This is the error surface, which is again a paraboloid in the three-dimensional space with axes ξ, w_0, and w_1. We locate the optimum weight vector, $\mathbf{W}^*$, corresponding with the minimum mean-square error, $\xi_{\min}$, from (2.17):

$$\mathbf{RW}^* = \mathbf{P}$$

or

$$\begin{bmatrix} 2 & 1 \\ 1 & 2 \end{bmatrix}\begin{bmatrix} w_0^* \\ w_1^* \end{bmatrix} = \begin{bmatrix} 7 \\ 8 \end{bmatrix}$$

Therefore,

$$[w_0^* \quad w_1^*] = [2 \quad 3] \tag{3.45}$$

Using (3.45) in (3.44), we obtain $\xi_{\min} = 4$. If we define $\mathbf{V} = \mathbf{W} - \mathbf{W}^*$ as before, we can then write (3.34) in terms of the translated coordinates, v_0 and v_1:

$$\xi = [v_0 \quad v_1]\begin{bmatrix} 2 & 1 \\ 1 & 2 \end{bmatrix}\begin{bmatrix} v_0 \\ v_1 \end{bmatrix} + 4 \tag{3.46}$$

Contours of constant ξ are shown in both coordinate systems in Figure 3.2.

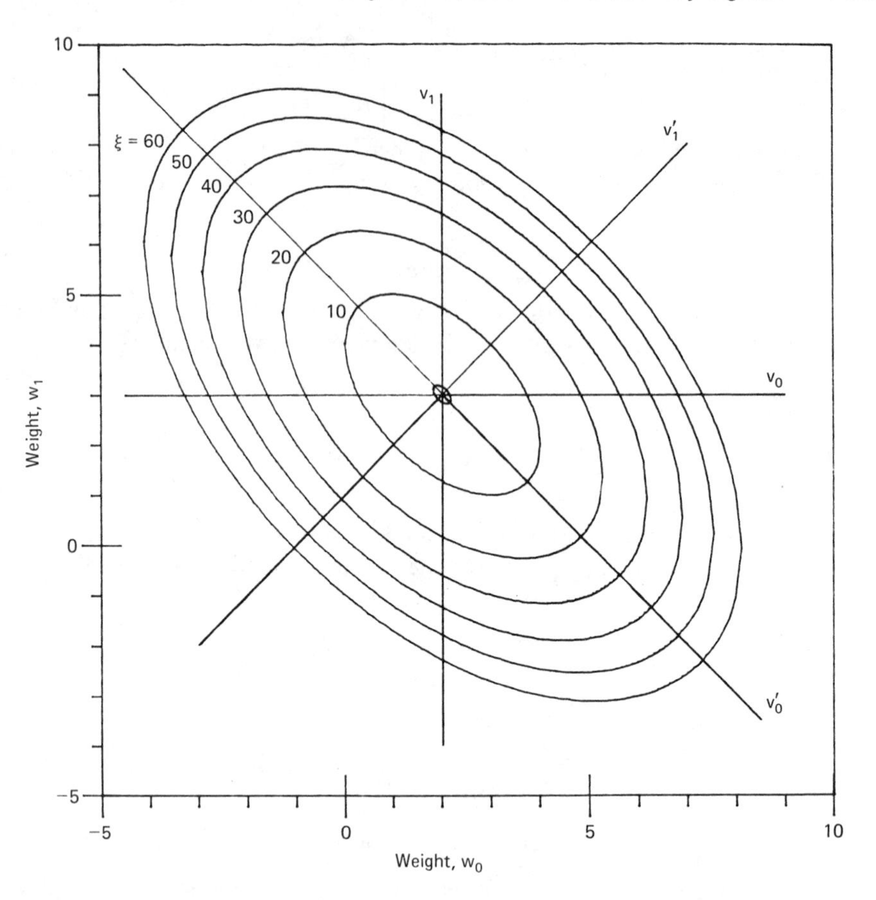

Figure 3.2 Contours of constant mean-square error in (3.44), showing translated (v_0–v_1) coordinate system and principal axes (v_0', v_1'). Eigenvectors, $\mathbf{Q}_0$ and $\mathbf{Q}_1$, point in positive v_0' and v_1' directions.

Similar to (3.22) in the previous example, the eigenvalues of **R** in this example are found as follows:

$$\lambda^2 - 4\lambda + 3 = 0$$

Therefore,

$$\lambda_0, \lambda_1 = 1, 3 \tag{3.47}$$

Also, similar to (3.21) and (3.22), the eigenvectors are determined with arbitrary constants from (3.1):

$$\begin{bmatrix} 1 & 1 \\ 1 & 1 \end{bmatrix} \begin{bmatrix} q_{00} \\ q_{10} \end{bmatrix} = \mathbf{0} \qquad \begin{bmatrix} q_{00} \\ q_{10} \end{bmatrix} = \begin{bmatrix} c_1 \\ -c_1 \end{bmatrix} \tag{3.48}$$

$$\begin{bmatrix} -1 & 1 \\ 1 & -1 \end{bmatrix} \begin{bmatrix} q_{01} \\ q_{11} \end{bmatrix} = \mathbf{0} \qquad \begin{bmatrix} q_{01} \\ q_{11} \end{bmatrix} = \begin{bmatrix} c_2 \\ c_2 \end{bmatrix} \tag{3.49}$$

These are the same eigenvectors as in the preceding example, so the normalized matrix of eigenvectors is the same as in (3.25), that is,

$$\mathbf{Q} = \frac{1}{\sqrt{2}}\begin{bmatrix} 1 & 1 \\ -1 & 1 \end{bmatrix} \tag{3.50}$$

These eigenvectors are shown on the error contour plot in Figure 3.2, colinear with the principal axes. Notice how $\mathbf{Q}_0$ and $\mathbf{Q}_1$, the columns of $\mathbf{Q}$ in (3.50), are unit vectors in the positive direction along the principal axes, v_0' and v_1'. As indicated by Exercise 15 at the end of this chapter, the Q-matrix is always of this form for the single-input adaptive linear combiner with two weights.

Notice also in Figure 3.2 how the eigenvalues indicate the steepness of the error surface along the principal axes. The surface is steeper in the v_1' direction with second derivative $2\lambda_1 = 6$ than in the v_0' direction with $2\lambda_0 = 2$.

EXERCISES

1. **(a)** Prove that Equation (3.30) represents an ellipse when there are two weights.
(b) What type of curve is represented when there is just one weight?

2. Starting with the definition of the gradient, give a detailed derivation of Equation (3.31).

3. Write the characteristic equation for **R** in terms of a polynomial, if:

(a) $R = \begin{bmatrix} a & b \\ b & a \end{bmatrix}$ **(b)** $\mathbf{R} = \begin{bmatrix} a & b & c \\ b & a & b \\ c & b & a \end{bmatrix}$

4. Find the eigenvalues of $\mathbf{R} = \begin{bmatrix} 3 & 2 \\ 2 & 3 \end{bmatrix}$.

5. Find the eigenvalues of $\mathbf{R} = \begin{bmatrix} 3 & 1 \\ 1 & 3 \end{bmatrix}$.

6. Write the characteristic equation for **R** in terms of a polynomial, if:

(a) $\mathbf{R} = \begin{bmatrix} a & b \\ b & c \end{bmatrix}$ **(b)** $\mathbf{R} = \begin{bmatrix} a & b & c \\ b & d & e \\ c & e & f \end{bmatrix}$

7. Which of the four input correlation matrices in Exercises 3 and 6 would apply to single-input adaptive linear combiners? Which could apply to multiple-input combiners?

8. Find the eigenvalues of $\mathbf{R} = \begin{bmatrix} 2 & 1 \\ 1 & 3 \end{bmatrix}$.

9. Find the eigenvalues of $\mathbf{R} = \begin{bmatrix} 4 & 1 \\ 1 & 3 \end{bmatrix}$.

10. Find the eigenvalues of $\mathbf{R} = \begin{bmatrix} 4 & 3 & 0 \\ 3 & 6 & 2 \\ 0 & 2 & 4 \end{bmatrix}$.

11. Find the eigenvalues of $\mathbf{R} = \begin{bmatrix} 4 & 2 & 0 \\ 2 & 4 & 3 \\ 0 & 3 & 6 \end{bmatrix}$.

12. Find the normalized eigenvectors:
   (a) In Exercise 4. **(b)** In Exercise 5.
   (c) In Exercise 8. **(d)** In Exercise 9.
   (e) In Exercise 10. **(f)** In Exercise 11.

13. Demonstrate that the eigenvectors are mutually orthogonal:
   (a) In Exercise 12(a).
   (b) In Exercise 12(c).
   (c) In Exercise 12(f).

14. Consider an adaptive linear combiner in the form of Figure 2.4 with two weights (i.e., $L = 1$). The signals x and d have the following properties: $E[x_{0k}^2] = 2$, $E[x_{1k}^2] = 3$, $E[x_{0k}x_{1k}] = 1$, $E[x_{0k}d_k] = 6$, $E[x_{1k}d_k] = 4$, $E[d_k^2] = 36$.
   (a) Write an expression for the mean-square error.
   (b) What is the optimal weight vector, $\mathbf{W}^*$?
   (c) What is the minimum mean-square error?
   (d) Find the eigenvalues and eigenvectors.
   (e) Make a plot similar to Figure 3.2.

15. Show that the eigenvectors of any single-input adaptive linear combiner with two weights are given by Equation (3.50).

ANSWERS TO SELECTED EXERCISES

1. **(a)** *Hint*: The general quadratic form, $Ax^2 + Bxy + Cy^2 + Dx + Ey + F = 0$, represents an ellipse if $B^2 - 4AC < 0$.

3. **(a)** $\lambda^2 - 2a\lambda + a^2 - b^2 = 0$
   (b) $\lambda^3 - 3a\lambda^2 + (3a^2 - 2b^2 - c^2)\lambda - a^3 + 2b^2(a - c) + ac^2 = 0$

4. $\lambda_0, \lambda_1 = 1, 5$

5. $\lambda_0, \lambda_1 = 2, 4$

6. **(a)** $\lambda^2 - (a + c)\lambda + ac - b^2 = 0$
   (b) $\lambda^3 + (a + d + f)\lambda^2 + (b^2 + c^2 + e^2 - ad - af - df)\lambda + adf + 2bce - ae^2 - b^2f - c^2d = 0$

7. Single-input: $3a, b$ (also $6a$ if $a = c$; $6b$ if $a = d = f$, $b = e$) Multiple-input: $6a, b$ ($3a, b$ could also apply)

8. $\lambda_0, \lambda_1 = 1.382, 3.618$

10. $\lambda_0, \lambda_1, \lambda_2 = 2.1716, 4, 7.8284$

12. **(a)** Same as Equation (3.50)
   (b) Same as Equation (3.50)
   (c) $\mathbf{Q}_0^T = [0.851 \quad -0.526]$; $\mathbf{Q}_1^T = [0.526 \quad 0.851]$

REFERENCES AND ADDITIONAL READINGS

1. F. R. Gantmacher, *The Theory of Matrices*, vol. 1. New York: Chelsea, 1960, p. 308.
2. J. N. Franklin, *Matrix Theory*. Englewood Cliffs, N.J.: Prentice-Hall, 1968, Sec. 4.6.

3. J. H. Wilkinson, *The Algebraic Eigenvalue Problem*. London: Oxford University Press, 1965.

4. L. G. Kelly, *Handbook of Numerical Methods and Applications*. Reading, Mass.: Addison-Wesley, 1967, Chap. 9.

5. H. Anton, *Elementary Linear Algebra*. New York: Wiley, 1973, Chap. 6.

6. D. K. Faddeev and V. N. Faddeeva, *Computational Methods of Linear Algebra*. San Francisco: W. H. Freeman, 1963.

chapter 4

Searching the Performance Surface

We have seen that the mean-square-error performance surface for the adaptive linear combiner is a quadratic function of the weights when the input signals and the desired response are statistically stationary. In many applications of interest the parameters of this quadratic performance surface are unknown, and an analytical description of it is not available. The location of points on the surface, however, can be measured or estimated by averaging the squared error over a period of time. The problem is to develop systematic procedures or algorithms capable of searching the performance surface and finding the optimal weight vector when only measured or estimated data are available. Most practical procedures of this type do not search exhaustively but find the optimal or nearly optimal solution by testing representative adjustments near it.

METHODS OF SEARCHING THE PERFORMANCE SURFACE

In this chapter we shall be concerned with developing algorithms for two well-known methods of searching a performance surface: Newton's method and the method of steepest descent. These methods entail the use of gradient estimates to indicate the direction in which the minimum of the surface lies. They are thus referred to as "descent methods." We will show that these methods are applicable to the quadratic performance surface in particular, and also to other types of performance surfaces.

Newton's method has fundamental mathematical significance, although it is frequently difficult to implement in practice. It is a method of gradient search that causes all components of the weight vector to be changed at each step in the search procedure or at what we shall later define as each iteration cycle. The changes are always in the direction of the minimum of the performance surface, provided that the surface is quadratic.

The method of steepest descent is readily implemented and has proven its value in a wide variety of practical applications. It is a method of gradient search that also causes all components of the weight vector to be changed at each step or iteration cycle. In this case, however, the changes are in the direction of the negative gradient of the performance surface. They are thus not necessarily in the direction of the minimum, since the negative gradient tends toward the minimum only when its origin lies on one of the principal axes of the surface, as discussed in Chapter 3 (see Figure 3.1 or 3.2, for example).

In subsequent chapters we shall describe additional methods of searching a performance surface and develop algorithms based on them. One is a gradient search method that uses a very rough estimate of the gradient at each step in the search procedure. Another is a method that does not employ gradients and has particular application to problems that we shall also introduce later with performance surfaces that are not quadratic. These methods are designated, respectively, "LMS" and "random search."

BASIC IDEAS OF GRADIENT SEARCH METHODS

To introduce the basic concepts of gradient search methods, including those of the recursive algorithm and convergence, we first consider the simplest case where there is only one weight. For this case, which has limited practical significance, all methods of gradient search reduce to a single method.

The one-weight (univariable) performance surface, which is a parabola, is illustrated in Figure 4.1. This surface can be represented as in (3.41):

$$\xi = \xi_{\min} + \lambda(w - w^*)^2 \tag{4.1}$$

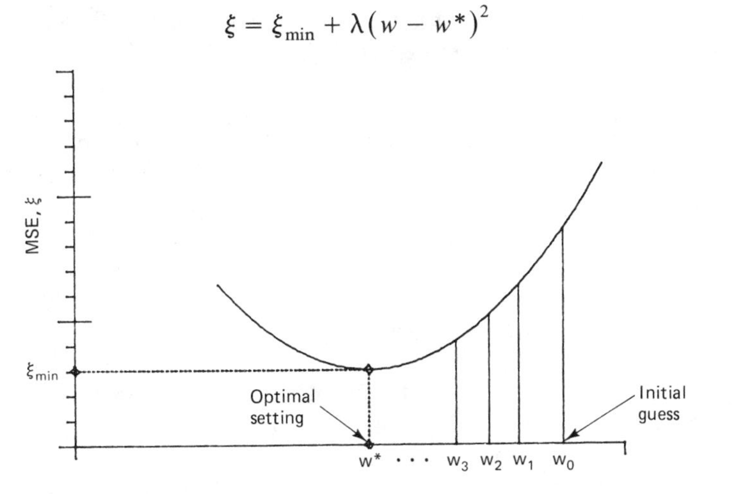

Figure 4.1 Gradient search of univariable performance surface.

We note that the eigenvalue, λ, equals r_{00} in the univariable case. The first derivative is

$$\frac{d\xi}{dw} = 2\lambda(w - w^*) \tag{4.2}$$

The second derivative, constant over the entire curve, is

$$\frac{d^2\xi}{dw^2} = 2\lambda \tag{4.3}$$

The problem is to find w^*, the weight adjustment that causes the mean-square error to be minimized. Not knowing the performance function, we begin with the arbitrary value w_0 and measure the slope of the curve at this point. We then choose a new value w_1 equal to the initial value w_0 plus an increment proportional to the negative of the slope.† Another new value, w_2, is then derived in the same way by measuring the slope of the curve at w_1. This procedure is repeated until the optimal value w^* is reached.

The value obtained by measuring the slope of the performance curve at the discrete intervals $w_0, w_1, w_2, \ldots$ is called the "gradient estimate." The method of making the measurement and the question of its accuracy are discussed in Chapter 5. For the purposes of the present chapter we simply assume that an exact value is available. Note that use of the negative of the gradient is necessary to proceed "downhill."

A SIMPLE GRADIENT SEARCH ALGORITHM AND ITS SOLUTION

With only a single weight, the repetitive or iterative gradient search procedure described above can be represented algebraically as

$$w_{k+1} = w_k + \mu(-\nabla_k) \tag{4.4}$$

where k is the step or iteration number. Thus w_k is the "present" adjustment value, while w_{k+1} is the "new" value. The gradient at $w = w_k$ is designated by ∇_k. The parameter μ is a constant that governs stability and rate of convergence; its choice is discussed later in this chapter.

The gradient ∇_k for the single-weight case is obtained from (4.2) as

$$\nabla_k = \left.\frac{d\xi}{dw}\right|_{w=w_k} = 2\lambda(w_k - w^*) \tag{4.5}$$

The dynamic or transient behavior of the iterative process, from initial value w_0 to optimal solution w^*, can be analyzed through the equation formed when (4.5) is substituted into (4.4):

$$w_{k+1} = w_k - 2\mu\lambda(w_k - w^*) \tag{4.6}$$

†Note that here, unlike Chapter 3, we are using the subscript to designate the iteration number rather than the weight number.

Rearranging terms, we obtain

$$w_{k+1} = (1 - 2\mu\lambda)w_k + 2\mu\lambda w^* \tag{4.7}$$

This equation is a linear, first-order, constant-coefficient, ordinary difference equation [1]. It can be solved by induction from the first few iterations. Starting with the initial guess w_0, the first three iterations of (4.7) yield

$$w_1 = (1 - 2\mu\lambda)w_0 + 2\mu\lambda w^* \tag{4.8}$$

$$w_2 = (1 - 2\mu\lambda)^2 w_0 + 2\mu\lambda w^*[(1 - 2\mu\lambda) + 1] \tag{4.9}$$

$$w_3 = (1 - 2\mu\lambda)^3 w_0 + 2\mu\lambda w^*\left[(1 - 2\mu\lambda)^2 + (1 - 2\mu\lambda) + 1\right] \tag{4.10}$$

From these results we can generalize to the kth iteration:

$$w_k = (1 - 2\mu\lambda)^k w_0 + 2\mu\lambda w^* \sum_{n=0}^{k-1} (1 - 2\mu\lambda)^n \tag{4.11}$$

$$= (1 - 2\mu\lambda)^k w_0 + 2\mu\lambda w^* \frac{1 - (1 - 2\mu\lambda)^k}{1 - (1 - 2\mu\lambda)} \tag{4.12}$$

$$= w^* + (1 - 2\mu\lambda)^k (w_0 - w^*) \tag{4.13}$$

This result gives us w_k explicitly at any point in the search procedure, and is thus a "solution" to the gradient search algorithm.

STABILITY AND RATE OF CONVERGENCE

In (4.13) the quantity $r = 1 - 2\mu\lambda$ is known as the "geometric ratio" because it is the ratio of successive terms in the geometric sum in (4.11). It is evident that r is the critical quantity in the one-weight iterative process. Equation (4.13) is defined to be "stable" if and only if

$$|r| = |1 - 2\mu\lambda| < 1 \tag{4.14}$$

This condition can also be expressed as

$$\frac{1}{\lambda} > \mu > 0 \tag{4.15}$$

If the condition in (4.14) or (4.15) is met, that is, if the algorithm in (4.13) is stable, the algorithm is seen to converge to the optimum solution:

$$\lim_{k \to \infty} [w_k] = w^* \tag{4.16}$$

The rate of convergence also obviously depends on the geometric ratio.

Figure 4.2 depicts typical action that takes place during the adjustment process for different values of the geometric ratio r. The lines have no physical significance but simply join the series of points representing the discrete values of w_k. Note that

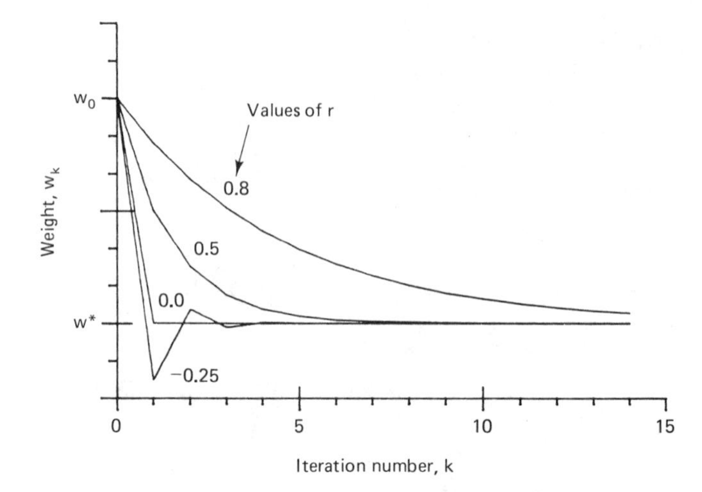

Figure 4.2 Weight adjustment for different values of the geometric ratio r. With $r = 0$ (Newton's method), w^* is reached in one iteration.

when the absolute value of r is less than 1, the rate of convergence increases as r decreases, reaching its maximum at $r = 0$, when the optimal solution is reached in a single step. Note also that for positive values of r of magnitude less than 1 there is no oscillation in the transient weight values, while for negative values the transient weight values overshoot the optimal setting and converge in a decaying oscillation. In the former case the process is said to be "overdamped" and in the latter "underdamped." When $r = 0$ the process is equivalent to Newton's method (discussed below) and is said to be "critically damped." When the absolute value of r is greater than or equal to 1, in accordance with (4.14), the process is unstable and does not converge.

The effects of the choice of μ on r and on the single-weight iterative process are summarized in Table 4.1.

TABLE 4.1 EFFECT OF μ ON CONVERGENCE OF THE SINGLE-WEIGHT GRADIENT SEARCH PROCESS

Stable (convergent)	$0 < \mu < \frac{1}{\lambda}$	$\lvert r\rvert < 1$
Overdamped	$0 < \mu < \frac{1}{2\lambda}$	$1 > r > 0$
Critically damped	$\mu = \frac{1}{2\lambda}$	$r = 0$
Underdamped	$\frac{1}{2\lambda} < \mu < \frac{1}{\lambda}$	$0 > r > -1$
Unstable (not convergent)	$\mu \geq \frac{1}{\lambda}$ and $\mu \leq 0$	$\lvert r\rvert > 1$

THE LEARNING CURVE

The effect of variations in the adjustment of the weight on the mean-square error can be observed from (4.1). If we let ξ_k be defined as the value of the mean-square error when the weight is fixed at w_k, then we can write from (4.1)

$$\xi_k = \xi_{\min} + \lambda(w_k - w^*)^2 \tag{4.17}$$

Substituting w_k from (4.13) into this expression yields

$$\xi_k = \xi_{\min} + \lambda(w_0 - w^*)^2(1 - 2\mu\lambda)^{2k} \tag{4.18}$$

It is evident that since w_k undergoes a geometric progression toward w^*, the mean-square error also undergoes a geometric progression toward $\xi_{\min}$. The geometric ratio of the mean-square-error progression is seen in (4.18) to be

$$r_{\text{mse}} = r^2 = (1 - 2\mu\lambda)^2 \tag{4.19}$$

Since this ratio can never be negative, the mean-square-error progression will never be oscillatory. Stability is assured as before by satisfaction of condition (4.14).

For the single-weight system, Figure 4.3 shows the relaxation of the mean-square error from its initial value ξ_0 toward the optimal value $\xi_{\min}$. The instance shown represents a value of $r_{\text{mse}} = 0.5$, corresponding to $r = 0.707$. The curve once again

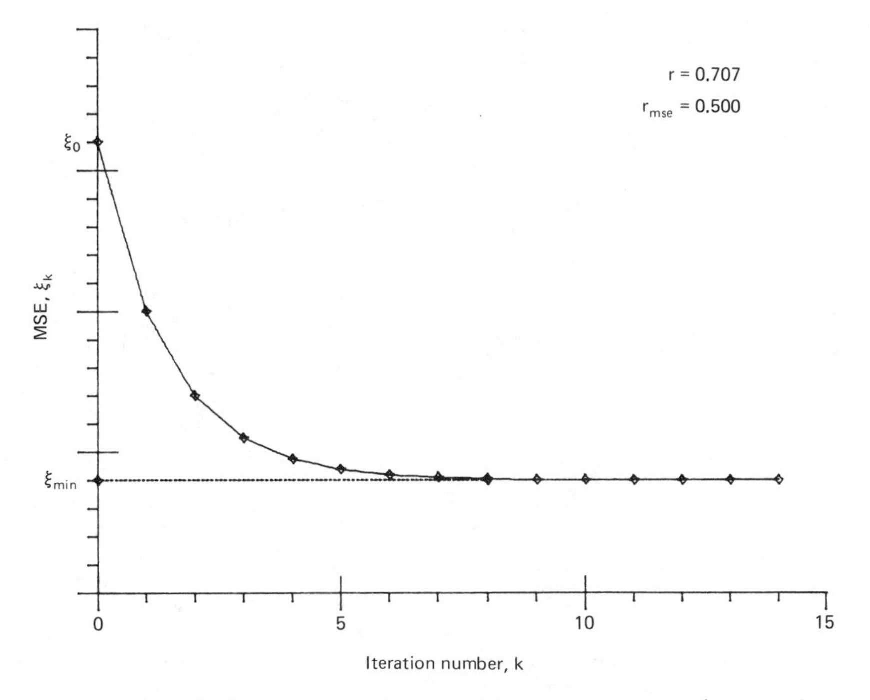

Figure 4.3 The "learning curve" is a plot of the mean-square error, ξ_k, versus k.

has no physical significance between integer values of k; it simply connects the discrete transient values of the error. The curve is called the "learning curve" and indicates the reduction of mean-square error during the iterative process.

GRADIENT SEARCH BY NEWTON'S METHOD

We have indicated that the univariable gradient search process is critically damped when $r = 0$, that is, where

$$r = 1 - 2\mu\lambda = 0 \qquad \mu = \frac{1}{2\lambda} \tag{4.20}$$

In this case the process converges in one step with quadratic mean-square-error functions and is called Newton's method because of its relationship to a method of elementary calculus for finding the roots of a polynomial. Let us now investigate the application of this method to particular single-weight (or univariable) functions and then extend its application to the multivariable performance surface.

Newton's method [2] is primarily a method for finding the zeros of a function, say $f(w)$, that is, a method for finding solutions to the equation $f(w) = 0$. The method consists of starting with an initial guess, w_0, and then using the first derivative, $f'(w_0)$, to compute the next estimate, w_1. As shown in Figure 4.4, w_1 is found where the tangent at $f(w_0)$ intersects the w-axis. Thus, using the geometry in Figure 4.4,

$$f'(w_0) = \frac{f(w_0)}{w_0 - w_1}$$

or

$$w_1 = w_0 - \frac{f(w_0)}{f'(w_0)} \tag{4.21}$$

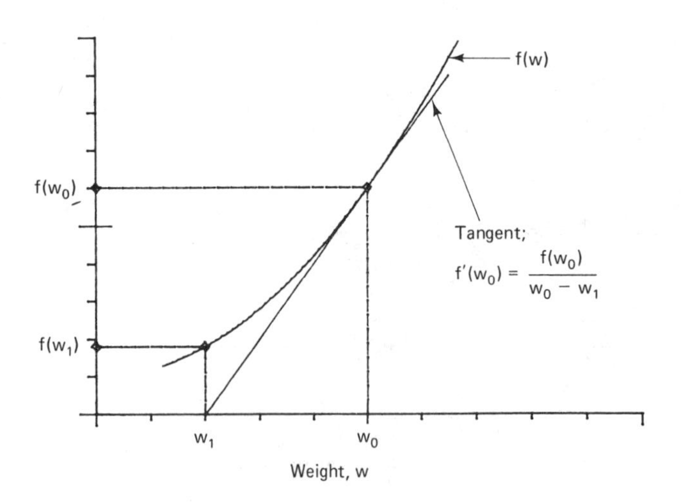

Figure 4.4 Newton's method for finding a zero of $f(w)$ proceeds from w_0 to w_1 using the tangent to $f(w)$.

The next point, w_2, is computed using w_1 as the initial guess, and so on. In general, then,

$$w_{k+1} = w_k - \frac{f(w_k)}{f'(w_k)}; \qquad k = 0, 1, \ldots \tag{4.22}$$

The convergence of Newton's method obviously depends on the initial guess, w_0, and on the nature of $f(w)$, but it is known to converge rapidly for a large class of functions.

Equation (4.22) is called the *continuous form* of Newton's method because the continuous function, $f(w)$, and its derivative are used explicitly. A *discrete form* also exists, and is used where $f'(w)$ must be estimated. Assuming that $f(w)$ is known or can be estimated accurately, we can estimate $f'(w)$ using a backward difference formula:

$$f'(w_k) \approx \frac{f(w_k) - f(w_{k-1})}{w_k - w_{k-1}} \tag{4.23}$$

Using this approximation in (4.22), the discrete form of Newton's method may be expressed as

$$w_{k+1} = w_k - \frac{f(w_k)(w_k - w_{k-1})}{f(w_k) - f(w_{k-1})}; \qquad k = 0, 1, \ldots \tag{4.24}$$

Note that in (4.24) as well as in (4.22), steps must be taken to assure that the denominator does not vanish at any iteration. For the present, we will use the continuous form of Newton's approximation.

To apply Newton's method in searching performance surfaces, we must begin with an equation of the form $f(w) = 0$, as above. This of course would be $\xi'(w) = 0$, or $\nabla = 0$ in general, as in (2.16), because we want to find the minimum of $\xi(w)$. Thus for the univariable performance surface, we use

$$f(w) = \xi'(w) \tag{4.25}$$

so that, for example, the continuous form in (4.22) becomes

$$w_{k+1} = w_k - \frac{\xi'(w_k)}{\xi''(w_k)}; \qquad k = 0, 1, \ldots \tag{4.26}$$

Note that the discrete form of the algorithm using (4.23) now involves an approximation to the second derivative of ξ.

When the performance surface is quadratic, as it has been in all of our previous discussions, the application of Newton's method leads to a one-step solution as shown in Figure 4.2. Using (4.2) in the algorithm (4.26) with $k = 0$ gives

$$w_1 = w_0 - \frac{2\lambda(w_0 - w^*)}{2\lambda} = w^* \tag{4.27}$$

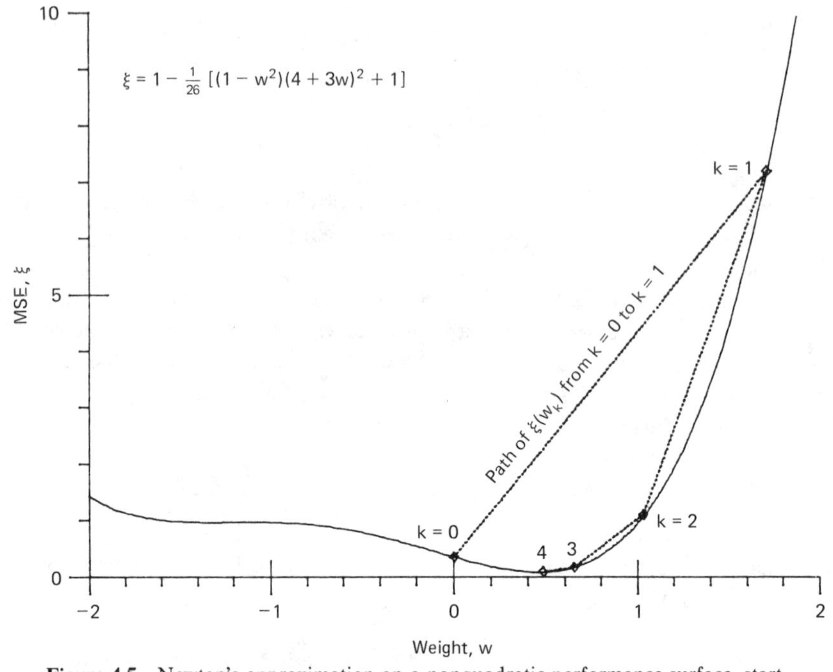

Figure 4.5 Newton's approximation on a nonquadratic performance surface, starting at $w = 0$.

Thus Newton's method works simply in the univariate case where the performance surface is quadratic and specified for all values of w.

There are two ways in which Newton's method can become more complicated in the univariate case. First, ξ' and ξ'' in (4.26) must be estimated if $\xi(w)$ is not known exactly. The subject of gradient estimation is addressed in Chapter 5. Second, the performance surface may not be quadratic. So far we have discussed only the adaptive linear combiner, for which ξ is quadratic, but we will later introduce other adaptive structures with nonquadratic performance surfaces. An example is shown in Figure 4.5 using a performance surface for a particular recursive adaptive filter. Note that even though the performance surface is nonquadratic, when the initial weight is at $w = 0$, Newton's method goes almost to the optimum, $w^* = 0.448$, after only four iterations. For some initial weight values, however, Newton's method will not find the optimum in this example, further development of which is left to Exercises 7 through 9.

NEWTON'S METHOD IN MULTIDIMENSIONAL SPACE

We have seen that Newton's method finds the optimum weight, w^*, in a single step when there is one weight and the performance surface is quadratic. We extend Newton's method to the multivariable case with many weights simply by defining it as a method which will perform similarly, that is, which will go to the optimum in one step on a quadratic performance surface.

Recall that in (2.17) the optimum weight vector is given by

$$\mathbf{W}^* = \mathbf{R}^{-1}\mathbf{P} \tag{4.28}$$

and that the gradient vector in (2.13) is

$$\nabla = 2\mathbf{RW} - 2\mathbf{P} \tag{4.29}$$

We can multiply (4.29) on the left by $\frac{1}{2}\mathbf{R}^{-1}$ and then combine these two equations to obtain

$$\mathbf{W}^* = \mathbf{W} - \tfrac{1}{2}\mathbf{R}^{-1}\nabla \tag{4.30}$$

We change this result into an adaptive algorithm as follows:†

$$\mathbf{W}_{k+1} = \mathbf{W}_k - \tfrac{1}{2}\mathbf{R}^{-1}\nabla_k \tag{4.31}$$

The subscript (k) on the gradient vector implies that the gradient is measured at step k, where the weight vector is $\mathbf{W}_k$.

Equation (4.31) is thus Newton's method for the multivariable case. When the error surface is quadratic, the method proceeds to the optimum solution in one step, as in (4.30). The two-weight, quadratic case is illustrated in Figure 4.6. In this "perfect" setting the weights jump from any initial setting, $\mathbf{W}_0 = (w_{00}, w_{10})$, to the optimum setting $\mathbf{W}^* = (w_0^*, w_1^*)$ in a single step.

As shown in Figure 4.6 and in (4.31), the steps in Newton's method do not proceed in the direction of the gradient. To do so, the weight path in Figure 4.6 would need to be perpendicular to each contour line. We observe that this would be the case only if $\mathbf{W}_0$ describes a point on one of the principal axes.

Notice in (4.31) that we could generalize Newton's method by reintroducing the constant, μ, introduced previously in (4.4) to regulate the convergence rate. If we change (4.31) to

$$\mathbf{W}_{k+1} = \mathbf{W}_k - \mu\mathbf{R}^{-1}\nabla_k \tag{4.32}$$

then we obtain the one-step formula by setting $\mu = \frac{1}{2}$. Otherwise, we could choose any other value of μ in the stable range seen in (4.35) below, that is,

$$0 < \mu < 1 \tag{4.33}$$

However, reasons for desiring overdamped operation and a smaller step size with μ less than $\frac{1}{2}$ are discussed in the next section. In (4.32), μ is dimensionless.

Furthermore, we can obtain a solution to (4.32) on the quadratic performance surface by substituting (4.29) for the gradient term and then using (4.28) to obtain

$$\mathbf{W}_{k+1} = (1 - 2\mu)\mathbf{W}_k + 2\mu\mathbf{W}^* \tag{4.34}$$

Now we have an equation in the form of (4.7), and we can obtain a solution

†Equation (4.31) can also be derived by approximating ξ with a truncated Taylor series (see Luenberger [3]).

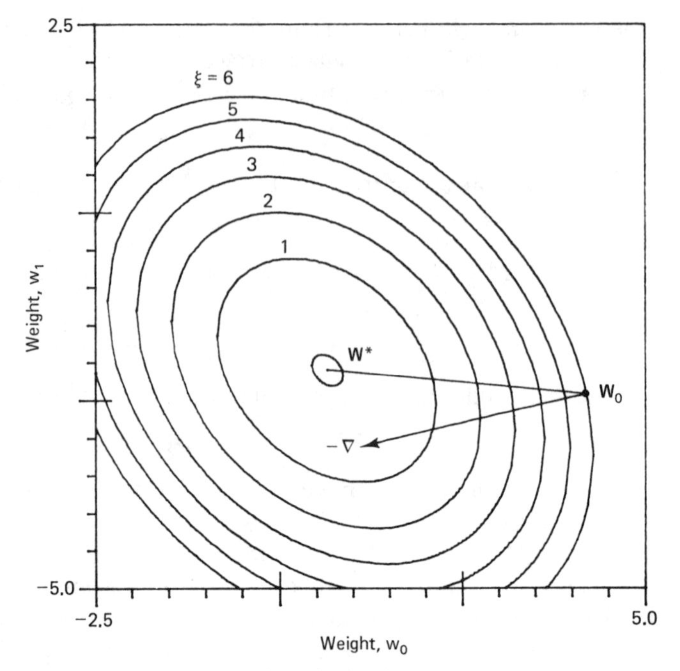

Figure 4.6 Illustration of Newton's method with $\mu = 1$ and 2 weights. The quadratic performance surface is the same as in Figure 3.1.

inductively just as (4.13) was obtained from (4.7). The corresponding solution here is

$$\mathbf{W}_k = \mathbf{W}^* + (1 - 2\mu)^k(\mathbf{W}_0 - \mathbf{W}^*) \tag{4.35}$$

As a check on this solution, observe that the one-step algorithm, $\mathbf{W}_1 = \mathbf{W}^*$, results when we set $\mu = \frac{1}{2}$, and that $\mathbf{W}_\infty = \mathbf{W}^*$ results when the condition (4.33) holds.

GRADIENT SEARCH BY THE METHOD OF STEEPEST DESCENT

The second principal search method to be discussed in this chapter is called the method of steepest descent, because in this method, unlike Newton's method, the weights are adjusted in the direction of the gradient at each step. An example is given in Figure 4.7 using the same quadratic performance surface as in Figure 4.6. Small steps are used in Figure 4.7 so that the path of steepest descent can be seen. This is unlike Figure 4.6, where convergence occurs in one large step.

Convergence in one step is a source of satisfaction to a numerical analyst who would like to minimize the number of iterations necessary to accomplish a surface search. To an adaptive system designer, however, it is generally too fast and not really desirable. In numerical analysis one can expect to be given the function to be

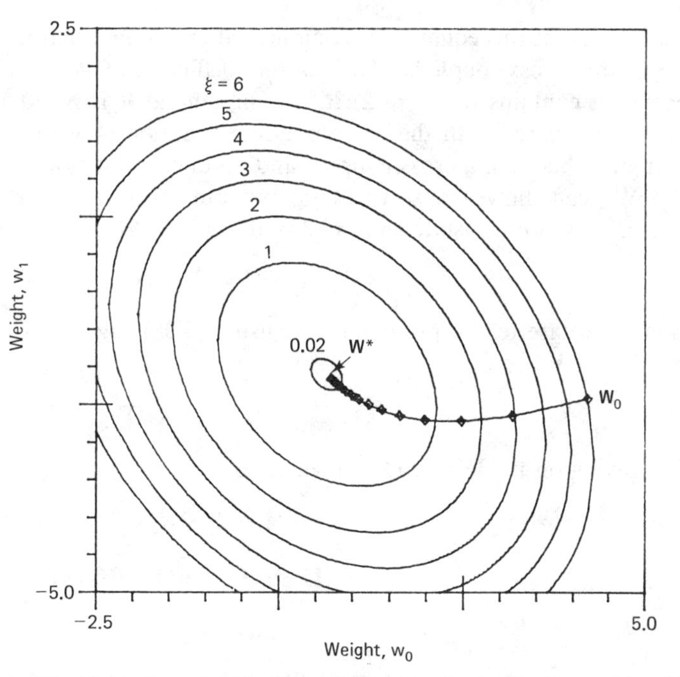

Figure 4.7 Illustration of the steepest-descent method with two weights. The quadratic performance surface is the same as in Figures 4.6 and 3.1, and $\mu = 0.3$.

searched, but in many practical adaptive system applications the performance function is unknown and must be measured or estimated on the basis of stochastic input data. Slow adaptation provides a filtering process that ameliorates the effects of gradient measurement noise. Newton's method is thus not as useful for devising practical algorithms as certain other methods, among which the method of steepest descent has thus far proven to be the most widely applicable.

From its definition, we can see that the method of steepest descent is expressed by the following algorithm, in which μ is a constant that regulates the step size and has dimensions of reciprocal signal power:

$$\mathbf{W}_{k+1} = \mathbf{W}_k + \mu(-\nabla_k) \tag{4.36}$$

Recall that we had a one-dimensional version of (4.36) in (4.4). The nature of the relaxation process resulting when this algorithm is used to search a quadratic performance surface can be determined by substituting (4.29) for the gradient term and then using (4.28):

$$\begin{aligned} \mathbf{W}_{k+1} &= \mathbf{W}_k - 2\mu\mathbf{R}\mathbf{V}_k \\ &= \mathbf{W}_k + 2\mu\mathbf{R}(\mathbf{W}^* - \mathbf{W}_k) \end{aligned} \tag{4.37}$$

Rearranging yields

$$\mathbf{W}_{k+1} = (\mathbf{I} - 2\mu\mathbf{R})\mathbf{W}_k + 2\mu\mathbf{R}\mathbf{W}^* \tag{4.38}$$

The solution of this equation is complicated by the fact that the various components of $\mathbf{W}_k$ are "cross-coupled." The matrix coefficient of $\mathbf{W}_k$ in (4.38) is not diagonal because it contains the term $2\mu\mathbf{R}$, and in general $\mathbf{R}$ is not diagonal. The equation may be compared with the corresponding equation (4.34) for Newton's method to distinguish between a cross-coupled and an uncoupled system.

We can, however, solve (4.38) by transforming to the principal coordinate system. First we translate as in (3.37), using $\mathbf{V} = \mathbf{W} - \mathbf{W}^*$, so that (4.38) becomes

$$\mathbf{V}_{k+1} = (\mathbf{I} - 2\mu\mathbf{R})\mathbf{V}_k \tag{4.39}$$

Then we rotate to the principal axes using (3.38) together with (3.13); that is, we have $\mathbf{V} = \mathbf{Q}\mathbf{V}'$, so

$$\mathbf{Q}\mathbf{V}'_{k+1} = (\mathbf{I} - 2\mu\mathbf{R})\mathbf{Q}\mathbf{V}'_k \tag{4.40}$$

Multiplying on the left by $\mathbf{Q}^{-1}$ then gives

$$\begin{aligned}\mathbf{V}'_{k+1} &= \mathbf{Q}^{-1}(\mathbf{I} - 2\mu\mathbf{R})\mathbf{Q}\mathbf{V}'_k \\ &= (\mathbf{Q}^{-1}\mathbf{I}\mathbf{Q} - 2\mu\mathbf{Q}^{-1}\mathbf{R}\mathbf{Q})\mathbf{V}'_k \\ &= (\mathbf{I} - 2\mu\mathbf{\Lambda})\mathbf{V}'_k\end{aligned} \tag{4.41}$$

Now the eigenvalue matrix, $\mathbf{\Lambda}$, is diagonal as in (3.4), so (4.41) represents a set of $L + 1$ equations of the form in (4.7). Thus it is clear that there is no cross-coupling in the principal-coordinate system. Furthermore, reasoning inductively, the solution to (4.41) must be

$$\mathbf{V}'_k = (\mathbf{I} - 2\mu\mathbf{\Lambda})^k\mathbf{V}'_0 \tag{4.42}$$

The result in (4.42) shows that the steepest-descent algorithm is stable and convergent when

$$\lim_{k\to\infty} (\mathbf{I} - 2\mu\mathbf{\Lambda})^k = \mathbf{0} \tag{4.43}$$

Since the product of two diagonal matrices is just the matrix of products of corresponding elements, (4.43) may also be written

$$\begin{bmatrix} \lim_{k\to\infty}(1 - 2\mu\lambda_0)^k & & & \\ & \lim_{k\to\infty}(1 - 2\mu\lambda_1)^k & & \\ & & \ddots & \\ & & & \lim_{k\to\infty}(1 - 2\mu\lambda_L)^k \end{bmatrix} = \mathbf{0} \tag{4.44}$$

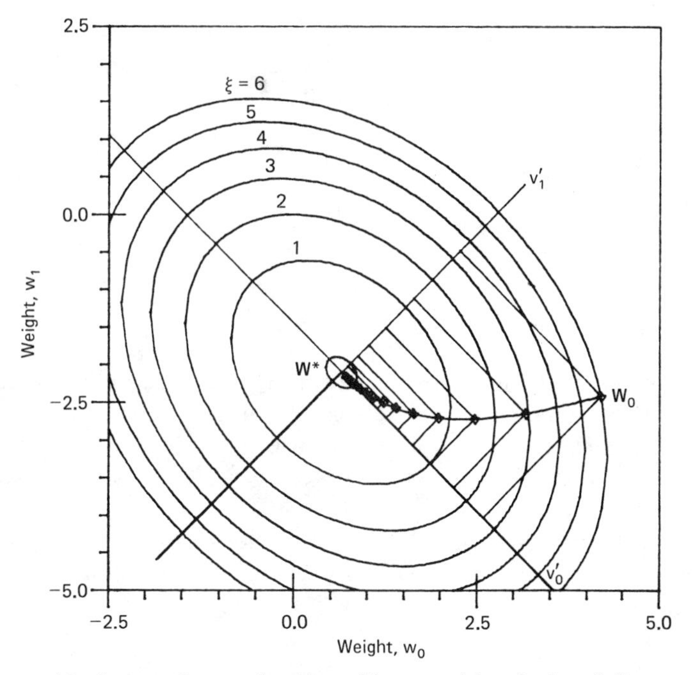

Figure 4.8 Steepest-descent algorithm with two weights. Ratios of distances are constant along principal axes v_0' and v_1', in accordance with (4.48).

In this form, we see that the convergence condition is satisfied by choosing μ so that

$$0 < \mu < \frac{1}{\lambda_{\max}} \tag{4.45}$$

where $\lambda_{\max}$ is the largest eigenvalue of $\mathbf{R}$. Condition (4.45) is necessary and sufficient for convergence of the steepest-descent algorithm with a quadratic performance surface. If this condition is satisfied, it follows that

$$\lim_{k\to\infty} \mathbf{V}_k' = \mathbf{0} \tag{4.46}$$

If we substitute $\mathbf{V}' = \mathbf{Q}^{-1}\mathbf{V} = \mathbf{Q}^{-1}(\mathbf{W} - \mathbf{W}^*)$ into (4.46) and transform back to natural coordinates, we find that

$$\lim_{k\to\infty} \mathbf{W}_k = \mathbf{W}^* \tag{4.47}$$

So the steepest-descent method is in general stable and convergent if and only if the condition in (4.45) is met.

Another example of the steepest-descent algorithm with a two-dimensional quadratic surface is illustrated in Figure 4.8. The principal axes, v_0' and v_1', are included. In accordance with (4.42), convergence takes place independently along each of the principal axes. The rate of convergence on each axis is governed by a unique geometric ratio. As seen in (4.44), these ratios are as follows:

$$\begin{aligned} r_0 &= 1 - 2\mu\lambda_0 \\ r_1 &= 1 - 2\mu\lambda_1 \\ &\vdots \\ r_L &= 1 - 2\mu\lambda_L \end{aligned} \tag{4.48}$$

That is, as the iterative process advances, the sequence of projections of $\mathbf{W}_k$ along each of the primed axes is purely geometric according to a ratio determined by the corresponding eigenvalue. The sequence of projections of $\mathbf{W}_k$ along the original unprimed coordinates is a sum of geometric progressions and is thus more complicated.

A description of the dynamic behavior of the iterative process in the natural coordinates can be obtained by reexpressing (4.42) in terms of $\mathbf{W}_k$. We first premultiply both sides of (4.42) by $\mathbf{Q}$ to obtain

$$\mathbf{QV}_k' = \mathbf{Q}(\mathbf{I} - 2\mu\mathbf{\Lambda})^k\mathbf{V}_0' \tag{4.49}$$

Again we transform to natural coordinates using $\mathbf{V}' = \mathbf{Q}^{-1}(\mathbf{W} - \mathbf{W}^*)$ and obtain

$$\mathbf{W}_k = \mathbf{W}^* + \mathbf{Q}(\mathbf{I} - 2\mu\mathbf{\Lambda})^k\mathbf{Q}^{-1}(\mathbf{W}_0 - \mathbf{W}^*) \tag{4.50}$$

Next, we use the following interesting relationship:

$$\begin{aligned} (\mathbf{QAQ}^{-1})^k &= \mathbf{QAQ}^{-1}\mathbf{QAQ}^{-1} \cdots \mathbf{QAQ}^{-1} \\ &= \mathbf{QA}^k\mathbf{Q}^{-1} \end{aligned} \tag{4.51}$$

where $\mathbf{A}$ is any matrix for which these products exist. Using $\mathbf{A} = \mathbf{I} - 2\mu\mathbf{\Lambda}$ in (4.50), and also (3.5) again, we have

$$\begin{aligned} \mathbf{W}_k &= \mathbf{W}^* + (\mathbf{QIQ}^{-1} - 2\mu\mathbf{Q\Lambda Q}^{-1})^k(\mathbf{W}_0 - \mathbf{W}^*) \\ &= \mathbf{W}^* + (\mathbf{I} - 2\mu\mathbf{R})^k(\mathbf{W}_0 - \mathbf{W}^*) \end{aligned} \tag{4.52}$$

This is the "solved" version of the steepest-descent algorithm in natural coordinates.

In summary, we have seen that the steepest-descent algorithm proceeds always in the direction of the gradient of the performance surface. The general form of the algorithm was given in (4.36), and its alternative or "solved" form in (4.52). The algorithm was seen to be stable if the condition in (4.45) is met.

COMPARISON OF LEARNING CURVES

It is interesting to compare the method of steepest descent with Newton's method by comparing learning curves. To derive formulas for the two learning curves, we begin with the quadratic mean-square-error function in (2.33) and (3.34), namely

$$\xi = \xi_{\min} + \mathbf{V}^{\mathrm{T}}\mathbf{R}\mathbf{V} \tag{4.53}$$

Into this we substitute the solution obtained in (4.35) for Newton's method, translated in accordance with (3.37). We can thus write

$$\xi_k = \xi_{\min} + (1 - 2\mu)^{2k}\mathbf{V}_0^{\mathrm{T}}\mathbf{R}\mathbf{V}_0 \tag{4.54}$$

This is a simple geometric progression having a single geometric ratio of

$$r_{\text{mse}} = r^2 = (1 - 2\mu)^2 \tag{4.55}$$

A typical learning curve for Newton's method, showing mean-square error as a function of number of iterations, is presented in Figure 4.9. This is a plot of a pure exponential function having a single time constant.

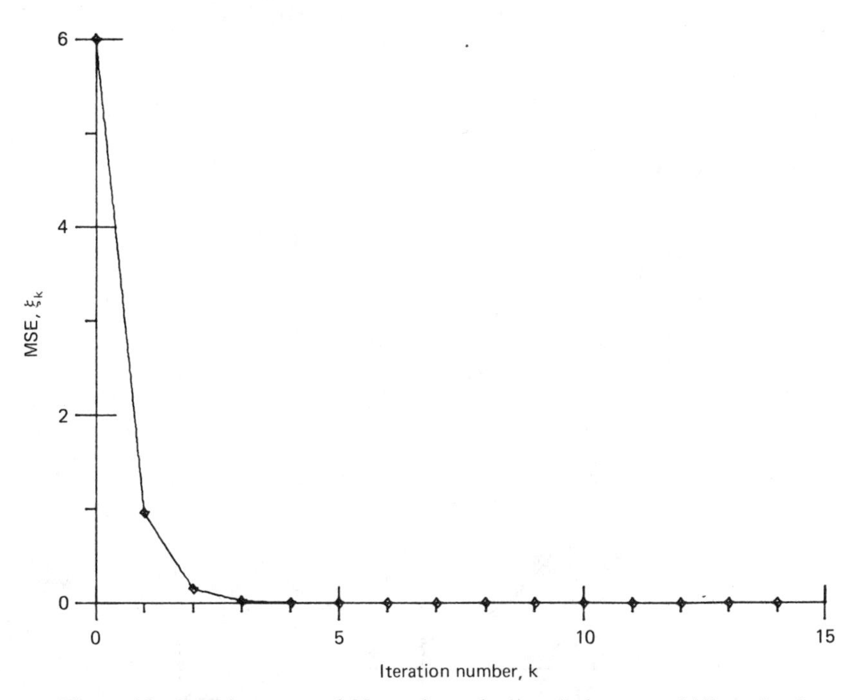

Figure 4.9 Learning curve of Newton's method applied to a multidimensional quadratic performance function. The ratio of each successive pair of ordinates is $(1 - 2\mu)^2$. In this example, the error surface is given by (2.24), its contours are in Figure 4.6, and $\mu = 0.3$.

The learning curve for the steepest-descent method is obtained similarly. We begin this time with the principal-coordinate version of (4.53), which is given in (3.35) as

$$\xi = \xi_{\min} + \mathbf{V}'^{\mathrm{T}}\mathbf{\Lambda}\mathbf{V}' \tag{4.56}$$

When the steepest-descent solution in (4.42) is substituted, (4.56) becomes

$$\begin{aligned}\xi_k &= \xi_{\min} + \left[(\mathbf{I} - 2\mu\mathbf{\Lambda})^k\mathbf{V}_0'\right]^{\mathrm{T}}\mathbf{\Lambda}(\mathbf{I} - 2\mu\mathbf{\Lambda})^k\mathbf{V}_0' \\ &= \xi_{\min} + \mathbf{V}_0'^{\mathrm{T}}\left[(\mathbf{I} - 2\mu\mathbf{\Lambda})^k\right]^{\mathrm{T}}\mathbf{\Lambda}(\mathbf{I} - 2\mu\mathbf{\Lambda})^k\mathbf{V}_0' \end{aligned} \tag{4.57}$$

Now the matrices $(\mathbf{I} - 2\mu\mathbf{\Lambda})$ and $\mathbf{\Lambda}$ are of course diagonal and we know that products of diagonal matrices are commutative. Therefore,

$$\xi_k = \xi_{\min} + \mathbf{V}_0'^{\mathrm{T}}(\mathbf{I} - 2\mu\mathbf{\Lambda})^{2k}\mathbf{\Lambda}\mathbf{V}_0' \tag{4.58}$$

$$= \xi_{\min} + \sum_{n=0}^{L} v_{0n}'^2\lambda_n(1 - 2\mu\lambda_n)^{2k} \tag{4.59}$$

The derivation of the final result in (4.59) is left to Exercise 18.

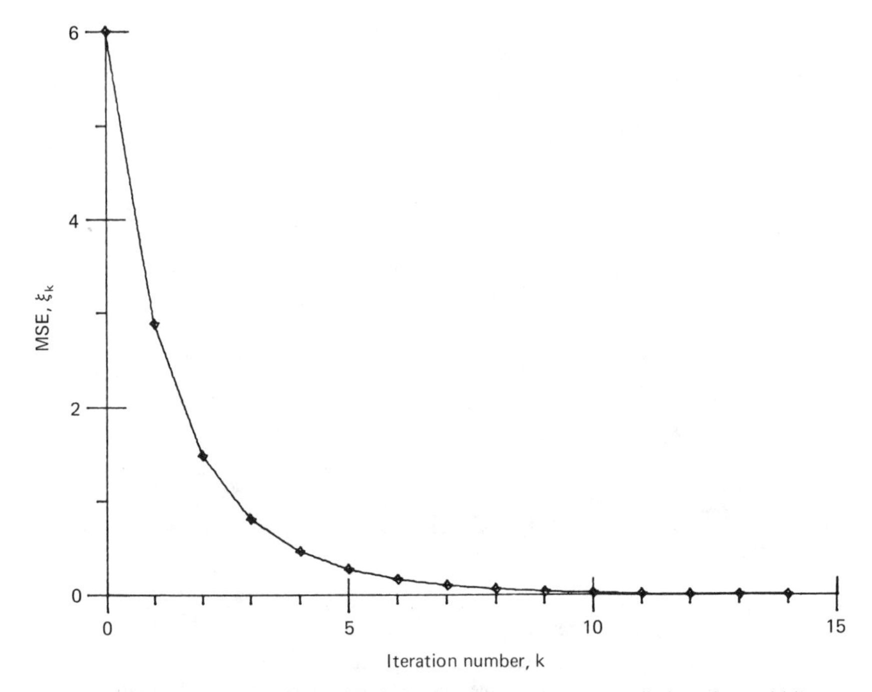

Figure 4.10 Learning curve of method of steepest descent applied to the multidimensional quadratic performance function given by (2.24). Contours are shown in Figure 4.7, and $\mu = 0.3$.

So we see that the learning curve for the steepest-descent method is the sum of decaying geometric progressions with geometric ratios of the form

$$(r_{mse})_n = r_n^2 = (1 - 2\mu\lambda_n)^2 \tag{4.60}$$

Figure 4.10 shows a learning curve for the method of steepest descent as applied to a quadratic performance surface. This curve is like a sum of exponentials having as many modes as there are weights.

Comparing Figures 4.9 and 4.10, we note that Newton's method seems to converge more rapidly than the steepest-descent method, given that μ and other factors are the same in both cases. Newton's method does indeed generally converge more rapidly, because it uses the information in the R matrix to find the direct path on the error surface to ξ_{min}. As suggested by Figures 4.7 and 4.8, the steepest-descent algorithm generally finds a longer path to ξ_{min}. This discussion of the convergence of the two methods is continued later, in Chapter 8.

Newton's method and the method of steepest descent may be regarded from another point of view as feedback processes: that is, as examples of what we have previously termed "performance feedback." We will show later how the performance feedback concept can be applied to analyze the gradient search of a quadratic performance surface.

EXERCISES

1. A single-weight performance surface has the parameters $\lambda = 0.1$, $\xi_{min} = 0$, and $w^* = 2$. Write an expression for this performance surface.
2. In Exercise 1, if the initial guess at the optimum weight is $w = 0$ and the convergence parameter is $\mu = 4$, what are the first five choices of w using the simple gradient search algorithm discussed at the beginning of this chapter?
3. Do Exercise 2 with $\mu = 8$.
4. For the univariable performance surface given by

$$\xi = 0.4w^2 + 4w + 11$$

   what range of values of the convergence parameter will provide an overdamped weight-adjustment curve?
5. Write an expression for and plot the learning curve for the performance surface in Exercise 4, given an initial value $w_0 = 0$ and a convergence parameter $\mu = 1.5$.
6. Derive a discrete form of Newton's algorithm similar to Equation (4.26) but with differences instead of derivatives.
7. Derive a weight-adjustment formula to apply Newton's method to the performance surface in Figure 4.5.
8. Beginning with $w_0 = 0$, what are the first seven weight values in Figure 4.5, to four decimal places?

9. In Figure 4.5, find w_1 to four decimal places when $w_0 = 0, -0.08, -0.14, -1.20$, and -1.30. Explain the results and show why Newton's method may not apply to non-quadratic error surfaces.
10. Write an algorithm in the form of Equation (4.31) specifically for the performance surface in Figure 3.2. Demonstrate one-step convergence from any starting point.
11. Using the modified Newton method in Equation (4.32) with convergence parameter μ set at 0.1, what are the first five weight vectors beginning from $\mathbf{W}_0 = (5, 2)$ in Figure 3.2? What is $\mathbf{W}_{20}$?
12. Suppose that the inverse matrix, $\mathbf{R}^{-1}$, and the gradient vector, ∇, are specified in the following form:

$$\mathbf{R}^{-1} = \begin{bmatrix} \rho_0 & \rho_1 \\ \rho_1 & \rho_0 \end{bmatrix} \qquad \nabla = \begin{bmatrix} \alpha \\ \beta \end{bmatrix}$$

Write explicit weight-adjustment formulas for Newton's method and for the steepest-descent algorithm.

13. Suppose that we have the following specifications:

$$\mathbf{R} = \begin{bmatrix} 2 & 1 \\ 1 & 2 \end{bmatrix} \qquad \mathbf{P} = \begin{bmatrix} 7 \\ 8 \end{bmatrix}$$

Write explicit weight-adjustment formulas for the Newton and steepest-descent algorithms, using Equations (4.34) and (4.38). Use the results to explain the notion of cross coupling.

14. Using the steepest-descent method with $\mu = 0.1$, what are the first five weight vectors beginning from $\mathbf{W}_0 = (5, 2)$ in Figure 3.2? What is $\mathbf{W}_{20}$?
15. Given the error surface in Equation (3.44), plot the learning curve for Newton's method when the initial weights are zero and the convergence parameter is $\mu = 0.05$.
16. Write the difference equations that describe the learning curves for Newton's method and the steepest-descent method in terms of natural coordinates.
17. Given the error surface in Equation (3.44), plot the learning curve for the steepest-descent method when the initial weights are zero and $\mu = 0.05$.
18. Derive Equation (4.59) from Equation (4.58) by carrying out the matrix products, element by element.
19. Explain why μ is dimensionless in (4.32) and has dimensions of reciprocal signal power in (4.36).

ANSWERS TO SELECTED EXERCISES

1. $\xi = 0.1(w - 2)^2$
2. $w_0 = 0, w_1 = 1.6, w_2 = 1.92, \ldots, w_5 = 1.99936$
4. $0 < \mu < 1.25$
5. $\xi_k = 1 + 10(-0.2)^{2k}$
7. $w_{k+1} = w_k - \dfrac{(3w_k + 4)(6w_k^2 + 4w_k - 3)}{54w_k^2 + 72w_k + 7}$

8. $w_0 = 0, w_1 = 1.7143, w_2 = 1.0347, \ldots, w_6 = w_7 = 0.4484$

9. 0.4484, 0.4484, -1.3335, -1.1151, -1.3333

11. $\mathbf{W}_{20} = [2.0346 \quad 2.9885]^{\mathrm{T}}$

14. $\mathbf{W}_{20} = [2.0231 \quad 2.9769]^{\mathrm{T}}$

15. $\xi_k = 4 + 38(0.9)^{2k}$

17. $\xi_k = 4 + 0.5[(0.9)^{2k} + 75(0.7)^{2k}]$

REFERENCES AND ADDITIONAL READINGS

1. B. Gold and C. M. Rader, *Digital Processing of Signals*. New York: McGraw-Hill, 1969, Chap. 2.
2. G. B. Thomas, Jr., *Calculus and Analytic Geometry*, 4th ed. Reading, Mass.: Addison-Wesley, 1968, Sec. 10.3.
3. D. G. Luenberger, *Introduction to Linear and Nonlinear Programming*. Reading, Mass.: Addison-Wesley, 1973, Sec. 7.7.
4. P. Eykhoff, *System Identification*. New York: Wiley, 1974, Chap. 5.

chapter 5

Gradient Estimation and Its Effects on Adaptation

In Chapter 4 we assumed that an exact measurement of the gradient vector required by the adaptive process was available at each iteration. In most applications, however, an exact measurement is not available, and an estimate based on a limited statistical sample must be used. Such an estimate is "noisy" and can be considered as a true gradient plus additive noise.

The purpose of this chapter is to describe a general method of estimating the gradient vector and to show the effects of noise in the estimate on the adaptive process. The method presented, called "derivative measurement," is equivalent to square-wave dithering of the weight vector. Methods based on sinusoidal dithering have also been successfully employed. They differ analytically from square-wave dithering, but their effects on the adaptive process are basically the same.

Besides derivative measurement, there is an "instantaneous method" of estimating the gradient that does not depend on dithering the weight vector. This method is the basis for the LMS algorithm and is discussed in Chapter 6. It is less general than derivative measurement because it requires specific knowledge about the nature of the performance surface. Derivative measurement requires only very general knowledge of the performance surface, and is therefore discussed first in order to provide a general introduction to the subject of gradient estimation.

GRADIENT COMPONENT ESTIMATION BY DERIVATIVE MEASUREMENT

A single component of the gradient vector can be measured in the straightforward manner illustrated in Figure 5.1. The parabolic mean-square-error function of a single variable was given in (3.41) and (4.1). In terms of the coordinate $v = w - w^*$, we have

$$\xi = \xi_{\min} + \lambda v^2 \tag{5.1}$$

Similar to (4.2) and (4.3), the derivatives of this function are

$$\frac{d\xi}{dv} = 2\lambda v \qquad \frac{d^2\xi}{dv^2} = 2\lambda \tag{5.2}$$

We recall that only the first derivative is required to implement the method of steepest descent, but that the first and second derivatives are both needed in Newton's method.

As suggested in Figure 5.1, the derivatives in (5.2) are estimated numerically by taking "central differences" [1]. Thus

$$\frac{d\xi}{dv} \approx \frac{\xi(v+\delta) - \xi(v-\delta)}{2\delta} \tag{5.3}$$

$$\frac{d^2\xi}{dv^2} \approx \frac{\xi(v+\delta) - 2\xi(v) + \xi(v-\delta)}{\delta^2} \tag{5.4}$$

These approximations become exact as δ approaches zero. They are also exact even for finite values of δ when the performance function is quadratic in v. For quadratic

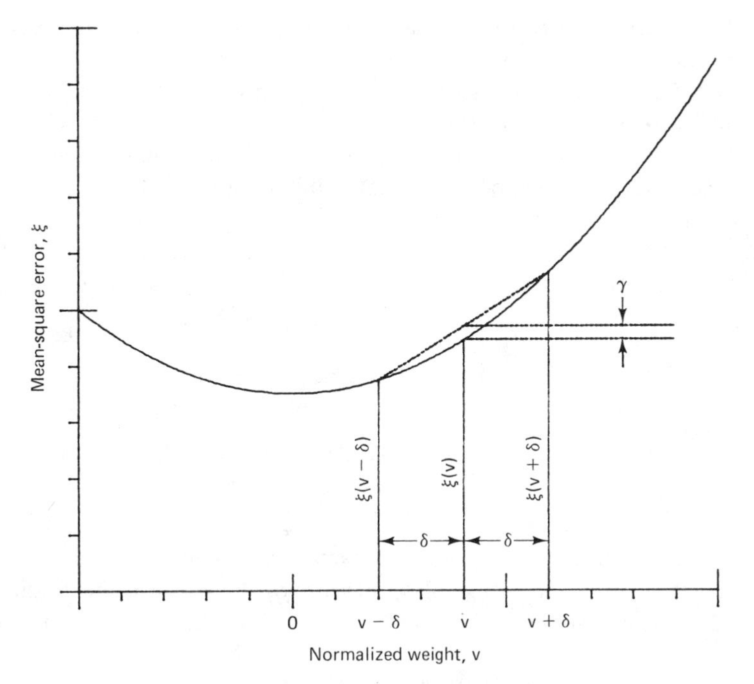

Figure 5.1 Derivative measurement.

performance surfaces, using (5.1), we have

$$
\begin{aligned}
\frac{\xi(v+\delta)-\xi(v-\delta)}{2\delta} &= \frac{\lambda(v+\delta)^2-\lambda(v-\delta)^2}{2\delta} \\
&= \frac{\lambda}{2\delta}(v^2+\delta^2+2v\delta-v^2-\delta^2+2v\delta) \\
&= 2\lambda v \\
&= \frac{d\xi}{dv}
\end{aligned}
\tag{5.5}
$$

$$
\begin{aligned}
\frac{\xi(v+\delta)-2\xi(v)+\xi(v-\delta)}{\delta^2} &= \frac{\lambda(v+\delta)^2-2\lambda(v)^2+\lambda(v-\delta)^2}{\delta^2} \\
&= \frac{2\lambda\delta^2}{\delta^2} \\
&= 2\lambda \\
&= \frac{d^2\xi}{dv^2}
\end{aligned}
\tag{5.6}
$$

THE PERFORMANCE PENALTY

The general procedure illustrated in Figure 5.1 requires that the weight adjustment be altered while the gradient measurement is being made. In order to estimate the first derivative, let us assume that the adaptive system spends time with the weight adjusted to $v-\delta$ and to $v+\delta$, in order to collect samples of the error and thus estimate $\xi(v-\delta)$ and $\xi(v+\delta)$. Suppose further that N samples of the error are accumulated at each point, so that equal amounts of time are spent at $v-\delta$ and at $v+\delta$, with no time being spent at v.

The "performance penalty" is defined as the increase, γ, brought about by misadjusting the weight and not leaving it set at v (see Figure 5.1). Thus

$$\gamma = \tfrac{1}{2}[\xi(v-\delta)+\xi(v+\delta)] - \xi(v) \tag{5.7}$$

For the one-weight quadratic performance function, the increase in ξ can be found explicitly by substituting (5.1) into (5.7). We have

$$
\begin{aligned}
\gamma &= \tfrac{1}{2}\left[2\xi_{\min}+\lambda(v-\delta)^2+\lambda(v+\delta)^2\right]-\left(\xi_{\min}+\lambda v^2\right) \\
&= \lambda\delta^2
\end{aligned}
\tag{5.8}
$$

In this result we see that γ is constant over a given performance function; it is not a function of v. A dimensionless measure of the effect of the gradient estimate on the adaptive adjustment, called the "perturbation" P, can further be defined in terms of

γ as follows:

$$P = \frac{\gamma}{\xi_{\min}} = \frac{\lambda\delta^2}{\xi_{\min}} \tag{5.9}$$

This expression gives the average increase in mean-square error normalized with respect to the minimum achievable mean-square error.

In the case of Newton's method, measurement of the second derivative using (5.4) requires an additional expenditure of time with v unperturbed in order to obtain the term $2\xi(v)$. This would obviously reduce the average perturbation, and thus the performance penalty. Since the second derivative in the case of quadratic performance surfaces is constant, however, it is measured only infrequently, and its effect on the perturbation can for practical purposes be ignored.

DERIVATIVE MEASUREMENT AND PERFORMANCE PENALTIES WITH MULTIPLE WEIGHTS

Derivative measurement to estimate a two-dimensional gradient is illustrated in Figure 5.2. With two weights, the quadratic performance function is derived from

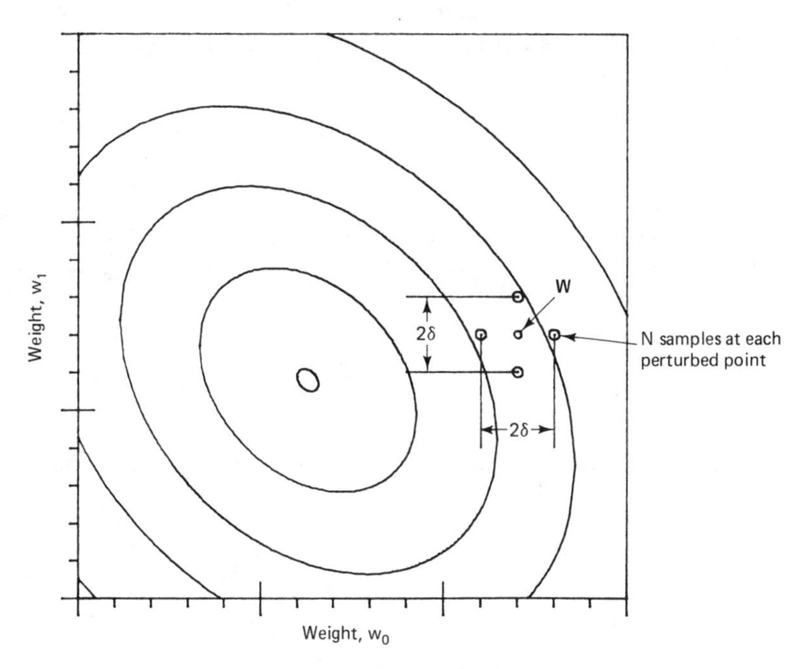

Figure 5.2 Two-dimensional derivative measurement.

(2.33) to be

$$\begin{aligned}\xi &= \xi_{\min} + \mathbf{V}^{\mathrm{T}}\mathbf{R}\mathbf{V}\\ &= \xi_{\min} + [v_0 \quad v_1]\begin{bmatrix} r_{00} & r_{01}\\ r_{10} & r_{11}\end{bmatrix}\begin{bmatrix} v_0\\ v_1\end{bmatrix}\\ &= \xi_{\min} + r_{00}v_0^2 + r_{11}v_1^2 + 2r_{01}v_0v_1 \qquad (5.10)\end{aligned}$$

When the partial derivative of this performance surface along coordinate v_0 is measured, the normalized performance penalty in terms of perturbation P_0, similar to (5.9), becomes

$$P_0 = \frac{r_{00}\delta^2}{\xi_{\min}} \qquad (5.11)$$

Similarly, the perturbation for derivative measurement along coordinate v_1 is

$$P_1 = \frac{r_{11}\delta^2}{\xi_{\min}} \qquad (5.12)$$

Assuming that equal time is required for the measurement of each gradient component (i.e., that $2N$ data samples are used for each measurement as before), the average perturbation during the entire measurement is given by

$$\begin{aligned}P &= \tfrac{1}{2}(P_0 + P_1)\\ &= \frac{\delta^2}{\xi_{\min}}\,\frac{r_{00} + r_{11}}{2} \qquad (5.13)\end{aligned}$$

Let us now define a general perturbation for $L + 1$ weights as the average of the perturbations of the individual gradient component measurements as follows:

$$P = \frac{\delta^2}{\xi_{\min}}\,\frac{\mathrm{trace}[\mathbf{R}]}{L + 1} \qquad (5.14)$$

Since the trace of $\mathbf{R}$ is equal to the sum of its eigenvalues as well as the sum of its diagonal elements, (5.14) can be reexpressed as

$$P = \frac{\delta^2}{\xi_{\min}}\,\frac{\sum_{n=0}^{L}\lambda_n}{L + 1} \qquad (5.15)$$

Further, since the sum of the eigenvalues divided by the number of eigenvalues is the average of the eigenvalues, (5.15) can be reexpressed as

$$P = \frac{\delta^2 \lambda_{av}}{\xi_{min}} \tag{5.16}$$

This is a convenient general expression for the perturbation with any number of weights, when the gradient is measured as in Figure 5.2.

VARIANCE OF THE GRADIENT ESTIMATE

Gradients estimated in the manner of Figures 5.1 and 5.2 are noisy because they are based on differences in noisy measurements of ξ, the mean-square error. The first step in determining the variance of the gradient estimate is thus to determine the variance in the estimate of ξ, where $\xi = E[\varepsilon_k^2]$ as in (2.13). Our estimate of ξ, denoted $\hat{\xi}$, is assumed to be based on N samples of ε_k^2.

For a review of the fundamental principles of variance, moments, and so on, the reader may refer to any elementary text on statistical or probability theory [2–5, 7]. To begin our derivation of the variance of $\hat{\xi}$ and of the gradient estimate, we first define an unbiased estimate of the rth moment of ε_k as

$$\hat{\alpha}_r = \frac{1}{N} \sum_{k=1}^{N} (\varepsilon_k)^r \tag{5.17}$$

From this definition we see that the true value or mean of $\hat{\alpha}_r$ is the expected value of ε_k^r, that is,

$$\alpha_r = E[\hat{\alpha}_r] = E[\varepsilon_k^r] \tag{5.18}$$

As an example of (5.18), let us derive the expected fourth moment, α_4, under the assumption that ε_k is Gaussian with zero mean; that is, the probability density of ε, denoted $p(\varepsilon)$ in general, is normal with mean equal to zero and standard deviation equal to σ. The true fourth moment is then

$$\begin{aligned} \alpha_4 &= \int_{-\infty}^{\infty} \varepsilon^4 p(\varepsilon)\, d\varepsilon \\ &= \int_{-\infty}^{\infty} \frac{\varepsilon^4}{\sigma\sqrt{2\pi}} e^{-\varepsilon^2/2\sigma^2}\, d\varepsilon = 3\sigma^4 \end{aligned} \tag{5.19}$$

Given $p(\varepsilon)$, we can find any true moment similarly. Note that when $r = 2$, we have in general

$$\alpha_2 = E[\varepsilon_k^2] = \xi \tag{5.20}$$

Having defined the true moment, we next define the variance of the moment estimate in (5.17) as the expected squared deviation from the mean, that is,

$$\text{var}[\hat{\alpha}_r] = E\left[(\hat{\alpha}_r - \alpha_r)^2\right] \tag{5.21}$$

Using (5.17) and (5.18) in this result, we have

$$\begin{aligned} \text{var}[\hat{\alpha}_r] &= E\left[\alpha_r^2 + \hat{\alpha}_r^2 - 2\alpha_r\hat{\alpha}_r\right] \\ &= E\left[\hat{\alpha}_r^2\right] - \alpha_r^2 \\ &= \frac{1}{N^2}\sum_{k=1}^{N}\sum_{l=1}^{N} E\left[(\varepsilon_k\varepsilon_l)^r\right] - \alpha_r^2 \end{aligned} \tag{5.22}$$

To simplify (5.22), we need to separate the N terms with $k = l$ from the rest of the terms in the summation. With $k = l$, each term is of the form $E[\varepsilon_k^{2r}]$. With k not equal to l, we assume ε_k and ε_l to be independent samples, and thus the expected product is the product of expected values. Hence

$$E\left[(\varepsilon_k\varepsilon_l)^r\right] = \begin{cases} E\left[\varepsilon_k^{2r}\right] = \alpha_{2r} & k = l \\ E\left[\varepsilon_k^r\right]E\left[\varepsilon_l^r\right] = \alpha_r^2 & k \neq l \end{cases} \tag{5.23}$$

We now use (5.23) in (5.22), noting that there are N^2 terms in the summation, N of which have $k = l$. The result is

$$\begin{aligned} \text{var}[\hat{\alpha}_r] &= \frac{1}{N^2}\left[N\alpha_{2r} + (N^2 - N)\alpha_r^2\right] - \alpha_r^2 \\ &= \frac{\alpha_{2r} - \alpha_r^2}{N} \end{aligned} \tag{5.24}$$

Accordingly, since $\alpha_2 = \xi$ as in (5.20), the variance of $\hat{\xi}$ is

$$\text{var}[\hat{\xi}] = \text{var}[\hat{\alpha}_2] = \frac{\alpha_4 - \alpha_2^2}{N} \tag{5.25}$$

The values of α in (5.25) depend, of course, on how ε_k is distributed, since each α is just the mean of a power of ε_k as in (5.18). For example, suppose that ε_k is distributed normally with zero mean and with a variance of σ_ε^2 [3, 5]. Then the mean fourth moment, α_4, was found in (5.19) to be $3\sigma_\varepsilon^4$, and the mean second moment, α_2, is just σ_ε^2. So in this case, (5.23) becomes

$$\text{var}[\hat{\xi}] = \frac{3\sigma_\varepsilon^4 - \sigma_\varepsilon^4}{N} = \frac{2\sigma_\varepsilon^4}{N} = \frac{2\xi^2}{N} \tag{5.26}$$

Thus, when ε_k is normally distributed with zero mean, the variance of the estimate of ε_k^2 is proportional to ξ^2 and inversely proportional to N.

From this result for the normal distribution of ε_k, we might anticipate that in general the variance of $\hat{\xi}$ could be expressed as

$$\text{var}[\hat{\xi}] = \frac{K\xi^2}{N} \tag{5.27}$$

We have shown that K is 2 when ε_k is distributed normally with zero mean. When the distribution is normal but with nonzero mean, K is somewhat less than 2. When the distribution is not Gaussian, K is also generally somewhat less than 2. A nonzero mean in a non-Gaussian ε_k may cause K to increase or decrease, depending on the distribution. Table 5.1 provides a summary of var$[\hat{\xi}]$ and the range of K for several different distributions of ε_k. We observe that $K = 2$ is either exact or

TABLE 5.1 VARIANCE OF THE MEAN-SQUARE-ERROR ESTIMATE

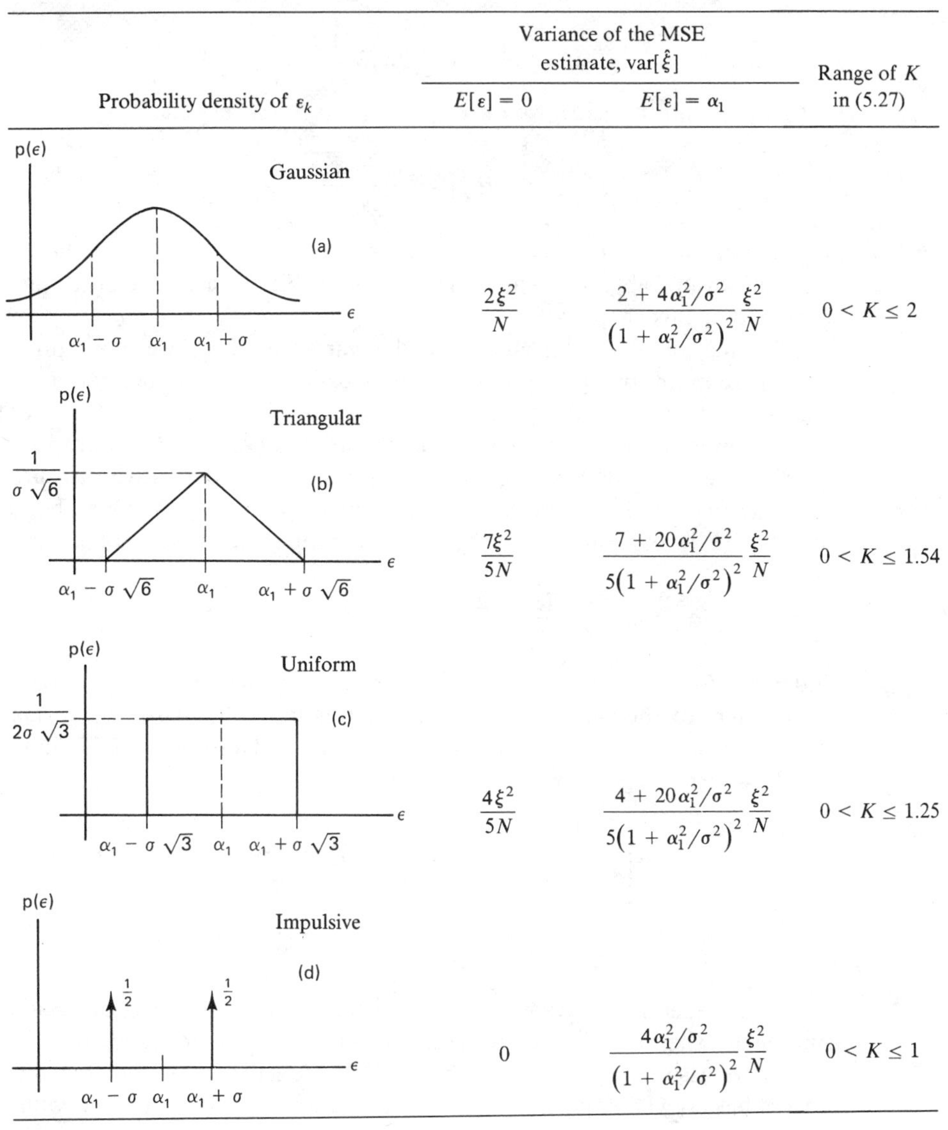

Probability density of ε_k	Variance of the MSE estimate, var$[\hat{\xi}]$: $E[\varepsilon] = 0$	Variance of the MSE estimate, var$[\hat{\xi}]$: $E[\varepsilon] = \alpha_1$	Range of K in (5.27)
Gaussian (a)	$\dfrac{2\xi^2}{N}$	$\dfrac{2 + 4\alpha_1^2/\sigma^2}{(1 + \alpha_1^2/\sigma^2)^2} \dfrac{\xi^2}{N}$	$0 < K \leq 2$
Triangular (b)	$\dfrac{7\xi^2}{5N}$	$\dfrac{7 + 20\alpha_1^2/\sigma^2}{5(1 + \alpha_1^2/\sigma^2)^2} \dfrac{\xi^2}{N}$	$0 < K \leq 1.54$
Uniform (c)	$\dfrac{4\xi^2}{5N}$	$\dfrac{4 + 20\alpha_1^2/\sigma^2}{5(1 + \alpha_1^2/\sigma^2)^2} \dfrac{\xi^2}{N}$	$0 < K \leq 1.25$
Impulsive (d)	0	$\dfrac{4\alpha_1^2/\sigma^2}{(1 + \alpha_1^2/\sigma^2)^2} \dfrac{\xi^2}{N}$	$0 < K \leq 1$

conservatively high for all cases in Table 5.1, so we will be using $K = 2$ for subsequent discussions.

To illustrate a non-Gaussian case in Table 5.1, suppose that ε_k is uniformly distributed with zero mean and with a standard deviation of σ. Then, similar to (5.19), the expected moments for even values of r are

$$\begin{aligned}\alpha_r &= \int_{-\infty}^{\infty} \varepsilon^r p(\varepsilon)\, d\varepsilon \\ &= \int_{-\sigma\sqrt{3}}^{\sigma\sqrt{3}} \varepsilon^r \left(\frac{1}{2\sigma\sqrt{3}} \right) d\varepsilon = \frac{\sigma^r 3^{r/2}}{r+1}\end{aligned} \tag{5.28}$$

Using this result in (5.25), we have

$$\mathrm{var}[\hat{\xi}] = \frac{\alpha_4 - \alpha_2^2}{N} = \frac{4\sigma^4}{5N} \tag{5.29}$$

Since the mean of ε_k is zero in this case we have $\xi = \sigma^2$, so (5.29) is the same as the result shown in Table 5.1. The other cases in Table 5.1 are covered in the exercises at the end of this chapter.

Having obtained formulas for the variance of $\hat{\xi}$, we now proceed to a determination of the gradient estimate, recalling that the method of derivative measurement involves taking differences between values of $\hat{\xi}$. For this determination we will continue to assume that the error samples (values of ε_k) are independent. We can then conclude that the values of $\hat{\xi}$ used in the derivative measurement are also independent. Let v be any component of $\mathbf{V} = (\mathbf{W} - \mathbf{W}^*)$ as before. Then, similar to (5.3), our estimate of the corresponding gradient component is

$$\widehat{\frac{\partial \xi}{\partial v}} = \frac{1}{2\delta}\hat{\xi}(v+\delta) - \frac{1}{2\delta}\hat{\xi}(v-\delta) \tag{5.30}$$

Under the foregoing assumption of independence, the variance of this estimate is just the sum of the variances of the two terms in (5.30). Also, the variance of $c\hat{\xi}$, where c is a constant, is c^2 times the variance of $\hat{\xi}$. Thus, using (5.27) in (5.30) with $K = 2$, we get

$$\begin{aligned}\mathrm{var}\left[\widehat{\frac{\partial \xi}{\partial v}}\right] &= \frac{1}{4\delta^2}\mathrm{var}[\hat{\xi}(v+\delta)] + \frac{1}{4\delta^2}\mathrm{var}[\hat{\xi}(v-\delta)] \\ &= \frac{1}{2N\delta^2}\left[\xi^2(v+\delta) + \xi^2(v-\delta)\right]\end{aligned} \tag{5.31}$$

This is the general (conservative) result for the variance of the estimate of a gradient component when the individual measurements of ε_k are independent.

If we assume that the perturbation δ in (5.31) is small and that the adaptive process has converged to a weight-vector solution near $\mathbf{W}^*$, the two values of ξ in

(5.31) are approximately equal to $\xi_{\min}$. In this case, (5.31) simplifies to

$$\operatorname{var}\left[\widehat{\frac{\partial \xi}{\partial v}}\right] = \frac{\xi_{\min}^2}{N\delta^2} \tag{5.32}$$

Since the values of N and δ are the same for the estimates of all components of the gradient vector, and since the samples of ε_k used in all estimates were assumed to be independent, the errors in all estimates are independent and have the same variance. The covariance matrix of the estimated gradient vector at the kth iteration is accordingly given by

$$\begin{aligned} \operatorname{cov}[\hat{\nabla}_k] &\triangleq E\left[(\hat{\nabla}_k - \nabla_k)(\hat{\nabla}_k - \nabla_k)^{\mathrm{T}}\right] \\ &= \frac{\xi_{\min}^2}{N\delta^2}\mathbf{I} \end{aligned} \tag{5.33}$$

EFFECTS ON THE WEIGHT-VECTOR SOLUTION

Having derived formulas for the variance of the gradient estimate, we now examine the effect of a "noisy" gradient estimate on the weight vector during the adaptation process. We will see that adaptation based on noisy gradient estimates results in noise in the weight-vector solution and a loss in performance. The exact nature of these effects varies according to the method of adaptation. In this section we describe the propagation of noise into the weight vector for Newton's method and the method of steepest descent in preparation for calculating the excess mean-square error and "misadjustment" in the sections following. For purposes of analysis we define the gradient estimation noise vector at the kth iteration, that is, at $\mathbf{W} = \mathbf{W}_k$, as $\mathbf{N}_k$. ($\mathbf{N}$ is thus a vector of size $L + 1$, and should not be confused with the scalar N, used previously to denote the number of observations of the error, ε_k.) The gradient estimate at the kth iteration, $\hat{\nabla}_k$, is then equal to the true gradient at $\mathbf{W} = \mathbf{W}_k$ plus the gradient estimation noise and can thus be written

$$\hat{\nabla}_k = \nabla_k + \mathbf{N}_k \tag{5.34}$$

We examine the effect of this noisy gradient estimate on the weight vector solution, first with Newton's method and then with the steepest-descent method.

Newton's method in its ideal form was derived in (4.32) as

$$\mathbf{W}_{k+1} = \mathbf{W}_k - \mu\mathbf{R}^{-1}\nabla_k \tag{5.35}$$

When the gradient is estimated and is therefore noisy as in (5.34), this expression becomes

$$\begin{aligned} \mathbf{W}_{k+1} &= \mathbf{W}_k - \mu\mathbf{R}^{-1}\hat{\nabla}_k \\ &= \mathbf{W}_k - \mu\mathbf{R}^{-1}\nabla_k - \mu\mathbf{R}^{-1}\mathbf{N}_k \end{aligned} \tag{5.36}$$

We can also write this expression for the weight vector in terms of the weight deviation $\mathbf{V}$ [Equation (2.32)] as

$$\mathbf{V}_{k+1} = \mathbf{V}_k - \mu \mathbf{R}^{-1} \nabla_k - \mu \mathbf{R}^{-1} \mathbf{N}_k \tag{5.37}$$

As in (2.35), we can now substitute $\nabla = 2\mathbf{RV}$ and obtain a difference equation in $\mathbf{V}$:

$$\begin{aligned} \mathbf{V}_{k+1} &= \mathbf{V}_k - 2\mu \mathbf{V}_k - \mu \mathbf{R}^{-1} \mathbf{N}_k \\ &= (1 - 2\mu)\mathbf{V}_k - \mu \mathbf{R}^{-1} \mathbf{N}_k \end{aligned} \tag{5.38}$$

So we have here a system of difference equations giving the weight deviation vector, $\mathbf{V}$, in terms of a "driving function" given by the gradient noise vector, $\mathbf{N}$. The system is cross-coupled as described in connection with (4.38), because here $\mathbf{N}$ is multiplied by $\mathbf{R}^{-1}$. We can uncouple the driving function by rotating to the principal-coordinate system, just as we did with (4.38). We use $\mathbf{V} = \mathbf{QV}'$ from (3.38) as well as $\mathbf{R}^{-1} = \mathbf{Q}\Lambda^{-1}\mathbf{Q}^{-1}$ from (3.5) in the result above and obtain

$$\mathbf{QV}'_{k+1} = (1 - 2\mu)\mathbf{QV}'_k - \mu \mathbf{Q}\Lambda^{-1}\mathbf{Q}^{-1}\mathbf{N}_k$$

or

$$\mathbf{V}'_{k+1} = (1 - 2\mu)\mathbf{V}'_k - \mu\Lambda^{-1}(\mathbf{Q}^{-1}\mathbf{N}_k) \tag{5.39}$$

In this result the quantity in parentheses is just the gradient noise projected into the principal-coordinate system. If we define this as $\mathbf{Q}^{-1}\mathbf{N} = \mathbf{N}'$, we can write

$$\mathbf{V}'_{k+1} = (1 - 2\mu)\mathbf{V}'_k - \mu\Lambda^{-1}\mathbf{N}'_k \tag{5.40}$$

Here, since Λ^{-1} is diagonal, (5.40) represents an uncoupled set of difference equations in the weight vector $\mathbf{V}'_k$. We can obtain a solution for $\mathbf{V}'_k$ similar to (4.35) by induction. Using the first three iterations of (5.40), we reason inductively as follows:

$$\begin{aligned} \mathbf{V}'_1 &= (1 - 2\mu)\mathbf{V}'_0 - \mu\Lambda^{-1}\mathbf{N}'_0 \\ \mathbf{V}'_2 &= (1 - 2\mu)^2\mathbf{V}'_0 - \mu\Lambda^{-1}\left[(1 - 2\mu)\mathbf{N}'_0 + \mathbf{N}'_1\right] \\ \mathbf{V}'_3 &= (1 - 2\mu)^3\mathbf{V}'_0 - \mu\Lambda^{-1}\left[(1 - 2\mu)^2\mathbf{N}'_0 + (1 - 2\mu)\mathbf{N}'_1 + \mathbf{N}'_2\right] \\ &\vdots \\ \mathbf{V}'_k &= (1 - 2\mu)^k\mathbf{V}'_0 - \mu\Lambda^{-1}\sum_{n=0}^{k-1}(1 - 2\mu)^n\mathbf{N}'_{k-n-1} \end{aligned} \tag{5.41}$$

Thus we have a solution as in (4.35) for Newton's method, except that here we are in the principal-coordinate system and the gradient noise, $\mathbf{N}'_k$, is included at each step. In (4.35) we could let k approach infinity and obtain the optimum solution, $\mathbf{W} = \mathbf{W}^*$, or $\mathbf{V}' = \mathbf{0}$, but here there is a residual error due to the gradient noise. If we let k approach infinity in (5.41) and assume that μ is in the stable range from 0 to $\frac{1}{2}$ as in (4.33), the factor $(1 - 2\mu)^k$ will become negligible and we will obtain the following "steady-state" solution:

$$\mathbf{V}_k' = -\mu\Lambda^{-1}\sum_{n=0}^{\infty}(1-2\mu)^n\mathbf{N}_{k-n-1}' \tag{5.42}$$

This gives us the steady-state error in the weight vector for Newton's method in terms of the input eigenvalues in Λ^{-1} and the gradient noise in the series of values given by $\mathbf{N}' = \mathbf{Q}^{-1}\mathbf{N}$. Note that Λ^{-1} is a diagonal matrix whose elements are $1/\lambda_0, 1/\lambda_1, \ldots, 1/\lambda_L$.

Having looked at Newton's method, let us now turn to the method of steepest descent and again examine the effects of gradient noise. Without noise, the steepest-descent method is described in Chapter 4. The difference formula in (4.36) is

$$\mathbf{W}_{k+1} = \mathbf{W}_k - \mu\nabla_k \tag{5.43}$$

In this we substitute the translated coordinate vector, $\mathbf{V} = \mathbf{W} - \mathbf{W}^*$, and use the noisy gradient vector, $\hat{\nabla} = 2\mathbf{RV} + \mathbf{N}$ as above, to obtain

$$\begin{aligned}\mathbf{V}_{k+1} &= \mathbf{V}_k - \mu(2\mathbf{RV}_k + \mathbf{N}_k)\\ &= (\mathbf{I} - 2\mu\mathbf{R})\mathbf{V}_k - \mu\mathbf{N}_k \end{aligned}\tag{5.44}$$

This result is similar to (5.38) for Newton's method and we can solve for the steady-state error in the same way as before. We first rotate to principal coordinates using $\mathbf{V} = \mathbf{QV}'$:

$$\mathbf{QV}_{k+1}' = (\mathbf{I} - 2\mu\mathbf{R})\mathbf{QV}_k' - \mu\mathbf{N}_k$$

Therefore,

$$\begin{aligned}\mathbf{V}_{k+1}' &= (\mathbf{I} - 2\mu\mathbf{Q}^{-1}\mathbf{RQ})\mathbf{V}_k' - \mu\mathbf{Q}^{-1}\mathbf{N}_k\\ &= (\mathbf{I} - 2\mu\Lambda)\mathbf{V}_k' - \mu\mathbf{N}_k' \end{aligned}\tag{5.45}$$

Again $\mathbf{N}' = \mathbf{Q}^{-1}\mathbf{N}$ is the gradient noise projected into the principal-coordinate system. Here we have a result similar to (5.40), and we can obtain a solution just as (5.41) was obtained by induction. The result in this case is

$$\mathbf{V}_k' = (\mathbf{I} - 2\mu\Lambda)^k\mathbf{V}_0' - \mu\sum_{n=0}^{k-1}(\mathbf{I} - 2\mu\Lambda)^n\mathbf{N}_{k-n-1}' \tag{5.46}$$

Again, if we assume μ is in the stability range given by (4.45), the first term becomes negligible for large k, and the steady-state solution is

$$\mathbf{V}_k' = -\mu\sum_{n=0}^{\infty}(\mathbf{I} - 2\mu\Lambda)^n\mathbf{N}_{k-n-1}' \tag{5.47}$$

This result should be compared with (5.42) for Newton's method.

Thus we have in (5.42) and (5.47) the steady-state effects of gradient noise on the weight vector solution. Now if we are given the gradient "noise" in terms of the covariance matrix derived in (5.33) in the preceding section, we can also derive covariance formulas for the weight vector. The covariance matrix of the weight vector is given by

$$\operatorname{cov}[\mathbf{V}_k'] = E\left[\mathbf{V}_k'\mathbf{V}_k'^{\mathrm{T}}\right] \tag{5.48}$$

with the expectation going over k, the iteration number. In the case of Newton's method, we can find the product $\mathbf{V}_k'\mathbf{V}_k'^{\mathrm{T}}$ from (5.39) as follows:

$$\begin{aligned}\mathbf{V}_k'\mathbf{V}_k'^{\mathrm{T}} &= (1-2\mu)^2\mathbf{V}_{k-1}'\mathbf{V}_{k-1}'^{\mathrm{T}} + \mu^2\Lambda^{-1}\mathbf{N}_{k-1}'\mathbf{N}_{k-1}'^{\mathrm{T}}(\Lambda^{-1})^{\mathrm{T}} \\ &\quad -\mu(1-2\mu)\left[\mathbf{V}_{k-1}'\mathbf{N}_{k-1}'^{\mathrm{T}}(\Lambda^{-1})^{\mathrm{T}} + \Lambda^{-1}\mathbf{N}_{k-1}'\mathbf{V}_{k-1}'^{\mathrm{T}}\right]\end{aligned} \tag{5.49}$$

Now Λ^{-1} is diagonal and so equal to its transpose, and the weight vector $\mathbf{V}_k'$ and the noise $\mathbf{N}_k'$ are assumed to be independent with zero mean, so the expectation of (5.49) is

$$\begin{aligned}\operatorname{cov}[\mathbf{V}_k'] &= (1-2\mu)^2\operatorname{cov}[\mathbf{V}_k'] + \mu^2(\Lambda^{-1})^2\operatorname{cov}[\mathbf{N}_k'] \\ &= \frac{\mu(\Lambda^{-1})^2}{4(1-\mu)}\operatorname{cov}[\mathbf{N}_k']\end{aligned} \tag{5.50}$$

In the steepest-descent case, a similar development from (5.45) results in

$$\begin{aligned}\mathbf{V}_k'\mathbf{V}_k'^{\mathrm{T}} &= (\mathbf{I}-2\mu\Lambda)\mathbf{V}_{k-1}\mathbf{V}_{k-1}'^{\mathrm{T}}(\mathbf{I}-2\mu\Lambda)^{\mathrm{T}} + \mu^2\mathbf{N}_{k-1}'\mathbf{N}_{k-1}'^{\mathrm{T}} \\ &\quad -\mu\left[(\mathbf{I}-2\mu\Lambda)\mathbf{V}_{k-1}'\mathbf{N}_{k-1}'^{\mathrm{T}} + \mathbf{N}_{k-1}'\mathbf{V}_{k-1}'^{\mathrm{T}}(\mathbf{I}-2\mu\Lambda)^{\mathrm{T}}\right]\end{aligned} \tag{5.51}$$

Again we note that $\mathbf{I} - 2\mu\Lambda$ is diagonal and that the expected cross products will vanish. So in this case the expectation of both sides produces

$$\begin{aligned}\operatorname{cov}[\mathbf{V}_k'] &= (\mathbf{I}-2\mu\Lambda)^2\operatorname{cov}[\mathbf{V}_k'] + \mu^2\operatorname{cov}[\mathbf{N}_k'] \\ &= \frac{\mu}{4}(\Lambda-\mu\Lambda^2)^{-1}\operatorname{cov}[\mathbf{N}_k']\end{aligned} \tag{5.52}$$

Thus we have in (5.50) and (5.52) the covariance of $\mathbf{V}_k'$ in terms of the covariance of the gradient noise. To relate these to the results of the preceding section, we first note from (5.33) and (5.34) that

$$\begin{aligned}\operatorname{cov}[\mathbf{N}_k'] &= E\left[\mathbf{N}_k'\mathbf{N}_k'^{\mathrm{T}}\right] = \mathbf{Q}^{-1}E\left[\mathbf{N}_k\mathbf{N}_k^{\mathrm{T}}\right]\mathbf{Q} \\ &= \mathbf{Q}^{-1}E\left[(\hat{\nabla}_k-\nabla_k)(\hat{\nabla}_k-\nabla_k)^{\mathrm{T}}\right]\mathbf{Q} \\ &= \mathbf{Q}^{-1}\operatorname{cov}[\hat{\nabla}_k]\mathbf{Q} \\ &= \frac{\xi_{\min}^2}{N\delta^2}\mathbf{Q}^{-1}\mathbf{I}\mathbf{Q} = \frac{\xi_{\min}^2}{N\delta^2}\mathbf{I}\end{aligned} \tag{5.53}$$

Notice again that the scalar N denotes the number of gradient observations, while the vector $\mathbf{N}$ stands for the gradient noise. Combining (5.53) with (5.50) and (5.52), we find that $\text{cov}[\mathbf{V}'_k]$ is diagonal, and can be expressed in the two cases as follows:

$$\textit{Newton's method:} \quad \text{cov}[\mathbf{V}'_k] = \frac{\mu(\Lambda^{-1})^2 \xi^2_{\min}}{4N\delta^2(1-\mu)} \tag{5.54}$$

$$\textit{Steepest descent:} \quad \text{cov}[\mathbf{V}'_k] = \frac{\mu(\Lambda - \mu\Lambda^2)^{-1} \xi^2_{\min}}{4N\delta^2} \tag{5.55}$$

Now to map these results into the unprimed coordinate system, we first note that, from (3.38),

$$\begin{aligned}\text{cov}[\mathbf{V}_k] &= E[\mathbf{V}_k\mathbf{V}_k^{\mathrm{T}}] \\ &= \mathbf{Q}E[\mathbf{V}'_k\mathbf{V}'^{\mathrm{T}}_k]\mathbf{Q}^{-1} = \mathbf{Q}\,\text{cov}[\mathbf{V}'_k]\mathbf{Q}^{-1}\end{aligned} \tag{5.56}$$

Using this result with (5.54) and (5.55), we can derive the following final expressions for $\text{cov}[\mathbf{V}]$. The details are left to Exercise 18.

Weight-vector covariance (Newton's method):

$$\text{cov}[\mathbf{V}_k] = \frac{\mu\xi^2_{\min}(\mathbf{R}^{-1})^2}{4N\delta^2(1-\mu)} \tag{5.57}$$

Weight-vector covariance (steepest-descent method):

$$\text{cov}[\mathbf{V}_k] = \frac{\mu\xi^2_{\min}(\mathbf{R} - \mu\mathbf{R}^2)^{-1}}{4N\delta^2} \tag{5.58}$$

where μ = adaptive gain constant

$\xi^2_{\min}$ = minimum mean-square error

$\mathbf{R}$ = input correlation matrix

N = number of independent error measurements at each perturbed weight value

δ = deviation of weights used to measure gradient

Thus in these expressions for the covariance of the weight vector in terms of the variables involved in error measurement, we have the principal result of this section. In the next section, we proceed to look at the excess mean-square error in terms of these same variables.

EXCESS MEAN-SQUARE ERROR AND TIME CONSTANTS

Without noise in the adaptive process Newton's method and the method of steepest descent, as well as other adaptive methods, would cause the weight vector to converge to a steady-state solution at the minimum point of the mean-square-error surface. The covariance of $\mathbf{V}$ would be zero, and the mean-square error would thus be equal to $\xi_{\min}$. Noise in the adaptive process, however, causes the steady-state weight vector solution to vary randomly about the minimum point—that is, to "climb the sides of the bowl." The result is "excess" mean-square error and a steady-state value of ξ greater than $\xi_{\min}$.

The mean-square error was defined in Chapter 2 as the expected value of the error with the weight vector fixed at $\mathbf{W}$. If the weight vector is not fixed, the instantaneous mean-square error ξ_k at the kth iteration is defined as the expected value of the error with $\mathbf{W} = \mathbf{W}_k$. See equations such as (4.17), (4.54), and so on, for examples. To derive an expression for *excess* mean-square error we must now further define an average taken over an arbitrarily large number of iterations of ξ_k. That is, we must allow the weight vector to be "noisy" and vary statistically as in the preceding section, and assess the resulting effects on ξ_k. Using the relationship between ξ and $\mathbf{V}$ in (4.53), we can define the excess mean-square error as

$$\begin{aligned}\text{excess MSE} &= E[\xi_k - \xi_{\min}] \\ &= E\left[\mathbf{V}_k^{\mathrm{T}}\mathbf{R}\mathbf{V}_k\right] \qquad (5.59)\end{aligned}$$

with the expectation as before going over k, the iteration index. This definition is applicable only where $\mathbf{V}_k$ undergoes a statistically stationary process over k—that is, where the noise $\mathbf{N}_k$ as well as the input signal vector $\mathbf{X}_k$ and desired response d_k are statistically stationary. Thus it is applicable only in steady state, after adaptive transients have died out.

Figure 5.3 illustrates the physical meaning of these definitions. Random weight variations about the optimum cause an increase in mean-square error. The average of these increases is the excess mean-square error. Note that the excess MSE here does *not* include the performance penalty discussed earlier, which was due to deliberate weight variations.

The formula for excess MSE in (5.59) is similar to the covariance of $\mathbf{V}_k'$ in (5.48). To derive results similar to (5.57) and (5.58) we must substitute for $\mathbf{V}_k$ in (5.59), and so separate derivations are again required for Newton's method and for the steepest-descent method. Since our previous solutions have been for $\mathbf{V}_k'$ rather than for $\mathbf{V}_k$, we modify (5.59) before beginning by using the relationship $\mathbf{V}^{\mathrm{T}}\mathbf{R}\mathbf{V} = \mathbf{V}'^{\mathrm{T}}\boldsymbol{\Lambda}\mathbf{V}'$, as expressed, for example, in (3.34) and (3.35). We have then as our starting point

$$\text{excess MSE} = E\left[\mathbf{V}_k'^{\mathrm{T}}\boldsymbol{\Lambda}\mathbf{V}_k'\right] \qquad (5.60)$$

Proceeding now in the case of Newton's method, we use (5.42) for $\mathbf{V}_k'$ in (5.60). To simplify the expression, we use r for the geometric ratio:

$$r = 1 - 2\mu \tag{5.61}$$

The substitution of (5.42) into (5.60) then yields

$$\begin{aligned}\text{excess MSE} &= \mu^2 E\left[\sum_{n=0}^{\infty} r^n \mathbf{N}_{k-n-1}'^{\mathrm{T}} \Lambda^{-1} \Lambda \Lambda^{-1} \sum_{m=0}^{\infty} r^m \mathbf{N}_{k-m-1}'\right] \\ &= \mu^2 \sum_{n=0}^{\infty} \sum_{m=0}^{\infty} r^{n+m} E\left[\mathbf{N}_{k-n-1}'^{\mathrm{T}} \Lambda^{-1} \mathbf{N}_{k-m-1}'\right]\end{aligned} \tag{5.62}$$

We have previously assumed that the gradient noise, $\mathbf{N}$, which results from independent errors in the measurement of the error, is independent from one iteration to the next. Thus all terms in (5.62) with $m \neq n$ must vanish, and thus

$$\text{excess MSE} = \mu^2 \sum_{n=0}^{\infty} r^{2n} E\left[\mathbf{N}_{k-n-1}'^{\mathrm{T}} \Lambda^{-1} \mathbf{N}_{k-n-1}'\right] \tag{5.63}$$

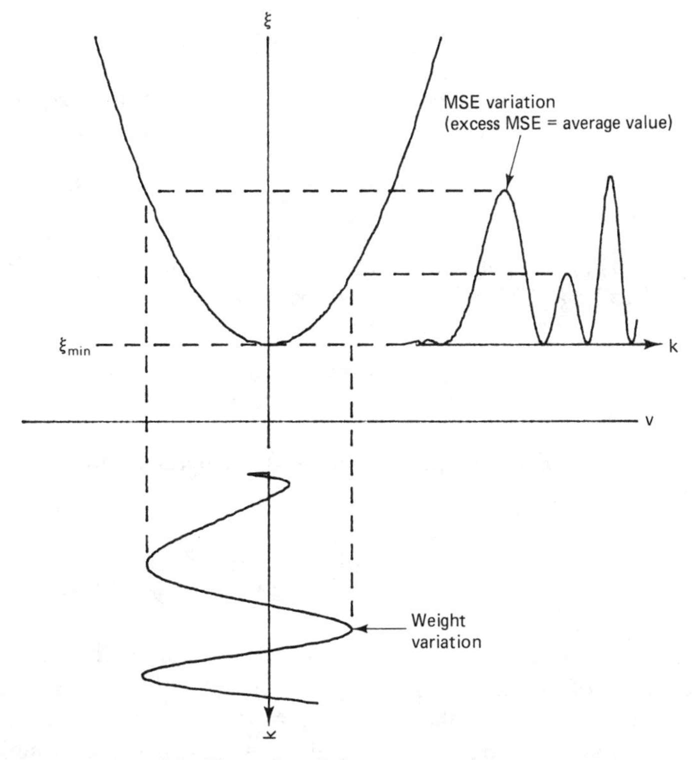

Figure 5.3 Illustration of the excess mean-square error.

Also, the gradient noise is assumed to be a stationary process, so the expectation in (5.63) goes over all subscripts of $\mathbf{N}'$. Thus

$$\text{excess MSE} = \mu^2 E\left[\mathbf{N}_k'^{\mathrm{T}} \Lambda^{-1} \mathbf{N}_k'\right] \sum_{n=0}^{\infty} r^{2n}$$

$$= \frac{\mu^2}{1 - r^2} E\left[\mathbf{N}_k'^{\mathrm{T}} \Lambda^{-1} \mathbf{N}_k'\right] \tag{5.64}$$

Here we have also assumed that the process is stable with r less than 1. Now, for $\mathbf{N}_k'^{\mathrm{T}} \Lambda^{-1} \mathbf{N}_k'$, we have

$$\mathbf{N}_k'^{\mathrm{T}} \Lambda^{-1} \mathbf{N}_k' = \left[n'_{0k} \quad \cdots \quad n'_{Lk}\right] \begin{bmatrix} \lambda_0^{-1} & & 0 \\ & \ddots & \\ 0 & & \lambda_L^{-1} \end{bmatrix} \begin{bmatrix} n'_{0k} \\ \vdots \\ n'_{Lk} \end{bmatrix}$$

$$= \sum_{m=0}^{L} \lambda_m^{-1} n'^2_{mk} \tag{5.65}$$

where n'_{mk} is a component of $\mathbf{N}_k'$. Using this result in (5.64), we obtain

$$\text{excess MSE} = \frac{\mu^2}{1 - r^2} \sum_{m=0}^{L} \lambda_m^{-1} E\left[n'^2_{mk}\right] \tag{5.66}$$

But $E[n'^2_{mk}]$ is a diagonal element of the covariance matrix of $\mathbf{N}'$, and was shown in (5.53) to be $\xi^2_{\min}/N\delta^2$. Thus (5.66) becomes

$$\text{excess MSE} = \frac{\xi^2_{\min} \mu^2}{N\delta^2 (1 - r^2)} \sum_{m=0}^{L} \frac{1}{\lambda_m} \tag{5.67}$$

Again using $r = 1 - 2\mu$, we obtain the general result for the excess MSE with Newton's method:

$$\text{excess MSE} = \frac{\xi^2_{\min} \mu}{4N\delta^2 (1 - \mu)} \sum_{m=0}^{L} \frac{1}{\lambda_m} \tag{5.68}$$

For practical reasons, it is convenient to express this result in terms of time constants of the adaptive process. To determine the number of time constants equal to a given value of the geometric ratio r, as in Figure 5.4, we construct an exponential envelope through the geometric sequence of samples. We then let the envelope be described by $\exp(-t/\tau)$, where t represents time and τ represents the time constant. If one unit of time corresponds to one iteration, we can write

$$\exp\left(-\frac{1}{\tau}\right) = r \tag{5.69}$$

which can be expanded as

$$r = \exp\left(-\frac{1}{\tau}\right) = 1 - \frac{1}{\tau} + \frac{1}{2!\tau^2} - \frac{1}{3!\tau^3} + \cdots \tag{5.70}$$

Since in most applications τ is large (10 or greater) and r is small (less than but near 1), the following is a useful approximation:

$$r \cong 1 - \frac{1}{\tau} \qquad (\text{large } \tau) \tag{5.71}$$

From the relationship $r = 1 - 2\mu$ for Newton's method, we also have

$$\mu \approx \frac{1}{2\tau} \tag{5.72}$$

so (5.68) becomes

$$\begin{aligned}\text{excess MSE} &\approx \frac{\xi_{\min}^2}{4N\delta^2(2\tau - 1)} \sum_{m=0}^{L} \frac{1}{\lambda_m} \\ &\approx \frac{\xi_{\min}^2}{8N\delta^2\tau} \sum_{m=0}^{L} \frac{1}{\lambda_m} \qquad (\text{large } \tau)\end{aligned} \tag{5.73}$$

We now further simplify (5.73) by expressing the sum in terms of an average,

$$\sum_{m=0}^{L} \frac{1}{\lambda_m} = (L + 1)\left(\frac{1}{\lambda}\right)_{\text{av}} \tag{5.74}$$

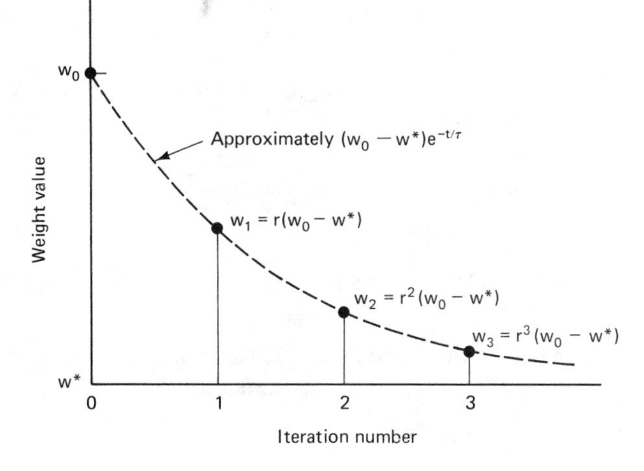

Figure 5.4 Exponential approximation to the geometric sequence of weight values.

and also by using the perturbation, that is, the average increase in MSE normalized with respect to $\xi_{\min}$, given in (5.16) as $P = \delta^2\lambda_{av}/\xi_{\min}$. With these substitutions, (5.73) becomes

$$\text{excess MSE} \approx \frac{(L+1)\xi_{\min}\lambda_{av}(1/\lambda)_{av}}{8NP\tau} \tag{5.75}$$

To derive the excess mean-square error for the method of steepest descent, we proceed in a similar manner. This time, instead of (5.42), we substitute (5.47) into (5.60). Using the fact that Λ is diagonal, we obtain

$$\begin{aligned}\text{excess MSE} &= E\left[\mathbf{V}_k'^{\mathrm{T}}\Lambda\mathbf{V}_k'\right]\\ &= \mu^2\sum_{n=0}^{\infty}\sum_{m=0}^{\infty}E\left[\mathbf{N}_{k-n-1}'^{\mathrm{T}}\Lambda(\mathbf{I}-2\mu\Lambda)^{n+m}\mathbf{N}_{k-m-1}'\right]\end{aligned} \tag{5.76}$$

The cross-terms vanish as they did in (5.62) and $\mathbf{N}'$ is again assumed to be statistically stationary as in (5.63), so (5.76) becomes

$$\begin{aligned}\text{excess MSE} &= \mu^2\sum_{n=0}^{\infty}E\left[\mathbf{N}_k'^{\mathrm{T}}\Lambda(\mathbf{I}-2\mu\Lambda)^{2n}\mathbf{N}_k'\right]\\ &= \mu^2 E\left[\mathbf{N}_k'^{\mathrm{T}}\Lambda\left(\sum_{n=0}^{\infty}(\mathbf{I}-2\mu\Lambda)^{2n}\right)\mathbf{N}_k'\right]\end{aligned} \tag{5.77}$$

with the expectation going over k as before. Now since the matrices in the sum in this result are diagonal, each element of the sum may be computed as a power series. Therefore, we can write

$$\sum_{n=0}^{\infty}(\mathbf{I}-2\mu\Lambda)^{2n} = \frac{1}{4\mu}(\Lambda-\mu\Lambda^2)^{-1} \tag{5.78}$$

(See Exercise 20 for more details on this result.) Using (5.78) together with Exercise 1(c) of Chapter 2 in (5.77) gives

$$\begin{aligned}\text{excess MSE} &= \mu^2 E\left[\mathbf{N}_k'^{\mathrm{T}}\Lambda\frac{1}{4\mu}(\Lambda-\mu\Lambda^2)^{-1}\mathbf{N}_k'\right]\\ &= \frac{\mu}{4}E\left[\mathbf{N}_k'^{\mathrm{T}}(\mathbf{I}-\mu\Lambda)^{-1}\mathbf{N}_k'\right]\end{aligned} \tag{5.79}$$

We have now arrived at a result that is the same as (5.64) if we use $\mu/4$ in place of $\mu^2/(1-r^2)$ and $(\mathbf{I}-\mu\Lambda)$ in place of Λ. Making these same replacements in (5.67), our result in (5.79) now becomes

$$\text{excess MSE} = \frac{\mu\xi_{\min}^2}{4N\delta^2}\sum_{m=0}^{L}\frac{1}{1-\mu\lambda_m}$$
$$= \frac{\mu\xi_{\min}\lambda_{\text{av}}}{4NP}\sum_{m=0}^{L}\frac{1}{1-\mu\lambda_m} \tag{5.80}$$

where P is the perturbation caused by gradient measurement, equal to $\delta^2\lambda_{\text{av}}/\xi_{\min}$ as before.

In the preceding development for Newton's method we noted that it is convenient to express excess mean-square error in terms of time constant and we defined the time constant τ pertaining to the relaxation of the weight vector. We now wish to define two other forms of time constant useful in describing the efficiency and rate of convergence of the adaptive process, the time constant of the learning curve, which we will call τ_{mse}, and what may be called for lack of a better term the "time constant of adaptation," which we will call T_{mse}. Both of these, as we shall see, are just scaled values of the weight relaxation time constant, τ.

The weight relaxation time constant τ is related by (5.71) to the geometric ratio r of the weights. The geometric ratio of the learning curve, on the other hand, from (4.19) is

$$r_{\text{mse}} = r^2 \tag{5.81}$$

The corresponding time constant τ_{mse} is obtained using (5.69) as follows:

$$\exp\left(-\frac{1}{\tau_{\text{mse}}}\right) = \exp\left(-\frac{2}{\tau}\right) = r^2 = r_{\text{mse}}$$

Therefore,

$$\tau_{\text{mse}} = \frac{\tau}{2} \tag{5.82}$$

This time constant is useful in describing the learning time of adaptive systems.

The basic unit of τ_{mse} is the iteration number. The basic unit of T_{mse}, the time constant of adaptation, on the other hand, is the data sample. Since, as illustrated in Figure 5.2, $2N$ samples are required for each gradient component measurement, $2(L+1)N$ samples are required during each iteration. The time constant of adaptation is accordingly defined as

$$T_{\text{mse}} \triangleq 2(L+1)N\tau_{\text{mse}} = N(L+1)\tau \tag{5.83}$$

This time constant is readily related to real time if the sampling rate is known.

The nth geometric ratio of the method of steepest descent was shown in (4.48) to be

$$r_n = 1 - 2\mu\lambda_n \tag{5.84}$$

The nth geometric ratio can be related to the nth weight relaxation time constant through (5.71)

$$r_n \approx 1 - \frac{1}{\tau_n} \qquad (\text{large } \tau_n) \tag{5.85}$$

Combining (5.84) and (5.85) yields

$$\mu\lambda_n \approx \frac{1}{2\tau_n} \tag{5.86}$$

As in (5.82), we define the nth time constant of the mean-square-error learning curve as half the time constant of the weight relaxation process:

$$(\tau_{\text{mse}})_n = \frac{\tau_n}{2} \tag{5.87}$$

Defining $(T_{\text{mse}})_n$ similarly and using (5.87) in (5.83), we also have

$$\begin{aligned}(T_{\text{mse}})_n &= N(L+1)\tau_n \\ &= 2N(L+1)(\tau_{\text{mse}})_n\end{aligned} \tag{5.88}$$

Using the first line of this result in (5.86), we have

$$\mu\lambda_n \approx \frac{N(L+1)}{2(T_{\text{mse}})_n} \tag{5.89}$$

On both sides of this result we can average over n to obtain

$$\mu\lambda_{\text{av}} \approx \frac{N(L+1)}{2}\left(\frac{1}{T_{\text{mse}}}\right)_{\text{av}} \tag{5.90}$$

The excess mean-square error for the method of steepest descent may now be found by substituting this result together with (5.86) into (5.80):

$$\text{excess MSE} \approx \frac{(L+1)\xi_{\min}}{8P}\left(\frac{1}{T_{\text{mse}}}\right)_{\text{av}} \sum_{m=0}^{L} \frac{1}{1 - 1/(2\tau_m)} \tag{5.91}$$

We have assumed that the steady-state weight-vector solution remains close to the optimal solution (hovers near the bottom of the bowl). In accordance with this assumption the adaptive process converges slowly, with small μ and large values of τ_n. We can accordingly write

$$\frac{1}{1 - 1/(2\tau_n)} \approx 1; \qquad n = 0, 1, \ldots, L \tag{5.92}$$

This approximation further simplifies the excess mean-square-error approximation:

$$\text{excess MSE} \approx \frac{(L+1)^2\xi_{\min}}{8P}\left(\frac{1}{T_{\text{mse}}}\right)_{\text{av}} \tag{5.93}$$

To summarize the results of this section, we restate the approximations for the steady-state excess mean-square error:

$$\text{excess MSE (Newton's method)} \approx \frac{(L+1)\xi_{\min}\lambda_{av}(1/\lambda)_{av}}{8NP\tau}$$

$$\approx \frac{(L+1)^2\xi_{\min}\lambda_{av}(1/\lambda)_{av}}{8PT_{\text{mse}}} \tag{5.94}$$

$$\text{excess MSE (steepest descent)} \approx \frac{(L+1)^2\xi_{\min}}{8P}\left(\frac{1}{T_{\text{mse}}}\right)_{av} \tag{5.95}$$

where $L+1$ = number of weights

$\xi_{\min}$ = minimum mean-square error

N = number of observations of the error ($2N$ observations for each gradient estimate)

P = perturbation (normalized increase in ξ caused by estimating the gradient) [see (5.9) through (5.16)]

τ = time constant of weight adjustment process (see Figure 5.4)

T_{mse} = time constant of adaptation [see (5.83) through (5.90)]

λ = eigenvalue of input correlation matrix

MISADJUSTMENT

Excess mean-square error is the average mean-square error less the minimum mean-square error. It provides a measure of the difference between actual and optimal performance averaged over time. Another measure of the difference between actual and optimal performance, particularly useful in the design of adaptive processes, is the misadjustment M, defined as the excess mean-square error divided by the minimum mean-square error:

$$M \triangleq \frac{(\text{excess MSE})}{\xi_{\min}} \tag{5.96}$$

The misadjustment is a dimensionless measure of the difference between adaptive and optimal Wiener performance as a result of gradient estimation noise. In other words, it is a normalized measure of the cost of adaptability. Note that M does not include the perturbation, P, which is due to deliberate weight variations rather than noise.

In the case of Newton's method, the misadjustment follows from (5.94):

$$M \approx \frac{(L+1)\lambda_{\text{av}}(1/\lambda)_{\text{av}}}{8NP\tau} \tag{5.97}$$

Alternatively, a preferable expression from (5.94) is

$$\boxed{M \approx \frac{(L+1)^2\lambda_{\text{av}}(1/\lambda)_{\text{av}}}{8PT_{\text{mse}}}} \tag{5.98}$$

Equation (5.98) provides a ready means of evaluating the performance of Newton's method. We can see that misadjustment decreases as the perturbation and time constant increase. Greater perturbation permits more accurate gradient measurements with the same amount of data, while a greater adaptive time constant allows more data to be averaged in arriving at the weight vector solution. We can also see that the misadjustment increases with the square of the number of weights, so that a substantial price must be paid for a small increase in adaptability. The eigenvalues can be seen to affect misadjustment significantly only when a great disparity exists among them. A separate examination of λ_{av} and $(1/\lambda)_{\text{av}}$ is presented below.

From (5.94) it can be seen that the excess mean-square error of Newton's method remains the same if the time constant is doubled and N simultaneously halved. One can double τ by halving μ and one can reduce N by taking fewer error samples per iteration. It is clear that small steps with small amounts of data per step are equivalent to large steps with large amounts of data per step. The important factor, as shown by (5.98), is the amount of data processed per time constant of adaptation.

The algorithm based on Newton's method described in this and the preceding chapter could be made somewhat more efficient by altering the gradient measurement procedure illustrated in Figures 5.1 and 5.2. The altered procedure would consist of making a measurement of ξ at the nominal setting of $\mathbf{W} = \mathbf{W}_k$, taking the difference between this common measurement and other measurements of ξ, and changing one weight at a time by δ. The amount of data per gradient vector estimate would be altered by a factor of $(L + 2)/2(L + 1) \approx 1/2$. The analysis of this procedure, however, would be complicated because the gradient noise components would be mutually correlated. In any case, the misadjustment would continue to increase with the square of the number of weights.

For the steepest-descent method, the misadjustment is given by substituting (5.95) into (5.96):

$$M \approx \frac{(L+1)^2}{8P}\left(\frac{1}{T_{\text{mse}}}\right)_{\text{av}} \tag{5.99}$$

This expression is similar to the corresponding expression (5.98) for Newton's method. The latter contains the term $\lambda_{\text{av}}(1/\lambda)_{\text{av}}$, which does not appear in (5.99). It also contains the term $1/T_{\text{mse}}$, whereas (5.99) contains the term $(1/T_{\text{mse}})_{\text{av}}$. Newton's method has only one time constant of adaptation, while the method of steepest descent has up to $L + 1$ separate T_{mse} values. It is clear that the two expressions for misadjustment would be identical if all eigenvalues of $\mathbf{R}$ were equal. Under this condition the performance surface would have circular symmetry, and the negative gradient would always lie in the direction of the minimum. Newton's method and the method of steepest descent would, in fact, be identical.

COMPARATIVE PERFORMANCE OF NEWTON'S AND STEEPEST-DESCENT METHODS

For a given perturbation and number of weights both Newton's method and the method of steepest descent show an increase in misadjustment with an increase in speed of adaptation (i.e., a decrease in the time constant of adaptation). The two methods, however, do not necessarily produce the same misadjustment for the same speed of adaptation. In a given situation the better algorithm will be the one that has the greater speed of adaptation for a fixed level of misadjustment or the smaller misadjustment for a fixed speed of adaptation.

To compare the two methods in terms of speed of adaptation and misadjustment, we must first express (5.99) in a more convenient form. Adaptation with the method of steepest descent involves multiple time constants, and the speed of adaptation is thus governed by the rate of convergence of the slowest transient mode. Let the adaptive time constant of this mode be designated $(T_{\text{mse}})_{\text{max}}$. From (5.88), we then have

$$(T_{\text{mse}})_{\text{max}} = \frac{(L+1)N}{2\mu\lambda_{\text{min}}} \tag{5.100}$$

From the same relationship (5.88) we can also obtain a formula for the average eigenvalue, λ_{av}, in terms of average reciprocals of the adaptation time constants:

$$\lambda_{av} = \frac{(L+1)N}{2\mu}\left(\frac{1}{T_{mse}}\right)_{av} \tag{5.101}$$

Combining this result with (5.100), we have

$$\left(\frac{1}{T_{mse}}\right)_{av} = \frac{\lambda_{av}}{\lambda_{min}(T_{mse})_{max}} \tag{5.102}$$

The misadjustment expression (5.99) for the steepest descent method can therefore be rewritten as

$$M \approx \frac{(L+1)^2\lambda_{av}}{8P\lambda_{min}(T_{mse})_{max}} \tag{5.103}$$

If we now let $(T_{mse})_{max}$ in (5.103) be equal to T_{mse} in (5.98) and further let L and P be the same in both expressions, the misadjustment of Newton's method and the method of steepest descent are the same except for the term $(1/\lambda)_{av}$ in the former case and $1/\lambda_{min}$ in the latter. Thus the misadjustment of the former (i.e., Newton's method) will generally be lower, since

$$\frac{1}{\lambda_{min}} \geq \left(\frac{1}{\lambda}\right)_{av} \tag{5.104}$$

Consider the following specific case. If the eigenvalues are 1 through 10, $1/\lambda_{min}$ is equal to 1 and $(1/\lambda)_{av}$ is approximately equal to 0.3. The misadjustment of the method of steepest descent would thus be about three times that of Newton's method.

When there is a large number of eigenvalues, it may be possible in some instances to represent them as being uniformly distributed between λ_{min} and λ_{max}. In this case the average value of $1/\lambda$ would be

$$\left(\frac{1}{\lambda}\right)_{av} = \frac{1}{\lambda_{max} - \lambda_{min}} \int_{\lambda_{min}}^{\lambda_{max}} \frac{d\lambda}{\lambda}$$

$$= \frac{\ln(\lambda_{max}/\lambda_{min})}{\lambda_{max} - \lambda_{min}} \tag{5.105}$$

If we let $\lambda_{max} = 10\lambda_{min}$, we can see that $(1/\lambda)_{av}$ is about 0.26 times $(1/\lambda_{min})$, and thus that the steepest-descent misadjustment is about four times the misadjustment for Newton's method. Similarly, if λ_{max} is $100\lambda_{min}$, the steepest-descent misadjustment is about 20 times the Newton misadjustment.

TOTAL MISADJUSTMENT AND OTHER PRACTICAL CONSIDERATIONS

Equations (5.98) and (5.99) show that misadjustment is inversely proportional to the perturbation P. Thus it might seem that the misadjustment could be made as small as desired by increasing the perturbation. We will now show that an unlimited increase in the perturbation is not practicable.

The perturbation has been defined as a dimensionless measure of the extent to which mean-square-error performance is affected by gradient component measurement. Equation (5.16) shows the perturbation as the excess mean-square error normalized with respect to the minimum mean-square error. The perturbation is much like misadjustment and is in fact a form of misadjustment due to gradient measurement in an "on-line" adaptive system. The "total" misadjustment of such a system can thus be defined as the sum of two misadjustments, one due to a random disturbance and the other to an independent deterministic disturbance of the steady-state weight vector:

$$M_{\text{tot}} \triangleq M + P \tag{5.106}$$

For the two methods discussed in this chapter, (5.98) and (5.99) give the total misadjustment as

$$\textit{Newton's method:} \quad M_{\text{tot}} \approx \frac{(L+1)^2 \lambda_{\text{av}} (1/\lambda)_{\text{av}}}{8PT_{\text{mse}}} + P \tag{5.107}$$

$$\textit{Steepest descent:} \quad M_{\text{tot}} \approx \frac{(L+1)^2 (1/T_{\text{mse}})_{\text{av}}}{8P} + P \tag{5.108}$$

We note that both (5.107) and (5.108) give M_{tot} in the form $P + A/P$, where A is not a function of P. If we set the derivative of this form with respect to P equal to zero, we can find P_{opt}, the value of P that minimizes M_{tot}. In either case, this value is

$$P_{\text{opt}} \approx \frac{A}{P_{\text{opt}}} \approx M_{\text{tot}} - P_{\text{opt}}$$

$$\approx \tfrac{1}{2} M_{\text{tot}} \tag{5.109}$$

The total adjustment is thus minimized when the perturbation is equal to about half of the total misadjustment.

A numerical example is useful in showing the application of these results. Suppose that a misadjustment of 10 percent is considered acceptable in a particular instance. Suppose further that the adaptive system contains an adaptive filter with 10 weights and that all eigenvalues of the **R**-matrix are equal, so that the misadjustments of Newton's method and of the method of steepest descent are equal. The best choice of P is thus 5 percent, and the adaptive time constant is found as follows:

$$M_{\text{tot}} \approx \frac{(L+1)^2}{8PT_{\text{mse}}} + P$$

$$\approx \frac{(10)^2}{8(0.05)T_{\text{mse}}} + 0.05 = 0.1$$

Therefore,

$$T_{\text{mse}} \approx 5000 \text{ samples} \tag{5.110}$$

If we consider the convergence time of the adaptive process to be on the order of four time constants, adaptive transients will have died out after 20,000 data samples have been taken. Given a 10-weight filter with 10 percent misadjustment, this amount of data is excessive. It nevertheless represents the best performance achievable with an on-line system and the given method of gradient estimation.

However, in designing a system to implement the adaptive process, there is at least one possibility for improvement. If circumstances permit, better performance could be achieved by measuring the gradient vector in an off-line system that could be perturbed without effect on the on-line system. With this procedure, illustrated in Figure 5.5, Newton's method would yield a misadjustment given by (5.98), and the method of steepest descent would yield a misadjustment given by (5.99). The 10-weight system described above with a specified misadjustment of 10 percent and a specified perturbation of 20 percent would have an adaptive time constant calculated as follows:

$$M = \frac{(L+1)^2}{8PT_{\text{mse}}} = \frac{(10)^2}{8(0.2)T_{\text{mse}}} = 0.1$$

Therefore,

$$T_{\text{mse}} = 625 \text{ samples} \tag{5.111}$$

The time constant in this case is greatly improved but still excessive.

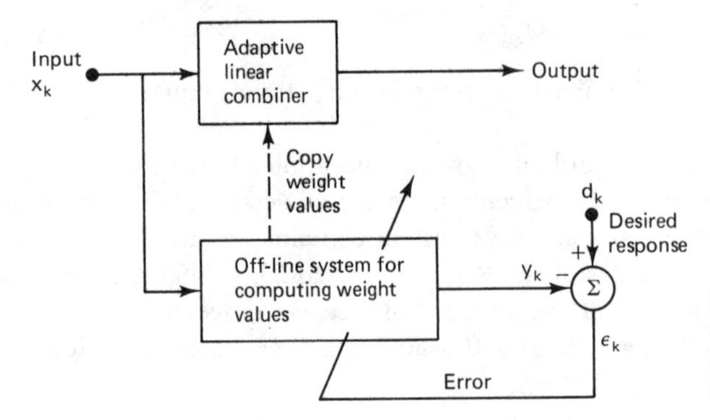

Figure 5.5 Off-line adaptation to alleviate effects of perturbation for gradient estimation. Output is not affected by gradient estimation process.

Since perturbing the off-line system does not contribute to the on-line system's misadjustment in Figure 5.5, it would seem reasonable to make the perturbation arbitrarily large to achieve the most accurate gradient estimate. Making P very large, however, violates the conditions assumed in deriving (5.33) from (5.32). These conditions were that P is small and that the system has adapted so that the weight vector solution is near the optimum. Making P large would thus eventually cause misadjustment to increase.

Another way to increase efficiency using the procedure illustrated in Figure 5.5 may be mentioned in concluding this discussion of practical considerations. If it is possible to operate at a rate much faster than real time, the input data can be repeated and all components of the gradient vector measured with the same data each iteration. Although an analysis will not be presented here, it can be shown that misadjustment in this case increases only as the first power rather than the square of the number of weights. An alternative or equivalent to repeating data is to use an array of off-line processors, each measuring one component of the gradient vector at each iteration.

EXERCISES

1. Explain why the central difference formulas (5.3) and (5.4) are exact for quadratic performance functions.
2. An adaptive system with a single weight has a performance surface given by

$$\xi = 5w^2 - 20w + 23$$

   Make a plot of ξ versus w, and on it show the values of $\xi_{\min}$, w^*, and λ. Also illustrate the performance penalty, γ, caused by perturbing w around $w = 2.5$, with a perturbation of $\pm\delta = \pm 1.0$. What is the value of γ?
3. Is a negative performance penalty possible with quadratic performance functions? Why or why not? Sketch a one-weight performance function with a negative performance penalty.
4. What is the perturbation in the case of Exercise 2?
5. Suppose that we have a transversal filter with the performance surface

$$\xi = 2w_0^2 + 2w_1^2 + 2w_0w_1 - 14w_0 - 16w_1 + 42$$

   and with a stationary random input signal, x, having samples correlated such that $E[x_k x_k] = 2$ and $E[x_k x_{k-1}] = 1$. What is the perturbation, P, in the mean-square error if there is a weight perturbation equal to δ?
6. Suppose that a weight is added to a given transversal filter. What in general is the effect on the perturbation (P)?
7. Suppose that in a particular case the error, ε_k, is distributed randomly such that all values from 1 to 3 are equally likely. What is the fourth moment, α_4, in this case?

8. Suppose that ε_k is distributed normally with the same mean and variance as in Exercise 7. How does the fourth moment compare with α_4 in Exercise 7?
9. In a certain situation, the error is distributed normally with zero mean and variance equal to 3. If the mean-square error is estimated on the basis of 10 independent samples of the error, what is the variance of the estimate?
10. Referring to Equation (5.25), show that K is somewhat less than 2 when the error, ε_k, is distributed normally with nonzero mean.
11. Derive the following results given in Table 5.1:
    (a) Row 2, column 2. **(b)** Row 4, column 2.
    (c) Row 1, column 3. **(d)** Row 2, column 3.
    (e) Row 3, column 3. **(f)** Row 4, column 3.
    (g) Row 2, column 4. **(h)** Row 3, column 4.
12. In Table 5.1, row 1, what must be true near the limits $K = 0$ and $K = 2$?
13. In Table 5.1, what ratio α_1/σ corresponds with the upper limit of K:
    (a) In row 2?
    (b) In row 3?
    (c) In row 4?
14. Given the conditions in Exercise 2, suppose that the gradient estimate is based on five observations of the error at each perturbed weight setting. What is the variance of the gradient estimate if ε_k is distributed normally?
15. Given the conditions in Exercise 5, assume that the gradient estimate is based on 50 observations of the error at each perturbed weight setting. Give the covariance matrix of the gradient estimate, assuming that ε_k is distributed normally.
16. Two time series, x_k and y_k, are specified to be zero for k less than zero, and are related by

$$x_k = ax_{k-1} + by_k$$

    Using induction, find a solution for x_k that does not involve a recursion relationship.
17. What is the variance of v, the translated weight vector, in the single-weight case, for Newton's method? How would your answer be affected if we changed to the steepest-descent method?
18. Derive the weight-vector covariance formulas (5.57) and (5.58), given the corresponding principal-axis formulas (5.54) and (5.55). Show each derivation in detail, one step at a time. *Hint*: Use Equation (3.38) together with Exercise 1(c) in Chapter 2.
19. For the conditions specified in Exercise 5, what is the weight-vector covariance matrix, cov[**V**]? Assume a steepest-descent algorithm with μ equal to $\frac{1}{2}$ its maximum stable value, and with $N = 10$ error observations.
20. In Equation (5.78), a geometric sum of diagonal matrices was found. In such a sum, each matrix element can be computed as a separate geometric sum. Use this fact to prove that

$$\sum_{n=0}^{\infty} \mathbf{D}^n = (\mathbf{I} - \mathbf{D})^{-1}$$

    where **D** is a diagonal matrix. What are the conditions for convergence? Is this result ever true when **D** is not diagonal? If so, when?

21. For the conditions in Exercise 5, what is the excess mean-square error, assuming that μ is one-half its maximum stable value [see Equations (4.33) and (4.45)] and that $N = 10$ error observations:
(a) With Newton's method?
(b) With the steepest-descent method?
In your answer, compare Equation (5.68) with (5.94) and Equation (5.80) with (5.95). Explain any discrepancies.

22. In a given adaptive process, there is only a single weight, the adaptive gain constant is $\mu = 0.01$, and the average squared input signal value is 2. What are the time constants for weight adjustment and for the learning curve:
(a) With Newton's method?
(b) With the steepest-descent method?

23. In Exercise 21, if a *total* of 10 observations is made of the error in order to adjust the weight at each iteration, what is the time constant of adaptation, T_{mse}, in each case?

24. Given the conditions in Exercise 21 with $N = 5$ error observations and a 5 percent perturbation, what is the excess mean-square error in each case?

25. An adaptive transversal filter with two weights is processing an input signal, x, having $E[x_k^2] = 3$ and $E[x_k x_{k-1}] = 2$. At each iteration, a *total* of 80 observations of ε_k is made. The corresponding weight perturbation causes P to be 0.05. The adaptive gain constant is $\mu = 0.01$. What is the misadjustment:
(a) With Newton's method?
(b) With the steepest-descent method?

ANSWERS TO SELECTED EXERCISES

5. $P = \delta^2/2$
7. $\alpha_4 = 24.2$
8. $\alpha_4 = 24.33$ (slightly larger)
9. $\text{var}[\hat{\xi}] = \sigma^4/5$
12. Near $K = 0$, $|\alpha_1/\sigma| \to \infty$; near $K = 2$, $|\alpha_1/\sigma| \to 0$
15. $\text{cov}[\hat{\nabla}] = (8/25\delta^2)\mathbf{I}$

REFERENCES AND ADDITIONAL READINGS

1. L. G. Kelly, *Handbook of Numerical Methods and Applications*. Reading, Mass.: Addison-Wesley, 1967, Chap. 2.
2. B. W. Lindgren, *Statistical Theory*. New York: Macmillan, 1962, Chap. 2.
3. W. Feller, *An Introduction to Probability Theory and Its Applications*, vol. 1. New York: Wiley, 1957, Chap. 9.

4. B. V. Gnedenko, *The Theory of Probability*. New York: Chelsea, 1967, Chap. V.
5. J. V. Uspensky, *Introduction to Mathematical Probability*. New York: McGraw-Hill, 1937, Chap. 12.
6. B. Widrow and J. M. McCool, "A comparison of adaptive algorithms based on the methods of steepest descent and random search," *IEEE Trans. Antennas Propag.*, vol. AP-24, no. 5, pp. 615–637, Sept. 1976.
7. A. Papoulis, *Probability, Random Variables, and Stochastic Processes*. New York: McGraw-Hill, 1965.

part III

ADAPTIVE ALGORITHMS AND STRUCTURES

(Chapters 6 – 8)

OBJECTIVES OF PART III

From Parts I and II we have most of the concepts necessary to the derivation and analysis of practical adaptive algorithms. These will be algorithms that work in nonstationary as well as stationary signal environments, algorithms that are meant to "track" as well as "seek" the minimum point on the performance surface.

The primary objective of Part III is to help the reader understand the LMS algorithm, which is the simplest scheme for adjusting the weights in a linear adaptive processor. The LMS algorithm, which is covered in Chapter 6, is widely applicable to all of the types of adaptive systems described in Part IV. A central objective of this text is to describe the LMS algorithm and its applications.

A second objective in Part III is to introduce types of adaptive algorithms other than the LMS algorithm, some of which have important advantages. To facilitate this, we introduce certain frequency-domain analysis techniques in Chapter 7. These are standard techniques of digital signal analysis, and the reader familiar with this subject may wish to skim Chapter 7.

The final objective in Part III is to acquaint the reader with lattice structures, which are introduced in the last part of Chapter 8. The lattice and its use in adaptive systems is a large subject, and only some basic ideas are covered in Chapter 8.

chapter 6

The LMS Algorithm

In Chapter 4 we introduced two algorithms, Newton's method and the steepest-descent method, for descending toward the minimum on the performance surface. Both of these algorithms required an estimation of the gradient at each iteration, so in Chapter 5 we discussed general methods for estimating the gradient. The gradient estimation methods in Chapter 5 are general because they are based on taking differences between estimated points on the performance surface, that is, differences between estimates of ξ.

In this chapter we introduce another algorithm for descending on the performance surface, known as the least-mean-square algorithm, or LMS algorithm. The LMS algorithm uses a special estimate of the gradient that is valid for the adaptive linear combiner described in Chapter 2.[†] Thus the LMS algorithm is more restricted in its use than the algorithms in Chapter 4.

On the other hand, the LMS algorithm is important because of its simplicity and ease of computation, and because it does not require off-line gradient estimations or repetitions of data. If the adaptive system is an adaptive linear combiner, and if the input vector $\mathbf{X}_k$ and the desired response d_k are available at each iteration, the LMS algorithm is generally the best choice for many different applications of adaptive signal processing.

DERIVATION OF THE LMS ALGORITHM

We recall that the adaptive linear combiner of Chapter 2 was applied in two basic ways, depending on whether the input is available in parallel (multiple inputs) or serial (single input) form. These two ways are shown together in Figure 6.1. In both

[†] However, it is possible to extend the LMS algorithm to recursive adaptive filters (see Chapter 8).

cases we have the combiner output, y_k, as a linear combination of the input samples. As in (2.8), we have

$$\varepsilon_k = d_k - \mathbf{X}_k^{\mathrm{T}} \mathbf{W}_k \tag{6.1}$$

where $\mathbf{X}_k$ is the vector of input samples in either of the two configurations in Figure 6.1.

To develop an adaptive algorithm using the previous methods, we would estimate the gradient of $\xi = E[\varepsilon_k^2]$ by taking differences between short-term averages of ε_k^2. Instead, to develop the LMS algorithm, we take ε_k^2 itself as an estimate of ξ_k. Then, at each iteration in the adaptive process, we have a gradient estimate of the form

$$\hat{\nabla}_k = \begin{bmatrix} \frac{\partial \varepsilon_k^2}{\partial w_0} \\ \vdots \\ \frac{\partial \varepsilon_k^2}{\partial w_L} \end{bmatrix} = 2\varepsilon_k \begin{bmatrix} \frac{\partial \varepsilon_k}{\partial w_0} \\ \vdots \\ \frac{\partial \varepsilon_k}{\partial w_L} \end{bmatrix} = -2\varepsilon_k \mathbf{X}_k \tag{6.2}$$

The derivatives of ε_k with respect to the weights follow directly from (6.1).

With this simple estimate of the gradient, we can now specify a steepest-descent type of adaptive algorithm. From (4.36), we have

$$\begin{aligned} \mathbf{W}_{k+1} &= \mathbf{W}_k - \mu \hat{\nabla}_k \\ &= \mathbf{W}_k + 2\mu \varepsilon_k \mathbf{X}_k \end{aligned} \tag{6.3}$$

This is the LMS algorithm [3, 4]. As before, μ is the gain constant that regulates the speed and stability of adaptation. Since the weight changes at each iteration are based on imperfect gradient estimates, we would expect the adaptive process to be noisy; that is, it would not follow the true line of steepest descent on the performance surface.

From its form in (6.3), we can see that the LMS algorithm can be implemented in a practical system without squaring, averaging, or differentiation and is elegant in its simplicity and efficiency. As noted above, each component of the gradient vector is obtained from a single data sample without perturbing the weight vector. Without averaging, the gradient components do contain a large component of noise, but the noise is attenuated with time by the adaptive process, which acts as a low-pass filter in this respect. We now proceed to discuss some of the properties of the LMS algorithm and to illustrate its use in some examples.

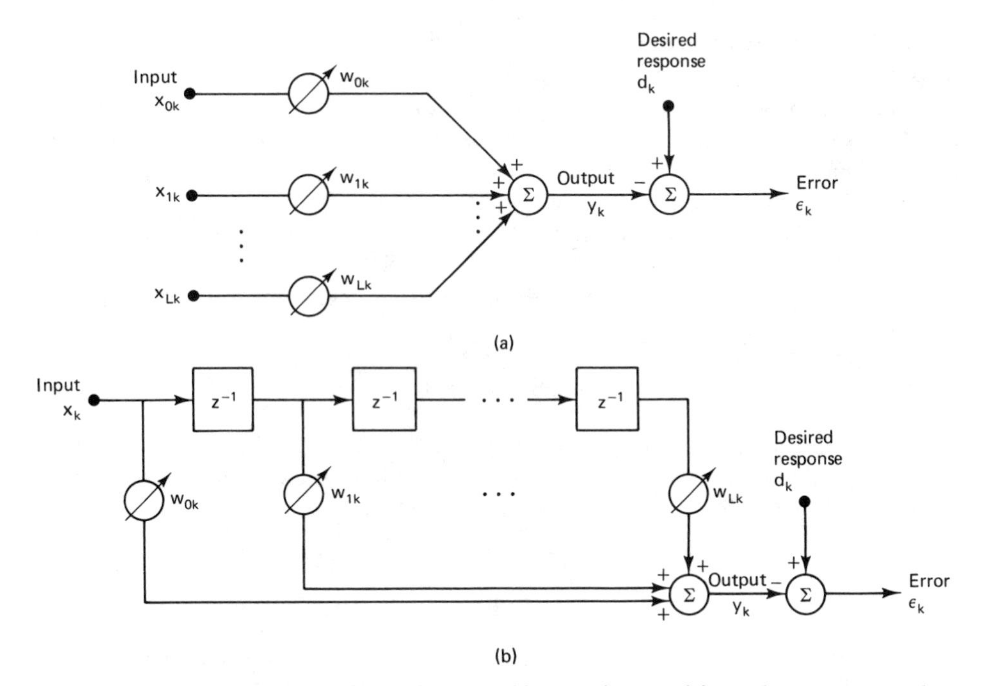

Figure 6.1 The adaptive linear combiner: (a) in general form; (b) as a transversal filter.

CONVERGENCE OF THE WEIGHT VECTOR

As with all adaptive algorithms, a primary concern with the LMS algorithm is its convergence to the optimum weight vector solution, where $E[\varepsilon_k^2]$ is minimized. To examine LMS convergence, we first note that the gradient estimate in (6.2) can readily be shown to be unbiased when the weight vector is held constant. The expected value of (6.2) with $\mathbf{W}_k$ held equal to $\mathbf{W}$ is

$$\begin{aligned} E[\hat{\nabla}_k] &= -2E[\varepsilon_k \mathbf{X}_k] \\ &= -2E\left[d_k \mathbf{X}_k - \mathbf{X}_k \mathbf{X}_K^{\mathrm{T}} \mathbf{W}\right] \\ &= 2(\mathbf{RW} - \mathbf{P}) = \nabla \end{aligned} \tag{6.4}$$

The second line of (6.4) follows from (6.1) plus the fact that ε_k is a scalar and can thus be commuted. The last line follows from the definitions of **P** and **R** in Chapter 2, and from (2.15). Since the mean value of $\hat{\nabla}_k$ is equal to the true gradient ∇, $\hat{\nabla}_k$ must be an unbiased estimate.

Seeing that the gradient estimate is unbiased, we could make the LMS algorithm into a true steepest-descent algorithm, at least in the limiting case, by estimating ∇ at each step as in (6.2) but not adapting the weights until many steps have occurred. In this way $\hat{\nabla}_k$ could be made to approach ∇_k, and (6.3) to approach

(4.36). With the weight vector changing at each iteration, we need to examine the weight vector convergence in a different manner, as follows.

From (6.3) we can see that the weight vector $\mathbf{W}_k$ is a function only of the past input vectors $\mathbf{X}_{k-1}, \mathbf{X}_{k-2}, \ldots, \mathbf{X}_0$. If we assume that successive input vectors are independent over time, $\mathbf{W}_k$ is independent of $\mathbf{X}_k$. For stationary input processes meeting this condition, the expected value of the weight vector $E[\mathbf{W}_k]$ after a sufficient number of iterations can be shown as follows to converge to the Wiener optimal solution given by (2.17), that is, to $\mathbf{W}^* = \mathbf{R}^{-1}\mathbf{P}$.

Taking the expected value of both sides of (6.3) yields the difference equation

$$\begin{aligned} E[\mathbf{W}_{k+1}] &= E[\mathbf{W}_k] + 2\mu E[\varepsilon_k \mathbf{X}_k] \\ &= E[\mathbf{W}_k] + 2\mu\left(E[d_k \mathbf{X}_k] - E\left[\mathbf{X}_k \mathbf{X}_k^T \mathbf{W}_k\right]\right) \end{aligned} \tag{6.5}$$

Using the foregoing assumption that $\mathbf{X}_k$ and $\mathbf{W}_k$ are independent, we have the expected products as in (6.4). Also, from (2.17) we have the optimum weight vector given as $\mathbf{W}^* = \mathbf{R}^{-1}\mathbf{P}$. Thus (6.5) becomes

$$\begin{aligned} E[\mathbf{W}_{k+1}] &= E[\mathbf{W}_k] + 2\mu(\mathbf{P} - \mathbf{R}E[\mathbf{W}_k]) \\ &= (\mathbf{I} - 2\mu\mathbf{R})E[\mathbf{W}_k] + 2\mu\mathbf{R}\mathbf{W}^* \end{aligned} \tag{6.6}$$

But now we have just the expected form of (4.38), which was solved (i.e., made nonrecursive) by changing to the principal-axis coordinate system. Using expected values, the solution from (4.42) is

$$E\left[\mathbf{V}_k'\right] = (\mathbf{I} - 2\mu\Lambda)^k \mathbf{V}_0' \tag{6.7}$$

where $\mathbf{V}'$ is the weight vector, $\mathbf{W}$, in the principal-axis system, Λ is the diagonal eigenvalue matrix of $\mathbf{R}$, and $\mathbf{V}_0'$ is the initial weight vector in the principal-axis system.

Thus, as k increases without bound, we see that the expected weight vector in (6.7) reaches the optimum solution (i.e., zero in the principal-axis system) only if the right side of the equation converges to zero. We saw in (4.45) that such convergence is guaranteed only if

$$\frac{1}{\lambda_{\max}} > \mu > 0 \tag{6.8}$$

where $\lambda_{\max}$ is the largest eigenvalue, that is, the largest diagonal element in Λ.

So, in (6.8), we have bounds on μ for convergence of the weight vector mean to the optimum weight vector. Within these bounds, the speed of adaptation and also the noise in the weight vector solution are determined by the size of μ. We also note that $\lambda_{\max}$ cannot be greater than the trace of $\mathbf{R}$, which is the sum of the diagonal elements of $\mathbf{R}$, that is,

$$\begin{aligned} \lambda_{\max} \le \operatorname{tr}[\Lambda] &= \sum (\text{diagonal elements of } \Lambda) \\ &= \sum (\text{diagonal elements of } \mathbf{R}) = \operatorname{tr}[\mathbf{R}] \end{aligned} \tag{6.9}$$

Furthermore, with a transversal adaptive filter, (2.11) gives tr[**R**] as just $(L + 1)E[x_k^2]$, or $L + 1$ times the input signal power. Thus convergence of the weight-vector mean is assured by:

$$\textit{In general:}\quad 0 < \mu < \frac{1}{\text{tr}[\mathbf{R}]}$$

$$\textit{Transversal filter:}\quad 0 < \mu < \frac{1}{(L + 1)(\text{signal power})} \tag{6.10}$$

This is a more restrictive bound on μ than (6.8), but is much easier to apply because the elements of **R** and the signal power can generally be estimated more easily than the eigenvalues of **R**.

The assumptions of decorrelation and stationarity of the input vector used to derive the results in this section are not necessary conditions for convergence of the LMS algorithm but have been adopted in this chapter for analytic convenience. Convergence with certain correlated and nonstationary inputs is demonstrated in the literature on the LMS algorithm [2, 6]. Under these conditions the analysis becomes much more complex. We know of no unconditional proof of convergence of the LMS algorithm.

AN EXAMPLE OF CONVERGENCE

To illustrate the convergence of the mean weight vector when employing the LMS algorithm in a specific situation, we again consider the simple two-weight combiner used as an example in Figure 2.6. We modify the input by adding a random signal as in Figure 6.2, in order to introduce a statistical fluctuation into the adaptive process. The input correlation matrix **R**, which was determined in (2.22), must be modified to include the random signal as follows. Let ϕ represent the average

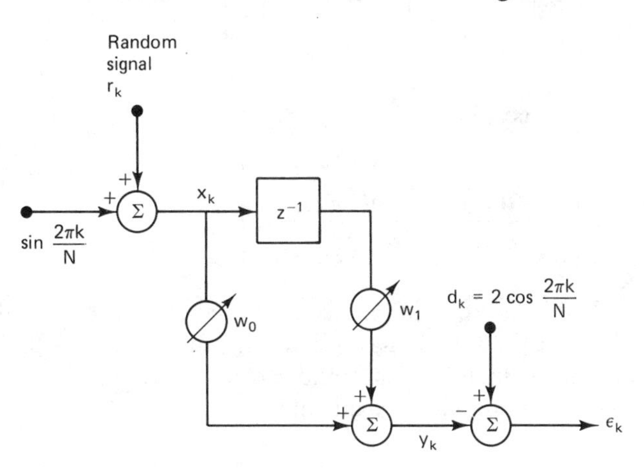

Figure 6.2 The adaptive linear combiner example from Chapter 2 with a random signal added at the input.

random signal power, that is,

$$\phi = E\left[r_k^2\right] \tag{6.11}$$

and suppose the random samples are independent of each other. Then

$$E\left[x_k^2\right] = E\left[\left(\sin\frac{2\pi k}{N} + r_k\right)^2\right] = 0.5 + \phi$$

$$E[x_k x_{k-1}] = E\left[\left(\sin\frac{2\pi k}{N} + r_k\right)\left(\sin\frac{2\pi(k-1)}{N} + r_{k-1}\right)\right]$$

$$= 0.5\cos\frac{2\pi}{N} \tag{6.12}$$

Thus the modified version of **R** in (2.22) becomes

$$\mathbf{R} = 0.5\begin{bmatrix} 1 + 2\phi & \cos\frac{2\pi}{N} \\ \cos\frac{2\pi}{N} & 1 + 2\phi \end{bmatrix} \tag{6.13}$$

A similar modification in (2.24) gives the performance surface for this example as

$$\xi = (0.5 + \phi)\left(w_0^2 + w_1^2\right) + w_0 w_1 \cos\frac{2\pi}{N} + 2w_1 \sin\frac{2\pi}{N} + 2 \tag{6.14}$$

The optimum weight vector, $\mathbf{W}^*$, can be found by solving $\mathbf{RW}^* = \mathbf{P}$ as described in Chapter 2, with **P** given in (2.23). The result is

$$\begin{bmatrix} 1 + 2\phi & \cos\frac{2\pi}{N} \\ \cos\frac{2\pi}{N} & 1 + 2\phi \end{bmatrix}\begin{bmatrix} w_0^* \\ w_1^* \end{bmatrix} = 2\begin{bmatrix} 0 \\ -\sin\frac{2\pi}{N} \end{bmatrix}$$

$$\begin{bmatrix} w_0^* \\ w_1^* \end{bmatrix} = \begin{bmatrix} \dfrac{2\cos(2\pi/N)\sin(2\pi/N)}{(1 + 2\phi)^2 - \cos^2(2\pi/N)} \\ \dfrac{-2(1 + 2\phi)\sin(2\pi/N)}{(1 + 2\phi)^2 - \cos^2(2\pi/N)} \end{bmatrix} \tag{6.15}$$

The performance of the LMS algorithm in (6.3) is illustrated in Figure 6.3 with $N = 16$ samples per signal cycle, and with a random signal power $\phi = 0.01$. Contours of ξ in (6.14) are shown in this figure, and we note that (6.15) gives $w_0^* = 3.784$ and $w_1^* = -4.178$ as the optimum weight values.

Together with the contours of constant ξ in the figure, there are two weight-value tracks. The two tracks have the following characteristics:

	Starting weight values	Value of μ	Number of iterations
Upper track	0, 0	0.10	250
Lower track	4, −10	0.05	500

The nature of the steepest-descent algorithm can be seen in that both tracks are more or less normal to the ξ contours, similar to Figure 4.7.

Because of the noisy gradient estimate at each iteration, the weight tracks in Figure 6.3 are seen to be erratic; they do not always proceed in the direction of the true gradient. The track for the larger value of μ is more erratic because the weight adjustment is greater at each iteration, but it also arrives at about the same distance from the minimum of ξ in about half the iterations taken for the track with smaller μ. If the LMS algorithm in Figure 6.3 were made to continue, either track would wander around the "bottom of the bowl," in the vicinity of the minimum of ξ,

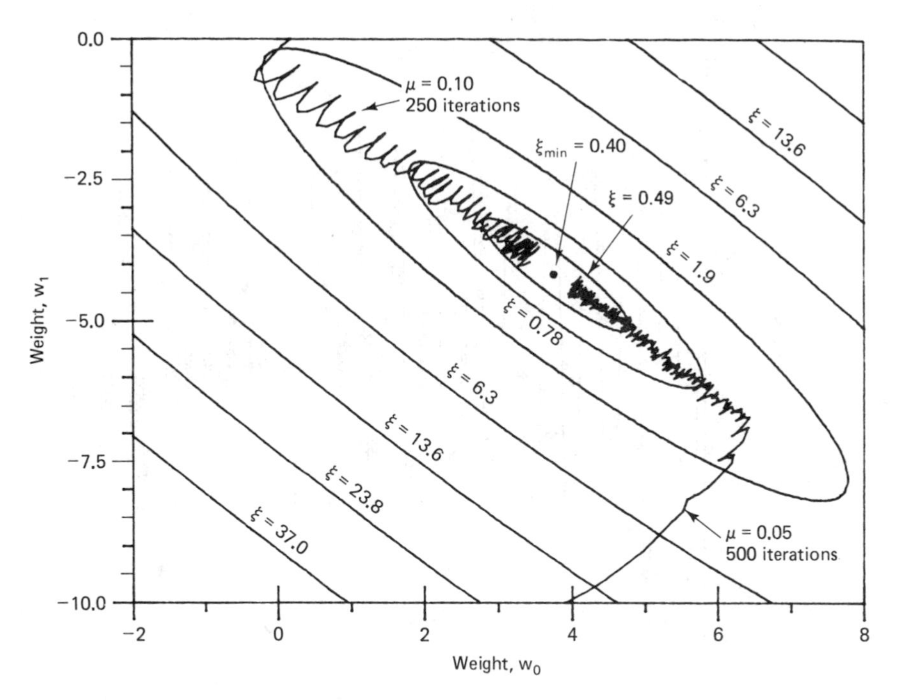

Figure 6.3 Performance surface contours and weight-value tracks for the LMS algorithm operating as in Figure 6.2 with $N = 16$ and $E[r_k^2] = 0.01$.

producing a noisy weight vector solution and an average misadjustment as discussed below.

We can observe the LMS convergence process in still another way, by plotting the error, ε_k, versus the iteration number. An example is given in Figure 6.4, with the error being the same as for the upper weight-value track in Figure 6.3. At first the error is quite sinusoidal, but as the adaptive filter learns to cancel the sinusoidal component, the error becomes increasingly random.

Finally, we note in the particular case of Figure 6.3 that the values of μ (i.e., 0.05 and 0.10) are well below the upper bound given by (6.10). With $\phi = 0.01$ we have from (6.9) and (6.13) the trace of **R** as

$$\text{tr}[\mathbf{R}] = 2(0.5 + \phi) = 1.02 \tag{6.16}$$

Therefore, (6.10) gives

$$0.98 > \mu > 0 \tag{6.17}$$

With the LMS algorithm, it is typical to use values of μ on the order of a tenth of the upper bound given in (6.10).

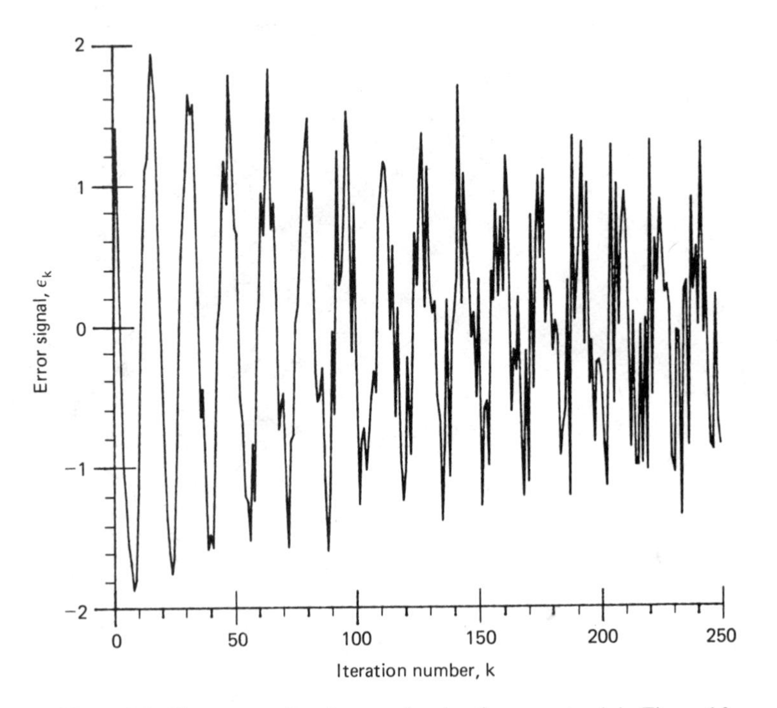

Figure 6.4 Error versus iteration number for the upper track in Figure 6.3.

LEARNING CURVE

In (4.59) we arrived at a formula for the learning curve, that is, the curve of ξ versus the iteration number k, for the steepest-descent method. The curve was theoretical in the sense that an exact knowledge of the gradient at each iteration was assumed. The theoretical learning curve was seen to decay in accordance with $L + 1$ geometric ratios of the form

$$r_n = 1 - 2\mu\lambda_n; \qquad n = 0, 1, \ldots, L \tag{6.18}$$

As a consequence of this result and of (5.86), the time constant for the exponential relaxation of the nth weight vector element to its optimal value is

$$\tau_n \approx \frac{1}{2\mu\lambda_n}; \qquad n = 0, 1, \ldots, L \tag{6.19}$$

From (5.87), the time constant associated with the nth mode of the learning curve is $\tau_n/2$, that is,

$$(\tau_{\text{mse}})_n \approx \frac{1}{4\mu\lambda_n}; \qquad n = 0, 1, \ldots, L \tag{6.20}$$

Furthermore, since each gradient estimate in the LMS algorithm is based on a single data observation, the time constant in terms of the input sample index, T_{mse}, is the same as τ_{mse}, the time constant in terms of algorithm iterations. Thus, for the LMS algorithm,

$$(T_{\text{mse}})_n = (\tau_{\text{mse}})_n \approx \frac{1}{4\mu\lambda_n}; \qquad n = 0, 1, \ldots, L \tag{6.21}$$

In some cases where the LMS algorithm is used, namely those where ε_k^2 at each iteration is a good approximation to $E[\varepsilon_k^2]$ for the current set of weight values, (6.21) gives a good approximation to the learning-curve time constants. But in general with the LMS algorithm, (6.21) will be somewhat optimistic, because ε_k^2 is not generally a good approximation to $E[\varepsilon_k^2]$, and the convergence process is erratic as in Figure 6.3. As an example we take the case illustrated in Figure 6.3, where the eigenvalues of $\mathbf{R}$ are found as follows, using (3.2) and (6.13) with $\phi = 0.01$ and $N = 16$:

$$\det \begin{bmatrix} 0.51 - \lambda & 0.5\cos\frac{\pi}{8} \\ 0.5\cos\frac{\pi}{8} & 0.51 - \lambda \end{bmatrix} = 0 \qquad \lambda_1, \lambda_2 = 0.972, 0.048 \tag{6.22}$$

For the lower weight track in the figure with $\mu = 0.05$, we have from (6.21)

$$(T_{\text{mse}})_1 \approx 5 \text{ iterations} \tag{6.23}$$

$$(T_{\text{mse}})_2 \approx 104 \text{ iterations} \tag{6.24}$$

For comparison with these values, a typical learning curve for this same situation is shown in Figure 6.5. Here ξ, or $E[\epsilon_k^2]$, is estimated by taking an average of 500 individual runs of ϵ_k^2 versus k. Each run had a different random sequence and a different starting point on the sinusoidal signals in Figure 6.2. We observe on the logarithmic plot in Figure 6.5 two different slopes of the learning curve, corresponding with the two directions of the lower path in Figure 6.3 and with the two time constants in (6.23) and (6.24). At first, the steeper slope is estimated to be approximately one decade in 13 iterations. Since a decade is a factor of 10 or $e^{2.30}$, a decade also represents a lapse of 2.30 time constants. Thus the initial time constant, $(T_{\text{mse}})_1$, is observed to be on the order of 13/2.30, or 6 iterations. After relaxation of this initial mode, the secondary time constant, $(T_{\text{mse}})_2$, is similarly observed to be approximately 265/2.3, or 115 iterations. In summary:

	Theoretical	Experimental
$(T_{\text{mse}})_1$	5	6
$(T_{\text{mse}})_2$	104	115

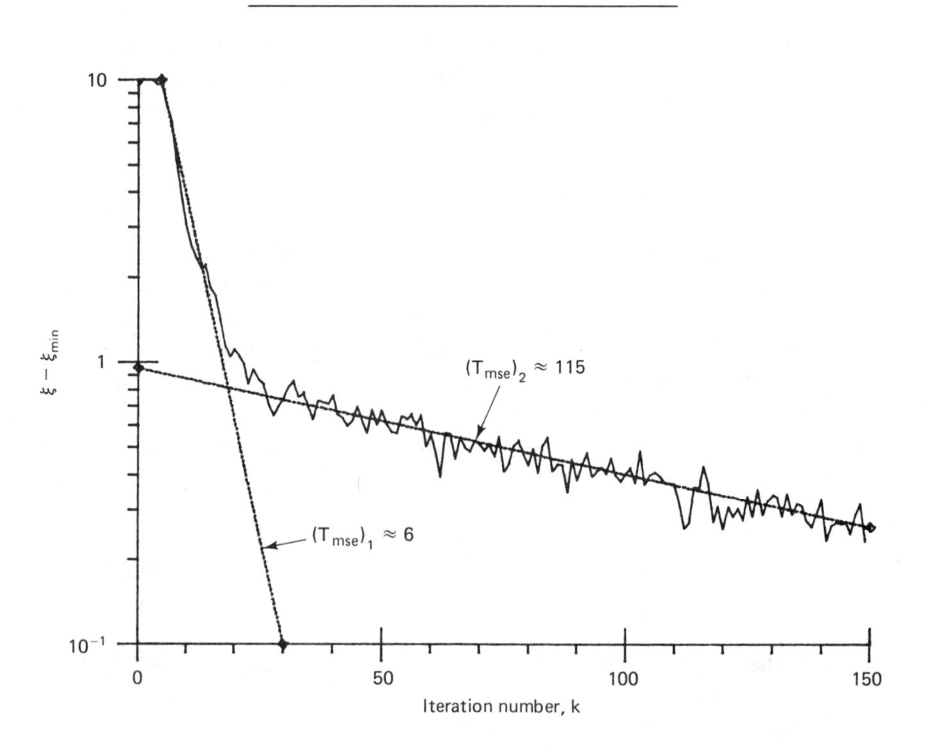

Figure 6.5 Learning curve for the lower track in Figure 6.3, with $\mu = 0.05$. ξ is estimated by averaging 500 runs as described in the text.

Thus, due to the noisy estimate of $E[\varepsilon_k^2]$ discussed previously, the experimental time constants are seen to be somewhat above the theoretical time constants. Such a result is typical with the LMS algorithm.

NOISE IN THE WEIGHT-VECTOR SOLUTION

In Chapter 5 we discussed the variance of the gradient estimate and the noise in the weight-vector solution under the assumption that the gradient estimate was based on weight perturbation. With the LMS algorithm, the gradient estimate as given by (6.2) is not based on weight perturbation, so we must reexamine its variance.

Just as in (5.34), let us define $\mathbf{N}_k$ as a vector of noise in the gradient estimate at the kth iteration. Thus, as in (5.34),

$$\hat{\nabla}_k = \nabla_k + \mathbf{N}_k \tag{6.25}$$

If we assume that the LMS process, using a small value of the adaptive gain constant μ, has converged to a steady-state weight vector solution near $\mathbf{W}^*$, then ∇_k in (6.25) will be close to zero. Then, in accordance with (6.2), the gradient noise is close to

$$\mathbf{N}_k = \hat{\nabla}_k = -2\varepsilon_k \mathbf{X}_k \tag{6.26}$$

The covariance of the noise is thus given by

$$\operatorname{cov}[\mathbf{N}_k] = E\left[\mathbf{N}_k\mathbf{N}_k^{\mathrm{T}}\right] = 4E\left[\varepsilon_k^2\mathbf{X}_k\mathbf{X}_k^{\mathrm{T}}\right] \tag{6.27}$$

If we assume that the weight vector, $\mathbf{W}_k$, remains near its optimum, $\mathbf{W}^*$, we conclude from (2.39) that ε_k^2 is approximately uncorrelated with the signal vector, so that (6.27) becomes

$$\begin{aligned}\operatorname{cov}[\mathbf{N}_k] &\approx 4E\left[\varepsilon_k^2\right]E\left[\mathbf{X}_k\mathbf{X}_k^{\mathrm{T}}\right]\\ &\approx 4\xi_{\min}\mathbf{R}\end{aligned} \tag{6.28}$$

To apply the results of Chapter 5, we need to transform (6.28) into the principal-axis coordinate system, as follows [see (3.38), (5.40), etc., for similar transformations]:

$$\begin{aligned}\operatorname{cov}\left[\mathbf{N}_k'\right] &= \operatorname{cov}\left[\mathbf{Q}^{-1}\mathbf{N}_k\right]\\ &= E\left[\mathbf{Q}^{-1}\mathbf{N}_k\left(\mathbf{Q}^{-1}\mathbf{N}_k\right)^{\mathrm{T}}\right]\\ &= \mathbf{Q}^{-1}E\left[\mathbf{N}_k\mathbf{N}_k^{\mathrm{T}}\right]\mathbf{Q}\\ &= \mathbf{Q}^{-1}\operatorname{cov}[\mathbf{N}_k]\mathbf{Q} \approx 4\xi_{\min}\mathbf{\Lambda}\end{aligned} \tag{6.29}$$

Now we can apply (5.52), which gives the weight vector covariance in the principal-axis coordinate system. The result is

$$\begin{aligned}\operatorname{cov}\left[\mathbf{V}_k'\right] &= \frac{\mu}{4}\left(\mathbf{\Lambda} - \mu\mathbf{\Lambda}^2\right)^{-1}\operatorname{cov}\left[\mathbf{N}_k'\right]\\ &\approx \mu\xi_{\min}\left(\mathbf{\Lambda} - \mu\mathbf{\Lambda}^2\right)^{-1}\mathbf{\Lambda}\end{aligned} \tag{6.30}$$

In practical situations the elements of $\mu\Lambda$ tend to be considerably less than 1, so we simplify the expression in (6.30) by neglecting the term $\mu\Lambda^2$ to obtain

$$\begin{aligned} \operatorname{cov}\left[\mathbf{V}_k'\right] &\approx \mu\xi_{\min}\Lambda^{-1}\Lambda \\ &\approx \mu\xi_{\min}\mathbf{I} \end{aligned} \tag{6.31}$$

Thus, transforming back to unprimed coordinates, we have the steady-state noise in the weight vector solution given approximately by

$$\begin{aligned} \operatorname{cov}[\mathbf{V}_k] &= \mathbf{Q}\operatorname{cov}\left[\mathbf{V}_k'\right]\mathbf{Q}^{-1} \\ &\approx \mu\xi_{\min}\mathbf{QIQ}^{-1} \\ &\approx \mu\xi_{\min}\mathbf{I} \end{aligned} \tag{6.32}$$

A further development of $\operatorname{cov}[\mathbf{V}_k]$ under less restrictive (nonstationary) conditions than those imposed to obtain (6.32) may be found in Farden [6]. In the example of Figure 6.3 the values of μ were 0.1 and 0.05 and $\xi_{\min}$ was about 0.4. Therefore, the diagonal elements of $\operatorname{cov}[\mathbf{V}_k]$ are 0.04 and 0.02 for the upper and lower tracks, respectively. These values correspond to rms weight fluctuations of 0.20 and 0.14, respectively.

MISADJUSTMENT

We recall from (5.96) that the misadjustment in an adaptive process is defined as the ratio of the excess mean-square error to the minimum mean-square error, and is thus a measure of how closely the adaptive process tracks the true Wiener solution, that is, a measure of the "cost of adaptability." The excess mean-square error is illustrated in Figure 5.3, and is given in (5.60) as

$$\text{excess MSE} = E\left[\mathbf{V}_k'^{\mathrm{T}}\Lambda\mathbf{V}_k'\right] \tag{6.33}$$

If we use the $L+1$ elements of $\mathbf{V}_k'$ and note that Λ is a diagonal matrix, we can express (6.33) in terms of a sum:

$$\text{excess MSE} = \sum_{n=0}^{L} \lambda_n E\left[v_{nk}'^2\right] \tag{6.34}$$

If we assume that the adaptive transient has died out and therefore that the squared error is near the "bottom of the bowl," we may assume that $E[v_{nk}'^2]$ in (6.34) is an element of $\operatorname{cov}[\mathbf{V}_k']$ in (6.31). Using (6.31) in (6.34) then gives the following approximation:

$$\begin{aligned} \text{excess MSE} &\approx \mu\xi_{\min}\sum_{n=0}^{L}\lambda_n \\ &\approx \mu\xi_{\min}\operatorname{tr}[\mathbf{R}] \end{aligned} \tag{6.35}$$

From this result we may compute the misadjustment, defined above and in (5.96), as

$$\boxed{\begin{aligned} M &= \frac{\text{excess MSE}}{\xi_{\min}} \\ &\approx \mu\,\mathrm{tr}[\mathbf{R}] \end{aligned}} \tag{6.36}$$

For example, in Figure 6.3, the values of μ were 0.1 and 0.05 and the trace of **R** was found in (6.16) to be 1.02, so we have for Figure 6.3,

$$M \approx \begin{cases} 0.1 & \text{(upper track)} \\ 0.05 & \text{(lower track)} \end{cases} \tag{6.37}$$

We can see in (6.36) that the misadjustment, M, is directly proportional to the adaptive gain constant, μ. Thus there is a trade-off between the misadjustment and the rate of adaptation. To see this trade-off more clearly, let us recall from (6.20) that the time constant for the nth mode of the learning curve is

$$(\tau_{\text{mse}})_n = \frac{1}{4\mu\lambda_n} \tag{6.38}$$

From this the trace of **R** can be written

$$\begin{aligned} \mathrm{tr}[\mathbf{R}] &= \sum_{n=0}^{L} \lambda_n \\ &= \frac{1}{4\mu} \sum_{n=0}^{L} \frac{1}{(\tau_{\text{mse}})_n} = \frac{L+1}{4\mu}\left(\frac{1}{\tau_{\text{mse}}}\right)_{\text{av}} \end{aligned} \tag{6.39}$$

Using this result in (6.36), we have

$$M \approx \frac{L+1}{4}\left(\frac{1}{\tau_{\text{mse}}}\right)_{\text{av}} \tag{6.40}$$

In the special case where all eigenvalues are equal, (6.40) reduces to

$$M \approx \frac{L+1}{4\tau_{\text{mse}}} \tag{6.41}$$

Experience shows this expression to be a good approximation to the relationship between misadjustment, time constant of the learning curve, and number of weights even when the eigenvalues are not equal. Such a relationship is needed in designing an adaptive system when the eigenvalues are unknown.

Since the trace of the R-matrix is the total power of the inputs to the weights, which is generally known, one can use (6.36) in choosing a value of μ that will produce a desired value of M. One can accordingly combine (6.41) and (6.36) to obtain a general expression for the time constant of the learning curve with equal eigenvalues:

$$\tau_{\text{mse}} \approx \frac{L+1}{4\mu\,\mathrm{tr}[\mathbf{R}]} \tag{6.42}$$

This expression is also a good approximation in many cases when the eigenvalues of the R-matrix are unequal.

Since adaptive transients die out or settle in approximately four time constants, (6.41) can be used to derive the following rule of thumb:

> Misadjustment equals number of weights divided by settling time, when the eigenvalues are equal.

Operation with 10 percent misadjustment, satisfactory in many applications, can generally be achieved with an adaptive settling time equal to 10 times the memory time span of the adaptive transversal filter.

PERFORMANCE

We have noted previously that the LMS algorithm differs from the algorithms discussed in Chapters 4 and 5 primarily in the way in which the gradient, ∇_k, is estimated at each time step. In effect, the LMS algorithm takes advantage of additional a priori information, namely that the performance surface is quadratic. We might expect that the LMS algorithm would have a performance advantage over the previous algorithms that used the difference method to estimate ∇_k, and in fact it does.

TABLE 6.1 MISADJUSTMENT AND TIME CONSTANTS OF STEEPEST-DESCENT AND LMS ALGORITHMS

	Steepest-descent algorithm	LMS algorithm
Misadjustment, M	$\dfrac{\mu(L+1)}{4N\delta^2}\xi_{\min}$	$\mu \operatorname{tr} \mathbf{R}$
	$=\dfrac{(L+1)^2}{8P}\left(\dfrac{1}{T_{\text{mse}}}\right)_{\text{av}}$	$=\dfrac{L+1}{4}\left(\dfrac{1}{T_{\text{mse}}}\right)_{\text{av}}$
Perturbation, P	$\dfrac{\delta^2\lambda_{\text{av}}}{\xi_{\min}}$	—
Total misadjustment, M_{tot}	$M+P$	M
Time constant of nth mode:		
In number of adaptive iterations, τ_{mse}	$\dfrac{1}{4\mu\lambda_n}$	$\dfrac{1}{4\mu\lambda_n}$
In number of data samples, T_{mse}	$\dfrac{N(L+1)}{2\mu\lambda_n}$	$\dfrac{1}{4\mu\lambda_n}$

The performance advantage of the LMS algorithm over the steepest-descent algorithm of Chapters 4 and 5 can readily be understood by comparing the expressions for misadjustment and time constant given in Table 6.1. The table entries originate from (5.16), (5.86) through (5.88), (5.99), (5.106), (6.20), and (6.40). It is apparent in both cases that misadjustment is reduced by slow adaptation; that is, by making the value of the time constant large. For a given time constant, however, the misadjustment of the LMS algorithm increases linearly with the number of weights rather than with the square of the number of weights. In typical circumstances much faster adaptation is thus possible.

The relative performance of the two algorithms is indicated in another way in Figure 6.6, which shows the time constant of adaptation T_{mse} as a function of the number of weights. For this comparison the misadjustment of the LMS algorithm was fixed at 10 percent. The *total* misadjustment of the steepest-descent algorithm was also fixed at 10 percent, with the perturbation P optimized according to (5.109). The eigenvalues of the R-matrix were further assumed to be equal. We obtain the curves in Figure 6.6 from Table 6.1 as follows:

$$\begin{aligned} &\textit{Steepest descent:} \quad && T_{\text{mse}} = \frac{(L+1)^2}{8MP} = 50(L+1)^2 \\ &\textit{LMS:} && T_{\text{mse}} = \frac{L+1}{4M} = 2.5(L+1) \end{aligned} \tag{6.43}$$

The figure illustrates that the LMS algorithm has a much shorter adaptation time, particularly when the number of weights is large.

In Chapter 5 we derived the time constant of adaptation of the method of steepest descent for an adaptive filter with ten weights and a total misadjustment of 10 percent with P optimized. The result obtained is given in (5.110) and is equal to 5000 data samples. A similar calculation for the LMS algorithm yields a time constant of adaptation of only 25 data samples. This value is much more favorable even than that of 625 data samples obtained in (5.111) for off-line adaptation to eliminate the effects of perturbing the weight vector. Since adaptive transients generally die out within approximately four time constants, the settling time would be approximately 100 sampling periods or iterations.

The efficiency of the LMS algorithm has been shown to approach a theoretical limit for adaptive algorithms when the eigenvalues of the R-matrix are equal or nearly equal [5]. With disparate eigenvalues, however, misadjustment is primarily determined by the fastest modes of adaptation, while settling time is limited by the slowest modes. Algorithms similar to the LMS algorithm but based on Newton's method rather than the method of steepest descent have been devised in an attempt to sustain efficiency under these conditions. In these algorithms the gradient estimate is premultiplied at each iteration by an estimate of the inverse of the R-matrix:

$$\mathbf{W}_{k+1} = \mathbf{W}_k + \mu\widehat{\mathbf{R}^{-1}}\hat{\nabla}_k \tag{6.44}$$

or

$$\mathbf{W}_{k+1} = \mathbf{W}_k + 2\mu\widehat{\mathbf{R}^{-1}}\varepsilon_k\mathbf{X}_k \tag{6.45}$$

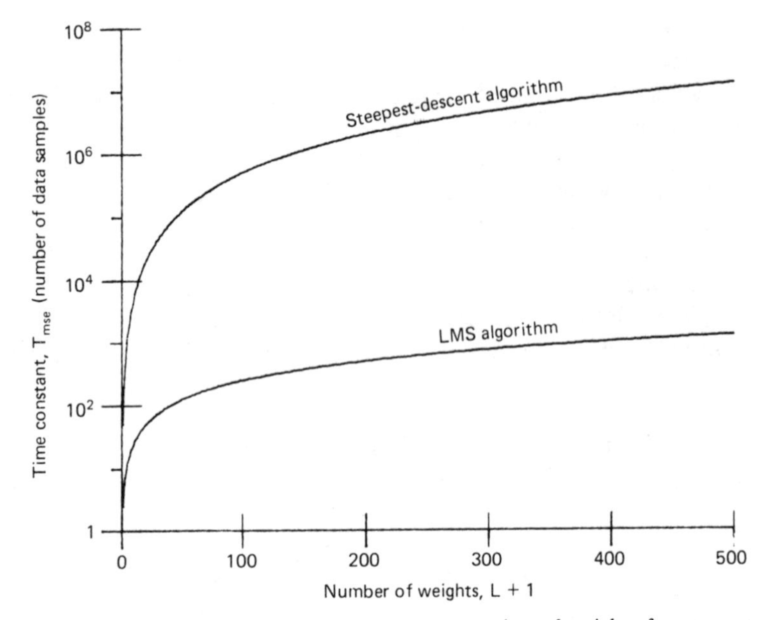

Figure 6.6 Time constant of adaptation versus number of weights for steepest-descent and LMS algorithms. Total misadjustment fixed at 10 percent.

This process causes all adaptive modes to have essentially the same time constant. Algorithms based on this principle are discussed in Chapter 8. They are potentially more efficient than the LMS algorithm but also typically more difficult to implement.

Attempts have also been made to devise algorithms more efficient than the LMS algorithm by using a time-varying value for the constant of adaptation μ. High values are first chosen to achieve rapid convergence; after convergence has been achieved low values are chosen to minimize misadjustment. This method works well as long as the input statistics remain stationary.

EXERCISES

1. Express the LMS algorithm for an individual weight in a single-input adaptive linear combiner.
2. If an adaptive transversal filter has an input signal power equal to p, what is a reasonable range for the adaptive gain constant, μ?
3. Describe the principal axes of the performance surface in Figure 6.3.
4. Explain why T_{mse} and τ_{mse} are identical in the case of the LMS algorithm.
5. In the example in Figure 6.3, what is the approximate covariance matrix for the gradient noise, assuming that the adaptive transient has died out?

6. What are the eigenvalues in the example in Figure 6.3?
7. What is the effect of adding a third weight to the adaptive transversal filter in Figure 6.2:
   (a) On the minimum mean-square error?
   (b) On the misadjustment?
   (c) On the learning curve?
8. Consider the adaptive signal-canceling situation in the diagram below.

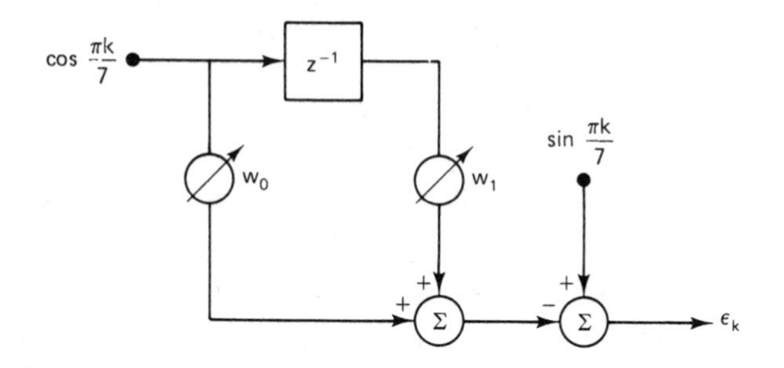

   (a) Write an expression for the performance surface.
   (b) Determine a range for the adaptive gain constant.
   (c) Express the LMS algorithm for this situation.
   (d) Using $\mu = 0.05$ and 0.005 with zero initial conditions, run the LMS algorithm and plot two learning curves similar to Figure 6.5 on the same graph. Estimate the two time constants.
9. Consider the adaptive predictor shown below.

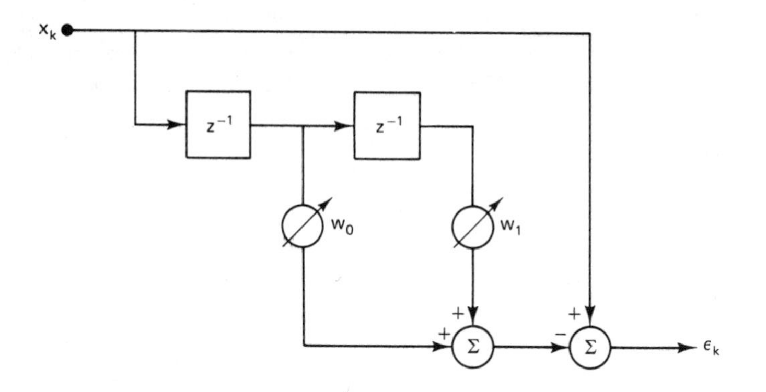

   (a) Given $\phi_{xx}(n) = E[x_k x_{k+n}]$, write an expression for the performance surface.
   (b) Give a specific expression for the performance surface when x_k is equal to $\sin(k\pi/5)$.
   (c) Write the LMS algorithm for the case where $x_k = \sin(k\pi/5)$, with μ at one-fifth of its maximum in (6.10).
   (d) Using the LMS algorithm in part (c), list the first 20 error values, ε_0 through ε_{19}. Assume that $\mathbf{W}_0 = \mathbf{0}$.

10. For the predictor in Exercise 9, generate 5000 samples of the input as follows, using zero initial conditions:

$$x_k = 0.95(r_k - 0.5) + 0.05x_{k-1}; \qquad k = 0,1,\ldots,4999$$

where r_k is the kth number in the random sequence in Appendix A. Then:

(a) Using the data, estimate the input correlation matrix, $\mathbf{R}$.

(b) Using (6.21), select μ to produce an average learning time constant equal to 100 data samples.

(c) Run the LMS algorithm with μ as above. Plot the first 5000 weight values, w_{0k} and w_{1k}, versus k.

(d) Plot a learning curve similar to that shown in Figure 6.5.

REFERENCES AND ADDITIONAL READINGS

1. B. Widrow and J. M. McCool, "A comparison of adaptive algorithms based on the methods of steepest descent and random search," *IEEE Trans. Antennas Propag.*, vol. AP-24, no. 5, pp. 615–637, Sept. 1976.
2. B. Widrow, J. M. McCool, M. G. Larimore, and C. R. Johnson, Jr., "Stationary and nonstationary learning characteristics of the LMS adaptive filter," *Proc. IEEE*, vol. 64, no. 8, pp. 1151–1162, Aug. 1976.
3. B. Widrow et al., "Adaptive noise cancelling: principles and applications," *Proc. IEEE*, vol. 63, no. 12, pp. 1692–1716, Dec. 1975.
4. B. Widrow, "Adaptive Filters," in *Aspects of Network and System Theory*, R. E. Kalman and N. De Claris (Eds.). New York: Holt, Rinehart and Winston, 1970, pp. 563–587.
5. B. Widrow and E. Walach, "On the statistical efficiency of the LMS algorithm with nonstationary inputs," *IEEE Trans. Information Theory—Special Issue on Adaptive Filtering*, vol. 30, no. 2, part 1, pp. 211–221, Mar. 1984.
6. D. C. Farden, "Tracking properties of adaptive signal processing algorithms," *IEEE Trans. Acoust. Speech Signal Process.*, vol. ASSP-29, p. 439, June 1981.

chapter 7

The z-Transform in Adaptive Signal Processing

In the preceding chapters we have described the adaptive linear combiner and its properties without recourse to the usual transform or frequency-domain analysis used with linear systems [1–4]. In this chapter we review some of these standard concepts and methods of digital signal processing and show how they apply particularly to the analysis of linear adaptive filters.

Specifically, we review the z-transform method of analyzing digital systems and show how the z-transform relates to the frequency response (gain and phase shift) of a system. We also reformulate the least-squares method and the LMS performance surface in terms of the z-transform. Our presentation is brief. There are many complete texts on the subject of digital signal processing [1–4, 7–9].

In the case of the adaptive linear combiner, the ideas of this chapter give us an alternative view of the design and analysis of adaptive systems. For other types of adaptive filters such as recursive adaptive filters, the transform methods discussed here are required, rather than alternatives.

THE z-TRANSFORM

As we have seen, the analysis of signal processing systems such as the adaptive linear combiner involves the use of sample sets, that is, ordered sequences of samples. These sample sets include various time series such as the input, desired, and error signals as well as the set of weight values.

The z-transform of any such sequence is defined as follows:

$$\begin{aligned} &\textit{Data sequence:} \quad [x_k] = [\cdots \quad x_{-1}x_0x_1x_2x_3 \quad \cdots] \\ &\textit{z-transform:} \quad X(z) = \sum_{k=-\infty}^{\infty} x_k z^{-k} \end{aligned} \tag{7.1}$$

In this form z is a continuous complex variable and $X(z)$ is called a "two-sided" z-transform because negative as well as positive values of the index k are involved. In fact, x_k must be known for all integer values of k in order to produce $X(z)$. Thus we could obtain a one-sided transform simply by assuming that x_k is zero for k less than zero.

As a simple example of a z-transform, let $[x_k]$ be the samples of an exponential function such that

$$x_k = \begin{cases} 0 & k < 0 \\ e^{-\alpha k} & k \geq 0; \quad \alpha > 0 \end{cases} \tag{7.2}$$

Then the z-transform is a simple rational function of α and z, expressed in Figure 7.1. Since x_k is zero for k less than zero, we have a one-sided transform in this case. The transform has a *zero* (i.e., has a value of zero) at $z = 0$ and has a *pole* (i.e., becomes infinite) at $z = e^{-\alpha}$.

As in the case of Figure 7.1, an infinite (or even a finite) geometric sum can always be written in closed rational form, so the z-transform allows us to express many sample sequences in a simple manner. Tabulations of z-transforms may be found in the literature [4].

Notice also that all of the information in the original sample sequence is conveyed by its z-transform. Each sample in (7.1) is associated with a unique power of z, so the complete sample set can be recovered from the z-transform. In other words, an inverse z-transform obviously exists. We will return shortly to a discussion of the inverse transform.

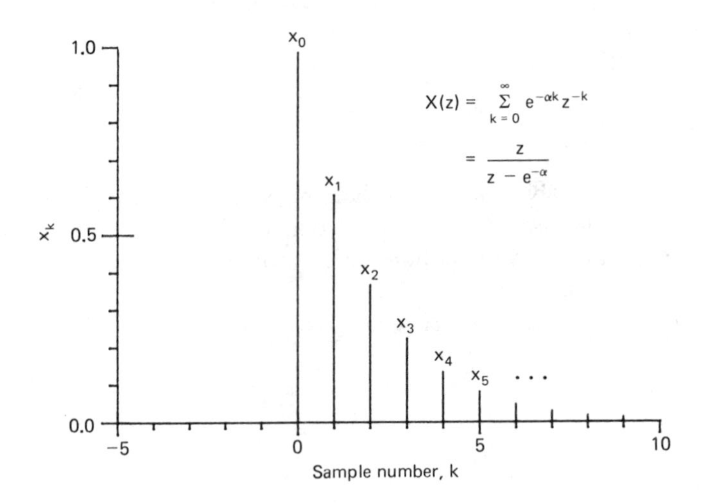

Figure 7.1 The sampled function $x_k = e^{-\alpha k}$ and its z-transform, which has a zero at $z = 0$ and a pole at $z = e^{-\alpha}$.

RIGHT- AND LEFT-HANDED SEQUENCES

The sample sequence in (7.2) is "right-handed" because the signal starts at $k = 0$ and proceeds to the right with k increasing. A "left-handed" sequence proceeds from $k = 0$ toward decreasing k, and a two-sided sequence goes in both directions from $k = 0$. All sequences can be transformed using (7.1) of course, but the region of z for which (7.1) converges varies. Consider the following examples:

Sample sequence:	*z-Transform*:
$x_k = e^{-\alpha k};\ k \geq 0;\ \alpha > 0$	$\sum_{k=0}^{\infty} (e^{-\alpha} z^{-1})^k = \frac{z}{z - e^{-\alpha}}$
$x_k = e^{\alpha k};\ k \leq 0;\ \alpha > 0$	$\sum_{k=-\infty}^{0} (e^{\alpha} z^{-1})^k = \frac{1}{1 - ze^{-\alpha}}$
$x_k = \cos \alpha k;\ k \geq 0$	$\sum_{k=0}^{\infty} \cos \alpha k z^{-k} = \frac{z(z - \cos \alpha)}{z^2 - 2z\cos\alpha + 1}$

Note in the first example that the summation converges only if $|z| > e^{-\alpha}$, so that the terms in the sum diminish with k. Similarly, the second transform converges only if $|z| < e^{\alpha}$ and the third transform[†] converges only if $|z| > 1$. Thus the following rules apply to these and other right- or left-handed sequences:

> If x_k is left-handed, $X(z)$ converges for $|z| < 1$ and has poles on or outside $|z| = 1$.
>
> If x_k is right-handed, $X(z)$ converges for $|z| > 1$ and has poles on or inside $|z| = 1$.

These rules are general and do not give the exact region of convergence of $X(z)$ on the z-plane.[‡] In the case of $x_k = e^{-\alpha k}$ for $k \geq 0$, for example, the region of convergence depends on α, but it always includes the region outside the unit circle defined by $|z| > 1$. We assume in all sequences that the samples $[x_k]$ are finite and do not grow without bound as $|k|$ increases.

There are cases, such as with correlation functions (later in this chapter) or noncausal filters (Chapter 10), where we may wish to transform a two-sided sequence, even though a region of convergence does not exist at all. In these cases it is generally convenient either not to sum the two-sided series [as in (7.70)], or else to treat the right- and left-handed portions of the sequence separately (as in Chapter 10). For example, the transform of $x_k = \cos \alpha k$ for $k < 0$ is minus the transform in

[†] The third series may be summed by writing $\cos \alpha k = (e^{j\alpha k} + e^{-j\alpha k})/2$, then summing the geometric series.

[‡] The complex z-plane contains points whose abscissas and ordinates are the real and imaginary parts of z. Every value of z corresponds to a point on the z-plane.

the example above,† where $x_k = \cos \alpha k$ for $k \geq 0$, but the two regions of convergence do not overlap. Thus the entire two-sided transform does not exist, even though the two-sided series may be used as in (7.70) to "represent" the entire sequence.

TRANSFER FUNCTIONS

Most of the readers of this book are already familiar with the transfer function concept, which is common to all types of linear systems analysis, continuous as well as digital. The transfer function is simply the transform of the output of a system divided by the transform of the input. With continuous systems we use the Laplace transform, and with digital systems we use the z-transform.

A general linear digital processing system (or algorithm) is illustrated in Figure 7.2. Compare this figure with Figure 2.2 and note that if we set the feedback coefficients (the b's) equal to zero, we have in Figure 7.2 the single-input adaptive linear combiner. The weights are shown without arrows in Figure 7.2 because for the present we want them to be fixed rather than adjustable.‡

Thus Figure 7.2 represents the general single-input linear combiner, or digital filter. Without the feedback portion, the filter is called "nonrecursive." With the feedback portion it is called "recursive." The expression for the output, y_k, in either case is

$$y_k = \sum_{n=0}^{L} a_n x_{k-n} + \sum_{n=1}^{L} b_n y_{k-n} \tag{7.3}$$

Thus, without the b's, we have essentially (2.3), that is, the nonrecursive version. The filter is also causal, because a_n is nonzero only for n positive.

To develop the transfer function, let us now take the z-transform of (7.3):

$$Y(z) = \sum_{k=-\infty}^{\infty} \sum_{n=0}^{L} a_n x_{k-n} z^{-k} + \sum_{k=-\infty}^{\infty} \sum_{n=1}^{L} b_n y_{k-n} z^{-k} \tag{7.4}$$

If we are to express the right side of this result in terms of transforms, we must define a_n and b_n for all values of n. As suggested above, we do this simply by letting a_n and b_n be zero when n designates a position "outside of the filter" in Figure 7.2. Thus we define

$$\begin{aligned} [a_n] &= [\cdots \quad 000 a_0 a_1 \quad \cdots \quad a_L 000 \quad \cdots] \\ [b_n] &= [\cdots \quad 000 b_1 \quad \cdots \quad b_L 000 \quad \cdots] \end{aligned} \tag{7.5}$$

† The reader may wish to verify this by summing the two series.

‡ When the weights are adapting, the time-varying filter does not have a transfer function. A transfer function implies fixed weights.

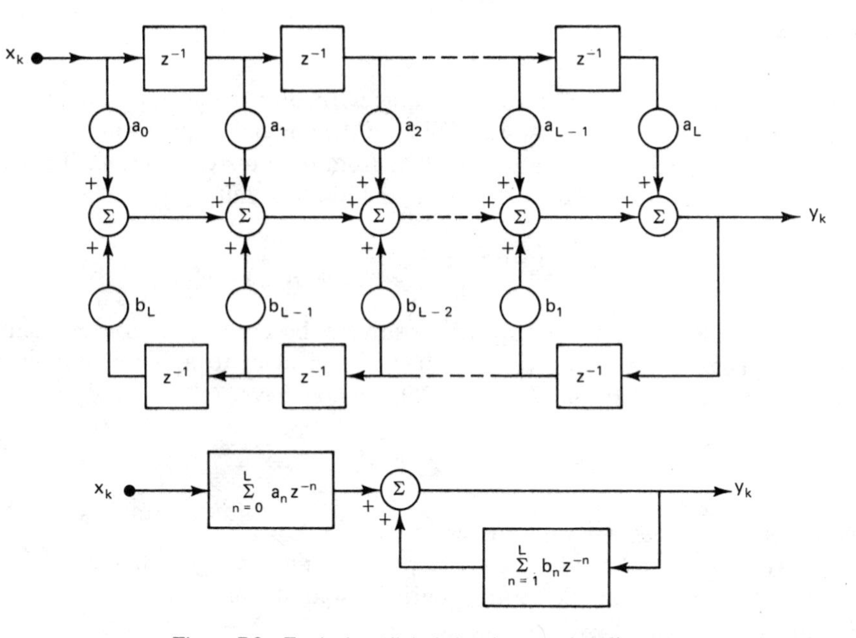

Figure 7.2 Equivalent digital signal processing diagrams.

Now we can write (7.4) with infinite limits and exchange the order of summation to obtain

$$Y(z) = \sum_{n=-\infty}^{\infty} a_n \sum_{k=-\infty}^{\infty} x_{k-n} z^{-k} + \sum_{n=-\infty}^{\infty} b_n \sum_{k=-\infty}^{\infty} y_{k-n} z^{-k} \tag{7.6}$$

Next, the right side can be written in terms of transforms simply by changing subscripts. If we let $m = k - n$ the limits on m are still $\pm$ infinity, and (7.6) becomes

$$\begin{aligned} Y(z) &= \left(\sum_{n=-\infty}^{\infty} a_n z^{-n} \right) \left(\sum_{m=-\infty}^{\infty} x_m z^{-m} \right) + \left(\sum_{n=-\infty}^{\infty} b_n z^{-n} \right) \left(\sum_{m=-\infty}^{\infty} y_m z^{-m} \right) \\ &= A(z)X(z) + Y(z)B(z) \end{aligned} \tag{7.7}$$

The transfer function, as mentioned above, is the output transform, $Y(z)$, divided by the input transform, $X(z)$. From (7.7), then,

$$\begin{aligned} \text{transfer function} = H(z) &= \frac{Y(z)}{X(z)} \\ &= \frac{A(z)}{1 - B(z)} \end{aligned} \tag{7.8}$$

Thus, with $B(z) = 0$, we have $Y(z) = A(z)X(z)$ for the nonrecursive case.

FREQUENCY RESPONSE

By simply replacing z with $e^{j\omega}$, where ω is the normalized frequency,[†] we are able to obtain the discrete Fourier transform (DFT) of the impulse response in (7.22) below, i.e., the frequency response of a linear filter, from its transfer function. To demonstrate this, we first divide $A(z)$ by $1 - B(z)$ in (7.8) to obtain

$$H(z) = \sum_{n=0}^{\infty} h_n z^{-n} = \frac{A(z)}{1 - B(z)} \tag{7.9}$$

Note here that h_n is nonzero only for positive n because a_n is nonzero only for positive n. Thus any causal recursive filter is equivalent to a causal nonrecursive filter of infinite length. From (7.3) and (7.9) we now have

$$y_k = \sum_{n=0}^{\infty} h_n x_{k-n} \tag{7.10}$$

Again, (7.10) describes *any* causal linear filter.

If we wish to find the frequency response of the filter given by $[h_k]$ in (7.10), we can let $[x_k]$ be the sample set of a unit sinusoid at some fixed frequency ω and compute $[y_k]$. So we let

$$x_k = e^{j\omega k} \tag{7.11}$$

Then, in (7.10), we have

$$\begin{aligned} y_k &= \sum_{n=0}^{\infty} h_n e^{j\omega(k-n)} \\ &= \left(\sum_{n=0}^{\infty} h_n e^{-j\omega n} \right) x_k \end{aligned} \tag{7.12}$$

Thus, since the sinusoid x_k is multiplied by the quantity in parentheses to obtain the sinusoid y_k, this quantity must represent the frequency response of the filter, that is, the gain and phase shift, at the frequency ω.

But the quantity in parentheses could have been obtained by substituting $e^{j\omega}$ for z in (7.9) or (7.8). Thus, for any linear filter in the form of Figure 7.2,

$$\begin{aligned} \text{frequency response} &= \frac{Y(e^{j\omega})}{X(e^{j\omega})} \\ &= \frac{A(e^{j\omega})}{1 - B(e^{j\omega})} \end{aligned} \tag{7.13}$$

We can see that the frequency response is a *periodic* function of ω, because $e^{j\omega}$ does

[†] That is, normalized so that the sampling frequency is at $\omega = 2\pi$. The reader should be aware that some authors use ω to represent an unnormalized frequency, which requires scaling by the time step T, as in (7.16).

not change as ω is increased by any multiple of 2π. Furthermore, if we use $2\pi - \omega$ in place of ω, we have

$$e^{j(2\pi-\omega)} = e^{-j\omega} \tag{7.14}$$

Thus, since the coefficients of $A(z)$ and $B(z)$ are real, we have

$$\frac{Y(e^{j(2\pi-\omega)})}{X(e^{j(2\pi-\omega)})} = \frac{Y(e^{-j\omega})}{X(e^{-j\omega})} \tag{7.15}$$

So the transfer function, Y/X, is unique only for $0 \le \omega < \pi$. This frequency range is called the *Nyquist interval*, with the *folding frequency* at $\omega = \pi$ and the sampling frequency at $\omega = 2\pi$. If we wanted to write x_k in (7.11) as a function of time instead of the sample number k, we would let

$$\begin{aligned} \omega &= \Omega T \ (\Omega \text{ in rad/s}) \\ &= 2\pi f T \ (f \text{ in hertz}) \end{aligned} \tag{7.16}$$

where T is the time step (distance between samples) in seconds, so that $t = kT$ would appear in the exponent. Then the folding frequency would be at π/T radians per second, or at $1/2T$ hertz, that is, at one-half the sampling frequency.

A specific example of frequency response is shown in Figure 7.3. Here we have the transfer function

$$H(z) = \frac{Y(z)}{X(z)} = \frac{0.27(z^2+1)}{z^2 - 1.27z + 0.81} \tag{7.17}$$

The frequency response, $H(e^{j\omega})$, is in this case

$$\begin{aligned} H(e^{j\omega}) &= \frac{0.27(e^{2j\omega}+1)}{e^{2j\omega} - 1.27e^{j\omega} + 0.81} \\ &= \frac{0.54\cos\omega(1.81\cos\omega - 1.27 - j0.19\sin\omega)}{(1.81\cos\omega - 1.27)^2 + (0.19\sin\omega)^2} \end{aligned} \tag{7.18}$$

The magnitude and phase of $H(e^{j\omega})$ are called the *amplitude gain* and *phase shift* of the filter. As in (7.18), we have

$$\begin{aligned} H(e^{j\omega}) &= \text{Real} + j\text{Imag} \\ \text{amplitude gain} &= |H(e^{j\omega})| = [\text{Real}^2 + \text{Imag}^2]^{1/2} \\ \text{phase shift} &= \tan^{-1}(\text{Imag/Real}) \end{aligned} \tag{7.19}$$

The amplitude gain and phase shift for the present example are plotted in Figure 7.3. Besides the amplitude gain, the *power gain* of a filter is also used. The power gain is the square of the amplitude gain and is sometimes specified in decibels (dB). Thus

$$\begin{aligned} \text{power gain} &= |H(e^{j\omega})|^2 \\ &= \text{Real}^2 + \text{Imag}^2 \\ \text{power gain in dB} &= 10\log_{10}|H(e^{j\omega})|^2 \end{aligned} \tag{7.20}$$

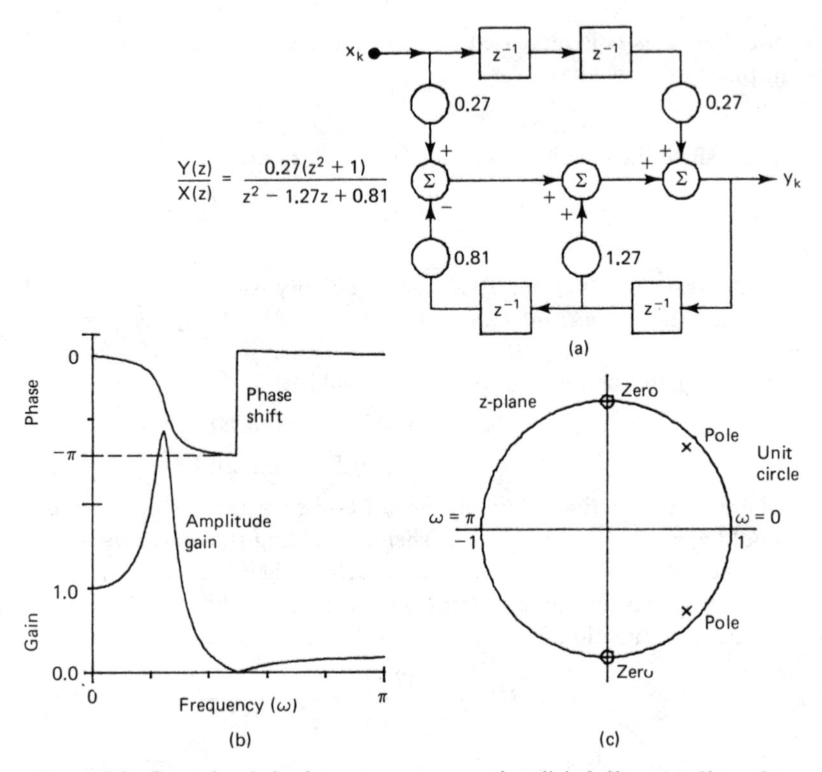

Figure 7.3 Example of the frequency response of a digital filter: (a) filter; (b) response; (c) pole–zero plot.

In the example of Figure 7.3 we can also note the effect of the *poles and zeros* of $H(z)$ on the gain and phase shift. To obtain the frequency response in (7.13) we set $z = e^{j\omega}$, and thus as ω changes from 0 to the folding frequency (π), z moves around the upper half of the unit circle on the z-plane. When ω is such that z is near a pole, as it is at $\omega = \pi/4$ in Figure 7.3, the gain is high. When z is near a zero of $H(z)$, as at $\omega = \pi/2$ in Figure 7.3, the gain is low. As z on the unit circle passes near or through a pole or zero, there is a sharp change in the phase characteristic, as also illustrated in Figure 7.3.

IMPULSE RESPONSE AND STABILITY

Suppose that we have an "impulse" sample set consisting of a single unit sample at $k = 0$, that is,

$$[x_k] = [\cdots \quad 00100 \quad \cdots] \tag{7.21}$$

By definition in (7.1), the z-transform of the impulse sample set is

$$X(z) = 1$$

So if $[x_k]$ is the input to a filter with transfer function $H(z)$, then from (7.8) the output must be the sequence whose transform is $H(z)$:

$$\text{impulse response} = Z^{-1}[H(z)] \tag{7.22}$$

where Z^{-1} stands for "inverse z-transform."

From this definition of the impulse response we can argue that nonrecursive digital filters such as the adaptive linear combiner are inherently "stable" in the sense that, as long as the weight values are finite, the impulse response is bounded and of finite length.

On the other hand, recursive filters have impulse responses of infinite length as suggested by (7.10). The causal recursive filter is stable only if the poles are inside the unit circle, as in Figure 7.3. We can show the stability requirement in the following way. Let the transfer function, $H(z)$, be a ratio of polynomials in z^{-1} as in (7.8). If we expand $H(z)$ into a sum of partial fractions, $H(z)$ will contain terms of the form $z^{-n}/(1 - z_0 z^{-1})$, where z_0 is a pole location. Thus we write $H(z)$ in the form

$$H(z) = \frac{Az^{-n}}{1 - z_0 z^{-1}} + G(z) \tag{7.23}$$

where $G(z)$ represents the rest of $H(z)$, and A is a constant. Since the input occurs at $k = 0$, we know that the response [i.e., the inverse of $H(z)$] must be right-handed. Therefore, we can express the first term in (7.23) as a right-handed series, and reexpress $H(z)$ as

$$H(z) = A \sum_{k=n}^{\infty} z_0^{k-n} z^{-k} + G(z) \tag{7.24}$$

From this result we have the impulse response as

$$\begin{aligned} h_k &= Z^{-1}[H(z)] \\ &= Az_0^{k-n} + Z^{-1}[G(z)] \qquad k \geq n \end{aligned} \tag{7.25}$$

Although the pole location z_0 is in general complex, we can see that the impulse response will grow without bound unless the modulus of z_0 is less than 1, that is, unless z_0 is inside the unit circle. (We define a filter with a pole *on* the unit circle to be "conditionally" stable.)

To summarize this section, we have the following tabulation of results for filters of finite length:

Filter type	Impulse response	Stability conditions
Nonrecursive	Finite (FIR)	Finite coefficients
Recursive	Infinite (IIR)	Finite coefficients; poles inside $\lvert z\rvert = 1$ for causal filter

The term "infinite" of course means infinite in extent, as in (7.10) or (7.25). The initials FIR and IIR are often used to specify the two classes of filters.

THE INVERSE z-TRANSFORM

As suggested by (7.24) and (7.25), the inverse z-transform of any rational function can be found via a partial fraction expansion and conversion to geometric series. For our analysis of least-squares systems, however, we need a more convenient form for the inverse. This is found in the realm of complex-variable theory and is

$$x_k = \frac{1}{2\pi j}\oint X(z)z^{k-1}\,dz \tag{7.26}$$

The path of the integration is taken to be counterclockwise on a circle centered at the origin on the z-plane, on which $X(z)$ in (7.1) converges.

The validity of (7.26) is easily proved by substituting the forward transform in (7.1). We have

$$\begin{aligned} x_k &= \frac{1}{2\pi j}\oint \sum_{n=-\infty}^{\infty} x_n z^{k-n-1}\,dz \\ &= \sum_{n=-\infty}^{\infty} \frac{1}{2\pi j}\oint x_n z^{k-n-1}\,dz \end{aligned} \tag{7.27}$$

The integral theorem of Cauchy [5] can be adapted to give the following general result:

$$\frac{1}{2\pi j}\oint z^m\,dz = \begin{cases} 0 & m \neq -1 \\ 1 & m = -1 \end{cases} \tag{7.28}$$

In (7.27) x_n is just a constant, so only one term in the sum, where $n = k$, is nonzero, and this term is x_k. So (7.27) becomes an identity and the validity of (7.26) is proved.

It is also useful to note that the substitution

$$z = e^{j\omega} \tag{7.29}$$

in (7.26) gives an inverse frequency transformation. The substitution causes dz to be replaced by $jz\,d\omega$, and a path of integration may be taken on the unit circle from $z = e^{-j\pi}$ counterclockwise to $z = e^{j\pi}$ (i.e., once around the circle). Thus (7.26) becomes

$$x_k = \frac{1}{2\pi}\int_{-\pi}^{\pi} X(e^{j\omega})e^{jk\omega}\,d\omega \tag{7.30}$$

and we have the sample set in terms of the spectral function, $X(e^{j\omega})$. This is essentially the inverse Fourier transform formula.

We turn now to the question of how to evaluate integrals like (7.26) or, equivalently, (7.30) in practice. If $X(z)$ is a polynomial in z, then (7.26) gives x_k simply as one of the coefficients of the polynomial, depending on k. But if $X(z)$ is a ratio of polynomials, the *residue theorem* [10] is helpful. The theorem states that

$$x_k = \frac{1}{2\pi j}\oint X(z)z^{k-1}\,dz$$
$$= \sum_n \text{Res}\left[X(z)z^{k-1} \text{ at pole } z_n\right] \tag{7.31}$$

That is, x_k is the sum of the residues of the integrand at all of its poles inside the path of integration. Given that $X(z)z^{k-1}$ is a rational function of z, each of these residues is found as follows. Let z_n be a pole of $X(z)z^{k-1}$, repeated r times. We write $X(z)z^{k-1}$ as

$$X(z)z^{k-1} = \frac{V(z)}{(z-z_n)^r} \tag{7.32}$$

and the residue at z_n is

$$\text{Res}\left[X(z)z^{k-1} \text{ at } z_n\right] = \frac{1}{(r-1)!}\left[\frac{d^{r-1}V(z)}{dz^{r-1}}\right]_{z=z_n} \tag{7.33}$$

In the case of $r = 1$, we have for a

$$\text{simple pole:}\quad \text{Res}\left[X(z)z^{k-1} \text{ at } z_n\right] = V(z_n) \tag{7.34}$$

Thus, using (7.31) through (7.34), we have a method for finding the inverse transforms of rational functions of z. [Note especially in (7.31) that when $k = 0$, there is usually an extra pole at $z = 0$.]

To illustrate, let us again examine the case of Figure 7.1. Here we had $X(z) = z/(z - e^{-a})$ with x_k being defined for $k \geq 0$, and so (7.31) becomes

$$x_k = \frac{1}{2\pi j}\oint X(z)z^{k-1}\,dz \qquad k \geq 0$$
$$= \frac{1}{2\pi j}\oint \frac{z^k\,dz}{z-e^{-a}} \qquad k \geq 0$$
$$= \text{Res}\left[\frac{z^k}{z-e^{-a}} \text{ at } e^{-a}\right] \tag{7.35}$$

Since we have a simple pole in this case, (7.34) applies and we have

$$x_k = V(e^{-a}) = e^{-ak} \qquad k \geq 0 \tag{7.36}$$

just as in Figure 5.1. Note that the range of k for which the inverse transform is valid is not carried within the integral formula itself and must therefore be known externally.

In the foregoing example, $[x_k]$ was right-handed and $k \geq 0$ was assumed. For left-handed sequences, the factor z^{k-1} in (7.26) makes using (7.33) difficult for the residue at $z = 0$. The easiest approach for left-handed sequences is to let $u = z^{-1}$ in the forward and inverse transforms (7.1) and (7.26). This substitution reverses the "handedness" of any sequence. For example, the "u-transform" of $x_k = e^{ak}$ for $k \leq 0$ would be $u/(u - e^{-a})$, and the inverse could be found as in (7.35) through (7.36) and then reflected to become a left-handed sequence. For further examples, see Exercises 16 through 18, 26, and 27.

CORRELATION FUNCTIONS AND POWER SPECTRA

Most of the analysis of adaptive filters assumes that the input signal or signals have statistical properties that are constant over the period of analysis, even though this may not be exactly true. Thus it is convenient to discuss signals that have either periodic or stationary random sample sequences. Such signals can be described in terms of *correlation functions*, which we define just as in (2.11) and (2.12), for the single-input adaptive linear combiner:

$$\textit{Cross-correlation:} \quad \phi_{xy}(n) = E[x_k y_{k+n}] \tag{7.37}$$

$$\textit{Autocorrelation:} \quad \phi_{xx}(n) = E[x_k x_{k+n}] \tag{7.38}$$

$$-\infty < n < \infty$$

The expectation goes over k, and we note that the autocorrelation function is just a special case of the more general cross-correlation function, with $y = x$ in (7.38). If the averages do not change with k, then $x_k y_{k+n}$ and $x_{k-n} y_k$ represent the same relative shift of x and y, and

$$\begin{aligned}\phi_{yx}(n) &= E[y_k x_{k+n}] \\ &= E[y_{k-n} x_k] = \phi_{xy}(-n)\end{aligned} \tag{7.39}$$

Thus the autocorrelation function is an even function, i.e.,

$$\phi_{xx}(n) = \phi_{xx}(-n) \tag{7.40}$$

We now define a *discrete power spectrum* as the z-transform of either of the correlation functions in (7.37) and (7.38). Thus

$$\textit{Cross-power spectrum:} \quad \Phi_{xy}(z) = \sum_{n=-\infty}^{\infty} \phi_{xy}(n) z^{-n} \tag{7.41}$$

$$\textit{Auto-power spectrum:} \quad \Phi_{xx}(z) = \sum_{n=-\infty}^{\infty} \phi_{xx}(n) z^{-n} \tag{7.42}$$

Again, we note that $\Phi_{xx}(z)$ is just a special case of $\Phi_{xy}(z)$ with $y = x$. And just as above in the section on frequency response, we can substitute $e^{j\omega}$ for z and obtain the power spectrum in terms of frequency:

$$\Phi_{xy}(e^{j\omega}) = \sum_{n=-\infty}^{\infty} \phi_{xy}(n) e^{-jn\omega} \tag{7.43}$$

In this formulation we have essentially the discrete Fourier transform of $\phi_{xy}(n)$, giving the distribution of the sample products, $[x_k y_{k+n}]$, over frequency, with $\omega = \pi$ at one-half the sampling rate as above.

An important property of the power spectrum is the symmetry property that arises from (7.39). By reversing the order of the signals x and y, we have the equivalent of the time reversal in (7.39):

$$\Phi_{yx}(z) = \sum_{n=-\infty}^{\infty} \phi_{xy}(-n) z^{-n}$$

$$= \sum_{m=-\infty}^{\infty} \phi_{xy}(m) z^{m} = \Phi_{xy}(z^{-1}) \tag{7.44}$$

The substitution of z^{-1} for z here implies the conjugate when z is on the unit circle, which is always true when we discuss frequency response.

We now examine the relationships between power spectra and transfer functions. Suppose that $[y_k]$ and $[x_k]$ are related through a linear transfer function, $H(z)$, as in Figure 7.4, which is in turn equivalent to Figure 7.2. Here we have $H(z)$ given in the form of (7.10), so that the actual filter could be either recursive or nonrecursive. If we use first (7.37) and then (7.10) in (7.41), we obtain

$$\Phi_{xy}(z) = \sum_{n=-\infty}^{\infty} E[x_k y_{k+n}] z^{-n}$$

$$= \sum_{n=-\infty}^{\infty} E\left[x_k \sum_{l=0}^{\infty} h_l x_{k+n-l}\right] z^{-n} \tag{7.45}$$

Since the expectation of any sum is the corresponding sum of expectations, we can move the "E" operator in (7.45) inside the inner sum. If we also exchange the order of summation, we get

$$\Phi_{xy}(z) = \sum_{l=0}^{\infty} h_l \sum_{n=-\infty}^{\infty} E[x_k x_{k+n-l}] z^{-n} \tag{7.46}$$

Now a simple change of index, with $m = n - l$, produces the desired transfer-function relationship:

$$\Phi_{xy}(z) = \sum_{l=0}^{\infty} h_l z^{-l} \sum_{m=-\infty}^{\infty} E[x_k x_{k+m}] z^{-m}$$

$$= \sum_{l=0}^{\infty} h_l z^{-l} \sum_{m=-\infty}^{\infty} \phi_{xx}(m) z^{-m}$$

$$= H(z)\Phi_{xx}(z) \tag{7.47}$$

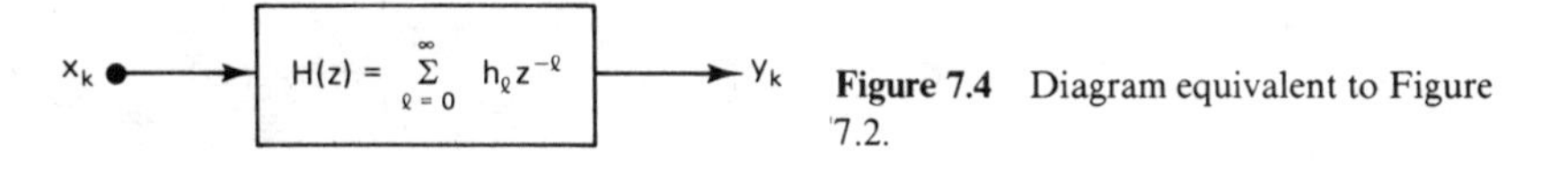

Figure 7.4 Diagram equivalent to Figure 7.2.

Proceeding in a similar manner, we can obtain a transfer-function relationship between Φ_{xx} and Φ_{yy} as follows:

$$
\begin{aligned}
\Phi_{yy}(z) &= \sum_{n=-\infty}^{\infty} \phi_{yy}(n) z^{-n} \\
&= \sum_{n=-\infty}^{\infty} E[y_k y_{k+n}] z^{-n} \\
&= \sum_{n=-\infty}^{\infty} E\left[\sum_{l=0}^{\infty} h_l x_{k-l} \sum_{m=0}^{\infty} h_m x_{k+n-m}\right] z^{-n} \\
&= \sum_{l=0}^{\infty} h_l z^{l} \sum_{m=0}^{\infty} h_m z^{-m} \sum_{n=-\infty}^{\infty} E[x_{k-l} x_{k+n-m}] z^{-(n-m+l)} \\
&= \sum_{l=0}^{\infty} h_l z^{l} \sum_{m=0}^{\infty} h_m z^{-m} \sum_{n=-\infty}^{\infty} \phi_{xx}(n-m+l) z^{-(n-m+l)} \\
&= H(z^{-1}) H(z) \Phi_{xx}(z) \\
&= |H(z)|^2 \Phi_{xx}(z)
\end{aligned}
\tag{7.48}
$$

Note that in the fourth line of this development, the two samples of x are separated by a total of $n - m + l$ sample points, and thus the expected product is $\phi_{xx}(n - m + l)$. Then, as in (7.46) and (7.47), a simple change of index gets us to the final result. On the last line, we have again assumed that z is on the unit circle to obtain the squared-magnitude result.

Furthermore, suppose that $[d_k]$ is a third sample sequence with stationary properties, similar to $[x_k]$ and $[y_k]$, which are still assumed to be related as in (7.10). Then the cross-spectrum $\Phi_{dy}(z)$ may be developed just as in (7.45) through (7.47):

$$
\begin{aligned}
\Phi_{dy}(z) &= \sum_{n=-\infty}^{\infty} E[d_k y_{k+n}] z^{-n} \\
&= \sum_{l=0}^{\infty} h_l z^{-l} \sum_{m=-\infty}^{\infty} \phi_{dx}(m) z^{-m} \\
&= H(z) \Phi_{dx}(z)
\end{aligned}
\tag{7.49}
$$

If we replace d with x in this result, we obtain the result in (7.47).

Besides these transfer-function relationships, we can also apply the inverse transform formula in (7.26) to get the correlation function in terms of the power spectrum. Since by definition in (7.41) Φ_{xy} is the transform of ϕ_{xy}, we have from (7.26)

$$
E[x_k y_{k+n}] = \phi_{xy}(n) = \frac{1}{2\pi j} \oint \Phi_{xy}(z) z^{n-1} dz \tag{7.50}
$$

$$
E[x_k x_{k+n}] = \phi_{xx}(n) = \frac{1}{2\pi j} \oint \Phi_{xx}(z) z^{n-1} dz \tag{7.51}
$$

In particular, with $n = 0$, we have

$$E[x_k y_k] = \phi_{xy}(0) = \frac{1}{2\pi j}\oint \Phi_{xy}(z)\frac{dz}{z} \tag{7.52}$$

$$E[x_k^2] = \phi_{xx}(0) = \frac{1}{2\pi j}\oint \Phi_{xx}(z)\frac{dz}{z} \tag{7.53}$$

The latter quantity, the mean-square value of x_k, is called the *total power* (or average power) in the sequence $[x_k]$. Note that, in terms of frequency, as (7.26) is the same as (7.30) so (7.53) is the same as

$$E[x_k^2] = \frac{1}{2\pi}\int_{-\pi}^{\pi} \Phi_{xx}(e^{j\omega})\,d\omega \tag{7.54}$$

Thus the total power, $E[x_k^2]$, is the integral of the discrete power spectrum.

To summarize this section, we list the important relationships together as follows:

If x, y, and d are stationary signals and if $Y(z) = H(z)X(z)$, then, with z on the unit circle,

$$\Phi_{xy}(z) = H(z)\Phi_{xx}(z) \tag{7.55}$$

$$\Phi_{dy}(z) = H(z)\Phi_{dx}(z) \tag{7.56}$$

$$\Phi_{yy}(z) = |H(z)|^2\,\Phi_{xx}(z) \tag{7.57}$$

$$\phi_{xy}(n) = \frac{1}{2\pi j}\oint \Phi_{xy}(z)z^{n-1}\,dz \tag{7.58}$$

$$E[x_k^2] = \phi_{xx}(0) = \frac{1}{2\pi j}\oint \Phi_{xx}(z)\frac{dz}{z} \tag{7.59}$$

THE PERFORMANCE FUNCTION

In previous chapters, beginning with Chapter 2, we have discussed the behavior of the adaptive linear combiner in terms of its performance function, ξ, the mean-square error. Let us now formulate this performance function in terms of the transfer function of the adaptive system and the signal power spectra.

We will assume that we have a *single-input* adaptive transversal filter. The nonrecursive single-input system (i.e., the single-input adaptive linear combiner) was illustrated first in Figure 2.2, and is shown again together with the desired output, d_k, and the error, ε_k, in Figure 7.5. We omit the subscript k from the weight values here because their dynamic behavior does not enter into the present discussion.

The performance surface of the nonrecursive system, that is, the mean-square error as a function of the weights, was given in (2.13) as

$$\xi = \phi_{dd}(0) + \mathbf{W}^{\mathrm{T}}\mathbf{R}\mathbf{W} - 2\mathbf{P}^{\mathrm{T}}\mathbf{W} \tag{7.60}$$

where

$$\mathbf{W} = [w_0 \quad w_1 \quad \cdots \quad w_L]^{\mathrm{T}} \tag{7.61}$$

$$\mathbf{R} = \begin{bmatrix} \phi_{xx}(0) & \phi_{xx}(1) & \phi_{xx}(2) & \cdots & \phi_{xx}(L) \\ \phi_{xx}(1) & \phi_{xx}(0) & \phi_{xx}(1) & \cdots & \phi_{xx}(L-1) \\ \phi_{xx}(2) & \phi_{xx}(1) & \phi_{xx}(0) & \cdots & \phi_{xx}(L-2) \\ \vdots & & & & \\ \phi_{xx}(L) & \phi_{xx}(L-1) & \phi_{xx}(L-2) & \cdots & \phi_{xx}(0) \end{bmatrix} \tag{7.62}$$

$$\mathbf{P} = [\phi_{dx}(0) \quad \phi_{dx}(-1) \quad \cdots \quad \phi_{dx}(-L)]^{\mathrm{T}} \tag{7.63}$$

Here we have replaced the expected value notation, E, of Chapter 2 with the correlation function, ϕ, used in this chapter. Using (7.61) through (7.63) in (7.60), we can express the performance function in terms of ϕ's:

$$\xi = \phi_{dd}(0) + \sum_{l=0}^{L} \sum_{m=0}^{L} w_l w_m \phi_{xx}(l-m) - 2\sum_{l=0}^{L} w_l \phi_{dx}(-l) \tag{7.64}$$

Having again an expression for the performance function of the nonrecursive adaptive system, we turn now to the more general case in Figure 7.6. The transfer function $H(z)$ is assumed to represent the type of digital filter shown in Figure 7.2. The weights (a's and b's in Figure 7.2) are now assumed to be adjustable, so that we can have ξ as a function of these weights. If the recursive weights (b's) are all zero, Figures 7.5 and 7.6 are equivalent. Using the results in the preceding section,

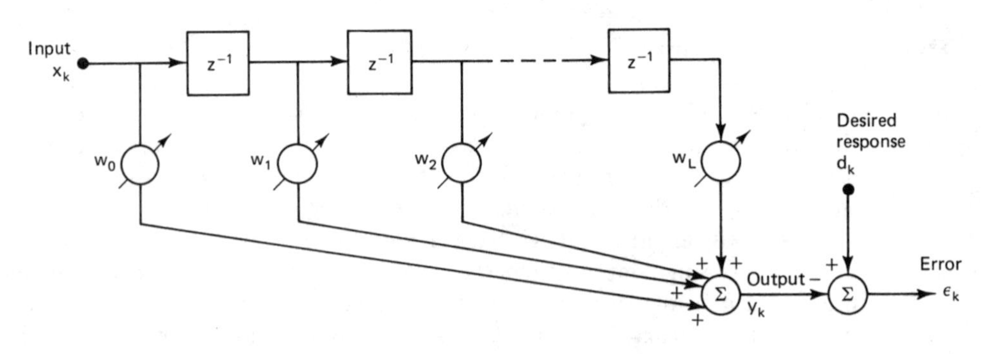

Figure 7.5 Single-input adaptive transversal filter.

particularly (7.56), (7.57), and (7.59), we have

$$\begin{aligned}
\xi &= E\left[\varepsilon_k^2\right] = E\left[(d_k - y_k)^2\right] \\
&= \phi_{dd}(0) + \phi_{yy}(0) - 2\phi_{dy}(0) \\
&= \phi_{dd}(0) + \frac{1}{2\pi j}\oint\left[\Phi_{yy}(z) - 2\Phi_{dy}(z)\right]\frac{dz}{z} \\
&= \phi_{dd}(0) + \frac{1}{2\pi j}\oint\left[H(z^{-1})\Phi_{xx}(z) - 2\Phi_{dx}(z)\right]H(z)\frac{dz}{z} \qquad (7.65)
\end{aligned}$$

In this result we have a general expression for the performance function of any single-input adaptive system.

We can show the equivalence of (7.65) and (7.64) in the nonrecursive case, that is, for the adaptive transversal filter. Using the notation in Figure 7.5, we have

$$H(z) = \sum_{l=0}^{L} w_l z^{-l} \qquad (7.66)$$

Using this nonrecursive form in (7.65), we have

$$\begin{aligned}
\xi &= \phi_{dd}(0) + \frac{1}{2\pi j}\oint\left[\sum_{l=0}^{L} w_l z^l \Phi_{xx}(z) - 2\Phi_{dx}(z)\right]\sum_{m=0}^{L} w_m z^{-m}\frac{dz}{z} \\
&= \phi_{dd}(0) + \sum_{l=0}^{L}\sum_{m=0}^{L} w_l w_m\left[\frac{1}{2\pi j}\oint\Phi_{xx}(z)z^{l-m-1}\,dz\right] \\
&\quad -2\sum_{m=0}^{L} w_m\left[\frac{1}{2\pi j}\oint\Phi_{dx}(z)z^{-m-1}\,dz\right] \\
&= \phi_{dd}(0) + \sum_{l=0}^{L}\sum_{m=0}^{L} w_l w_m \phi_{xx}(l-m) - 2\sum_{m=0}^{L} w_m \phi_{dx}(-m) \qquad (7.67)
\end{aligned}$$

To obtain the final result here, we used (7.58). Thus we have proved that (7.65) is equivalent to (7.64) for the adaptive linear combiner, and demonstrated that (7.65) is a general expression for the performance surface of a single-input adaptive system.

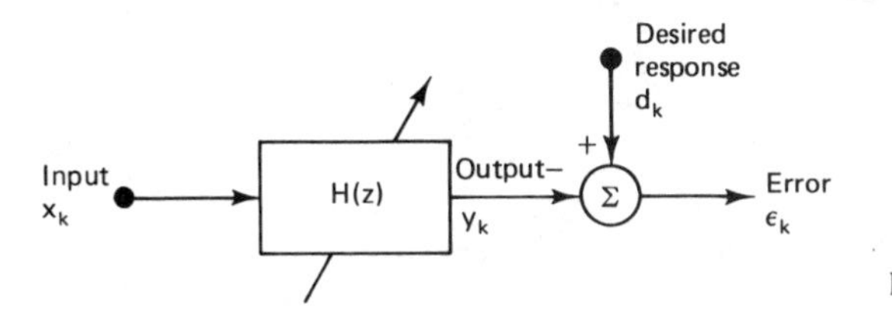

Figure 7.6 Single-input adaptive filter.

EXAMPLES OF PERFORMANCE SURFACES

In this section we develop two examples of performance surfaces using (7.65). The first example is the same as the example of the adaptive transversal filter in Figure 2.6. For this system there were two weights, w_0 and w_1. The input and desired signals were

$$x_k = \sin\frac{2\pi k}{N} \qquad \text{and} \qquad d_k = 2\cos\frac{2\pi k}{N} \tag{7.68}$$

The correlation functions, found essentially in (2.20) and (2.21), are

$$\begin{aligned} \phi_{xx}(n) &= 0.5\cos\frac{2\pi n}{N} \\ \phi_{dx}(n) &= \sin\frac{2\pi n}{N} \qquad -\infty \le n \le \infty \\ \phi_{dd}(0) &= 2 \end{aligned} \tag{7.69}$$

Thus the terms needed for (7.65) are

$$\begin{aligned} H(z) &= w_0 + w_1 z^{-1} \\ \Phi_{xx}(z) &= 0.5\sum_{n=-\infty}^{\infty}\cos\frac{2\pi n}{N}z^{-n} \\ \Phi_{dx}(z) &= \sum_{n=-\infty}^{\infty}\sin\frac{2\pi n}{N}z^{-n} \end{aligned} \tag{7.70}$$

It is important here that we keep Φ_{xx} and Φ_{dx} as sums, because the sums do not converge as described previously in the section "Right- and Left-Handed Sequences." In general, when the correlation function is *periodic*, the power spectrum is considered to be zero everywhere except at a single frequency where it is infinite, and the spectrum is represented correctly by the sums in (7.70).

If we substitute (7.70) into (7.65), we have the performance surface given by

$$\begin{aligned} \xi &= \phi_{dd}(0) + \frac{1}{2\pi j}\oint\Bigg[(w_0 + w_1 z)(0.5)\sum_{n=-\infty}^{\infty}\cos\frac{2\pi n}{N}z^{-n} \\ &\qquad -2\sum_{n=-\infty}^{\infty}\sin\frac{2\pi n}{N}z^{-n}\Bigg](w_0 + w_1 z^{-1})\frac{dz}{z} \\ &= 2 + \frac{1}{2}\sum_{n=-\infty}^{\infty}\cos\frac{2\pi n}{N}\left[\frac{1}{2\pi j}\oint[w_0^2 + w_1^2 + w_0 w_1(z + z^{-1})]\frac{dz}{z^{n+1}}\right] \\ &\qquad -2\sum_{n=-\infty}^{\infty}\sin\frac{2\pi n}{N}\left[\frac{1}{2\pi j}\oint(w_0 + w_1 z^{-1})\frac{dz}{z^{n+1}}\right] \\ &= 2 + 0.5(w_0^2 + w_1^2) + w_0 w_1\cos\frac{2\pi}{N} + 2w_1\sin\frac{2\pi}{N} \end{aligned} \tag{7.71}$$

Thus we have again the performance surface given in (2.24) and plotted in Figure 2.5 for $N = 5$ samples per cycle. We note again that this surface is quadratic in w_0 and w_1 and has a single global minimum.

For our second example of a performance surface, we consider the system shown in Figure 7.7. This is a type of "system identification" or "modeling" example in which the input, x_k, is a broadband signal (white noise in this case; see below) and is applied simultaneously to the adaptive filter and to a system to be modeled. The desired output, d_k, is the output of the system to be modeled, so when $E[\varepsilon_k^2]$ is minimized, the adaptive filter becomes the best possible model within its realm of adjustment. Note that the adaptive filter itself is in the form of Figure 7.2 with $a_0 = 1$, $a_1 = w_0$, and $b_1 = w_1$, and is recursive because of the feedback coefficient, w_1. From Figure 7.7, the transfer function of the adaptive filter, in accordance with (7.8), is

$$H(z) = \frac{1 + w_0 z^{-1}}{1 - w_1 z^{-1}} = \frac{z + w_0}{z - w_1} \tag{7.72}$$

For the input sequence, $[x_k]$ in Figure 7.7, we define "white noise" as a signal having the following properties:

Properties of white noise, $[x_k]$:

Uncorrelated samples:
$$\phi_{xx}(n) = \begin{cases} E[x_k^2] & n = 0 \\ 0 & n \neq 0 \end{cases} \tag{7.73}$$

Constant power:
$$\Phi_{xx}(z) = E[x_k^2] \tag{7.74}$$

Note that (7.74) follows from (7.73) because $\Phi_{xx}(z)$ is the z-transform of $\phi_{xx}(n)$, as in (7.42). Here, for convenience, we will let $\Phi_{xx}(z) = 1$.

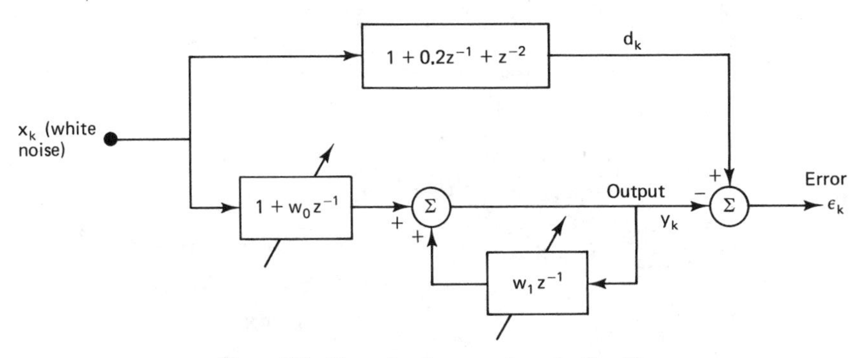

Figure 7.7 Example of a recursive adaptive filter.

With $\Phi_{xx}(z) = 1$, we can find $\Phi_{dx}(z)$ by using (7.55) and (7.44) with a change of subscript; that is, from Figure 7.7,

$$\Phi_{dx}(z) = \left[(1 + 0.2z^{-1} + z^{-2})\Phi_{xx}(z)\right]_{z \leftarrow z^{-1}}$$

$$= 1 + 0.2z + z^2 \tag{7.75}$$

Furthermore, from (7.59), (7.57), and (7.28), we have

$$\phi_{dd}(0) = \frac{1}{2\pi j}\oint |1 + 0.2z^{-1} + z^{-2}|^2 \frac{dz}{z}$$

$$= 2.04 \tag{7.76}$$

So we now have all of the functions needed to find the performance surface from (7.65), which becomes

$$\xi = 2.04 + \frac{1}{2\pi j}\oint \left[\frac{1 + w_0 z}{1 - w_1 z} - 2(1 + 0.2z + z^2)\right]\frac{z + w_0}{z - w_1}\frac{dz}{z} \tag{7.77}$$

In this integral there are poles at $z = 0$, w_1, and $1/w_1$. As discussed earlier in the section "Impulse Response and Stability," w_1 must be less than 1 for stability, so the pole at $1/w_1$ lies outside the unit circle. Thus, in accordance with (7.31), the integral in (7.77) must be the sum of the residues at $z = w_1$ and $z = 0$. These are found

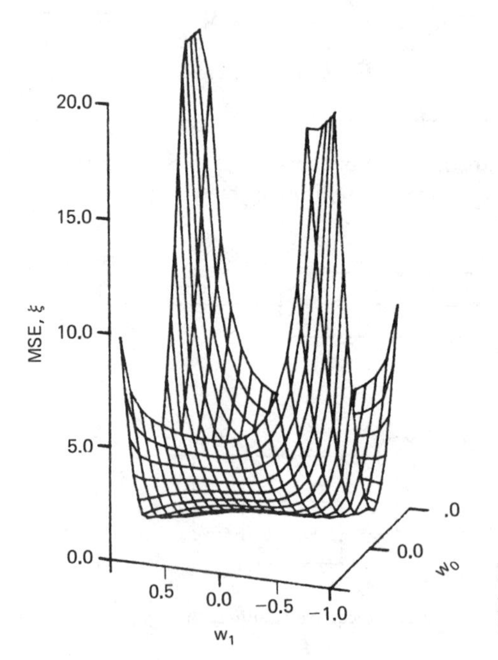

Figure 7.8 Mean-square-error performance surface for the system of Figure 7.7.

using (7.32) and (7.34) as follows:

$$\xi = 2.04 + \left[\frac{1 + w_0 w_1}{1 - w_1^2} - 2\left(1 + 0.2w_1 + w_1^2\right)\right]\frac{w_1 + w_0}{w_1} + \frac{w_0}{w_1}$$

$$= 2.04 + \frac{1 + 2w_0 w_1 + w_0^2}{1 - w_1^2} - 2[1 + (w_1 + 0.2)(w_0 + w_1)] \qquad (7.78)$$

In the first line of this result, the second term is the residue at $z = w_1$ and the third term, w_0/w_1, is the residue at $z = 0$.

Thus we have in (7.78) the performance surface for our second example. It is illustrated in Figure 7.8, where we note that in contrast with the first example above, ξ is quadratic in w_0 but is neither quadratic nor unimodal in w_1. Thus, adapting w_0 is generally straightforward, but adapting w_1 is not. Choices of w_1 beyond the stable range cause ξ to become infinite. An automatic system for adapting w_1 would have two stability problems: stability of the adaptive algorithm and stability of the filter being adapted.

EXERCISES

1. Find the z-transform of each of the following signals. Assume that each signal begins at $k = 0$.
   (a) Impulse function, $x_0 = 1$, $x_k = 0$; $k > 0$
   (b) Step function, $x_k = 1$, $k \geq 0$
   (c) Exponential function, $x_k = e^{-ak}$; $k \geq 0$
   (d) Ramp function, $x_k = ak$; $k \geq 0$
   (e) Cosine function, $x_k = A\cos(\lambda_0 k)$; $k \geq 0$
2. (Shifting Theorem) Find $Y(z)$ in terms of $X(z)$ if $y_k = x_{k-n}$ (i.e., y is a shifted version of x).
3. Find $Y(z)$ in terms of $F(z)$ if $y_k = e^{-ak}F_k$.
4. Prove that the z-transform of Ax_k is $AX(z)$.
5. Prove that the z-transform of $x_k + y_k$ is $X(z) + Y(z)$.
6. For the adaptive transversal filter with two weights shown below:
   (a) Find $H(z)$.
   (b) Write an expression for the amplitude gain.

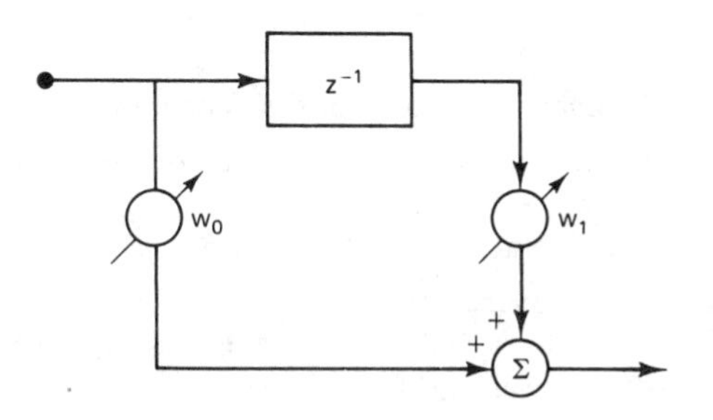

7. For the transversal filter shown below, plot the amplitude gain versus frequency for $w_1 = 0, 1$, and 10, to show how the gain changes with w_1.

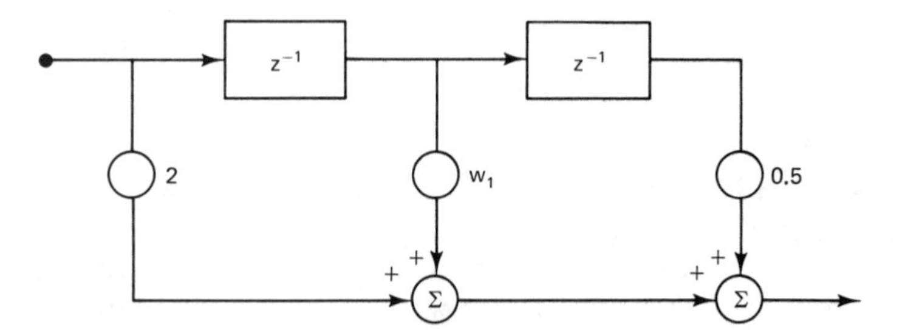

8. Plot the three phase-shift curves for Exercise 7.

9. A recursive linear system has the transfer function

$$H(z) = \frac{z^2 + 1}{z^2 - \sqrt{2}\,z + 1}$$

(a) Draw a filter diagram similar to Figure 7.2.
(b) Plot the gain and phase shift versus frequency.
(c) Explain the plots in terms of a pole–zero plot.

10. Plot the power gain versus frequency for the linear combiner in Exercise 7 with $w_1 = -1$.

11. Express the impulse response in Exercise 7.

12. Express the impulse response in Exercise 9. Note that the poles are on the unit circle; therefore, (7.26) may be applied by using a slightly larger contour. Also note that for $k = 0$, there is an extra pole at $z = 0$.

13. What are the ranges of w_0 and w_1 for which the system shown below is stable?

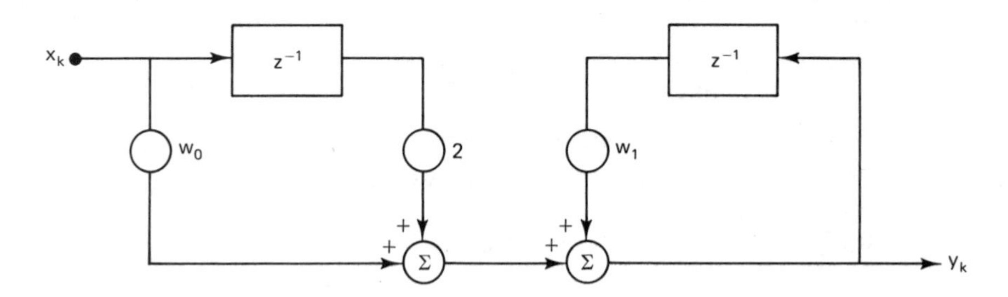

14. What is the impulse response of the system in Exercise 13?

15. Use residues to find the inverse z-transform. Assume that $x_k = 0$ for $k < 0$.
(a) $X(z) = z/(z - a)$.
(b) $X(z) = ze^{-\alpha}\sin\beta/(z^2 - 2ze^{-\alpha}\cos\beta + e^{-2\alpha})$.

16. A left-handed sequence is given by $x_k = e^{ak}$; $k < 0$, with $a > 0$. Using u in place of z^{-1}, find the u-transform and show that it converges outside $|u| < 1$. Then, using residues, show how to recover $[x_k]$ from $X(u)$.

17. Suppose that $[x_k]$ has an autocorrelation function given by $\phi_{xx}(n) = e^{-a|n|}$, defined for all n. Find the power spectrum of $[x_k]$ as a function of frequency (ω). (*Hint*: Treat the right and left portions of ϕ_{xx} separately.)

18. Suppose that $[x_k]$ is a right-handed sequence, nonzero only for $k \geq 0$, and let $[y_k]$ be the mirror image of $[x_k]$ (i.e., $y_{-k} = x_k$) for all k.

(a) What is $Y(z)$ in terms of $X(z)$?

(b) What is the amplitude spectrum of $[y_k]$ in terms of the amplitude spectrum of $[x_k]$?

(c) What is the phase spectrum of $[y_k]$ in terms of the phase spectrum of $[x_k]$?

19. For the diagram in Exercise 13, write expressions for

(a) $\Phi_{yy}(z)$ in terms of $\Phi_{xx}(z)$.

(b) $\Phi_{xy}(z)$ in terms of $\Phi_{xx}(z)$.

(c) $\Phi_{yx}(z)$ in terms of $\Phi_{xx}(z)$.

20. Suppose that RANDOM (1.) is a Fortran library function that returns independent random numbers, uniformly distributed from 0 to 1. Write a Fortran statement that will create a sample from a white-noise sequence having unit power.

21. Suppose that the desired output, $[d_k]$, is developed by passing $[x_k]$ through a "plant," $H_1(z)$, and that $H_2(z)$ is an adaptive filter, as shown below. Develop a general expression for the performance function in this case.

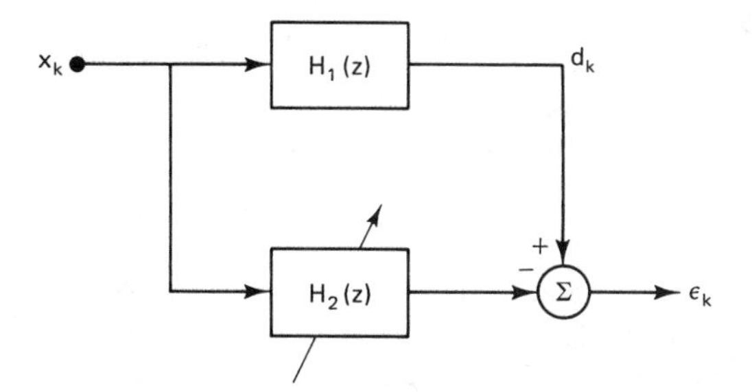

22. In Exercise 21, derive an expression for the performance surface when $[x_k]$ is unit white noise, $H_1(z) = z/(z - 0.5)$, and $H_2(z) = \Sigma_{n=0}^{L} w_n z^{-n}$.

23. In Exercise 21, derive an expression for the performance surface when $[x_k]$ is unit white noise, $H_1(z) = 1 - 2z^{-1}$, and $H_2(z) = w_0 z/(z - w_1)$.

24. Suppose that r_k is a sample of the uniform random variate described in Appendix A, with range from zero to 1. Suppose that a time series is generated from $x_k = a(r_k - 0.5)$, where a is a constant. What are the correlation coefficients, $\phi_{xx}(0)$ and $\phi_{xx}(1)$?

25. What are $\phi_{xx}(0)$ and $\phi_{xx}(1)$ if $x_k = a(r_k - 0.5) + bx_{k-1}$ where r_k is as in Exercise 24 and $|b| < 1$?

26. Use the u-transform concept to determine the inverse of (i.e., the left-handed sequence corresponding to) $X(z) = A/(1 - az^{-1})$, where a is any pole outside the unit circle, $|z| = 1$.

27. Use the u-transform concept together with Equation (7.57) to determine the correlation function $\phi_{yy}(n)$ shown below, assuming that x_k is white noise with unit power.

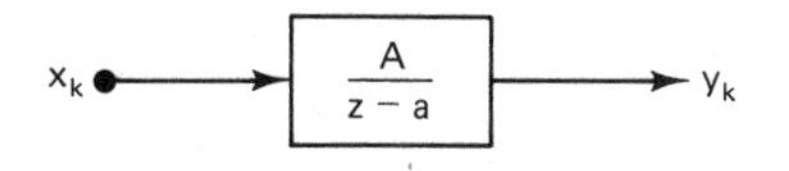

ANSWERS TO SELECTED EXERCISES

12. $\delta(k) + 2\sin(k\pi/4);\ k \geq 0$

16. $X(u) = u/(u - e^{-a})$

24. $a^2/12, 0$

25. $a^2/[12(1 - b^2)];\ a^2b/[12(1 - b^2)]$

26. $x_k = 0;\ k \geq 0$; and $x_k = -Aa^k;\ k < 0$

REFERENCES AND ADDITIONAL READINGS

1. R. W. Hamming, *Digital Filters*. Englewood Cliffs, N.J.: Prentice-Hall, 1977, Chapter 2.
2. A. V. Oppenheim and R. W. Schafer, *Digital Signal Processing*. Englewood Cliffs, N.J.: Prentice-Hall, 1975, Chaps. 1–4.
3. L. R. Rabiner and B. Gold, *Theory and Application of Digital Signal Processing*. Englewood Cliffs, N.J.: Prentice-Hall, 1975, Chap. 2.
4. S. D. Stearns, *Digital Signal Analysis*. Rochelle Park, N.J.: Hayden, 1975, Chaps. 8 and 9.
5. W. Kaplan, *Advanced Calculus*. Reading, Mass.: Addison-Wesley, 1952, Chap. 9.
6. S. Karni and W. J. Byatt, *Mathematical Methods in Continuous and Discrete Systems*. New York: Holt, Rienhart and Winston, 1981, Chaps. 2 and 5.
7. N. Ahmed and T. Natarajan, *Discrete-Time Signals and Systems*. Reston, Va.: Reston, 1983.
8. S. A. Tretter, *Introduction to Discrete-Time Signal Processing*. New York: Wiley, 1976.
9. A. Peled and B. Liu, *Digital Signal Processing*. New York: Wiley, 1976.
10. R. V. Churchill, *Complex Variables and Applications*. New York: McGraw-Hill, 1948, Chap. 7.

chapter 8

Other Adaptive Algorithms and Structures

In the introduction to Chapter 6 we remarked that the LMS algorithm, although simple and efficient, is restricted in its application, being applicable primarily to nonrecursive linear filters. We also saw that the LMS algorithm is a steepest-descent type of algorithm which does not necessarily descend by the most direct route to the minimum mean-square error.

In this chapter we introduce different types of algorithms that are not as simple as the LMS algorithm, but have other advantages. First, the sequential regression (SER) algorithm is introduced as an approximation to Newton's method. The SER algorithm does not usually follow the path of steepest descent, but proceeds on a more direct course to the optimum set of weights.

Next, we will extend the LMS algorithm to recursive adaptive filters, that is, to adaptive filters with poles. As we saw in the example in Figure 7.8, these filters can produce nonquadratic error surfaces with local minima and unstable regions, so special precautions must be taken when gradient search methods such as the LMS algorithm are used with recursive adaptive filters.

In addition to following the direction of the gradient (steepest-descent method) or taking the direct path (Newton's method) to the minimum mean-square error, there is the possibility of searching the error surface in a random pattern for its minimum. We will therefore introduce the random search type of algorithm, which has advantages in situations where the error surface is not unimodal, as well as in some other situations.

Besides these other types of algorithms, we will also introduce in this chapter the lattice structure, which is the major alternative to the direct form of the adaptive linear combiner used up to this point. We will show that the lattice structure has certain advantages over the direct form.

AN IDEAL: THE LMS / NEWTON ALGORITHM

Before introducing the SER algorithm, which approximates Newton's method, we consider here an ideal adaptive algorithm which we will call the "LMS/Newton" algorithm. This is an algorithm that is ideal in the sense that its performance can be approached but not equaled in practical situations, and thus it becomes a standard to which the performance of other algorithms may be compared.

To derive the LMS/Newton algorithm, we start with the form of Newton's algorithm given in (4.32):

$$\mathbf{W}_{k+1} = \mathbf{W}_k - \mu \mathbf{R}^{-1} \nabla_k \tag{8.1}$$

With this formula we saw that under ideal conditions, convergence to the optimum weight vector, $\mathbf{W}^*$, will occur in a single step; that is, starting with $\mathbf{W}_0$, we have $\mathbf{W}_1 = \mathbf{W}^*$. The ideal conditions are:

1. $\mu = \frac{1}{2}$
2. Exact knowledge of the gradient vector, ∇, at each iteration
3. Exact knowledge of the (unchanging) signal inverse correlation matrix, $\mathbf{R}^{-1}$

If the first of these conditions is removed and μ is made to be between 0 and $\frac{1}{2}$, the algorithm requires a larger number of steps, but still proceeds along a straight path to $\mathbf{W}^*$, as illustrated in Figure 8.1. The latter is similar to Figure 4.6, except that μ has been reduced from 0.5 to 0.05, thus requiring more than just one step to reach the optimum weights. (See also the learning curve in Figure 4.9.) The specific error surface used for Figure 8.1 is the same as that used in Figure 6.3, and given by (6.14) with $N = 16$ and $\phi = 0.01$.

Next, let us remove the second condition above and assume that we must use a noisy gradient estimate, $\hat{\nabla}$, in place of ∇. Then we have

$$\mathbf{W}_{k+1} = \mathbf{W}_k - \mu \mathbf{R}^{-1} \hat{\nabla}_k \tag{8.2}$$

The algorithm is now ideal only because of the third condition, that is, only because an exact knowledge of $\mathbf{R}^{-1}$ is still assumed. Without relaxing the third condition, we can now convert (8.2) into an "LMS" type of algorithm by using the same sort of gradient estimate used in Chapter 6, that is, by using ε_k^2 as an estimate for ξ. As in (6.2), this gradient estimate is given by

$$\hat{\nabla}_k = -2\varepsilon_k \mathbf{X}_k \tag{8.3}$$

where $\mathbf{X}_k$ is of course the input signal vector at the kth iteration. Using (8.3) in (8.2), we have

$$\mathbf{W}_{k+1} = \mathbf{W}_k + 2\mu \mathbf{R}^{-1} \varepsilon_k \mathbf{X}_k \tag{8.4}$$

Now this result is the same as the LMS algorithm in (6.3), except for the presence of $\mathbf{R}^{-1}$ in the weight increment term. We can increase the similarity between the two algorithms by noting that when $\mathbf{R}$ is diagonal with equal eigenvalues, we have

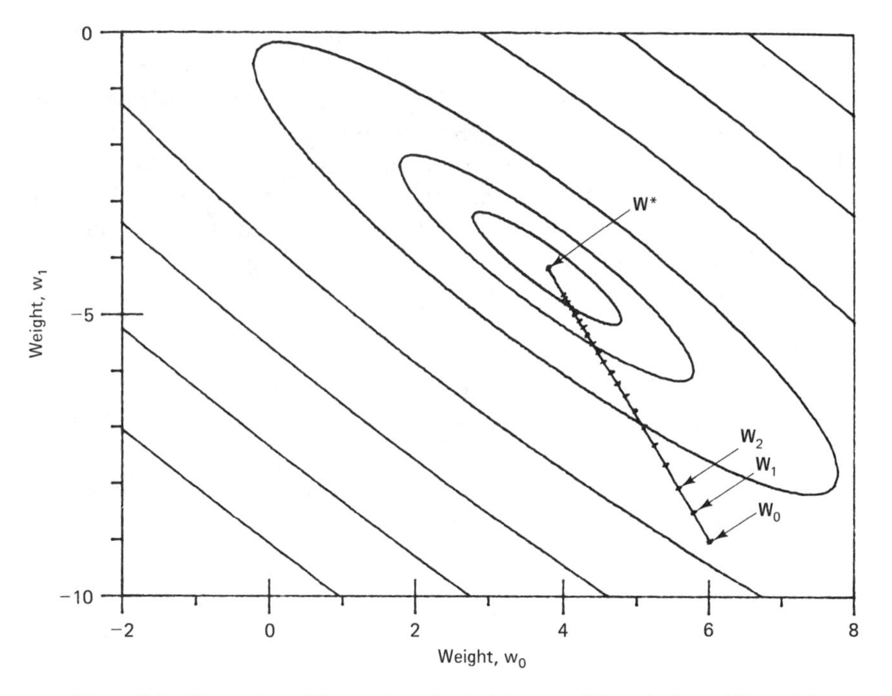

Figure 8.1 Illustration of Newton's method with two weights, similar to Figure 4.6 with μ reduced to 0.05. Error surface is given by (6.14) with $N = 16$ and $\phi = 0.01$. Note direct path from $\mathbf{W}_0$ to $\mathbf{W}^*$.

$\lambda_{av}\mathbf{R}^{-1} = \mathbf{I}$. Thus we give μ the same range of values as in Chapter 6 by using λ_{av} as a scaling factor in (8.4) to obtain

$$\mathbf{W}_{k+1} = \mathbf{W}_k + 2\mu\lambda_{av}\mathbf{R}^{-1}\varepsilon_k\mathbf{X}_k \tag{8.5}$$

We will call this the "LMS/Newton" algorithm. Note that the units of λ_{av} and $\varepsilon_k\mathbf{X}_k$ are units of power, the units of $\mathbf{R}^{-1}$ and μ are units of reciprocal power, and $\mathbf{W}$ is of course dimensionless, so that (8.5) is correct dimensionally. Notice also that the range of μ has now been scaled by λ_{av} and, from (6.8) and condition 1 above, we now have

$$\textit{For convergence:} \quad \frac{1}{\lambda_{max}} > \mu > 0 \tag{8.6}$$

$$\textit{For one-step convergence under noiseless conditions:} \quad \mu = \frac{1}{2\lambda_{av}} \tag{8.7}$$

The LMS/Newton algorithm in (8.5) is ideal in the sense that an exact knowledge of $\mathbf{R}^{-1}$ is assumed. We have already discussed how such knowledge is usually not available in adaptive situations, that is, $\mathbf{X}$ is usually nonstationary and $\mathbf{R}$ is considered to change slowly with time, in an unknown way. The LMS/Newton algorithm is also ideal because, under noiseless conditions on a parabolic error surface, the weight track is a direct path to $\mathbf{W}^*$ as in Figure 8.1.

In this sense, even under noisy conditions, the LMS/Newton algorithm is generally superior to the LMS algorithm, as illustrated in Figure 8.2. To draw Figure 8.2 we used again the error surface given in (6.14) with $N = 16$ and $\phi = 0.01$, and we used $(6, -9)$ as the starting weight vector, (w_{00}, w_{10}). The inverse $\mathbf{R}$ matrix, which is assumed to be known for the LMS/Newton case, is found by inverting the two-dimensional $\mathbf{R}$ matrix:

$$\mathbf{R}^{-1} = \begin{bmatrix} r_1 & r_2 \\ r_2 & r_1 \end{bmatrix}^{-1} = \frac{1}{r_1^2 - r_2^2}\begin{bmatrix} r_1 & -r_2 \\ -r_2 & r_1 \end{bmatrix} \tag{8.8}$$

We obtain the elements r_1 and r_2 from (6.13) with $N = 16$ and $\phi = 0.01$. Also, for the LMS/Newton algorithm in (8.5), we need λ_{av}. From (3.2), we have

$$\lambda_1, \lambda_2 = r_1 \pm r_2 = 0.51 \pm 0.4619$$
$$\lambda_{av} = r_1 = 0.51 \tag{8.9}$$

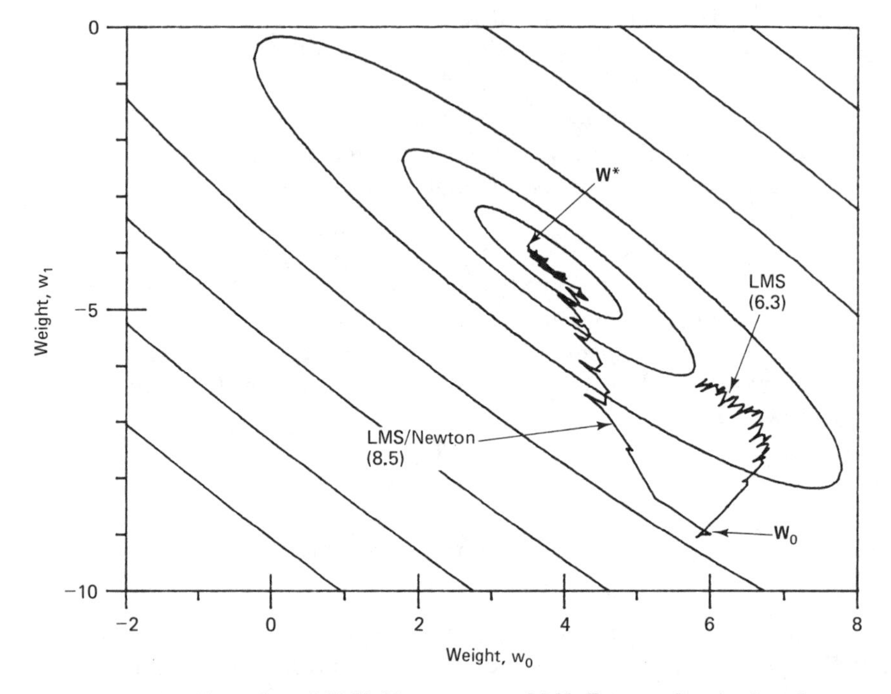

Figure 8.2 Illustration of LMS/Newton versus LMS. Error surface is given by (6.14) with $N = 16$, $\phi = 0.01$, and $\mu = 0.05$. Each track represents 100 iterations.

So in Figure 8.2 we see the tracks produced from the first 100 iterations of (6.3) and (8.5). The LMS track follows approximately the path of steepest descent and will eventually reach the optimum point at w_0^*, w_1^*, while the LMS/Newton track follows approximately the direct path and reaches the optimum in 100 iterations. Both tracks are noisy due to the noisy gradient estimate at each iteration, but the superiority of the LMS/Newton algorithm is clearly illustrated.

PROPERTIES OF THE LMS / NEWTON ALGORITHM

Since the LMS/Newton algorithm represents an ideal standard of performance, we are interested in its theoretical properties of convergence and its excess mean square error. In this section we derive these properties and compare them with those of the LMS algorithm.

Under ideal, noiseless conditions the LMS algorithm behaves in accordance with the steepest-descent formulas developed in Chapters 4 and 5, while the LMS/Newton algorithm behaves in accordance with the formulas developed for Newton's method. Remembering that we have now scaled μ by λ_{av} (i.e., replaced μ with $\mu\lambda_{av}$) in (8.5), we recall from (5.61) and (5.84) that the geometric ratio for the relaxation of the nth weight is

$$\textit{Newton}: \quad r = 1 - 2\mu\lambda_{av} \tag{8.10}$$

$$\textit{Steepest descent}: \quad r_n = 1 - 2\mu\lambda_n \qquad 0 \le n \le L \tag{8.11}$$

Thus the methods are equivalent when the eigenvalues of $\mathbf{R}$ are all equal. From these ratios, as in (6.21), we find that the time constant associated with the nth mode of the learning curve is

$$\textit{Newton}: \quad T_{mse} = \frac{1}{4\mu\lambda_{av}} \tag{8.12}$$

$$\textit{Steepest descent}: \quad (T_{mse})_n = \frac{1}{4\mu\lambda_n} \qquad 0 \le n \le L \tag{8.13}$$

Thus, as in the example in Figure 8.2, learning is slower for LMS than for LMS/Newton when the eigenvalues are unequal. For the specific case of Figure 8.2, when we use (8.9) in (8.12) and (8.13) we obtain $T_{mse} \approx 10$ iterations for the LMS/Newton learning curve, whereas the longer time constant for the LMS learning curve is $T_{mse} \approx 100$ iterations. This result accounts in a rough way for the shorter LMS track in Figure 8.2.

Besides the learning-curve time constant, which gives us a measure of the algorithm's speed of convergence to ξ_{min}, we are also interested in the excess mean-square error, which measures the algorithm's ability to remain near to ξ_{min}. In Chapter 6 we derived the excess MSE for the LMS case, and we now proceed to a similar development for the LMS/Newton case.

To derive the excess MSE, we require the covariance matrix of $\mathbf{V}_k'$, the weight vector in the principal-axis coordinate system, to use as in (6.34). In (5.50), we derived cov[$\mathbf{V}_k'$] for Newton's method with a noisy gradient estimate, so the result is the same for the LMS/Newton algorithm except that we must again use $\mu\lambda_{av}$ in place of μ, as we did above and in (8.5). Thus we have

$$\text{cov}\left[\mathbf{V}_k'\right] = \frac{\mu\lambda_{av}(\Lambda^{-1})^2}{4(1-\mu\lambda_{av})}\text{cov}\left[\mathbf{N}_k'\right] \tag{8.14}$$

Now $\mathbf{N}_k'$ is the gradient noise in the principal-axis coordinate system, and in (8.3) we used the same gradient estimate for LMS/Newton as for LMS in (6.2). Therefore, the gradient noise must be the same and the result for cov[$\mathbf{N}_k'$] in (6.29) is valid for the LMS/Newton algorithm:

$$\text{cov}\left[\mathbf{N}_k'\right] = 4\xi_{\min}\Lambda \tag{8.15}$$

Substituting (8.15) into (8.14), we obtain

$$\text{cov}\left[\mathbf{V}_k'\right] = \frac{\mu\lambda_{av}\xi_{\min}}{1-\mu\lambda_{av}}\Lambda^{-1} \tag{8.16}$$

We now use the diagonal elements of cov[$\mathbf{V}_k'$] to obtain the excess mean-square error as in (6.34). The result is

$$\begin{aligned}\text{excess MSE} &= \sum_{n=0}^{L}\lambda_n E\left[v_{nk}'^2\right]\\ &= \sum_{n=0}^{L}\lambda_n\frac{\mu\lambda_{av}\xi_{\min}\lambda_n^{-1}}{1-\mu\lambda_{av}}\\ &= \frac{(L+1)\mu\lambda_{av}\xi_{\min}}{1-\mu\lambda_{av}}\end{aligned} \tag{8.17}$$

We can simplify this result by noting that, as in (6.9), $L+1$ times λ_{av} is the same as the sum of the diagonal elements of $\mathbf{R}$ (i.e., the trace of $\mathbf{R}$) and also that, in the derivation of cov[$\mathbf{N}_k'$], μ was assumed to be small compared with the one-step convergence value in (8.7), that is,

$$\mu \ll \frac{1}{2\lambda_{av}} \tag{8.18}$$

Thus the denominator in (8.17) is assumed to be close to 1, and

$$\text{excess MSE} \approx \mu\xi_{\min}\,\text{tr}\,[\mathbf{R}] \tag{8.19}$$

Therefore, from (6.36) and (8.19) we conclude that the excess MSE and the misadjustment, M, are the same for the LMS and the LMS/Newton algorithms. In either case the misadjustment is

$$\begin{aligned}M &= \frac{\text{excess MSE}}{\xi_{\min}}\\ &= \mu\,\text{tr}\,[\mathbf{R}]\end{aligned} \tag{8.20}$$

To summarize, we tabulate the theoretical properties of the LMS/Newton and LMS algorithms:

	LMS/Newton	LMS
Longest learning-curve time constant, T_{mse}	$\dfrac{1}{4\mu\lambda_{av}}$	$\dfrac{1}{4\mu\lambda_{min}}$
Misadjustment, M	$\mu\,\mathrm{tr}[\mathbf{R}]$	$\mu\,\mathrm{tr}[\mathbf{R}]$

Since both algorithms have the same misadjustment for a given μ, we can observe that the LMS/Newton algorithm converges faster by the ratio of λ_{av} to λ_{min}. Thus, when $\mathbf{R}$ has highly unequal eigenvalues, the LMS algorithm and other steepest-descent algorithms fall far short of the ideal LMS/Newton algorithm. In the next section we develop an algorithm whose performance approaches more closely that of the LMS/Newton.

THE SEQUENTIAL REGRESSION ALGORITHM

Comparing the LMS and LMS/Newton algorithms in (6.3) and (8.5), we see that it is the knowledge of $\mathbf{R}^{-1}$ that allows $\mathbf{W}$ to take the direct path, rather than the path of steepest descent, to $\mathbf{W}^*$. To develop an algorithm more like the LMS/Newton, we might therefore think in terms of estimating $\mathbf{R}^{-1}$ at each step, and thus approaching the ideal in (8.5).

The sequential regression (SER) algorithm [1, 2] embodies precisely this sort of improvement. It computes an estimate of $\mathbf{R}^{-1}$ that generally improves with each iteration, and thus approaches (8.5). To develop the SER algorithm, let us look first at how we might estimate $\mathbf{R}$, which is a simpler problem than that of estimating $\mathbf{R}^{-1}$.

Using the notation in (7.38) and (7.62), we have the elements of $\mathbf{R}$ given by the input correlation function, where n is the distance from the main diagonal:

$$\phi_{xx}(n) = E[x_k x_{k+n}] \tag{8.21}$$

Thus we could also write

$$\mathbf{R} = E\left[\mathbf{X}_k \mathbf{X}_k^{\mathrm{T}}\right] \tag{8.22}$$

as we did when $\mathbf{R}$ was first introduced in (2.11). The latter form suits our present purpose better, because we can include both single- and multiple-input adaptive systems.

Instead of letting the expectation go over all values of k as in (8.22), suppose now that we have a finite number of observations of the signal $\mathbf{X}$, say $\mathbf{X}_0$ through $\mathbf{X}_k$. Under stationary conditions, our best unbiased estimate of $\mathbf{R}$ would then be

$$\hat{\mathbf{R}}_k = \frac{1}{k+1} \sum_{l=0}^{k} \mathbf{X}_l \mathbf{X}_l^{\mathrm{T}} \tag{8.23}$$

In adaptive situations where $\mathbf{X}$ is nonstationary, we can see that (8.23) would not be a good estimate of $\mathbf{R}$. Because of its infinite memory, this estimate would become insensitive to changes in $\mathbf{R}$ for large values of k.

To provide the effect of a short-term memory in the estimate of $\mathbf{R}$, consider the following function:

$$\mathbf{Q}_k = \sum_{l=0}^{k} \alpha^{k-l} \mathbf{X}_l \mathbf{X}_l^{\mathrm{T}} \tag{8.24}$$

Comparing this with (8.23), we see that we have here a measure similar to $\mathbf{R}_k$ times a scaling factor, except that current products are emphasized, giving the estimate a "fading" memory.† As a rule of thumb, we might let α be chosen such that the half-life of the exponential function is equal to the number of iterations over which $\mathbf{X}_l$ is stationary. Thus

$$\begin{gathered} 0 < \alpha < 1 \\ \alpha \approx 2^{-1/(\text{length of stationarity of } \mathbf{X})} \end{gathered} \tag{8.25}$$

Now the total value of this scaling factor over k iterations is

$$\sum_{l=0}^{k} \alpha^{k-l} = \frac{1 - \alpha^{k+1}}{1 - \alpha} \tag{8.26}$$

and thus our modified estimate of $\mathbf{R}$ at the kth iteration (which would be exact, for example, if $\mathbf{X}_k$ were constant for $k \geq 0$) is

$$\begin{aligned} \hat{\mathbf{R}}_k &= \frac{1 - \alpha}{1 - \alpha^{k+1}} \mathbf{Q}_k \\ &= \frac{1 - \alpha}{1 - \alpha^{k+1}} \sum_{l=0}^{k} \alpha^{k-l} \mathbf{X}_l \mathbf{X}_l^{\mathrm{T}} \end{aligned} \tag{8.27}$$

We can see that in the limiting case where $\mathbf{X}_l$ is stationary for all time, α approaches 1 in (8.25), and if we take the limit as α approaches 1 in (8.27), we get agreement with (8.23).

Having the estimate $\hat{\mathbf{R}}_k$, we are ready to begin the derivation of the SER algorithm. It will be more convenient to omit the scaling factor and use $\mathbf{Q}_k$ rather than $\hat{\mathbf{R}}_k$. To do this we begin with the formula for the optimum weight vector given first in (2.16):

$$\hat{\mathbf{R}}_k \mathbf{W}_k = \hat{\mathbf{P}}_k \tag{8.28}$$

† The reader is cautioned not to confuse $\mathbf{Q}_k$ here with its previous use to represent the eigenvectors in Chapter 3.

Here we have used the kth estimates in place of the true values used in Chapter 2. Let us assume that $\mathbf{P}$ is estimated as $\mathbf{R}$ is estimated in (8.27). Then, from the definition of $\mathbf{P}$ in (2.12), we obtain

$$\hat{\mathbf{P}}_k = \frac{1-\alpha}{1-\alpha^{k+1}} \sum_{l=0}^{k} \alpha^{k-l} d_l \mathbf{X}_l \tag{8.29}$$

Using (8.27) and (8.29) in (8.28), we can cancel the scaling factor and obtain

$$\mathbf{Q}_k \mathbf{W}_k = \sum_{l=0}^{k} \alpha^{k-l} d_l \mathbf{X}_l \tag{8.30}$$

The SER algorithm is now developed as follows, beginning with (8.30). This development is similar to that of Ahmed et al. [2], Graupe [1], Sage [32], and others. Let us assume that $\mathbf{W}_{k+1}$ (rather than $\mathbf{W}_k$) is to be computed in terms of $\hat{\mathbf{R}}_k$ and $\hat{\mathbf{P}}_k$. Then, from (8.28) through (8.30),

$$\begin{aligned} \mathbf{Q}_k \mathbf{W}_{k+1} &= \alpha \sum_{l=0}^{k-1} \alpha^{(k-1)-l} d_l \mathbf{X}_l + d_k \mathbf{X}_k \\ &= \alpha \mathbf{Q}_{k-1} \mathbf{W}_k + d_k \mathbf{X}_k \\ &= \left(\mathbf{Q}_k - \mathbf{X}_k \mathbf{X}_k^{\mathrm{T}}\right) \mathbf{W}_k + d_k \mathbf{X}_k \end{aligned} \tag{8.31}$$

Here we rewrote (8.30), then in the last line used the following relationship from (8.24):

$$\mathbf{Q}_k = \alpha \mathbf{Q}_{k-1} + \mathbf{X}_k \mathbf{X}_k^{\mathrm{T}} \tag{8.32}$$

Next we substitute (2.8) for the desired signal, d_k:

$$\begin{aligned} \mathbf{Q}_k \mathbf{W}_{k+1} &= \left(\mathbf{Q}_k - \mathbf{X}_k \mathbf{X}_k^{\mathrm{T}}\right) \mathbf{W}_k + \left(\varepsilon_k + \mathbf{X}_k^{\mathrm{T}} \mathbf{W}_k\right) \mathbf{X}_k \\ &= \mathbf{Q}_k \mathbf{W}_k + \varepsilon_k \mathbf{X}_k \end{aligned} \tag{8.33}$$

We now multiply on the left by $\mathbf{Q}_k^{-1}$ and obtain finally

$$\mathbf{W}_{k+1} = \mathbf{W}_k + \mathbf{Q}_k^{-1} \varepsilon_k \mathbf{X}_k \tag{8.34}$$

Since $\mathbf{Q}_k^{-1}$ is a scaled approximation to $\mathbf{R}^{-1}$, we have here the form of the LMS/Newton algorithm in (8.5). In fact, from (8.27), we have

$$\mathbf{Q}_k^{-1} = \frac{1-\alpha}{1-\alpha^{k+1}} \hat{\mathbf{R}}_k^{-1} \tag{8.35}$$

and if we consider the steady-state case where k is large enough to neglect α^{k+1} in

(8.35), we can make (8.34) an approximation to (8.5) as follows:

$$\mathbf{W}_{k+1} = \mathbf{W}_k + \frac{2\mu\lambda_{\text{av}}}{1-\alpha}\mathbf{Q}_k^{-1}\varepsilon_k\mathbf{X}_k \tag{8.36}$$

Note that under nonstationary conditions, λ_{av} is a changing quantity that may have to be adjusted during the adaptive process. Note also that omitting the factor $(1 - \alpha^{k+1})$ from the last term in (8.36) is equivalent to using a larger value of μ at first in (8.5). If startup conditions are important, one could include the factor and revise (8.36) to obtain

$$\mathbf{W}_{k+1} = \mathbf{W}_k + \frac{2\mu\lambda_{\text{av}}(1-\alpha^{k+1})}{1-\alpha}\mathbf{Q}_k^{-1}\varepsilon_k\mathbf{X}_k \tag{8.37}$$

In either of the latter two forms of the SER algorithm, we need to be able to compute $\mathbf{Q}_k^{-1}$ iteratively. The algorithm for doing this can be derived as above, beginning this time with the iterative formula for $\mathbf{Q}_k$ in (8.32). If (8.32) is premultiplied by $\mathbf{Q}_k^{-1}$ and postmultiplied by $\mathbf{Q}_{k-1}^{-1}$ and then by $\mathbf{X}_k$, we obtain

$$\mathbf{Q}_{k-1}^{-1} = \alpha\mathbf{Q}_k^{-1} + \mathbf{Q}_k^{-1}\mathbf{X}_k\mathbf{X}_k^{\mathrm{T}}\mathbf{Q}_{k-1}^{-1} \tag{8.38}$$

or

$$\begin{aligned}\mathbf{Q}_{k-1}^{-1}\mathbf{X}_k &= \alpha\mathbf{Q}_k^{-1}\mathbf{X}_k + \mathbf{Q}_k^{-1}\mathbf{X}_k\mathbf{X}_k^{\mathrm{T}}\mathbf{Q}_{k-1}^{-1}\mathbf{X}_k \\ &= \mathbf{Q}_k^{-1}\mathbf{X}_k\left(\alpha + \mathbf{X}_k^{\mathrm{T}}\mathbf{Q}_{k-1}^{-1}\mathbf{X}_k\right)\end{aligned} \tag{8.39}$$

We can now divide by the scalar factor in parentheses and multiply on the right by $\mathbf{X}_k^{\mathrm{T}}\mathbf{Q}_{k-1}^{-1}$ to get

$$\frac{\mathbf{Q}_{k-1}^{-1}\mathbf{X}_k\mathbf{X}_k^{\mathrm{T}}\mathbf{Q}_{k-1}^{-1}}{\alpha + \mathbf{X}_k^{\mathrm{T}}\mathbf{Q}_{k-1}^{-1}\mathbf{X}_k} = \mathbf{Q}_k^{-1}\mathbf{X}_k\mathbf{X}_k^{\mathrm{T}}\mathbf{Q}_{k-1}^{-1} \tag{8.40}$$

Substituting (8.38) for the right side of (8.40) and rearranging, we have

$$\mathbf{Q}_k^{-1} = \frac{1}{\alpha}\left[\mathbf{Q}_{k-1}^{-1} - \frac{(\mathbf{Q}_{k-1}^{-1}\mathbf{X}_k)(\mathbf{Q}_{k-1}^{-1}\mathbf{X}_k)^{\mathrm{T}}}{\alpha + \mathbf{X}_k^{\mathrm{T}}(\mathbf{Q}_{k-1}^{-1}\mathbf{X}_k)}\right] \tag{8.41}$$

So we have in (8.41) an iterative procedure for computing $\mathbf{Q}_k^{-1}$ in (8.34). Note that the vector

$$\mathbf{S}_k = \mathbf{Q}_{k-1}^{-1}\mathbf{X}_k \tag{8.42}$$

is used three times in (8.41) and would therefore be computed first in the algorithm, and also that the denominator term in (8.41) is a scalar and would therefore be computed separately.

Concerning the initial value of $\mathbf{Q}_k^{-1}$ in (8.41), Lee [5] has presented an argument for starting with $\mathbf{Q}_0^{-1} = q_0\mathbf{I}$, where q_0 is a large constant, for stationary stochastic situations. This choice would suffice in adaptive situations as well, although it may be preferable to set $\mathbf{Q}_0^{-1}$ closer to its correct value, if such can be

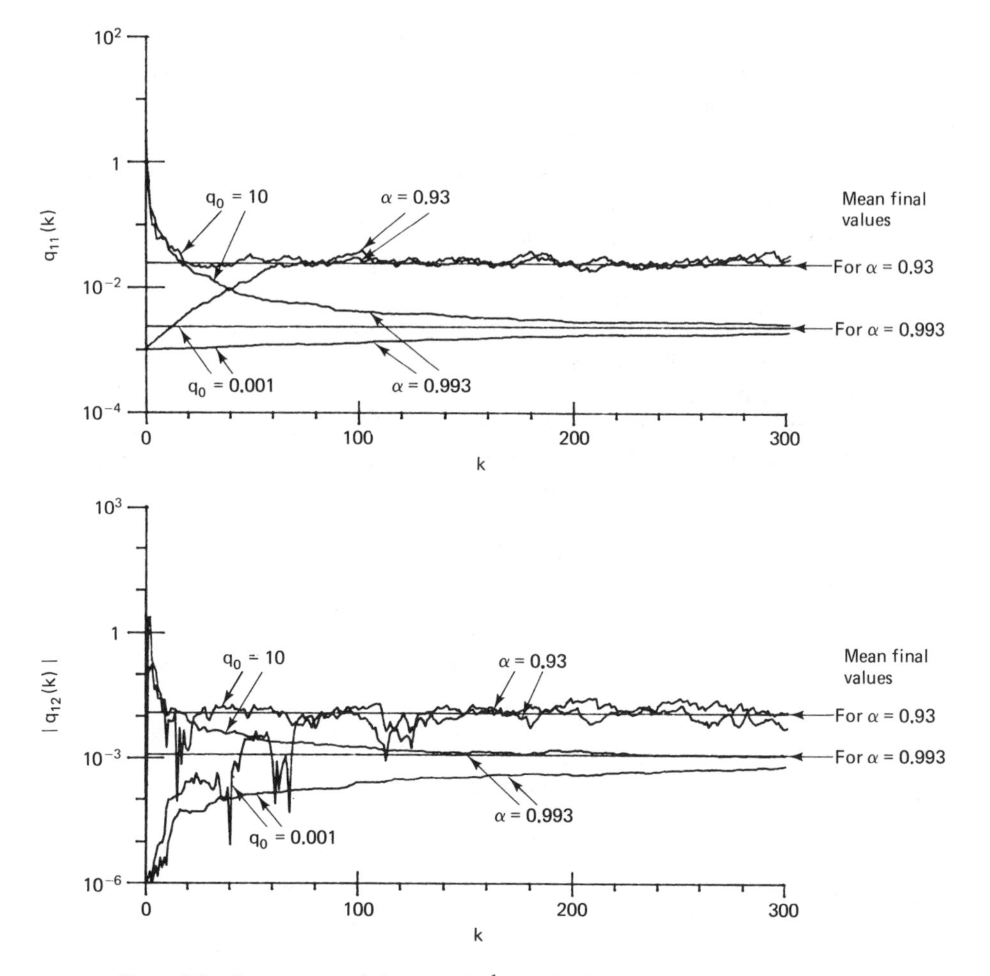

Figure 8.3 Convergence of elements $\mathbf{Q}_k^{-1}$; $q_{11}(k)$ in upper plot and magnitude of $q_{12}(k)$ in lower plot. Values of q_0 and α are as shown. Mean final values are from (8.43).

estimated. An example of the convergence of $\mathbf{Q}_k^{-1}$ from different initial conditions is given in Figure 8.3. For this example we used the following data:

$$\mathbf{R} = \begin{bmatrix} 4 & 2 \\ 2 & 4 \end{bmatrix} \qquad \mathbf{R}^{-1} = \frac{1}{6}\begin{bmatrix} 2 & -1 \\ -1 & 2 \end{bmatrix}$$

$$\mathbf{Q}_\infty^{-1} \approx \frac{1-\alpha}{6}\begin{bmatrix} 2 & -1 \\ -1 & 2 \end{bmatrix} \tag{8.43}$$

The input signal $\mathbf{X}_k$, with correlation matrix $\mathbf{R}$, was generated using the result of Exercise 25 in Chapter 7. The values in $\mathbf{R}^{-1}$ and $\mathbf{Q}_\infty^{-1}$ then follow from (8.8) and from letting k become large in (8.35). The upper curves in Figure 8.3 show the convergence of q_{11} for two values of α corresponding to sequence lengths of 10 and 100 in (8.25) and two values of the starting constant, q_0. The lower curves show the convergence of q_{12} for the same values of α and q_0. With q_0 nearer to the correct final value of q_{11}, the situation is seen to be better for q_{11} but somewhat worse for q_{12}, considering the logarithmic vertical scale. As expected, the estimates are noisier but converge more rapidly for the smaller value of α. As seen from (8.41), $\mathbf{Q}_k^{-1}$ remains symmetric if $\mathbf{Q}_0^{-1}$ is symmetric, as is the case here.

So we have completed our derivation of the SER algorithm. This algorithm has been developed similarly by Graupe [1], Ahmed et al. [2], Kailath [24], Lee [5], and others, usually with α set equal to 1. Another "finite-memory" form was also proposed by Ahmed et al. [4]. The finite-memory algorithm as developed here may be summarized as follows. As noted above, λ_{av} can be estimated from actual data and may require continual updating in nonstationary situations. When signal statistics are completely unavailable, note that the factor $\mu\lambda_{\text{av}}$ must always be between zero and 1. One may thus select a presumed "safe" value (e.g., 0.05 in a typical application) for this factor (see Exercise 14).

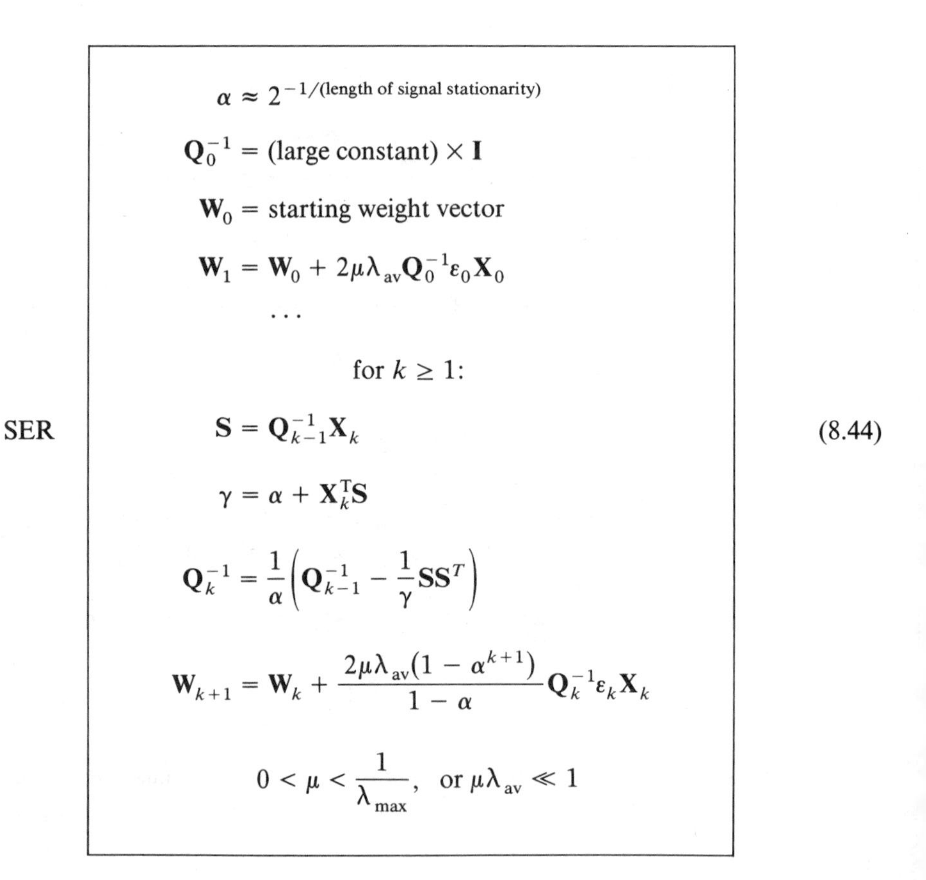

SER

$$\alpha \approx 2^{-1/(\text{length of signal stationarity})}$$

$$\mathbf{Q}_0^{-1} = (\text{large constant}) \times \mathbf{I}$$

$$\mathbf{W}_0 = \text{starting weight vector}$$

$$\mathbf{W}_1 = \mathbf{W}_0 + 2\mu\lambda_{\text{av}}\mathbf{Q}_0^{-1}\varepsilon_0\mathbf{X}_0$$

$$\cdots$$

for $k \geq 1$:

$$\mathbf{S} = \mathbf{Q}_{k-1}^{-1}\mathbf{X}_k \qquad (8.44)$$

$$\gamma = \alpha + \mathbf{X}_k^T\mathbf{S}$$

$$\mathbf{Q}_k^{-1} = \frac{1}{\alpha}\left(\mathbf{Q}_{k-1}^{-1} - \frac{1}{\gamma}\mathbf{S}\mathbf{S}^T\right)$$

$$\mathbf{W}_{k+1} = \mathbf{W}_k + \frac{2\mu\lambda_{\text{av}}(1-\alpha^{k+1})}{1-\alpha}\mathbf{Q}_k^{-1}\varepsilon_k\mathbf{X}_k$$

$$0 < \mu < \frac{1}{\lambda_{\max}}, \quad \text{or } \mu\lambda_{\text{av}} \ll 1$$

The subscript was omitted from $\mathbf{S}$ and γ to emphasize that these quantities need not be saved from one iteration to the next. The computation of $\mathbf{W}_k$ comes from the more accurate form in (8.37), $\mathbf{S}$ is from (8.42), and $\mathbf{Q}_k^{-1}$ and γ are from (8.41). The values of α and $\mathbf{Q}_0^{-1}$ need only be approximated roughly, as illustrated in Figure 8.3, and λ_{av} may be approximated with the input signal power.

The performance of the SER algorithm in (8.44) is illustrated in Figure 8.4. The situation is the same as in Figure 8.2 and the two figures may be compared directly. The value of $\alpha = 0.93$ was chosen to correspond with a stationary sequence length of 10 samples in the first line of (8.44). The SER algorithm performs better than the LMS and reaches the optimum weight vector, $\mathbf{W}^*$, in well under 100 iterations because, as seen in Figure 8.3, $\mathbf{Q}_k^{-1}$ approximates $\mathbf{R}^{-1}$ quite well after a few iterations. On the other hand, the SER paths in Figure 8.4 are not quite as direct as the LMS/Newton paths in Figure 8.2 because of the inaccuracies in $\mathbf{Q}_k^{-1}$ during the first few iterations.

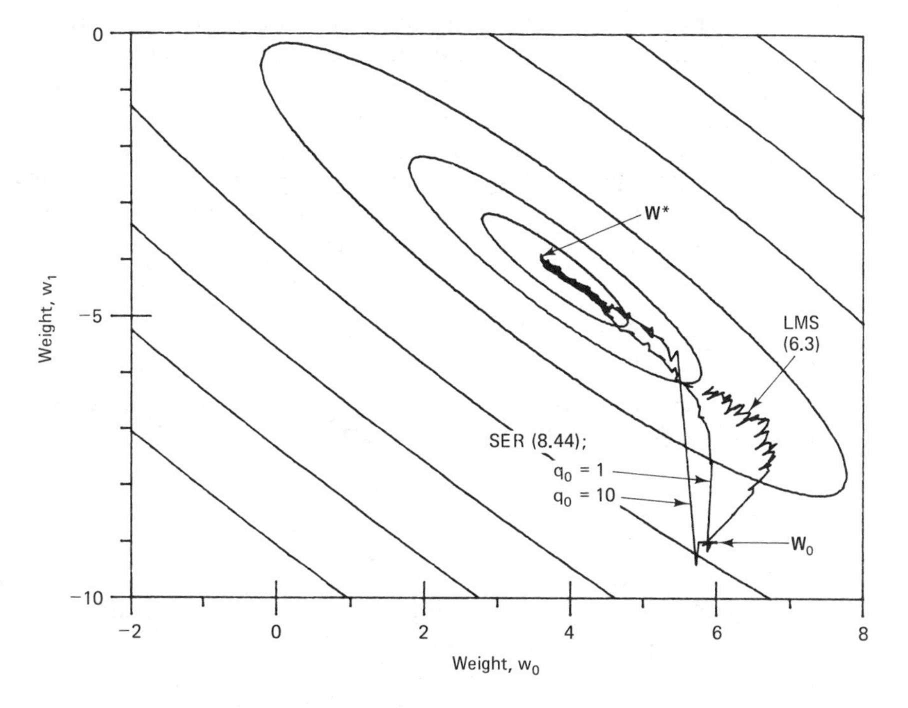

Figure 8.4 Illustration of SER versus LMS. Error surface same as in Figure 8.2. Parameters were $\mu = 0.05$, $\alpha = 0.93$, and q_0 as shown. Each track represents 100 iterations. In this example $\lambda_{max} = 0.97$, so μ is well within its acceptable range in (8.44).

ADAPTIVE RECURSIVE FILTERS

In Chapter 7 we discussed the possibility of using a recursive filter in place of the adaptive linear combiner in an adaptive system. The recursive filter, with poles as well as zeros, would offer the same advantages (resonance, sharp cutoff, etc.) that recursive filters offer in time-invariant applications [6].

On the other hand, we saw in Chapter 7 that recursive adaptive filters have two disadvantages that are not found in the adaptive linear combiner:

1. They become unstable if the poles move outside the unit circle during the adaptive process.
2. Their performance surfaces are generally nonquadratic and may even have local minima.

These are serious disadvantages, and consequently the recursive adaptive filter has had very limited application. Instability must be prevented by limiting the coefficients in some manner, and neither the steepest-descent method nor Newton's method will work correctly on a multimodal performance surface.

Regarding the second disadvantage, there is some indication at the time of this writing that if the recursive adaptive filter has a sufficient number of zeros and poles, the error surface will be unimodal [7]. Thus one may be able to remove the local minima simply by adding filter weights.

To develop algorithms for the recursive adaptive filter, let us place the recursive filter in Figure 7.2 in the standard adaptive mode, as illustrated in Figure 8.5. Here the vector $\mathbf{X}_k$ can represent either the multiple- or single-input situation, and of course y_k is a scalar. As in (7.3) we have

$$y_k = \sum_{n=0}^{L} a_n x_{k-n} + \sum_{n=1}^{L} b_n y_{k-n} \tag{8.45}$$

This form applies to the single-input case, which we assume here for convenience. A time-varying weight vector and a new signal vector, $\mathbf{W}_k$ and $\mathbf{U}_k$, are now defined as

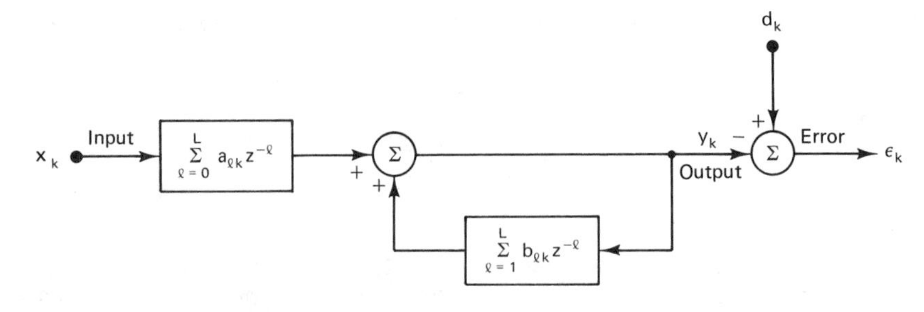

Figure 8.5 Recursive adaptive filter.

follows:

$$\mathbf{W}_k = [a_{0k} a_{1k} \quad \cdots \quad a_{Lk} b_{1k} \quad \cdots \quad b_{Lk}]^{\mathrm{T}} \tag{8.46}$$

$$\mathbf{U}_k = [x_k x_{k-1} \quad \cdots \quad x_{k-L} y_{k-1} \quad \cdots \quad y_{k-L}]^{\mathrm{T}} \tag{8.47}$$

From Figure 8.5 and from (8.45), we can write

$$\begin{aligned} \varepsilon_k &= d_k - y_k \\ &= d_k - \mathbf{W}_k^{\mathrm{T}} \mathbf{U}_k \end{aligned} \tag{8.48}$$

This is quite similar to the nonrecursive case [e.g., (2.8)], the main difference being that $\mathbf{U}_k$ contains values of y as well as x.

Considering first the LMS algorithm, we again use the gradient approximation in (6.2):

$$\begin{aligned} \hat{\nabla}_k &= \frac{\partial \varepsilon^2}{\partial \mathbf{W}_k} = 2\varepsilon \frac{\partial \varepsilon}{\partial \mathbf{W}_k} \\ &= 2\varepsilon_k \left[\frac{\partial \varepsilon_k}{\partial a_{0k}} \cdots \frac{\partial \varepsilon_k}{\partial a_{Lk}} \frac{\partial \varepsilon_k}{\partial b_{1k}} \cdots \frac{\partial \varepsilon_k}{\partial b_{Lk}} \right]^{\mathrm{T}} \\ &= -2\varepsilon_k \left[\frac{\partial y_k}{\partial a_{0k}} \cdots \frac{\partial y_k}{\partial a_{Lk}} \frac{\partial y_k}{\partial b_{1k}} \cdots \frac{\partial y_k}{\partial b_{Lk}} \right]^{\mathrm{T}} \end{aligned} \tag{8.49}$$

The derivatives in (8.49) present a special problem because y_k is now a recursive function. Using (8.45), we define

$$\begin{aligned} \alpha_{nk} &\triangleq \frac{\partial y_k}{\partial a_n} = x_{k-n} + \sum_{l=1}^{L} b_l \frac{\partial y_{k-l}}{\partial a_n} \\ &= x_{k-n} + \sum_{l=1}^{L} b_l \alpha_{n,k-l} \end{aligned} \tag{8.50}$$

$$\begin{aligned} \beta_{nk} &\triangleq \frac{\partial y_k}{\partial b_n} = y_{k-n} + \sum_{l=1}^{L} b_l \frac{\partial y_{k-l}}{\partial b_n} \\ &= y_{k-n} + \sum_{l=1}^{L} b_l \beta_{n,k-l} \end{aligned} \tag{8.51}$$

With the derivatives defined in this manner, we have

$$\hat{\nabla}_k = -2\varepsilon_k [\alpha_{0k} \quad \cdots \quad \alpha_{Lk} \beta_{1k} \quad \cdots \quad \beta_{Lk}]^{\mathrm{T}} \tag{8.52}$$

Similar to (6.3), we write the LMS algorithm as follows:

$$\mathbf{W}_{k+1} = \mathbf{W}_k - \mathbf{M} \hat{\nabla}_k \tag{8.53}$$

Here we have replaced the constant, μ, with the following diagonal matrix:

$$\mathbf{M} = \text{diag}[\mu \quad \cdots \quad \mu\nu_1 \quad \cdots \quad \nu_L] \tag{8.54}$$

With the nonquadratic error surface, we now have a convergence parameter μ for each a, and then a different convergence factor for each b. We may even wish to have these factors vary with time. Using current values of the b's in (8.50) and (8.51) the LMS algorithm for recursive adaptive filters is as follows:

IIR LMS

$$\begin{aligned}
y_k &= \mathbf{W}_k^{\mathrm{T}}\mathbf{U}_k \\
\alpha_{nk} &= x_{k-n} + \sum_{l=1}^{L} b_{lk}\alpha_{n,k-l} \qquad 0 \le n \le L \\
\beta_{nk} &= y_{k-n} + \sum_{l=1}^{L} b_{lk}\beta_{n,k-l} \qquad 1 \le n \le L \\
\hat{\nabla}_k &= -2(d_k - y_k)[\alpha_{0k} \quad \cdots \quad \alpha_{Lk}\beta_{1k} \quad \cdots \quad \beta_{Lk}]^{\mathrm{T}} \\
\mathbf{W}_{k+1} &= \mathbf{W}_k - \mathbf{M}\hat{\nabla}_k
\end{aligned} \tag{8.55}$$

Initialization is the same as discussed previously except that here, in addition, the α's and β's should be set initially at zero unless their values are known. Note that the signal vector, $\mathbf{U}_k$, is defined in (8.47) and that b_{lk} is one of the feedback weights of $\mathbf{W}_k$ in (8.46).

It is helpful to picture (8.55) in the form of a diagram. If we use the notation in Chapter 7 with

$$A_k(z) = \sum_{l=0}^{L} a_{lk} z^{-l} \quad \text{and} \quad B_k(z) = \sum_{l=1}^{L} b_{lk} z^{-l} \tag{8.56}$$

then, just as in (7.8), we can write the transfer function for either the second or the third line of (8.55) as

$$\text{transfer function} = \frac{z^{-n}}{1 - B_k(z)} \tag{8.57}$$

For example, the computation of α_{nk} is illustrated in Figure 8.6. Using this result,

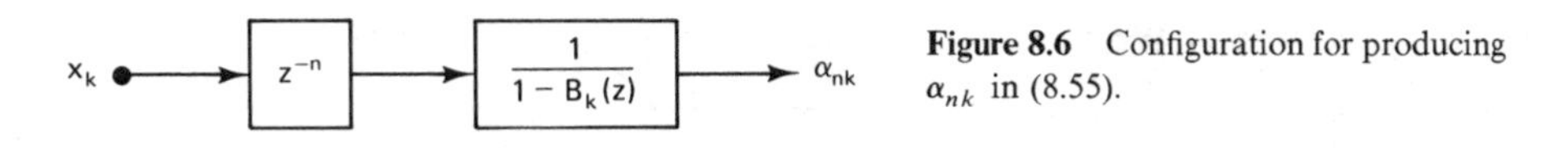

Figure 8.6 Configuration for producing α_{nk} in (8.55).

the entire adaptive filter is illustrated in Figure 8.7. It is interesting to picture the operation in this form, even though the weight updates in the last line of (8.55) are not shown. Similar illustrations have been given by Hsia [8] and by White, who originally proposed the recursive LMS algorithm [9]. Various simplifications and improvements to the recursive LMS algorithm have been proposed by Feintuch [14], David and Stearns [15], Hsia [8], and others.

An entire class of algorithms known as the HARF (hyperstable adaptive recursive filter) algorithms has been proposed by Larimore et al. [10]. The simplest member of this class, known as SHARF [10–13], can be described as follows. In (8.55), let α_{nk} and β_{nk} be approximated with x_{k-n} and y_{k-n}, respectively. Then, instead of using $\varepsilon_k = d_k - y_k$ in estimating ∇_k, we use a smoothed version of ε_k, obtained by filtering ε_k. The latter step is the key feature of the HARF class of

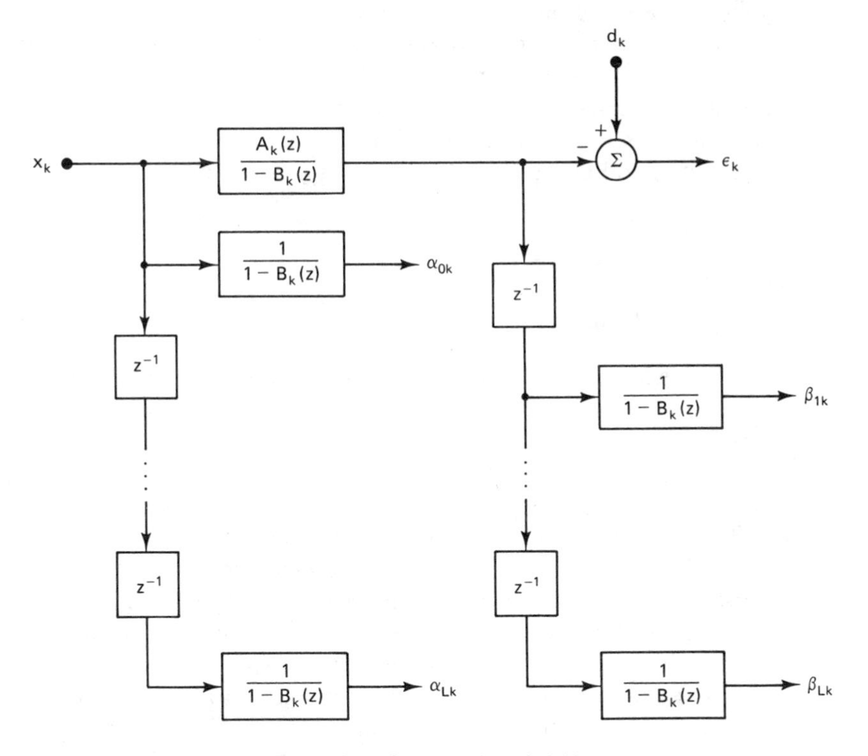

Figure 8.7 Configuration of (8.55).

algorithms. We have

IIR SHARF

$$\begin{aligned} y_k &= \mathbf{W}_k^{\mathrm{T}}\mathbf{U}_k \\ \varepsilon_k &= d_k - y_k \\ \nu_k &= \varepsilon_k + \sum_{n=1}^{N} c_n \varepsilon_{k-n} \\ \hat{\nabla}_k &= -2\nu_k [x_k \quad \cdots \quad x_{k-L} y_{k-1} \quad \cdots \quad y_{k-L}]^{\mathrm{T}} \\ \mathbf{W}_{k+1} &= \mathbf{W}_k - \mathbf{M}\hat{\nabla}_k \end{aligned} \tag{8.58}$$

The c's are constant coefficients used in smoothing ε_k to obtain ν_k. The SHARF algorithm is thus simpler than (8.55). Its convergence has been proved in some situations [25] and it has been applied in noise canceling [12] and in prediction [13]. The choice of the smoothing coefficients, $[c_n]$, is a subject of current research and is beyond the range of our discussion here.

Since LMS algorithms in general are hampered by the nonquadratic IIR error surface, we are again prompted to consider approximations to Newton's method, even though the latter may also be hampered for the same reason. We can easily develop an SER type of algorithm for recursive filters by replacing $\mathbf{X}$ with $\mathbf{U}$, the IIR signal vector, and using the IIR gradient estimate. Thus:

Nonrecursive:	*Recursive*:
$\varepsilon_k = d_k - \mathbf{W}_k^{\mathrm{T}}\mathbf{X}_k$	$\varepsilon_k = d_k - \mathbf{W}_k^{\mathrm{T}}\mathbf{U}_k$
$\hat{\nabla}_k = -2\varepsilon_k \mathbf{X}_k$	$\hat{\nabla}_k = -2\varepsilon_k [\alpha_{0k} \quad \cdots \quad \alpha_{Lk}\beta_{1k} \quad \cdots \quad \beta_{Lk}]^{\mathrm{T}}$

The easiest way to arrive at a recursive SER algorithm is to think of first replacing $\mathbf{X}$ with $\mathbf{U}$ in (2.11) and (2.12), thus redefining $\mathbf{R}$ and $\mathbf{P}$, noting that the dimension increases from $L + 1$ to $2L + 1$.

Having done this, we have the IIR error surface given by (2.13). Now the only simple approach to an SER algorithm is to take the gradient of (2.13) and to assume that $\mathbf{R}$ and $\mathbf{P}$ are not functions of $\mathbf{W}$. Although this assumption was valid for FIR adaptive filters and led to (2.16), it is not valid for IIR filters during convergence, because both $\mathbf{R}$ and $\mathbf{P}$ contain expected products involving y_k. However, after convergence with stationary inputs, $\mathbf{R}$ and $\mathbf{P}$ become constants even in the IIR case, so (2.16) may be said to be correct near $\mathbf{W} = \mathbf{W}^*$.

Using the revised $\mathbf{R}$ and $\mathbf{P}$ and assuming that (2.16) and (8.28) still represent the weight vector solution, we obtain the IIR SER algorithm simply by replacing $\mathbf{X}_k$ by $\mathbf{U}_k$ in (8.36), that is, all of the derivation is the same as in the FIR case. We must also replace the gradient estimate (8.3) used in (8.5) and subsequently in (8.37) with the recursive gradient estimate as above. When we do this, (8.37) becomes

$$\mathbf{W}_{k+1} = \mathbf{W}_k - \frac{\mathbf{M}\lambda_{\mathrm{av}}(1 - \alpha^{k+1})}{1 - \alpha}\mathbf{Q}_k^{-1}\hat{\nabla}_k \tag{8.59}$$

As in the recursive LMS algorithm, we have replaced μ with $\mathbf{M}$ in (8.54) to allow different convergence factors for the b's. The complete recursive SER algorithm is

IIR SER

$$\begin{aligned}
\mathbf{S} &= \mathbf{Q}_{k-1}^{-1}\mathbf{U}_k \\
\gamma &= \alpha + \mathbf{U}_k^{\mathrm{T}}\mathbf{S} \\
\mathbf{Q}_k^{-1} &= \frac{1}{\alpha}\left(\mathbf{Q}_{k-1}^{-1} - \frac{1}{\gamma}\mathbf{S}\mathbf{S}^{\mathrm{T}}\right) \\
\alpha_{nk} &= x_{k-n} + \sum_{l=1}^{L} b_{lk}\alpha_{n,k-l} \qquad 0 \le n \le L \\
\beta_{nk} &= y_{k-n} + \sum_{l=1}^{L} b_{lk}\beta_{n,k-l} \qquad 1 \le n \le L \\
\hat{\nabla}_k &= -2(d_k - y_k)[\alpha_{0k} \quad \cdots \quad \alpha_{Lk}\beta_{1k} \quad \cdots \quad \beta_{Lk}]^{\mathrm{T}} \\
\mathbf{W}_{k+1} &= \mathbf{W}_k - \frac{\mathbf{M}\lambda_{\mathrm{av}}(1 - \alpha^{k+1})}{1 - \alpha}\mathbf{Q}_k^{-1}\hat{\nabla}_k
\end{aligned} \tag{8.60}$$

Note that the unsubscripted α is the memory constant in (8.25), while the subscripted α's are the derivatives in (8.50). The SER initial conditions in (8.44) may be used here. Again, from (8.46) and (8.47), $\mathbf{S}$ and $\mathbf{Q}^{-1}$ are now dimensioned $2L + 1$. Convergence and stability are both potential problems and need to be considered in each application of (8.60).

To provide an example of the performance of these algorithms, consider the identification configuration in Figure 8.8. Beginning with $a_0 = b_1 = b_2 = 0$, the recursive adaptive filter is to search the error surface for the point where $a_0 = 1$, $b_1 = 1.2$, and $b_2 = -0.6$, and thus identify the "unknown" system. With a white-noise input, the three coefficients must converge to these exact values in order to reduce $\xi = E[\varepsilon_k^2]$ to zero.

Typical convergence tracks of the IIR LMS and SER algorithms, (8.55) and (8.60), are shown in Figure 8.9. Note that these are only examples from a wide variety of IIR algorithms and parameter settings. [A SHARF example could also

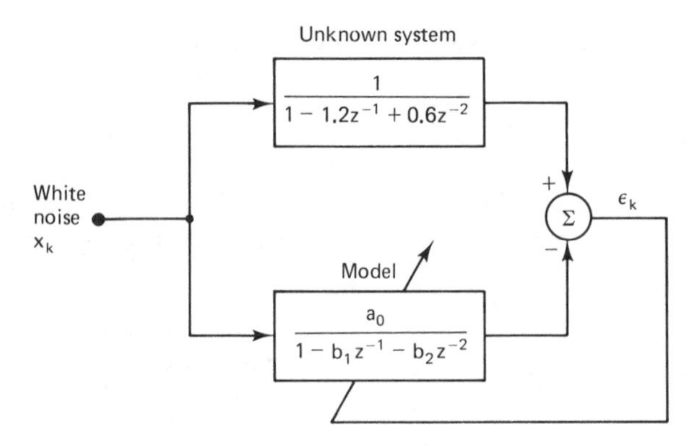

Figure 8.8 Configuration for the recursive performance examples in Figure 8.9: adaptive modeling of an unknown system.

have been included, with the appropriate choice of $[c_n]$ in (8.58).] However, several typical features of the IIR algorithms can be seen in these examples.

First, the error surface is found from (7.65) (see exercises 35 through 37). Just as in the example in Chapter 7, $\xi = E[\varepsilon_k^2]$ is quadratic in a_0 in this example. Therefore, we have shown the error contours only in the $b_1 b_2$ plane in Figure 8.9. The contours are drawn with a_0 optimized for each pair (b_1, b_2). However, the tracks in the figure include the adaption of a_0 as well as b_1 and b_2.

Second, as discussed in Chapter 7, the poles of the adaptive filter must lie inside the unit circle for stability. Thus there is a region of stability in the performance space, a triangle in Figure 8.9, and the b's should be constrained to stay within this region. The correspondence between the unit circle and the triangle is left for Exercise 12.

As mentioned above, the weights were started at $a_0 = b_1 = b_2 = 0$, and in both cases were allowed to reach their optimum values. The convergence parameters, $\mathbf{M}$, in (8.54) were

$$\begin{aligned} &LMS\colon \quad \mathbf{M} = \text{diag}\,[0.05 \quad 0.005 \quad 0.0025] \\ &SER\colon \quad \mathbf{M} = \text{diag}\,[0.5 \quad 0.1 \quad 0.05] \end{aligned} \tag{8.61}$$

The SER algorithm was started with $q_0 = 1$, and $\alpha = 0.93$ for 10-sample stationarity. Note that the LMS algorithm more or less follows the path of steepest descent but is erratic in the bottom of the valley leading down to $\xi_{\min}$ in the nonquadratic error surface. This type of performance is typical with IIR filters.

The SER algorithm, on the other hand, is not a good approximation to Newton's method, at least during the initial part of the learning process, because, as mentioned above, $\mathbf{Q}_k^{-1}$ is a function of $\mathbf{W}_k$ until $\mathbf{W}_k$ approaches the optimum point. Near this point, however, the SER track is smoother than the LMS track, and in this example the SER algorithm needed fewer iterations to reach the optimum at $b_1 = 1.2$, $b_2 = -0.6$.

The tracks in examples such as Figure 8.9 are sensitive to the choice of $\mathbf{M}$ as in (8.61), as well as q_0, α, and of course the starting weight values. The choice of $\mathbf{M}$ and the best form of the overall algorithm are still subjects of research in IIR adaptive filter theory.

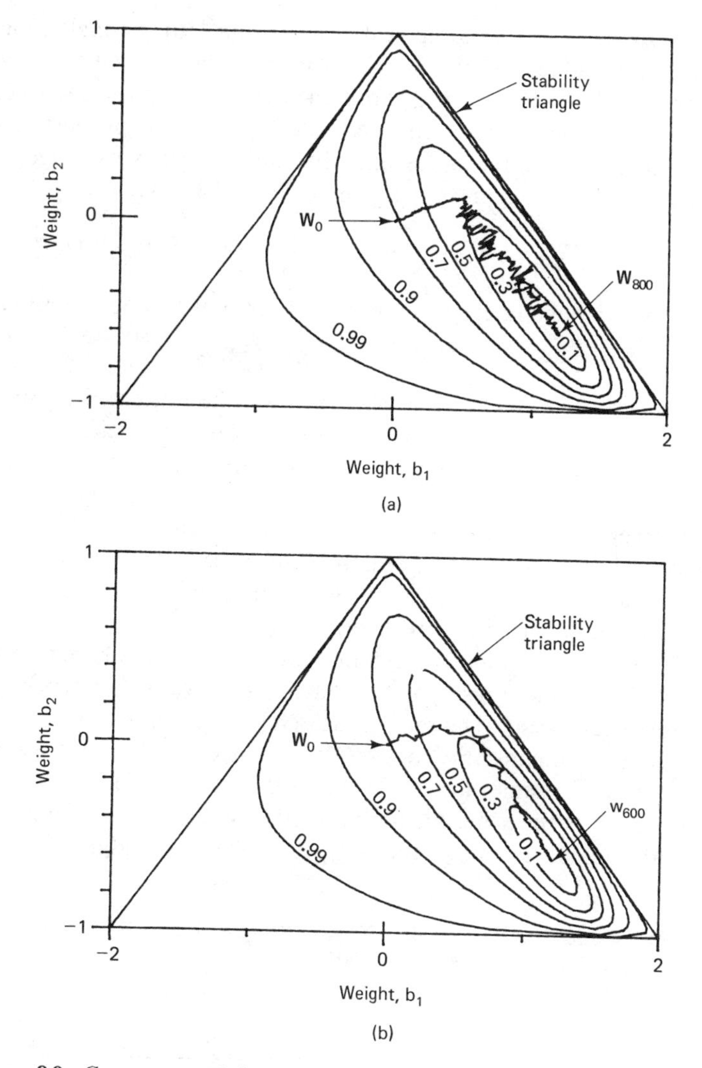

Figure 8.9 Convergence of the IIR LMS [(a), 800 iterations] and SER [(b), 600 iterations] algorithms. Normalized values of $E[\varepsilon_k^2]$ shown next to error contours, with $E[\varepsilon_k^2] = 0$ at $b_1 = 1.2$, $b_2 = -0.6$.

RANDOM-SEARCH ALGORITHMS

Besides Newton's method and the steepest-descent method, there is the possibility of searching the performance surface in a random fashion. In this section we briefly introduce two types of random-search algorithms. First, the linear random search algorithm is a type of algorithm that chooses a *random direction* in which to travel in the weight space, and second, the genetic optimizer algorithm is presented as a type of algorithm that selects a series of *random locations* in the weight space.

In the linear random-search algorithm [16], a small random change, $\mathbf{U}_k$, is tentatively added to the weight vector at the beginning of each iteration. The corresponding change in performance is observed as described in Chapter 5. A permanent weight vector change, from $\mathbf{W}_k$ to $\mathbf{W}_{k+1}$, proportional to the product of the change in performance and the initial tentative change, is then made. This procedure can be expressed algebraically as follows:

$$\mathbf{W}_{k+1} = \mathbf{W}_k + \frac{\mu}{\sigma^2}[\hat{\xi}(\mathbf{W}_k) - \hat{\xi}(\mathbf{W}_k + \mathbf{U}_k)]\mathbf{U}_k \tag{8.62}$$

For the adaptive linear combiner, $\mathbf{U}_k$ is assumed to be a random vector with

$$\text{cov}[\mathbf{U}_k] = \sigma^2\mathbf{I} \tag{8.63}$$

The functions $\hat{\xi}(\mathbf{W}_k)$ and $\hat{\xi}(\mathbf{W}_k + \mathbf{U}_k)$ are the estimated mean-square errors of the present and tentative weight vector values, respectively. Both the convergence parameter, μ, and σ^2 are constants that affect stability and rate of convergence.

The linear random-search algorithm has been analyzed for the adaptive linear combiner, and the results are summarized in Table 8.1 [16]. Table 8.1 is similar to Table 6.1, with the random-search results substituted for the LMS results in Chapter 6. As defined in Chapter 5, N is the number of observations used to make each estimate of the mean-square error.

Although the linear random-search algorithm is based on changing the weight vector in a random fashion, its behavior is shown in Reference 16 to be quite similar to steepest descent. The random-search weights relax geometrically toward $\mathbf{W}^*$ with the same time constants that result with steepest descent. We can see in Table 8.1 that μ has the same effect as in the steepest-descent algorithm, and in fact the stable range of μ is the same for both algorithms. For a given convergence rate, the LRS misadjustment is twice the steepest-descent misadjustment. Thus the steepest-

TABLE 8.1 MISADJUSTMENT AND TIME CONSTANTS OF LINEAR RANDOM-SEARCH (LRS) AND STEEPEST-DESCENT ALGORITHMS

	LRS algorithm	Steepest-descent algorithm
Misadjustment, M	$\frac{\mu(L+1)}{N\sigma^2}\xi_{\min} = \frac{(L+1)^2}{4P}\left(\frac{1}{T_{\text{mse}}}\right)_{\text{av}}$	$\frac{\mu(L+1)}{4N\delta^2}\xi_{\min} = \frac{(L+1)^2}{8P}\left(\frac{1}{T_{\text{mse}}}\right)_{\text{av}}$
Perturbation, P	$\frac{\sigma^2(L+1)\lambda_{\text{av}}}{2\xi_{\min}}$	$\frac{\delta^2\lambda_{\text{av}}}{\xi_{\min}}$
Total misadjustment, M_{tot}	$M + P$	$M + P$
Time constant of nth mode:		
In number of adaptive iterations, τ_{mse}	$\frac{1}{4\mu\lambda_n}$	$\frac{1}{4\mu\lambda_n}$
In number of data samples, T_{mse}	$\frac{N}{2\mu\lambda_n}$	$\frac{N(L+1)}{4\mu\lambda_n}$

descent algorithm outperforms the LRS algorithm. However, the LRS algorithm is simple to implement. It is also useful as a model for "natural selection," and for studying theories of the development of living systems.

The steepest-descent and linear random-search algorithms are much less efficient than the LMS algorithm in terms of data usage, and they have a higher misadjustment for a given speed of convergence. However, they are applicable where the LMS algorithm may not be usable, that is, where the input signals are unavailable or where the adjustable parameters are not signal weights.

The linear random-search algorithm is, again, an algorithm that chooses a random direction to test at each iteration. An algorithm that tests random *locations* in the performance space has been developed for adaptive systems by Etter and Masukawa [17]. Since its behavior brings to mind the processes of cell division and selective survival, it is called the "genetic optimizer" algorithm.

We will describe the operation of the genetic optimizer in terms of a single weight, w. The algorithm begins by selecting an initial population of M values of w, equally spaced over a predetermined range. Then $\xi(w)$ is estimated at each of these M values of w, and a revised and larger population of $N > M$ values is created. In this revised population, the M original values of w are replicated in inverse proportion to the corresponding values of $\xi(w)$. Thus, if the original population contains the values w_1 and w_2, and if $\xi(w_2) > \xi(w_1)$, the revised population will contain more values of w_1 than of w_2. The size of the total population is now N, with some values of w repeated several times.

Next, in the second generation, offspring are created as follows. Two values of w are chosen at random and, at an arbitrary point, the binary representations of the two values are split. The first segment of one binary string is concatenated with the second segment of the other to yield the offspring. For example, suppose that w is represented with 8 bits so that 256 values are possible, and that splitting takes place in the middle of each string. Then

Binary representation of parent 1: $a_1a_2a_3a_4a_5a_6a_7a_8$

Binary representation of parent 2: $b_1b_2b_3b_4b_5b_6b_7b_8$

Binary representation of offspring: $a_1a_2a_3a_4b_5b_6b_7b_8$

The second generation is formed by creating M offspring in this manner and replacing all of the parents in the first generation. Note that if both parents represent the same value of w, their offspring also represents this same value.

The population of the second generation is again increased from M to N as above, replicating values of w that have lower values of $\xi(w)$, and so on. The entire process converges as the offspring become more and more alike, and may be made to diverge or search the error space by introducing random mutations. Examples of the genetic optimizer algorithm with an adaptive delay have been given by Etter and Masukawa [17].

LATTICE STRUCTURES

In our discussion of adaptive algorithms so far we have given examples using only a direct realization of the adaptive system. For nonrecursive systems this amounts to the adaptive transversal filter introduced in Chapter 2, and for recursive systems we have the direct recursive form introduced in Figure 7.2. More generally there are at least four types of realizations or structures that are of potential interest in adaptive processing:

1. Direct form
2. Cascade form
3. Parallel form
4. Lattice structure

Suppose that we have a single-input, recursive filter with transfer function similar to (7.8):

$$H(z) = \frac{A(z)}{1 + B(z)} = \frac{a_0 + a_1 z^{-1} + \cdots + a_L z^{-L}}{1 + b_1 z^{-1} + \cdots + b_L z^{-L}} \tag{8.64}$$

This is the direct form.† The cascade form is obtained from the direct form by factoring $A(z)$ and $1 + B(z)$, usually into second-order sections, and the parallel from the cascade via a partial fraction expansion [20]. These two forms are not used extensively in adaptive signal processing, although the adaptive cascade structure has been a subject of research [21].

Various lattice structures can also be obtained from (8.64), and these are used in adaptive processing, particularly in linear prediction [22, 23]. Let us begin our introduction to adaptive lattices by reviewing how to convert (8.64) into a lattice type of structure. The following algorithm has been discovered by Gray and Markel [18]. We rewrite (8.64) as

$$H(z) = \frac{Y(z)}{X(z)} = \frac{A_L(z)}{B_L(z)} \tag{8.65}$$

$$= \frac{a_{L0} + a_{L1} z^{-1} + \cdots + a_{LL} z^{-L}}{b_{L0} + b_{L1} z^{-1} + \cdots + b_{LL} z^{-L}} \tag{8.66}$$

Here X and Y are the input and output signal transforms, the subscript L is used to

† For the sake of symmetry we let the b's here have a sign opposite to that in Chapter 7.

designate explicitly the filter size, and b_{L0} is understood to be always equal to 1. Starting with (8.66), we form a succession of shorter polynomials. The algorithm is:

For $l = L, L-1, \ldots, 1$:

$$zC_l(z) = z^{-l}B_l(z^{-1}) \tag{8.67}$$

$$\kappa_{l-1} = b_{ll} \tag{8.68}$$

$$B_{l-1}(z) = \frac{B_l(z) - \kappa_{l-1}zC_l(z)}{1 - \kappa_{l-1}^2} \tag{8.69}$$

$$\nu_l = a_{ll} \tag{8.70}$$

$$A_{l-1}(z) = A_l(z) - \nu_l zC_l(z) \tag{8.71}$$

Upon executing this algorithm, we have κ_{l-1} and ν_l for $l = 1, 2, \ldots, L$, and we also define $\nu_0 \triangleq a_{00}$. These values of κ and ν will become the coefficients used in the lattice. To show this, we note first that (8.71) can be substituted into itself to obtain

$$\begin{aligned} A_L(z) &= A_{L-1}(z) + \nu_L zC_L(z) \\ &= A_{L-2}(z) + \nu_{L-1}zC_{L-1}(z) + \nu_L zC_L(z) \\ &\;\;\vdots \\ &= A_0(z) + \nu_1 zC_1(z) + \cdots + \nu_L zC_L(z) \end{aligned} \tag{8.72}$$

But $A_0(z)$ is a_{00}, or ν_0 by the definition above, and from (8.67) we have

$$zC_0(z) = B_0(z^{-1}) = B_0(z) = b_{00} = 1 \tag{8.73}$$

Therefore, (8.72) may be written

$$A_L(z) = \sum_{l=0}^{L} \nu_l zC_l(z) \tag{8.74}$$

Using the signal transforms in (8.65), this result is also

$$Y(z) = \sum_{l=0}^{L} X(z)\nu_l \frac{zC_l(z)}{B_L(z)} \tag{8.75}$$

We will use this result to describe the lattice structure, but first we need one additional result. With $B_l(z)$ and $C_l(z)$ as defined above, Itakura and Saito [19] have shown that

$$\begin{bmatrix} B_l(z) \\ C_l(z) \end{bmatrix} = \begin{bmatrix} 1 & \kappa_{l-1} \\ z^{-1}\kappa_{l-1} & z^{-1} \end{bmatrix} \begin{bmatrix} B_{l-1}(z) \\ C_{l-1}(z) \end{bmatrix} \tag{8.76}$$

We will not repeat the proof of this result here, but will demonstrate it below in an example. A lattice element can be formed by rewriting (8.76), and then assembled into a lattice structure using (8.75). From (8.76) we have

$$B_l(z) = B_{l-1}(z) + \kappa_{l-1}C_{l-1}(z)$$

or

$$B_{l-1}(z) = B_l(z) - \kappa_{l-1}z^{-1}[zC_{l-1}(z)] \tag{8.77}$$
$$C_l(z) = z^{-1}\kappa_{l-1}B_{l-1}(z) + z^{-1}C_{l-1}(z)$$

or

$$zC_l(z) = \kappa_{l-1}B_{l-1}(z) + z^{-1}[zC_{l-1}(z)] \tag{8.78}$$

Using $X(z)/B_L(z)$ as an input signal transform, we can see that (8.77) and (8.78) are implemented in the lattice element in Figure 8.10. Note that all summing inputs are positive, and that we use the triangular weight symbol here to emphasize the signal direction.

We now assemble these lattice elements into a cascade structure, as in the example in Figure 8.11 with $L = 3$. On the right end of the structure we have the pair of nodes for which $l - 1 = 0$ in Figure 8.10, that is,

$$\text{right upper signal} = \frac{X(z)B_0(z)}{B_L(z)} = \frac{X(z)}{B_L(z)} \tag{8.79}$$

$$\text{right lower signal} = \frac{zX(z)C_0(z)}{B_L(z)} = \frac{X(z)}{B_L(z)} \tag{8.80}$$

These results follow from (8.73). Since these signals are the same, the right-hand nodes are always tied together as in Figure 8.11.

On the left end of the structure we have the nodes for which $l = L$ in Figure 8.10, so

$$\text{left upper signal} = \frac{X(z)B_L(z)}{B_L(z)} = X(z) \tag{8.81}$$

Thus we show the input signal, x_k, entering the structure at the upper left in Figure 8.11. In the lower part of the structure in Figure 8.11, we can see from Figure 8.10

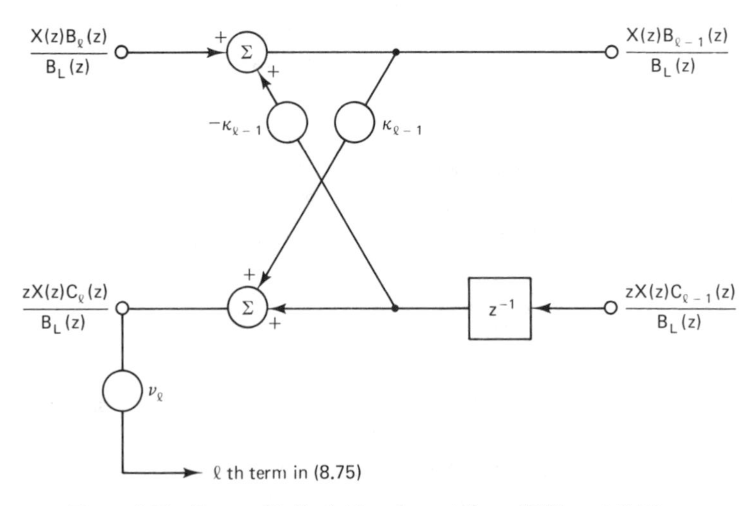

Figure 8.10 Two-multiplier lattice element, from (8.77) and (8.78).

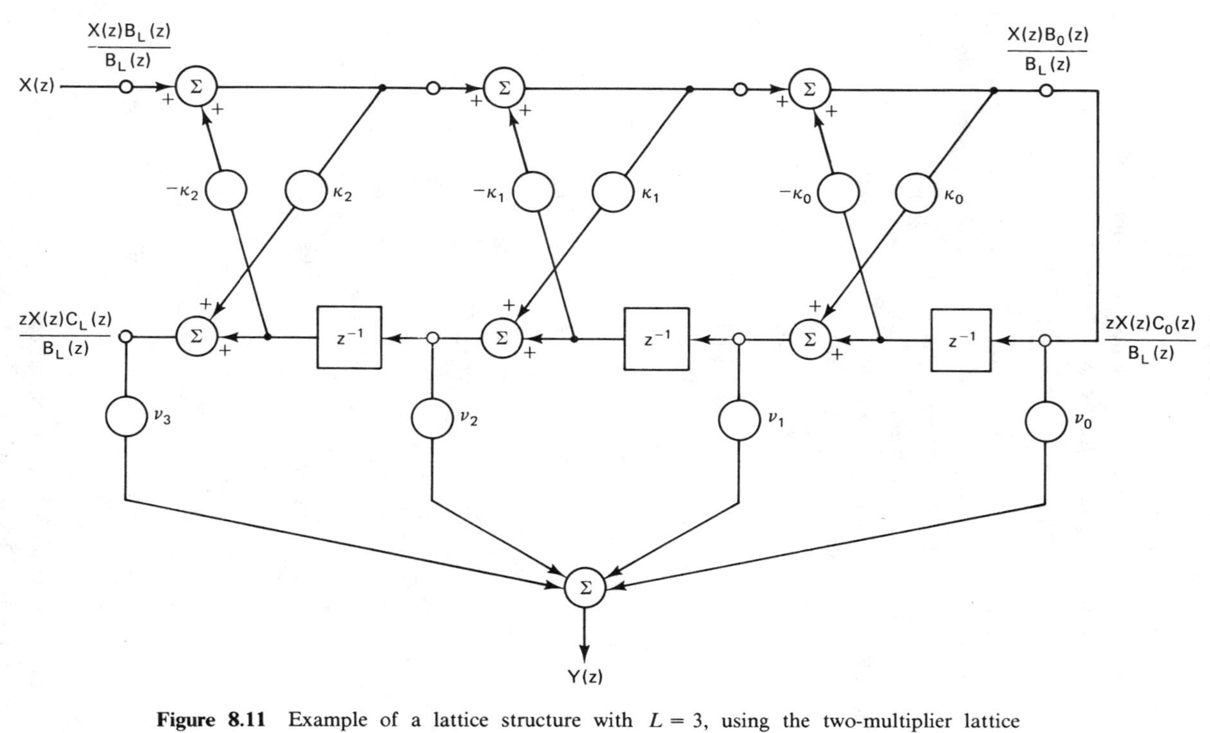

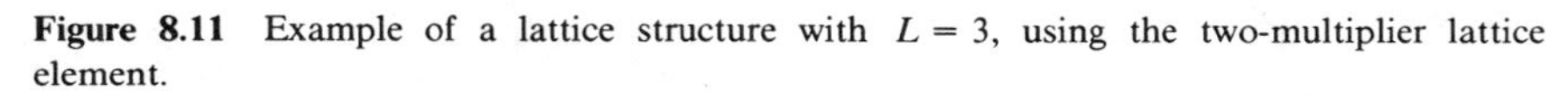
Figure 8.11 Example of a lattice structure with $L = 3$, using the two-multiplier lattice element.

how each node produces a term in the sum in (8.75), and thus the lattice structure is equivalent to the direct form in (8.64).

An important theorem by Jury [27] states that the poles of (8.64) will be inside the circle $|z| = 1$ if $|\kappa_l| < 1$ for all l. Thus, we have a "stable lattice" if $|\kappa_l| < 1$.

Before giving an example, we note a ladder form that is equivalent to Figures 8.10 and 8.11. First, substituting (8.77) into (8.78), we have

$$zC_l(z) = \kappa_{l-1}B_l(z) + z^{-1}(1 - \kappa_{l-1}^2)[zC_{l-1}(z)] \tag{8.82}$$

Equations (8.77) and (8.82) then describe the ladder element in Figure 8.12, which is equivalent to Figure 8.10. There is also a one-multiplier lattice element, with coefficients derived in Reference 18.

For a more specific example of the lattice structure, let us convert the general second-order filter into a lattice form. The translation algorithm [Equations (8.67) through (8.71)] gives

$$zC_2(z) = z^{-2} + b_1 z^{-1} + b_2$$

$$\kappa_1 = b_2 \tag{8.83}$$

$$B_1(z) = 1 + \frac{b_1 z^{-1}}{1 + b_2}$$

$$\nu_2 = a_2 \tag{8.84}$$

$$A_1(z) = a_0 - a_2 b_2 + (a_1 - a_2 b_1) z^{-1}$$

and

$$zC_1(z) = z^{-1} + \frac{b_1}{1 + b_2}$$

$$\kappa_0 = \frac{b_1}{1 + b_2} \tag{8.85}$$

$$B_0(z) = 1$$

$$\nu_1 = a_1 - a_2 b_1 \tag{8.86}$$

$$\nu_0 = A_0(z) = a_0 - a_2 b_2 - \frac{b_1(a_1 - a_2 b_1)}{1 + b_2} \tag{8.87}$$

From these results it is easy to solve for the direct coefficients in terms of the lattice coefficients. The results are listed in Table 8.2. Note that $|b_2|$ must be less than 1 for stability.

Let us now look at two special cases of the general lattice structure. The first, shown in Figure 8.13, is the "all-pole" version of Figure 8.11. If we set

$$a_0 = 1$$

$$a_1 = a_2 = \cdots = a_L = 0 \tag{8.88}$$

in (8.64), then we have an all-pole transfer function. If we use (8.88) in (8.70) and

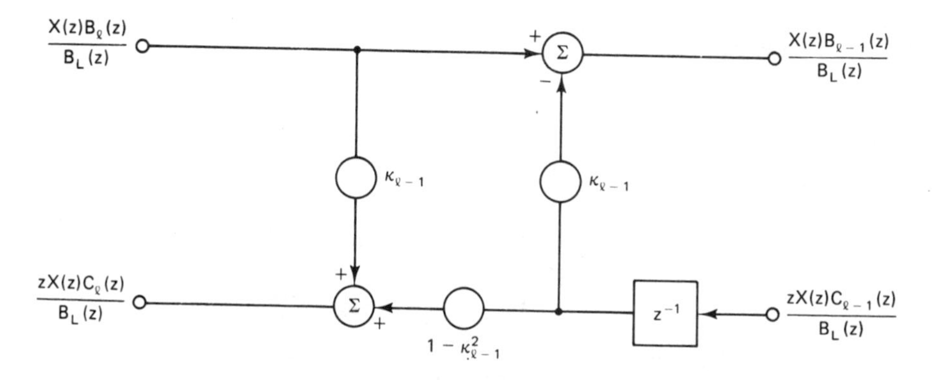

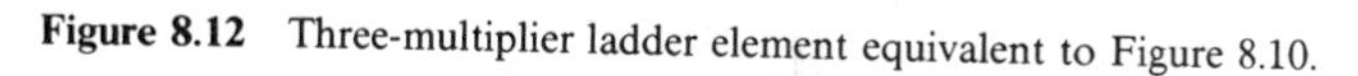

Figure 8.12 Three-multiplier ladder element equivalent to Figure 8.10.

TABLE 8.2 CONVERSIONS WITH $L = 2$

Direct to lattice:	$\nu_0 = a_0 - \dfrac{b_1(a_1 - a_2 b_1)}{1 + b_2} - a_2 b_2$
	$\nu_1 = a_1 - a_2 b_1$
	$\nu_2 = a_2$
	$\kappa_0 = \dfrac{b_1}{1 + b_2}$
	$\kappa_1 = b_2$
Lattice to direct:	$a_0 = \nu_0 + \nu_1\kappa_0 + \nu_2\kappa_1$
	$a_1 = \nu_1 + \nu_2\kappa_0(1 + \kappa_1)$
	$a_2 = \nu_2$
	$b_1 = \kappa_0(1 + \kappa_1)$
	$b_2 = \kappa_1$

(8.71) we find that

$$\nu_L = \nu_{L-1} = \cdots = \nu_1 = 0$$
$$\nu_0 = 1 \tag{8.89}$$

Thus Figure 8.13 is the all-pole version of Figure 8.11. It is easy to see, again from (8.73), that the output on the right is just $Y(z) = X(z)/B_L(z)$. The all-pole version is just a special case of the general lattice structure. Note, as for Figure 8.11, that the structure in Figure 8.13 is stable if $|\kappa_l| < 1$ for all l. The second special case is an "all-zero" version, obtained via a conversion of the all-pole lattice. Let us rewrite (8.77) and (8.78) as follows:

$$B_l(z) = B_{l-1}(z) + \kappa_{l-1} z^{-1}[zC_{l-1}(z)] \tag{8.90}$$

$$zC_l(z) = \kappa_{l-1} B_{l-1}(z) + z^{-1}[zC_{l-1}(z)] \tag{8.91}$$

This "reversion" of Figure 8.10 is shown in Figure 8.14 with $X(z)$ used as the input signal transform. Assembling these elements into a lattice, we have Figure 8.15 as an example, again with $L = 3$. On the left end of the lattice we have, again from (8.73),

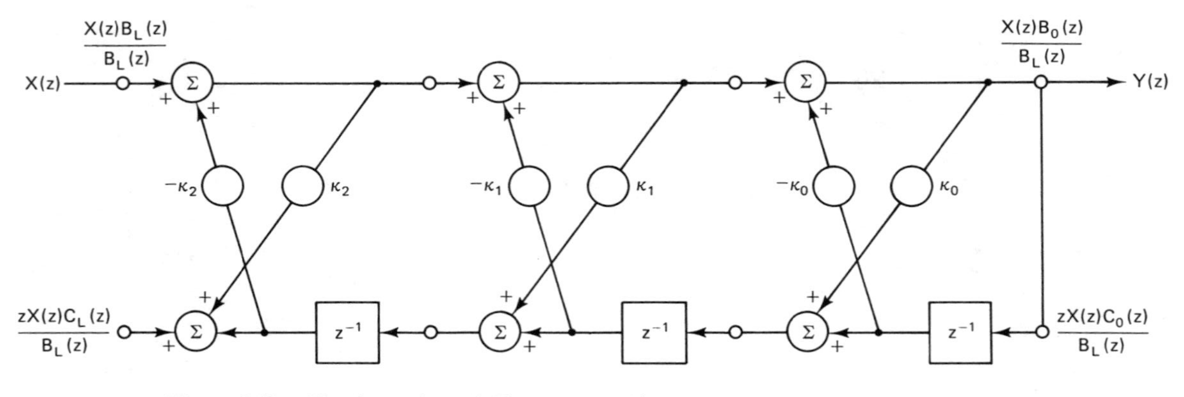

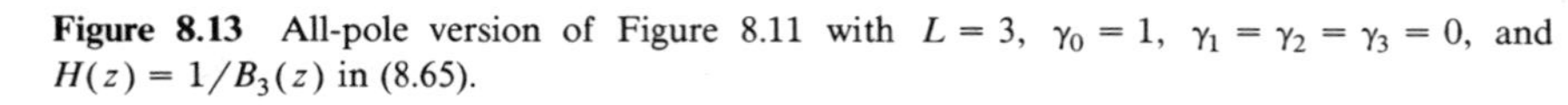

Figure 8.13 All-pole version of Figure 8.11 with $L = 3$, $\gamma_0 = 1$, $\gamma_1 = \gamma_2 = \gamma_3 = 0$, and $H(z) = 1/B_3(z)$ in (8.65).

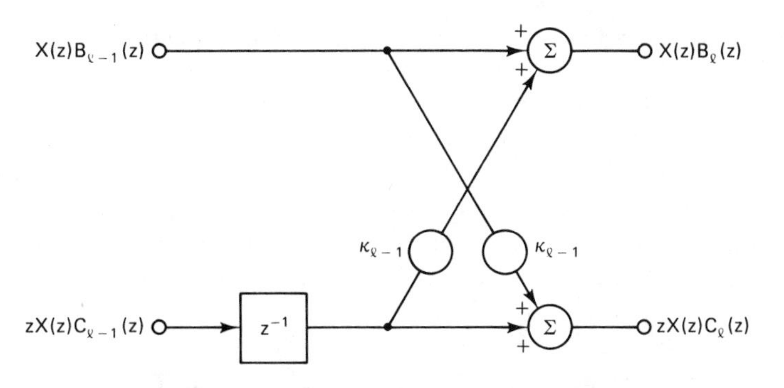

Figure 8.14 Reversion of the two-multiplier lattice element in Figure 8.10.

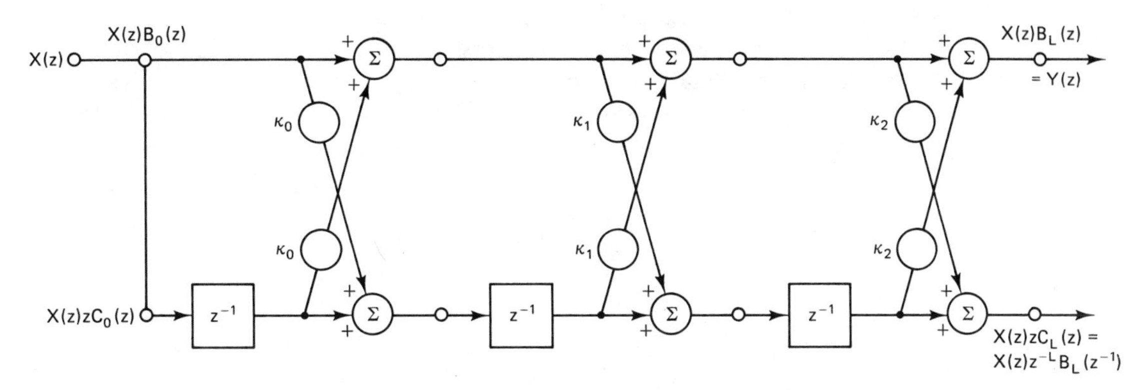

Figure 8.15 All-zero reversion of Figure 8.13 using the lattice element in Figure 8.14. $L = 3$, $H_f(z) = B_3(z)$, and $H_b(z) = z^{-3}B_3(z^{-1})$.

$X(z)$ at both input nodes, and on the upper right we have the desired all-zero output,†

$$Y(z) = B_L(z)X(z)$$

$$H_f(z) = \sum_{l=0}^{L} b_l z^{-l}; \qquad b_0 = 1 \tag{8.92}$$

Thus we have in Figure 8.15 the lattice version of the transversal filter in Chapter 2 or the nonrecursive filter in Chapter 7, with the initial weight equal to 1. Notice also that at the lower right in Figure 8.15 we obtain a secondary output with the associated transfer function

$$H_b(z) = z^{-L}B_L(z^{-1})$$

$$= \sum_{l=0}^{L} b_{L-l} z^{-l}; \qquad b_0 = 1 \tag{8.93}$$

† For simplicity, we write b_{Ll} simply as b_l.

With the transfer functions in (8.92) and (8.93), the lattice in Figure 8.15 can be seen to function as a *one-step predictor*. Consider the equivalent one-step predictor diagrams in Figure 8.16. In part (a) we have the type of system first illustrated in Figure 1.4, with the prediction delay, M, equal to 1 and the prediction filter an adaptive linear combiner with weights $[-b_l]$. Note that since l begins at 1, we have a one-step predictor in Figure 8.16(a). As shown by the (equivalent) second diagram in part (a) the upper transfer function is just $H_f(z)$ in (8.92), so the output in Figure 8.16(a) is called the *forward prediction error*, $E_f(z)$. The current sample, x_k, is being predicted in terms of the first L previous samples, x_{k-1} through x_{k-L}.

In Figure 8.16(b) we have, by the same argument, the *backward prediction error*. Here we have x_{k-L} being "predicted" by x_k through x_{k-L+1}. The weights in this case are $[-b_{L-l}]$ as shown. The bottom diagram in Figure 8.16(b) represents (8.93). To summarize, Figure 8.17 shows the complete case with $L = 3$, including the forward and backward prediction errors, ε_{fk} and ε_{bk}. The weights, κ_0 through κ_L, are variable in Figure 8.17 so that the predictor may be adaptive. The intermediate forward and backward signals are labeled s_{lk} and s'_{lk}, respectively, for use

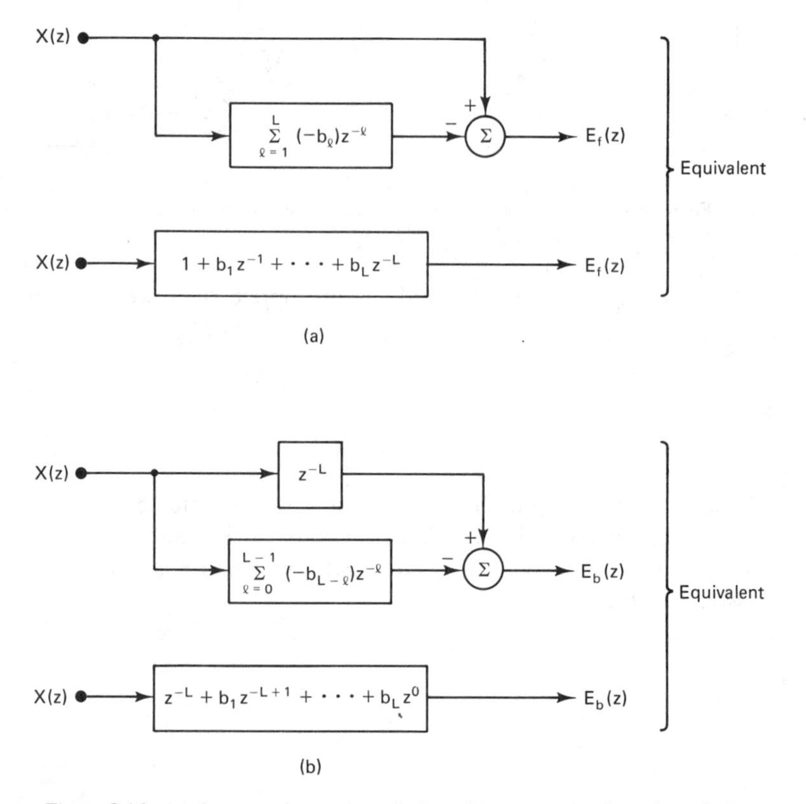

Figure 8.16 (a) One-step forward prediction; (b) one-step backward prediction.

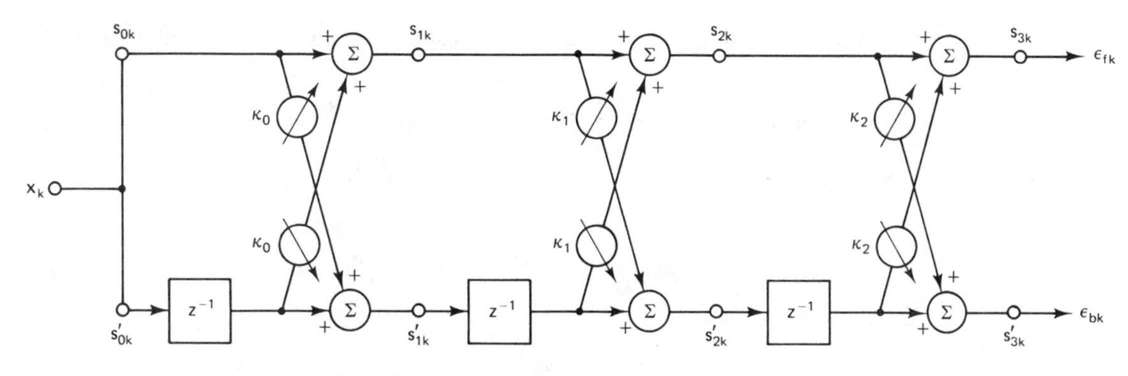

Figure 8.17 One-step adaptive lattice predictor, $L = 3$.

in the next section. Thus we have

$$s_{0k} = s'_{0k} = x_k$$

$$s_{l+1,k} = s_{lk} + \kappa_l s'_{l,k-1} \qquad 0 \le l \le L-1$$

$$s'_{l+1,k} = \kappa_l s_{lk} + s'_{l,k-1} \qquad 0 \le l \le L-1 \tag{8.94}$$

$$\varepsilon_{fk} = s_{Lk}$$

$$\varepsilon_{bk} = s'_{Lk}$$

THE ADAPTIVE LATTICE PREDICTOR

In the one-step predictor in Figure 8.17 we have arrived at a lattice structure whose coefficients, $[\kappa_l]$, are to be made time varying, or adaptive. The other lattice structures introduced in Figures 8.11 and 8.13 could be (and in fact have been) made adaptive [26], but the main applications of adaptive lattices involve the use of the adaptive lattice predictor, typically in speech-processing situations.

To adapt the lattice in Figure 8.17, we would think of adjusting all of the κ's to minimize the expected squared forward prediction error $E[\varepsilon_f^2]$. Instead, however, the best method [28, 29] involves adjusting each κ_l to minimize $E[s_{l+1}^2]$, the expected squared error at *each stage*, with $E[\varepsilon_f^2] = E[s_L^2]$ as in (8.94). We will derive a simple LMS type of algorithm for the adjustment of each κ_l, but first let us examine the implications of adjusting the κ's in this manner.

In Figure 8.17 and in (8.94) we have the forward and backward prediction errors at each stage given by s_{lk} and s'_{lk}. To denote the various mean-square values of these errors, let us adopt a correlation notation similar to that in Chapter 7:

$$\phi_l(n) \triangleq E[s_{lk}s_{l,k+n}]$$

$$\phi'_l(n) \triangleq E\left[s'_{lk}s_{l,k+n}\right] \tag{8.95}$$

$$\phi''_l(n) \triangleq E\left[s'_{lk}s'_{l,k+n}\right]$$

With these definitions we assume that the signals in the lattice are stationary.

Let us first examine $\phi_l(n)$ and $\phi_l''(n)$. Taking the mean-square value of (8.94), we have

$$\begin{aligned}\phi_{l+1}(n) &= E\left[\left(s_{lk} + \kappa_l s'_{l,k-1}\right)\left(s_{l,k+n} + \kappa_l s'_{l,k+n-1}\right)\right] \\ &= \kappa_l^2 \phi_l''(n) + \kappa_l\left[\phi_l'(1-n) + \phi_l'(1+n)\right] + \phi_l(n) \end{aligned} \tag{8.96}$$

$$\begin{aligned}\phi''_{l+1}(n) &= E\left[\left(\kappa_l s_{lk} + s'_{l,k-1}\right)\left(\kappa_l s_{l,k+n} + s'_{l,k+n-1}\right)\right] \\ &= \kappa_l^2 \phi_l(n) + \kappa_l\left[\phi_l'(1-n) + \phi_l'(1+n)\right] + \phi_l''(n) \end{aligned} \tag{8.97}$$

At the left end of the lattice we have

$$\phi_0(n) = \phi_0''(n) = E[x_k x_{k+n}] \tag{8.98}$$

Using (8.98) in (8.96) and (8.97), we can show that $\phi_1(n) = \phi_1''(n)$, then $\phi_2(n) = \phi_2''(n)$, and so on, so that we have

$$\begin{aligned}\phi_{l+1}(n) &= \phi''_{l+1}(n) \\ &= \left(\kappa_l^2 + 1\right)\phi_l(n) + \kappa_l\left[\phi_l'(1-n) + \phi_l'(1+n)\right] \end{aligned} \tag{8.99}$$

If κ_l is adjusted to minimize the forward prediction error, $E[s_{l+1,k}^2] = \phi_{l+1}(0)$, at each stage, we have, from (8.99),

$$\frac{\partial \phi_{l+1}(0)}{\partial \kappa_l} = 2\kappa_l \phi_l(0) + 2\phi_l'(1) = 0$$

or

$$\kappa_l^* = -\frac{\phi_l'(1)}{\phi_l(0)} \tag{8.100}$$

The asterisk is used here to denote the optimum value of κ_l. To prove that κ_l^* is the true optimum in the sense that not only $\phi_{l+1}(0)$ but also the final mean-square prediction error, $\phi_L(0)$, is minimized with respect to κ_l, we must return to the recursion formula (8.76) and find the optimum $B_l(z)$ at each stage. This involves essentially a derivation of Levinson's algorithm [30] for finding optimum filter weights, and Tretter [31] presents an excellent derivation of the algorithm for the type of situation that we have here. Rather than repeat the derivation, let us consider a two-stage lattice with just two parameters, κ_0 and κ_1.

For the two-stage lattice, (8.99) gives the mean-square output error as

$$\phi_2(0) = \left(\kappa_1^2 + 1\right)\phi_1(0) + 2\kappa_1\phi_1'(1) \tag{8.101}$$

In this result we could again express ϕ_1 using (8.99) and ϕ_1' from the following general relationship, which follows in turn from (8.95) and (8.99):

$$\begin{aligned}\phi'_{l+1}(n) &= E\left[\left(\kappa_l s_{lk} + s'_{l,k-1}\right)\left(s_{l,k+n} + \kappa_l s'_{l,k+n-1}\right)\right] \\ &= \kappa_l^2 \phi_l'(1-n) + 2\kappa_l\phi_l(n) + \phi_l'(n+1) \end{aligned} \tag{8.102}$$

In both cases, $\phi_1(0)$ and $\phi_1'(1)$ are not functions of κ_1, so we can minimize $\phi_2(0)$ with respect to κ_1 using (8.100). With $\kappa_1 = \kappa_1^* = -\phi_1'(1)/\phi_1(0)$, we have

$$[\phi_2(0)]_{\min} = \left(\frac{\phi_1'^2(1)}{\phi_1^2(0)} + 1\right)\phi_1(0) - 2\frac{\phi_1'^2(1)}{\phi_1(0)}$$

$$= \phi_1(0) - \frac{\phi_1'^2(1)}{\phi_1(0)} \tag{8.103}$$

Now we must find the value of κ_0 that minimizes this result. If κ_0 is found in accordance with (8.100), then $\phi_1(0)$ is minimized in (8.103) and its derivative with respect to κ_0 must be zero. Thus, for $[\phi_2(0)]_{\min}$ to be minimized with respect to κ_0 in this manner, the derivative of the remaining term in (8.103) must also equal zero. (Since the maximum is unbounded, this must lead to a minimum.) Therefore, from (8.102),

$$\frac{\partial \phi_1'(1)}{\partial \kappa_0} = 2\kappa_0\phi_0'(0) + 2\phi_0(1) = 0$$

$$\kappa_0 = -\frac{\phi_0(1)}{\phi_0'(0)} = -\frac{E[x_k x_{k+1}]}{E[x_k^2]} = -\frac{\phi_0'(1)}{\phi_0(0)} = \kappa_0^* \tag{8.104}$$

Thus we have shown that κ_0^* in (8.100) is the true optimum weight in the sense that it minimizes $\phi_2(0)$ as well as $\phi_1(0)$. We could have obtained the same result by finding the optimum adaptive linear combiner, then using Table 8.2 to convert from b's to κ's, as in Exercises 24 through 26.

Notice in the previous example that the global minimum is achieved only after both κ_0 and κ_1 are adjusted to their optimum values. Thus, in an adaptive lattice, we would think of convergence occurring approximately stage by stage, with κ_l being adjusted to minimize $\phi_{l+1}(0)$ first for $l = 0$, then for $l = 1$, and so on. Convergence does not occur in this manner in the adaptive linear combiner.

We can use the result in (8.100) to develop an LMS algorithm for the adaptive lattice predictor. Let us estimate the gradient of the mean-square error, $\phi_{l+1}(0)$, using as before the gradient of the square error itself:

$$\frac{\partial \widehat{\phi_{l+1}(0)}}{\partial \kappa_l} = \frac{\partial s_{l+1,k}^2}{\partial \kappa_l}$$

$$= 2 s_{l+1,k} \frac{\partial s_{l+1,k}}{\partial \kappa_l}$$

$$= 2 s_{l+1,k} s'_{l,k-1} \tag{8.105}$$

The final result is obtained by differentiating $s_{l+1,k}$ in (8.94). Then, just as in (6.3), we use the gradient estimate in the steepest-descent algorithm (4.36) to obtain the lattice LMS algorithm:

Lattice LMS

$$\begin{aligned}\kappa_{l,k+1} &= \kappa_{lk} - \mu_l \frac{\partial \widehat{\phi_{l+1}(0)}}{\partial \kappa_l} \\ &= \kappa_{lk} - 2\mu_l s_{l+1,k} s'_{l,k-1} \qquad 0 \le l \le L-1\end{aligned} \tag{8.106}$$

The signal values are of course computed in accordance with (8.94). The time-invariant convergence parameter, μ_l, should be allowed to differ from stage to stage according to Griffiths [28], who presented essentially this same algorithm.

Before presenting an application of (8.106) let us discuss briefly the range of μ_l at each stage of the lattice. Assuming a perfect gradient estimate, we use the derivative of (8.99) in (8.106) to obtain

$$\kappa_{l,k+1} = \kappa_{lk} - \mu_l\left[2\kappa_{lk}\phi_l(0) + 2\phi'_l(1)\right] \tag{8.107}$$

Now let us define a translated weight, δ_l, that bears the same relation to κ_l that **V** had to **W** in Chapter 3 [Equation (3.29)] and following chapters:

$$\delta_{lk} = \kappa_{lk} - \kappa_l^* \tag{8.108}$$

Using this definition and also (8.100) in (8.107), we have

$$\begin{aligned}\delta_{l,k+1} &= \delta_{lk} - 2\mu_l\left[\left(\delta_{lk} + \kappa_l^*\right)\phi_l(0) + \phi'_l(1)\right] \\ &= \delta_{lk} - 2\mu_l\left[\left(\delta_{lk} - \frac{\phi'_l(1)}{\phi_l(0)}\right)\phi_l(0) + \phi'_l(1)\right] \\ &= \delta_{lk}\left[1 - 2\mu_l\phi_l(0)\right] \\ &= \left[1 - 2\mu_l\phi_l(0)\right]^{k+1}\delta_{l0}\end{aligned} \tag{8.109}$$

Since δ_{lk} must converge to zero, we conclude that for convergence we must have

$$0 < \mu_l < \frac{1}{2\phi_l(0)} \tag{8.110}$$

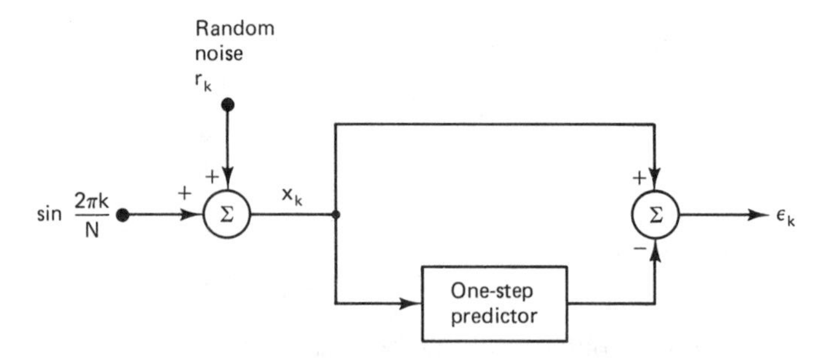

Figure 8.18 One-step predictor used as an example.

In a practical adaptive lattice we could estimate $\phi_l(0)$ by averaging s_{lk}^2 over its recent history, and thus use (8.110) to assure that μ_l is kept within its limits. As with the adaptive linear combiner, these convergence parameters affect both the learning curve and the excess mean-square error, or misadjustment.

An example of the performance of an adaptive lattice predictor is given in Figures 8.18 through 8.21. The predictor configuration is shown in Figure 8.18, and we see that the signal being predicted is the same sine wave plus noise that was used in previous examples. The predictor tries to predict the signal, x_k, one time step ahead, and thus we would expect it to eliminate the sine wave from x_k and leave the unpredictable white-noise component, r_k. We would then expect the output power, $E[\varepsilon_k^2]$, to equal the noise power, $E[r_k^2]$.

The autocorrelation matrix of x_k was derived previously in (2.20) and (6.13), and is

$$\mathbf{R} = \begin{bmatrix} \rho_0 & \rho_1 & \rho_2 \\ \rho_1 & \rho_0 & \rho_1 \\ \rho_2 & \rho_1 & \rho_0 \end{bmatrix} \qquad \begin{aligned} \rho_0 &= 0.5 + E\left[r_k^2\right] \\ \rho_1 &= 0.5\cos\frac{2\pi}{N} \\ \rho_2 &= 0.5\cos\frac{4\pi}{N} \end{aligned} \tag{8.111}$$

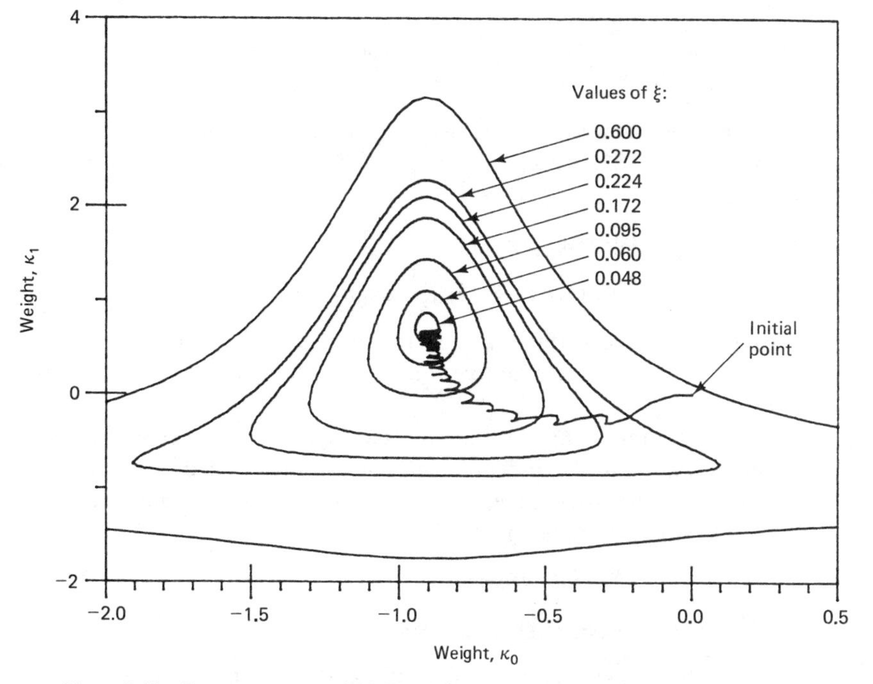

Figure 8.19 Error contours and LMS weight track for the two-stage adaptive lattice predictor used in Figure 8.18. Parameters are $N = 16$, $E[r_k^2] = 0.01$, $\mu_0 = 0.05$, $\mu_1 = 0.1$. Weight track consists of 200 iterations.

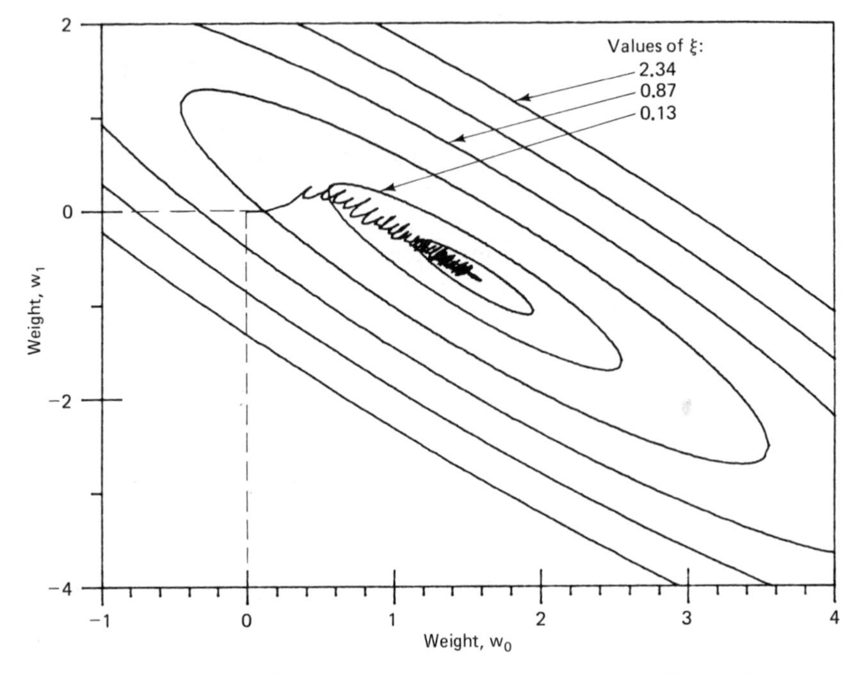

Figure 8.20 Same as Figure 8.19, but with an adaptive transversal filter in place of the adaptive lattice. Weight track consists of 300 iterations, and $\mu = 0.1$. Initial weights at $(0, 0)$.

For this example we will use a two-stage lattice with weights κ_0 and κ_1. From (8.98), (8.99), and (8.102) we can develop the following expression for the performance surface:

$$\begin{aligned}\xi = \phi_2(0) &= \left(\kappa_1^2 + 1\right)\phi_1(0) + 2\kappa_1\phi_1'(1) \\ &= \left(\kappa_1^2 + 1\right)\left[\left(\kappa_0^2 + 1\right)\rho_0 + 2\kappa_0\rho_1\right] + 2\kappa_1\left[\kappa_0^2\rho_0 + 2\kappa_0\rho_1 + \rho_2\right] \end{aligned} \tag{8.112}$$

This performance surface is seen to be a quadratic function of each weight, but unlike the case for the adaptive transversal filter, the contours of constant ξ are not elliptical.

The contours of (8.112) are plotted together with a typical convergence track in Figure 8.19. Note the independence of κ_0 discussed previously. For any value of κ_1, κ_0 can be adjusted independently to its optimum value, which is given below. The following parameters were used:

$$\begin{aligned} &\textit{Signal parameters:} && N = 16; \quad E\left[r_k^2\right] = 0.01 \\ &\textit{Convergence parameters:} && \mu_0 = 0.05, \quad \mu_1 = 0.1 \end{aligned} \tag{8.113}$$

With $\phi_0(0) = \rho_0 = 0.51$, and, of course, $\phi_1(0)$ less than this value as long as κ_0 is less than 1, these values of μ are well within the limits of (8.110). As with the adaptive linear combiner, it is typical to use values of μ on the order of 0.1 times the upper limit in (8.110). Notice that the track in Figure 8.19 follows approximately the path of steepest descent, as we would expect with the LMS algorithm in (8.106). The optimum weight values for this example are found from (8.99), (8.100), and (8.102) to be

$$\kappa_0^* = -\frac{\phi_0'(1)}{\phi_0(0)} = -\frac{\rho_1}{\rho_0} = -0.906$$

$$\kappa_1^* = -\frac{\phi_1'(1)}{\phi_1(0)} = \frac{\rho_1^2 - \rho_2\rho_0}{\rho_0^2 - \rho_1^2} = 0.708 \tag{8.114}$$

In less than 200 iterations, the weight track in Figure 8.10 reached the vicinity of these optimum values.

For comparison with Figure 8.19, Figure 8.20 shows the same adaptive process using the adaptive transversal filter instead of the adaptive lattice. Here we have the familiar elliptical error contours. The bowl is quite flat in the vicinity of $\xi_{\min}$, however, and with $\mu = 0.1$, about 300 iterations were required to reach the vicinity of $\xi_{\min}$. The advantage of having a different convergence parameter at each stage, computed as suggested after (8.110), generally allows the lattice predictor to converge faster than the ordinary LMS algorithm [34]. (The same random sequence was used in Figures 8.19 and 8.20.)

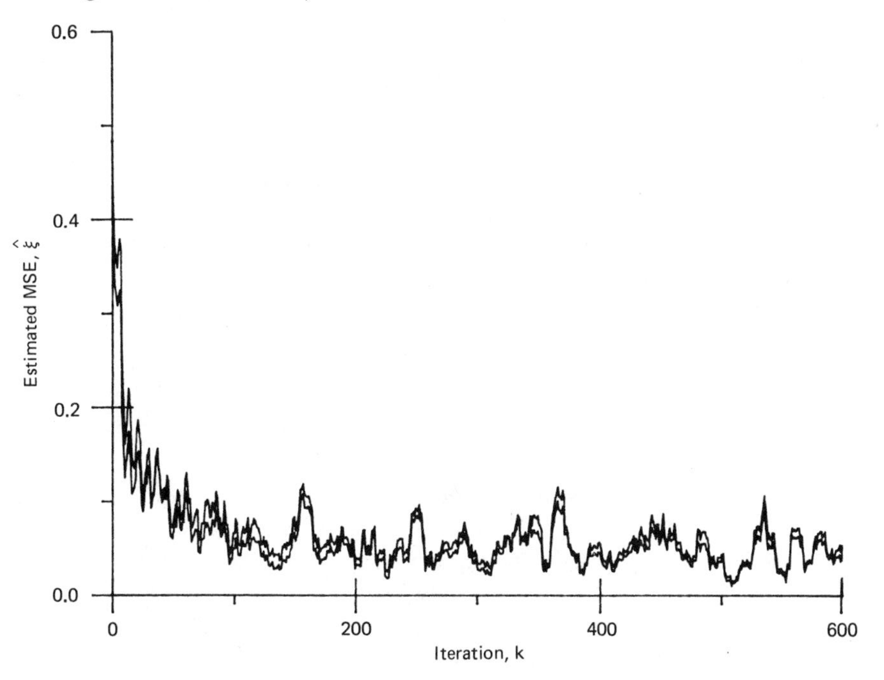

Figure 8.21 Learning curves for the LMS adaptive transversal predictor and for the adaptive lattice predictor. Curves are nearly identical.

Learning curves for the two types of predictors are compared in Figure 8.21, using the same parameters as in the previous two figures. The mean-square error was estimated by averaging each ε_k^2 with its 10 nearest neighbors. With the parameters μ_0, μ_1, and μ set in this manner, the two learning curves are nearly identical for the same random sequence. Note that the steady-state value of ξ appears to be around 0.05. Since $E[r_k^2] = 0.01$ and since the sine wave is predicted, we have $\xi_{\min} = 0.01$ and the misadjustment in this case must be about $M = 4$.

In a manner similar to that used above with the adaptive linear combiner, we can also develop a sequential regression algorithm (i.e., an approximation to Newton's method) for the adaptive lattice. From (8.99) the output mean-square error for the lth lattice stage is

$$\phi_{l+1}(0) = \left(\kappa_l^2 + 1\right)\phi_l(0) + 2\kappa_l\phi_l'(1) \tag{8.115}$$

We take the derivative of (8.115) and designate the result as the gradient with respect to κ_l:

$$\frac{\partial\phi_{l+1}(0)}{\partial\kappa_l} = 2\kappa_l\phi_l(0) + 2\phi_l'(1) \tag{8.116}$$

Next we solve (8.116) for $\phi_l'(1)$ and place the result into (8.100) to obtain

$$\begin{aligned} \kappa_l^* &= -\frac{1}{\phi_l(0)}\left[-\kappa_l\phi_l(0) + \frac{1}{2}\frac{\partial\phi_{l+1}(0)}{\partial\kappa_l}\right] \\ &= \kappa_l - \frac{1}{2}\frac{1}{\phi_l(0)}\frac{\partial\phi_{l+1}(0)}{\partial\kappa_l} \end{aligned} \tag{8.117}$$

Here we have a one-step Newton formula similar to (4.30) in Chapter 4, so we multiply the second term by μ as in (8.1). If we do this and also replace the gradient with its estimate in (8.105), we obtain an SER algorithm comparable to that obtained above:

$$\begin{aligned} \kappa_{l,k+1} &= \kappa_{lk} - \frac{\mu}{2}\frac{1}{\phi_l(0)}\frac{\partial\phi_{l+1}(0)}{\partial\kappa_l} \\ &= \kappa_{lk} - \frac{\mu}{\phi_l(0)}s_{l+1,k}s'_{l,k-1} \end{aligned} \tag{8.118}$$

Just as we had to estimate $\mathbf{R}$ for the adaptive linear combiner, so here we must estimate $\phi_l(0) = E[s_{lk}^2]$. For the nonstationary, adaptive environment we give the past history of s_{lk}^2 a decreasing emphasis as we did in (8.27), so that the estimate becomes

$$P_{lk} \triangleq \hat{\phi}_{lk}(0) = \frac{1-\alpha}{1-\alpha^{k+1}}\sum_{i=0}^{k}\alpha^{k-i}s_{li}^2 \tag{8.119}$$

By using (8.119) recursively we obtain the following recursion formula:

$$P_{lk} = \frac{1}{1-\alpha^{k+1}}\left[(1-\alpha)s_{lk}^2 + \alpha(1-\alpha^k)P_{l,k-1}\right] \tag{8.120}$$

Assuming zero initial conditions and α chosen as in (8.44), the complete SER algorithm thus becomes

Lattice SER (8.121)

Initially: $\alpha \approx 2^{-1/(\text{length of signal stationarity})}$

$0 < \mu < 1.0$

P_{l0} = estimated signal power; $0 \le l < L$

$\kappa_{l0} = 0; \quad 0 \le l < L$

For $0 \le l < L$:

$$P_{lk} = \frac{1}{1-\alpha^{k+1}}\left[(1-\alpha)s_{lk}^2 + \alpha(1-\alpha^k)P_{l,k-1}\right] \qquad k > 0$$

$$\kappa_{l,k+1} = \kappa_{lk} - \frac{\mu}{P_{lk}} s_{l+1,k}s'_{l,k-1} \qquad k \ge 0$$

As for the LMS algorithm in (8.106), the signal values are assumed to be computed in accordance with (8.94). In fact, observing (8.106) and (8.121) together, we see that the LMS and SER algorithms are essentially the same for the lattice, the only difference being that $\phi_l(0)$ is estimated at each step in (8.121). An application of the SER algorithm to the preceding example is shown in Figure 8.22, and we note that the weight track is improved slightly over the track in Figure 8.19.

Thus in this section we have seen how the adaptive lattice serves as a predictor. One of its main applications is in speech compression, where the speech is predicted by an adaptive lattice of length sufficient to produce a near-white output [35]. The lattice coefficient values are recorded at a relatively low sampling rate. To reconstruct the speech, a time-varying all-pole lattice is then formed with these coefficients as in Figure 8.13 and driven with locally generated noise sequences. The reconstructed speech is taken from the output of the all-pole lattice. The major advantage of the lattice in this application is that (as mentioned above) stability is guaranteed; that is, the zeros of the predictor and the poles of the all-pole inverse are all inside $|z| = 1$ provided that $|\kappa_l| < 1$ for $0 \le l < L$.

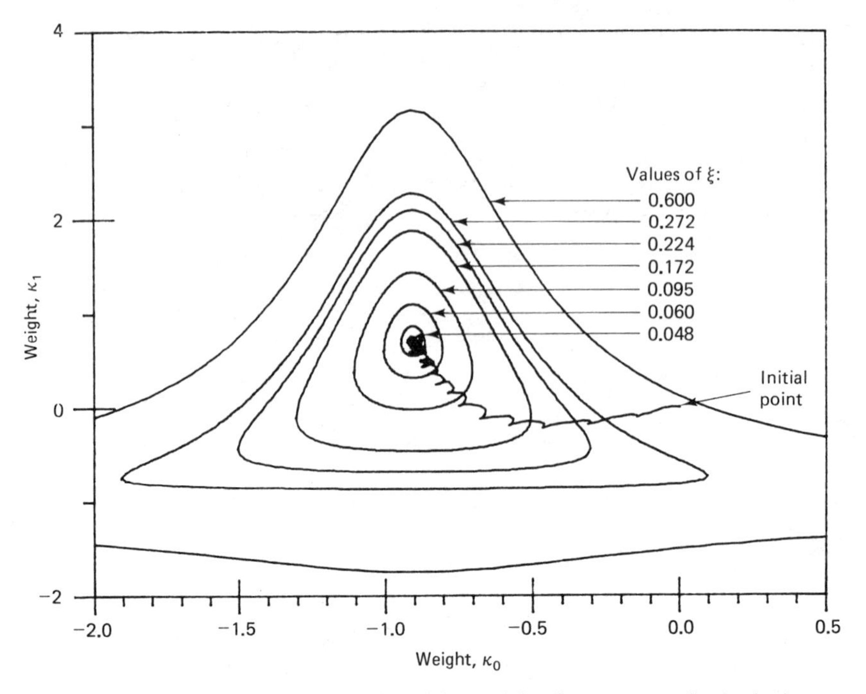

Figure 8.22 Error contours and SER weight track for the two-stage adaptive lattice predictor used in Figure 8.18. Parameters are $N = 16$, $E[r_k^2] = 0.01$, $\mu = 0.02$, $\alpha = 0.9$. Weight track consists of 200 iterations.

ADAPTIVE FILTERS WITH ORTHOGONAL SIGNALS

In the preceding section we saw how the weights of the adaptive lattice predictor could be adjusted independently, using local signals as in (8.106), to minimize the final prediction error. On account of this property, the signals within the lattice are mutually orthogonal. The error signal at each stage is uncorrelated with the other error signals. In this section we will discuss similar types of adaptive filters with orthogonal signals.

Our motivation for considering such adaptive structures is to retain the simplicity of the LMS algorithm and yet incorporate some of the benefits of the more complex algorithms such as the LMS-Newton and SER algorithms discussed previously. Of all algorithms, the LMS algorithm involves the least computation per iteration cycle and the least use of memory; moreover, it is the easiest to implement in software and the simplest to debug and to understand. We saw in Chapter 6 how it is used to adjust the weights of a transversal filter with the goal of minimizing the mean squared error. In cases of high eigenvalue disparity, however, we have seen that other algorithms can often achieve faster convergence than the LMS algorithm and should be used accordingly. For a given level of misadjustment, the use of Newton's method or of some form of orthogonalization of the inputs to the adaptive weights can result in more rapid adaptation than is possible with LMS alone when there is a large spread in the eigenvalues. The benefits of orthogonalization are explained in the literature [36].

Two orthogonalizing methods for preprocessing the signals before final weighting and summing will be described here. One of these methods is based on the use of lattice filters, and the other uses the discrete Fourier transform. Other methods using "Gram-Schmidt" orthogonalization have been applied to adaptive antennas by Brennan and Reed, whose work has been included in the text by Monzingo and Miller [37].

The first orthogonalization scheme to be considered is based on the discrete Fourier transform (DFT) and was devised by Narayan and Peterson [38]. Narayan's adaptive filter is shown in Figure 8.23. The input signal is applied to a series of delays with taps connecting to a multi-input, multi-output implementation of the DFT, which was described in Chapter 7. With each new input sample, the data slides one step down the delay line and a new DFT is computed. Each of the DFT outputs is associated with a specific frequency band. The DFT used in this manner can be considered as a means of implementing a bank of band-pass filters uniformly spaced in frequency between zero and half the sampling frequency.

The DFT outputs in Figure 8.23 are complex discrete functions of the sampling index, k. They are approximately uncorrelated with each other, being in different frequency bands. They are not perfectly uncorrelated because the DFT band-pass filters overlap somewhat [6], causing "leakage" of signal components from one band to another.

These complex output signals of the DFT are weighted in Figure 8.23 with complex adaptive weights to produce y_k, which is also complex. The real desired response is treated as a complex signal with a zero imaginary part. The error signal used in the adaptive process is therefore also complex. Although y_k is complex, its imaginary part will generally be small since the desired response is real. The weights are adapted in accordance with the "complex LMS algorithm" of Widrow, McCool, and Ball [39]:

$$\mathbf{W}_{k+1} = \mathbf{W}_k + 2\mu\varepsilon_k\overline{\mathbf{X}}_k.$$

The bar over $\mathbf{X}_k$ represents the complex conjugate. This algorithm causes the complex weight vector $\mathbf{W}_k$ to converge in the mean to a complex least-squares solution which minimizes the sum of the mean square of the real part of the error plus the mean square of the imaginary part of the error.

For example, suppose the input signal, x_k in Figure 8.23, is a pure sinusoidal signal with exactly N samples per cycle. Then the only nonzero DFT output is the input to w_{1k}, which is a complex sinusoidal signal with fixed amplitude and linearly changing phase. Suppose further that the desired output, d_k, is sinusoidal and has the same frequency as the input signal. Then, to drive ε_k to zero, w_{1k} will converge to a constant complex value that creates a real output signal, y_k, with amplitude and phase equal to the amplitude and phase of d_k, and the other weights will converge to zero. The reader may wish at this point to consider other examples of converged

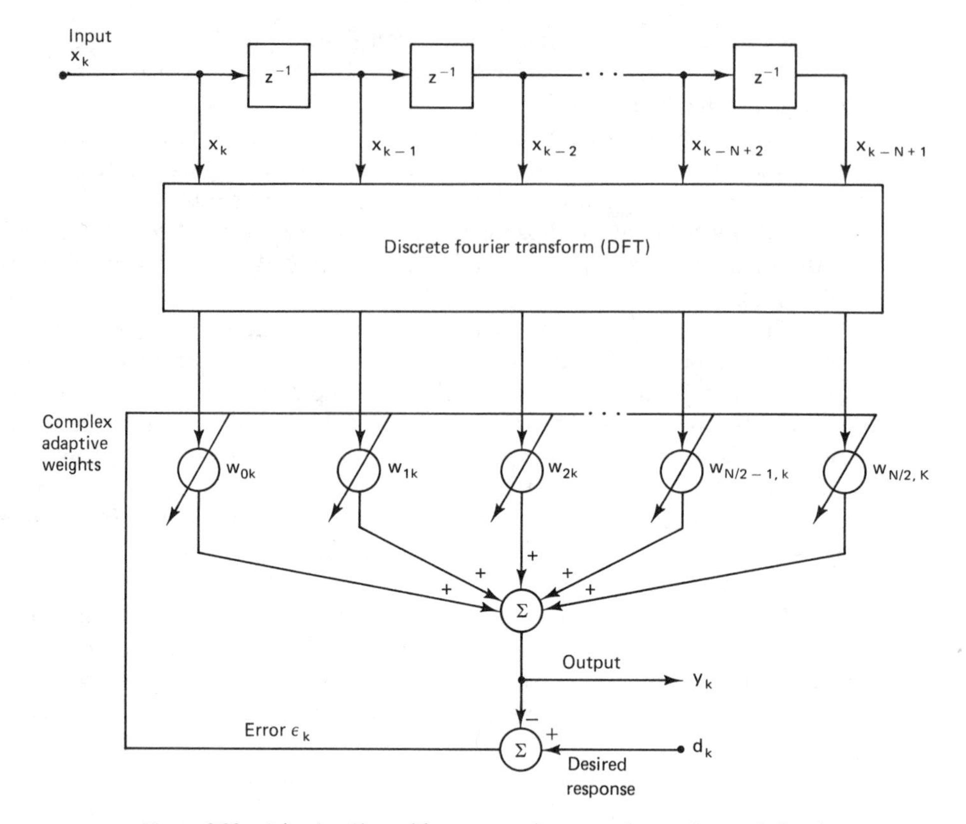

Figure 8.23 Adaptive filter with preprocessing to produce orthogonal signals.

weight vectors that drive ε_k to zero, in order to understand better how the system in Figure 8.23 works.

The adaptive process of Figure 8.23 can be improved by normalizing each of the DFT outputs to equal power levels. Coefficients having values that are inversely proportional to the square roots of the respective DFT average output powers could be incorporated into the signal paths. These power levels could be obtained by exponential time averaging, by uniform time averaging, or by some other weighting formula. The result is a practical algorithm that is highly effective with large eigenvalue disparity.

When using a conventional transversal adaptive filter, the tap weights directly determine the filter impulse response. On the other hand, when using Narayan's frequency-domain orthogonalization scheme of Figure 8.23, the weights directly determine the frequency response of the filter, both in magnitude and in phase.

The orthogonalization scheme of Figure 8.23 uses a fixed preprocessor consisting of the discrete Fourier transform. The orthogonalization scheme of Figure 8.24 uses a lattice filter which is adaptive. It is interesting to note that the adaptation of the weights of the lattice (the κ's) depends only on the input signal, and not on the

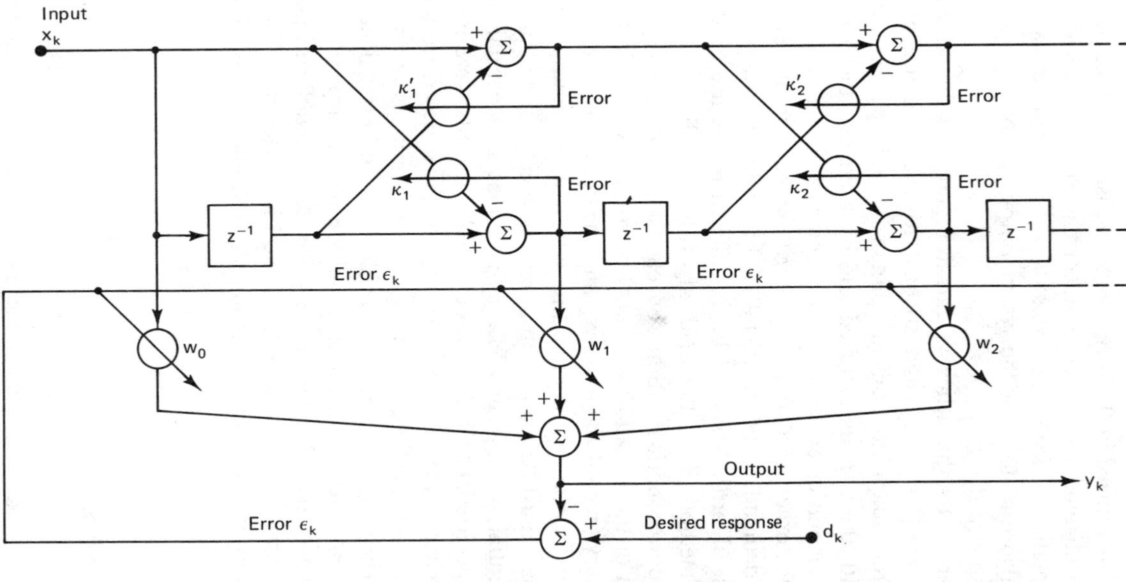

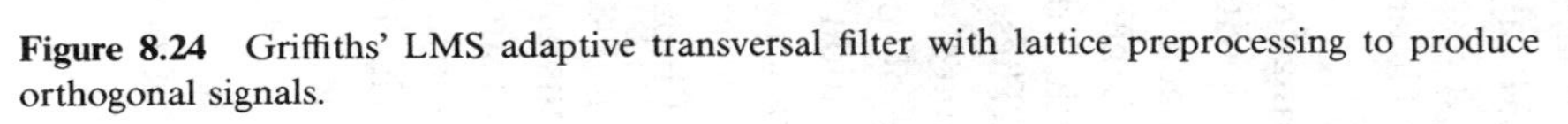

Figure 8.24 Griffiths' LMS adaptive transversal filter with lattice preprocessing to produce orthogonal signals.

desired response, but the adaptation of the output weights (the w's) depends on both the input signal and the desired response. All weights may be adapted by the LMS algorithm, as suggested by Griffiths [23].

When the lattice weights (the κ's) converge so that their corresponding error signals are minimized in the mean square sense, the lattice outputs (the inputs to the w's) are orthogonal and therefore uncorrelated, as discussed above.

Adaptation of the output weights (the w's) can be readily accomplished using the LMS algorithm to cause the system output y_k to be a best least-squares fit to the desired response d_k.

Theoretically, the lattice weights in Figure 8.24 should match as in Figure 8.17, that is, $\kappa_1 = \kappa_1'$, $\kappa_2 = \kappa_2'$, and so on. However, because of noise in the weights resulting from adaptation, this matching will not be perfect. By starting the adaptive process with the weights matched and by averaging the corresponding corrections with each adaptation cycle, perfect matching can be sustained.

Other "exact least-squares" algorithms have been devised by Morf and Lee, Kailath, Friedlander, and others [40–46] for adjusting the κ's and the w's in Figure 8.24. These algorithms should be as efficient in the use of statistical input data as Newton's method. With the LMS algorithm, Griffiths has emphasized the importance of power normalization (or choosing appropriate values of μ for LMS adaptation of the κ's and the w's). Because the signals in the lattice structure are nonstationary (even if the input x_k and the desired response d_k are stationary) due to the weight variations of preceding lattice stages, power estimates of the various weight inputs must be made with a moving average technique such as the one using an exponential window in (8.121). A similar exponential window was used by Griffiths and Medaugh [47]. Various time constants were used to produce performances similar to the exact least-squares methods described above.

Orthogonalizing algorithms and their applications are a subject of current research. They appear to be very useful when rapid transient adaptation is required and the eigenvalues of the input signal are highly disparate. Many situations exist, however, where the adaptive lattice provides no significant performance advantage over the transversal filter adapted by the LMS algorithm; for example, with certain types of nonstationary inputs such as those discussed by Widrow and Walach (see ref. 36).

EXERCISES

1. What two types of information must be known in order to have the perfect weight track in Figure 8.1? Why is this information not available for adaptive filters?
2. Explain the difference between the "Newton" weight track in Figure 8.1 and the "LMS/Newton" weight track in Figure 8.2.
3. What is the ratio of adaption time of the LMS/Newton algorithm to that of the LMS algorithm? When is this ratio equal to 1?

4. For the sequential regression algorithm, develop a formula for computing $\mathbf{Q}_k$ in terms of $\mathbf{Q}_{k-1}$. What is the equivalent transfer function, $H(z)$, from $\mathbf{X}_k \mathbf{X}_k^T$ to $\mathbf{Q}_k$?

Note: In some of the remaining exercises we will need the sequences $[r_k]$, $[s_k]$, and $[x_k]$. The sequences should go from $k = 0$ to $k = 1000$; however, they may be shortened as required to accommodate small computers. Generate the sequences as follows:

$$r_k = \sqrt{12}\,[\text{RANDOM}(1.) - 0.5]$$

$$s_k = \sin\left[\frac{2\pi k}{30 - k/50}\right]$$

$$x_k = \sum_{n=0}^{\infty} e^{(40n-k)/4} u(k - 40n)$$

The sequence $[r_k]$ is random white noise with unit power, generated using the RANDOM routine, version 1 or 2, in Appendix A. The sinusoidal sequence, $[s_k]$, has a slowly changing period. The sequence $[x_k]$ is a train of exponential pulses occurring at 40-sample intervals.

5. Given the one-step prediction situation illustrated below, write an LMS algorithm to adjust w_0 and w_1 at each iteration. What is the maximum useful value of μ?

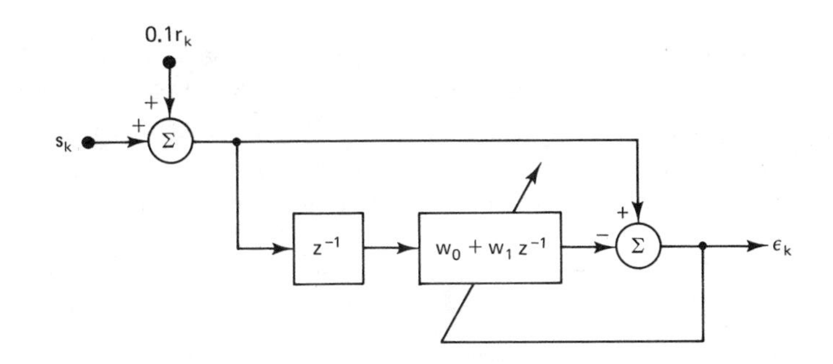

6. Using the result of Exercise 5 with $\mathbf{W}_0 = \mathbf{0}$ and $\mu = 0.02\mu_{\max}$, plot ε_k versus k. Also plot w_0 and w_1 versus k. Also plot a learning curve, estimating $E[\varepsilon_k^2]$ at each point by averaging ε_k^2 with its 20 nearest neighbors. (Use any procedure to take care of end effects.)
7. Using Equation (8.44), write an explicit SER algorithm for the predictor illustrated in Exercise 5. What is $\mu_{\max}$ in this case?
8. Using the SER algorithm from Exercise 7 with $\mu = 0.1\mu_{\max}$, plot ε_k versus k. Also plot w_0 and w_1 versus k. Also plot a learning curve, estimating $E[\varepsilon_k^2]$ by averaging ε_k^2 with its 20 nearest neighbors. (Use any procedure to take care of end effects.)
9. Give a modified version of the SER algorithm (8.44) for the completely stationary case with $\alpha = 1$. Give the corresponding limiting form of Equation (8.27). What difficulty would occur if the SER algorithm were allowed to go on in this form, with k unbounded?
10. Given the system identification configuration in Figure 8.8, write an explicit IIR LMS algorithm using Equation (8.55).

11. Using $\mu = 0.05$, $\nu_1 = 0.005$, and $\nu_2 = 0.0025$, run the IIR LMS algorithm for Figure 8.8 with $[r_k]$ as the input sequence. Plot ε_k versus k. Also plot a_{0k}, b_{1k}, and b_{2k} versus k. Also plot a learning curve as in Exercises 6 and 8.

12. For the second-order adaptive recursive filter, prove that (b_1, b_2) must lie inside the triangles shown in Figure 8.9 in order for the filter to be stable, that is, that the triangles correspond with the unit circle on the z-plane.

13. The configuration shown below illustrates the "equalizing" or "inverse filtering" problem. Starting with $\mathbf{W}_0 = \mathbf{0}$ the weights are to be adjusted so that the adaptive filter compensates for the channel, except for a propagation delay, z^{-5}. Write an LMS algorithm for this example and plot the learning curve, using $\mu = 0.04\mu_{\max}$. Estimate the residual error, $E[\varepsilon_k^2]_{\text{final}}$.

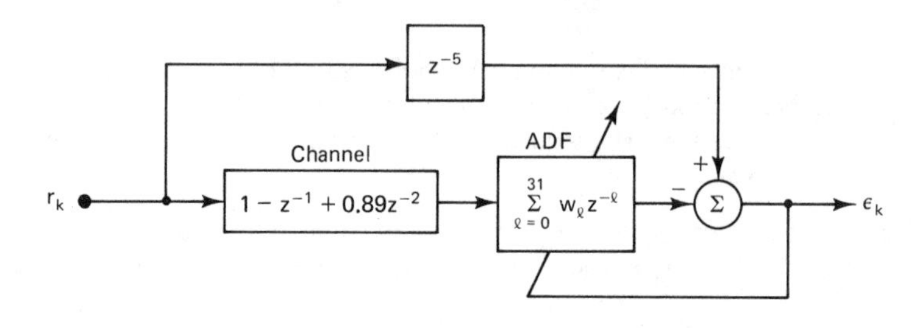

14. Do Exercise 13 using the SER algorithm with $\mathbf{W}_0 = \mathbf{0}$, $\alpha = 0.95$, $\mu\lambda_{\text{av}} = 0.05$, and $\mathbf{Q}_0^{-1} = 100\,\mathbf{I}$. Compare the learning curve with the LMS learning curve.

15. The equalizer below is similar to that above, but with an IIR adaptive filter. Write an LMS algorithm. Choose appropriate values for μ, ν_1, and ν_2, and plot the learning curve. Estimate the final mean value of ε_k^2 and compare with Exercise 13.

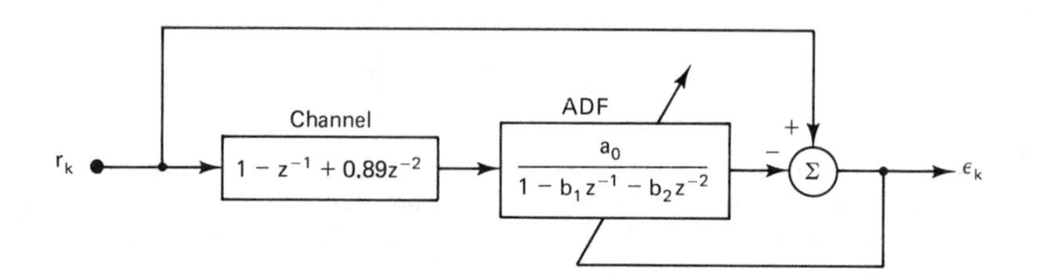

16. For Exercise 15, plot a_{0k}, b_{1k}, and b_{2k} versus k.

17. Work Exercise 13 using $[x_k]$, defined above, as the input data sequence. Explain any differences in your result.

18. Work Exercise 14 using $[x_k]$ as the input.

19. Work Exercise 15 using $[x_k]$ as the input.

20. Do Exercise 13 using the linear random-search algorithm. Use the same μ used in Exercise 13, and use $N = 5$ observations to get each $\hat{\xi}$. Adjust σ^2 so that the theoretical misadjustment will be one-tenth of $\xi_{\min}$.

21. Develop a lattice structure equivalent to the one-step predictor that follows.

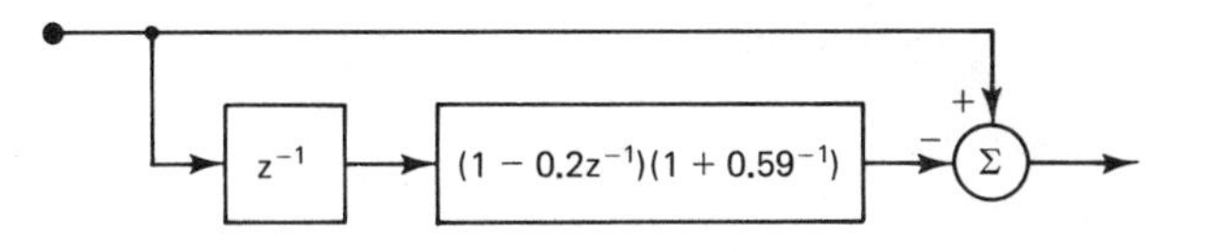

22. Sketch the lattice in Figure 8.13 using three-multiplier ladder elements.

23. Construct a lattice with transfer function $H(z) = (z^2 + 2z + 1)/(z^2 - z + 0.89)$. Verify the structure by computing and comparing the first five samples of the impulse response.

24. For the one-step predictor shown below, find the optimum values of b_1 and b_2 in terms of input correlation coefficients.

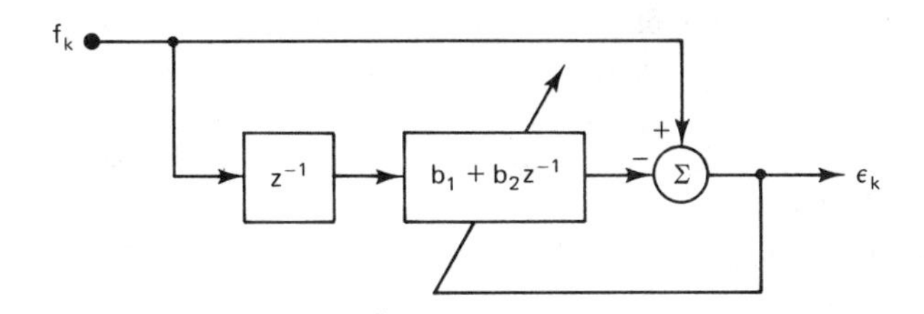

25. Convert the predictor shown in Exercise 24 into a lattice predictor. Give the lattice coefficients in terms of the input correlation coefficients using the answer to Exercise 24.

26. Given a two-stage lattice predictor, find the optimum weights in terms of input correlation coefficients, using Equations (8.100), (8.99), and (8.102). Compare with the answer to Exercise 25.

27. Illustrate the convergence of the three-stage adaptive lattice predictor shown below by plotting ε_k versus k for $k = 0, 1, \ldots, 500$. Use the algorithm in Equation (8.121) with $\alpha = 0.99$, $\mu = 0.1$, and estimated signal power equal to input signal power.

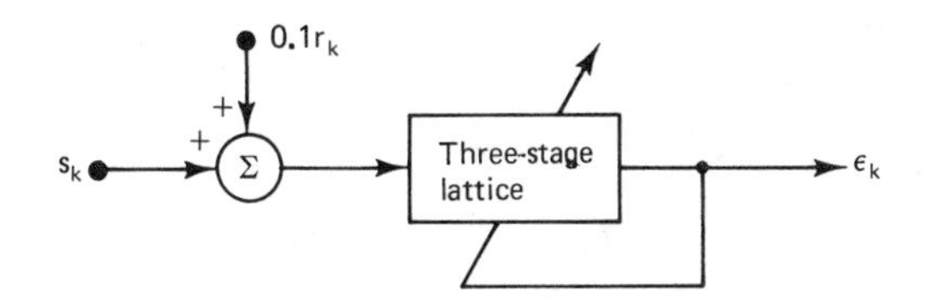

28. For the predictor described in Exercise 27, plot κ_0, κ_1, and κ_2 versus k for $k = 0, 1, \ldots, 1000$. Discuss briefly the behavior of these weights.

29. Derive an expression for the error surface for the foregoing predictor, as it exists in the vicinity of $k = 500$.

30. In the system described in Exercise 27, assume that $s_k = \sin(k\pi/10)$ and that there is a two-stage lattice with adaptive weights κ_0 and κ_1. Derive an expression for the error surface and make a contour plot similar to Figure 8.19.

31. What are the optimum weight values for the predictor in Exercise 27?

32. What are the optimum weight values in Exercise 27 if the sequence $[x_k]$ is used in place of $[s_k]$?

33. Select appropriate initial conditions for the situation depicted below and run the predictor for $k = 0, 1, \ldots, 1000$, using $\mu = 0.2$ times its maximum initial value in Equation (8.106). Plot a learning curve, estimating $E[\varepsilon_k^2]$ by averaging each ε_k^2 with its 10 nearest neighbors. Estimate the time constant.

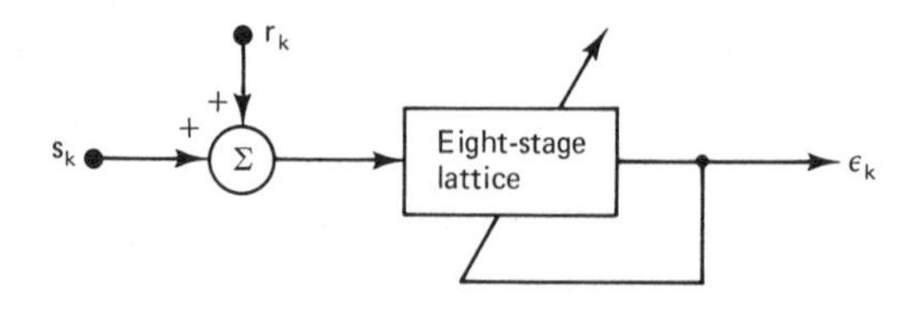

34. Using x_k in place of s_k, run the predictor shown in Exercise 33 until $E[\varepsilon_k^2]$ reaches a steady value. Explain the relationship between this value, $E[r_k^2]$, and $E[x_k^2]$.

35. In Figure 8.8, let $\Phi_{xx}(z) = 1$. What are $\phi_{dd}(0)$ and $\Phi_{dx}(z)$?

36. Find an expression for the performance surface, $\xi(a_0, b_1, b_2)$, in Figure 8.8. Begin with Equation (7.65).

37. Show that the performance surface for Figure 8.8 has a single global minimum within the stable range of (b_1, b_2).

REFERENCES AND ADDITIONAL READINGS

1. D. Graupe, *Identification of Systems*. New York: Van Nostrand Reinhold, 1972, Chap. 6.
2. N. Ahmed, D. L. Soldan, D. R. Hummels, and D. D. Parikh, "Sequential regression considerations of adaptive filtering," *Electron. Lett.*, p. 446, July 21, 1977.
3. D. Parikh and N. Ahmed, "A sequential regression algorithm for recursive filters," *Electron. Lett.*, p. 266, Apr. 27, 1978.
4. N. Ahmed, D. R. Hummels, M. Uhl, and D. L. Soldan, "A short term sequential regression algorithm," *IEEE Trans. Acoust. Speech Signal Process.*, vol. ASSP-27, p. 453, Oct. 1979.
5. R. C. K. Lee, *Optimal Estimation, Identification, and Control*. Cambridge, Mass.: MIT Press, 1966, Sec. 4.3.1.
6. S. D. Stearns, *Digital Signal Analysis*. Rochelle Park, N.J.: Hayden, 1975, Chaps. 7, 9 and 12.
7. S. D. Stearns, "Error surfaces of recursive adaptive filters," *IEEE Trans. Circuits Syst.*, vol. CAS-28, Special Issue on Adaptive Systems, June 1981.
8. T. C. Hsia, "A simplified adaptive recursive filter design," *Proc. IEEE*, vol. 69, p. 1153, Sept. 1981.
9. S. A. White, "An adaptive recursive digital filter," *Proc. 9th Asilomar Conf. Circuits Syst. Comput.*, p. 21, Nov. 1975.
10. M. G. Larimore, J. R. Treichler, and C. R. Johnson, Jr., "SHARF: an algorithm for adapting IIR digital filters," *IEEE Trans. Acoust. Speech Signal Process.*, vol. ASSP-28, p. 428, Aug. 1980.

11. J. R. Treichler, M. G. Larimore, and C. R. Johnson, Jr., "Simple adaptive IIR filtering," *Proc.* 1978 *ICASSP*, p. 118, Apr. 1978.
12. M. G. Larimore, C. R. Johnson, Jr., and J. R. Treichler, "Adaptive cancelling using SHARF," *Proc. 21st Midwest Symp. Circuits Syst.*, Aug. 1978.
13. M. G. Larimore, J. R. Treichler, and C. R. Johnson, Jr., "Multipath cancellation by adaptive recursive filtering," *Proc. 12th Asilomar Conf. Circuits Syst. Comput.*, Nov. 1978.
14. P. L. Feintuch, "An adaptive recursive LMS filter," *Proc. IEEE*, vol. 64, p. 1622, Nov. 1976.
15. R. A. David and S. D. Stearns, "Adaptive IIR algorithms based on gradient search," *Proc. 24th Midwest Symp. Circuits Syst.*, June 1981.
16. B. Widrow and J. M. McCool, "A comparison of adaptive algorithms based on the methods of steepest descent and random search," *IEEE Trans. Antennas Propag.*, vol. AP-24, p. 615, Sept. 1976.
17. D. M. Etter and M. M. Masukawa, "A comparison of algorithms for adaptive estimation of the time delay between sampled signals," *Proc. ICASSP-81*, p. 1253, Mar. 1981.
18. A. H. Gray, Jr., and J. D. Markel, "Digital lattice and ladder filter synthesis," *IEEE Trans. Audio Electroacoust.*, vol. AU-21, p. 491, Dec. 1973.
19. F. Itakura and S. Saito, "Digital filtering techniques for speech analysis and synthesis," *Proc. 7th Int. Conf. Acoust.*, vol. 3, Paper 25C-1, p. 261, 1971.
20. S. D. Stearns, *Digital Signal Analysis*. Rochelle Park, N.J.: Hayden, 1975, Chap. 9.
21. R. A. David, "IIR Adaptive Algorithms Based on Gradient Search Techniques," Stanford Univ., Stanford, Calif., Aug. 1981 (Ph.D. dissertation).
22. J. Makhoul and R. Viswanathan, "Adaptive lattice methods for linear prediction," *Proc. ICASSP-78*, p. 83.
23. L. J. Griffiths, "An adaptive lattice structure for noise-cancelling applications," *Proc. ICASSP-78*, p. 87.
24. T. Kailath, *Linear Systems*. Englewood Cliffs, N.J.: Prentice-Hall, 1980.
25. C. R. Johnson, Jr., M. G. Larimore, J. R. Treichler, and B. D. O. Anderson, "SHARF convergence properties," *IEEE Trans. Acoust. Speech Signal Process.*, vol. ASSP-29, p. 659, June 1981.
26. D. Parikh, N. Ahmed, and S. D. Stearns, "An adaptive algorithm for recursive filters," *IEEE Trans. Acoust. Speech Signal Process.*, vol. ASSP-28, p. 110, Feb. 1980.
27. E. I. Jury, "A note on the reciprocal zeros of a real polynomial with respect to the unit circle," *IEEE Trans. Commun. Technol.*, vol. CT-11, p. 292, June 1964.
28. L. J. Griffiths, "A continuously-adaptive filter implemented as a lattice structure," *Proc. ICASSP-77*, p. 683, May 1977.
29. M. D. Srinath and M. M. Viswanathan, "Sequential algorithm for identification of parameters of an autoregressive process," *IEEE Trans. Autom. Control*, vol. AC-20, p. 542, Aug. 1975.
30. N. Levinson, "The Wiener RMS error criterion in filter design and prediction," *J. Math. Phys.*, vol. 25, pp. 261–278, 1946.
31. S. A. Tretter, *Introduction to Discrete-Time Signal Processing*. New York: Wiley, 1976, Sec. 7.6.
32. A. P. Sage, *Optimum System Design*. Englewood Cliffs, N.J.: Prentice-Hall, 1968, p. 276.

33. G. C. Goodwin and R. L. Payne, *Dynamic System Identification: Experiment Design and Data Analysis*. New York: Academic Press, 1977, Chap. 7.
34. R. S. Medaugh and L. J. Griffiths, "A comparison of two fast linear predictors," *Proc. ICASSP-81*, p. 293, Mar. 1981.
35. J. I. Makhoul and L. K. Cosell, "Adaptive lattice analysis of speech," *IEEE Trans. Acoust. Speech Signal Process.*, vol. ASSP-29, p. 654, June 1981.
36. B. Widrow and E. Walach, "On the statistical efficiency of the LMS algorithm with nonstationary inputs," *IEEE Trans. Information Theory—Special Issue on Adaptive Filtering*, vol. 30, no. 2, part 1, pp. 211–221, Mar. 1984.
37. R. A. Monzingo and T. W. Miller, *Introduction to Adaptive Arrays*. New York: John Wiley, 1980, Sec. 8.3.
38. S. S. Narayan and A. M. Peterson, "Frequency domain least-mean-square algorithm," *Proc. IEEE*, vol. 69, no. 1, pp. 124–126, Jan. 1981.
39. B. Widrow, J. McCool, and M. Ball, "The complex LMS algorithm," *Proc. IEEE*, vol. 63, no. 4, pp. 719–720, Apr. 1975.
40. M. Morf, A. Vieira, and D. T. Lee, "Ladder forms for identification and speech processing," *Proc. 1977 IEEE Conf. Decision and Control*, New Orleans, LA, pp. 1074–1078, Dec. 1977.
41. M. Morf and D. T. Lee, "Recursive least-squares ladder forms for fast parameter tracking," *Proc. 1979 IEEE Conf. Decision and Control*, San Diego, CA, pp. 1362–1367, Jan.
42. D. T. Lee, M. Morf, and B. Friedlander, "Recursive least-squares ladder estimation algorithms," *IEEE Trans. Acoust. Speech Signal Process. Joint Special Issue on Adaptive Signal Processing*, vol. ASSP-29, no. 3, pp. 627–641, June 1981.
43. B. Porat, B. Friedlander and M. Morf, "Square-root covariance ladder algorithms," *IEEE Trans. on Acoust. Control*, vol. AC-27, no. 4, pp. 813–829, Aug. 1982.
44. B. Friedlander, "Lattice filters for adaptive processing," *Proc. IEEE*, vol. 70, no. 8, pp. 829–867, Aug. 1982.
45. B. Friedlander, "Lattice methods for spectral estimation," *Proc. IEEE*, vol. 70, no. 9, pp. 990–1017, Sept. 1982.
46. H. Lev-Ari, T. Kailath, and J. M. Cioffi, "Least-squares adaptive lattice and transversal filters: a unified geometric theory," *IEEE Trans. Information Theory—Special Issue on Adaptive Filtering*, vol. 30, no. 2, part 1, pp. 222–236, Mar. 1984.
47. L. J. Griffiths and R. S. Medaugh, "Convergence properties of an adaptive noise cancelling lattice structure," *Proc. 1978 IEEE Conf. Decision and Control*, San Diego, CA, pp. 1357–1361, Jan. 1979.

part IV

APPLICATIONS

(Chapters 9–14)

OBJECTIVES OF PART IV

The main objective of this final part of the text is to show how the theory developed in Parts I, II, and III is put into practice in several different areas. A secondary objective is to review some related topics such as spread-spectrum techniques, equalization, basic control theory, etc., in order to make the discussion more useful.

In these final six chapters there are only four basic adaptive configurations, which are used in different ways. First, *adaptive modeling* is introduced and applied to different situations in Chapter 9 and to adaptive control systems in Chapter 11. Second, *inverse modeling* (adaptive equalization) applications are introduced in Chapter 10 and again in Chapter 11. Third, *adaptive interference canceling* is discussed in Chapter 12 and applied to adaptive arrays in Chapers 13 and 14. A fourth basic configuration, *adaptive prediction*, has already been introduced and is used again in Chapter 12. We recall that these four configurations were illustrated in Chapter 1, Fig. 1.5.

chapter 9

Adaptive Modeling and System Identification

Modeling and system identification is a very broad subject, of great importance in the fields of control systems, communications, and signal processing. Modeling is also important outside the traditional engineering disciplines, as when one studies social systems, economic systems, or biological systems. We cannot treat such a broad range of subjects here. Our goal is a modest one, that of showing how simple adaptive filters can be used in system modeling and of demonstrating the application of adaptive system identification to several physical situations.

GENERAL DESCRIPTION

An adaptive filter can be used in modeling, that is, imitating the behavior of physical dynamic systems which may be regarded as unknown "black boxes" having one or more inputs and one or more outputs. Modeling a single-input, single-output dynamic system or "plant" (control systems terminology) is illustrated in Figure 9.1(a). Both the unknown system and the adaptive filter are driven by the same input. The adaptive filter adjusts itself with the goal of causing its output to match that of the unknown system, generally to cause its output to be a best least-squares fit to that of the unknown system. If enough flexibility resides in the adaptive system, if it is sufficiently adjustable and contains enough "degrees of freedom" (adjustable weights), a close fit or perhaps a perfect fit is possible. Upon convergence, the structure and parameter values of the adaptive system may or may not resemble those of the unknown system, but the input–output response relationships will match. In this sense, the adaptive system becomes a model of the unknown system. If the input is wideband and if the structure of the adaptive system is such that when its adjustable parameters are suitably chosen an exact match is possible, adapting to minimize the mean-square error will result in an exact match in every detail.

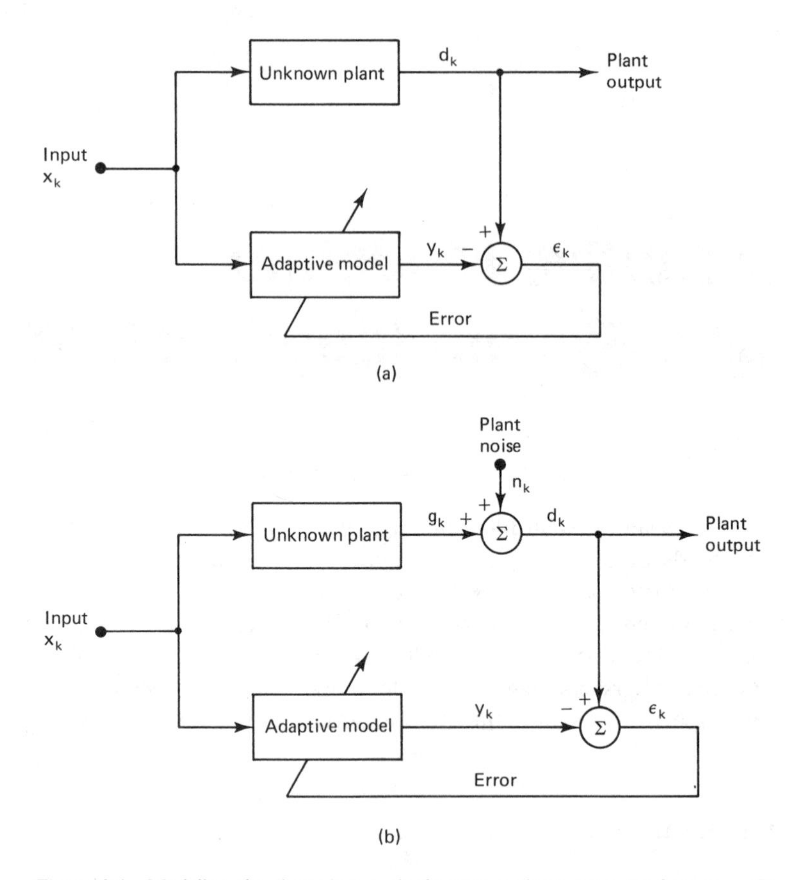

Figure 9.1 Modeling the single-input, single-output plant: (a) noise-free case; (b) noisy plant case.

In many practical cases, the plant to be modeled is noisy, that is, has internal random disturbing forces. As such, when the adaptive model has enough flexibility to match the dynamic response of the unknown plant, its output will perfectly match that of the unknown plant except for plant noise n_k, indicated in Figure 9.1(b) as a noise added to the plant output. Internal plant noise appears at the plant output and is commonly represented there as an additive noise. This noise is generally uncorrelated with the plant input. If this is the case and if the adaptive model is an adaptive linear combiner whose weights are adjusted to minimize mean-square error, it can be shown that the least-squares solution will be unaffected by the presence of the plant noise. This is not to say that the convergence of the adaptive process will be unaffected by plant noise, only that the expected weight vector of the adaptive model after convergence will be unaffected. (Exercise 6 considers this issue.) The least-squares solution will be determined primarily by the impulse response of the plant to be modeled. It could also be significantly affected by the statistical or spectral character of the plant input signal.

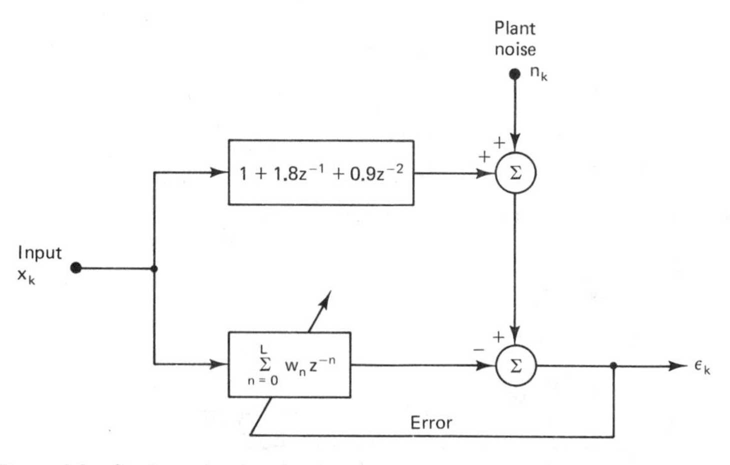

Figure 9.2 Configuration for the single-input system identification experiment in Figure 9.3.

We illustrate the single-input system identification process using the configuration in Figure 9.2. Here we have an all-zero plant, so the "perfect fit" described above is possible provided that $L \geq 2$. We assume that the signal, x_k, and the noise, n_k, are mutually uncorrelated white random sequences† and that the adaptive transversal filter uses the LMS algorithm, with

$$E\left[x_k^2\right] = \tfrac{1}{12}, \qquad E\left[n_k^2\right] = \tfrac{1}{12} \text{ or } 0, \qquad \mu = 0.5 \tag{9.1}$$

With these values we can determine the excess mean-square error and the learning-curve time constant using (6.35) and (6.38). With no noise and with $L \geq 2$, the excess MSE is

$$\text{excess MSE (no noise)} = \mu \xi_{\min} \text{tr}[\mathbf{R}] = 0 \tag{9.2}$$

That is, if the adaptive filter is of sufficient length, $\xi_{\min}$ is zero, so the excess MSE is also zero. When n_k is added in Figure 9.2, the adaptive filter with $L \geq 2$ still cancels the plant output, so $\xi_{\min}$ is due entirely to the noise power:

$$\xi_{\min} = E\left[n_k^2\right] = \tfrac{1}{12} \tag{9.3}$$

When the adaptive filter is of insufficient size with $L < 2$, then of course $\xi_{\min}$ is nonzero with or without plant noise.

For white random noise the input correlation matrix is diagonal with elements equal to $E[x_k^2]$, so the convergence time constant is

$$T_{\text{mse}} = \frac{1}{4\mu\lambda} = \frac{1}{4(0.5)(1/12)} = 6 \text{ iterations} \tag{9.4}$$

†Generated using RANDOM in Appendix A; for example, $x_k = \text{RANDOM}(1.) - 0.5$ at each iteration. The variance of the unit uniform variate is $E[x_k^2] = \frac{1}{12}$.

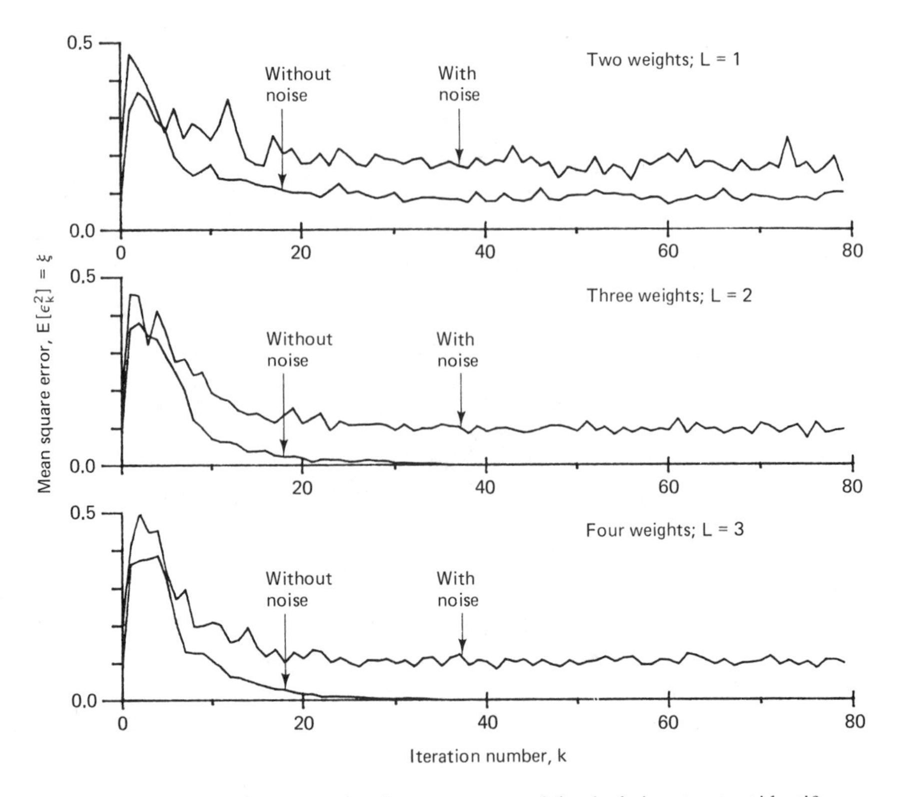

Figure 9.3 Learning curves showing convergence of the single-input system identification process in Figure 9.2. Each curve is the average of 100 runs.

The excess MSE in (9.2) and (9.3) as well as the time constant in (9.4) are illustrated approximately in Figure 9.3, which gives six experimental learning curves. Each curve is the average of 100 computer runs of the configuration in Figure 9.2. There are curves with and without noise for two, three, and four adaptive weights.

With only two weights we note a nonzero ξ_{min} even without noise, which indicates that the adaptive filter cannot completely identify the plant. With three or four weights we observe complete identification in the noiseless case and approximate verification of (9.3) in the noisy case. In all cases we observe approximate verification of the six-iteration time constant in (9.4).

A means of modeling multiple input–multiple output systems can be inferred from the two input–two output case, as illustrated in Figure 9.4. The plant of Figure 9.4 is a black box having two inputs and two outputs. Its internal structure is assumed to be unknown. The purpose of the adaptive filter is to determine the input–output characteristics of the unknown plant, and if possible, to unmask its detailed structure. Not knowing the plant structure but merely knowing that it is linear with two inputs and two outputs and is time invariant over the learning period, a general linear modeling structure can be chosen as shown in Figure 9.4.

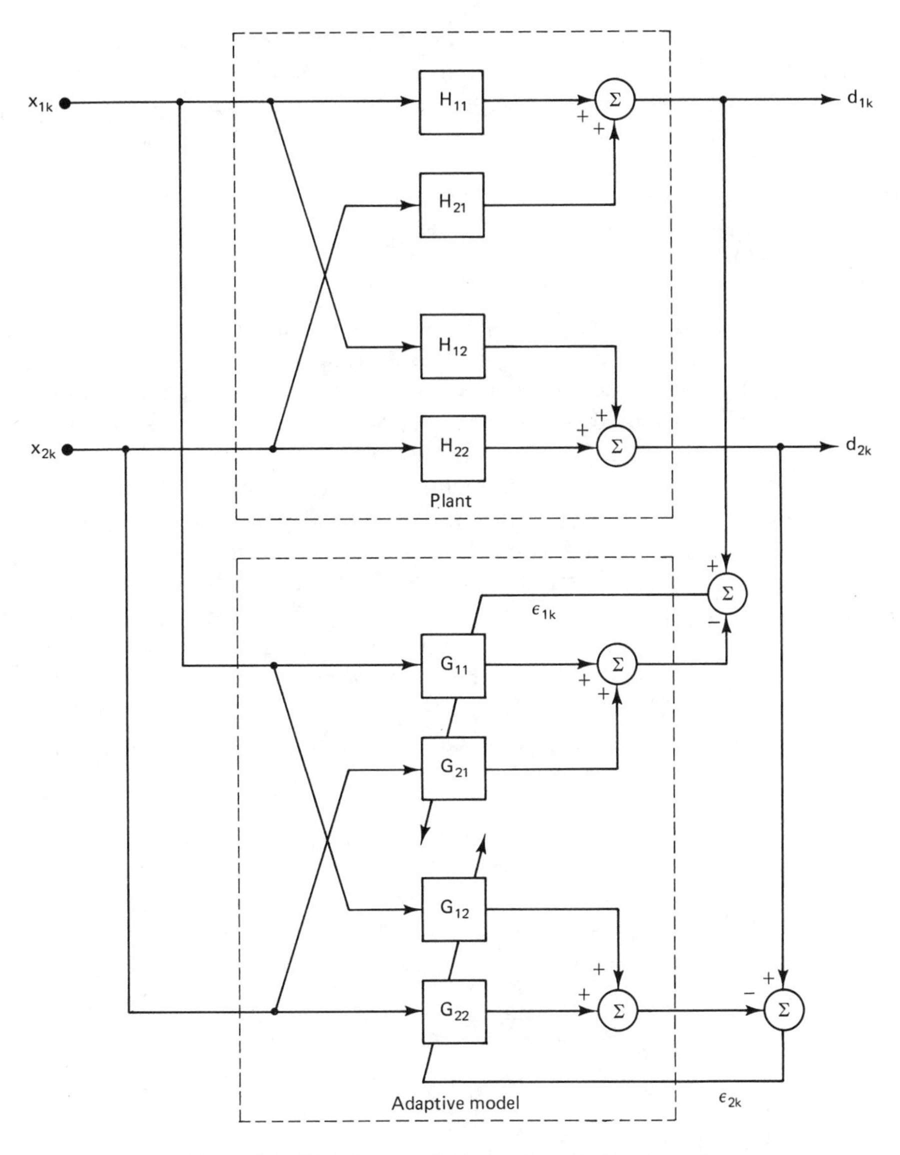

Figure 9.4 Modeling a multiple-input, multiple-output plant.

Assuming that each adaptive filter has a sufficient number of degrees of freedom, upon convergence each of the subadaptive filters will match the corresponding component filters of the plant (i.e., $G_{11} \rightarrow H_{11}$, $G_{22} \rightarrow H_{22}$, $G_{12} \rightarrow H_{12}$, $G_{21} \rightarrow H_{21}$). This will certainly occur if the inputs x_1 and x_2 are mutually uncorrelated. Good modeling may also occur when x_1 and x_2 are correlated, as long as x_1 cannot be perfectly generated by filtering x_2, and vice versa.

Three examples will now be presented which are representative of applications of adaptive modeling techniques to practical problems. The first application is to communications, the second application to a problem in geophysics, and the third is to the design of digital filters.

ADAPTIVE MODELING OF A MULTIPATH COMMUNICATION CHANNEL

The reader is probably familiar with the "multipath" type of communication problem illustrated in Figure 9.5. It is a problem in which the transmitted information takes more than one path to the receiver, thus producing interference and echoes in the received signal. To see and appreciate how adaptive modeling is used to identify the paths (i.e., the impulse response in Figure 9.5), we will first describe briefly a "spread-spectrum" technique [1, 13] for sending binary signals reliably in a high-noise, multipath environment.

In spread-spectrum communications each information bit, either a ONE or a ZERO, is transmitted as a string of coded bits. The ONE bit may be represented by a particular string which, for example, may be 32 bits long. The ZERO bit would be represented by a different 32-bit string. For detection, the receiver tries to correlate the strings, which are decoded as ONEs or ZEROs, depending on correlation maxima. The ONE and ZERO codes are "pseudorandom" and are designed as well as possible to be mutually orthogonal and individually to have autocorrelations that

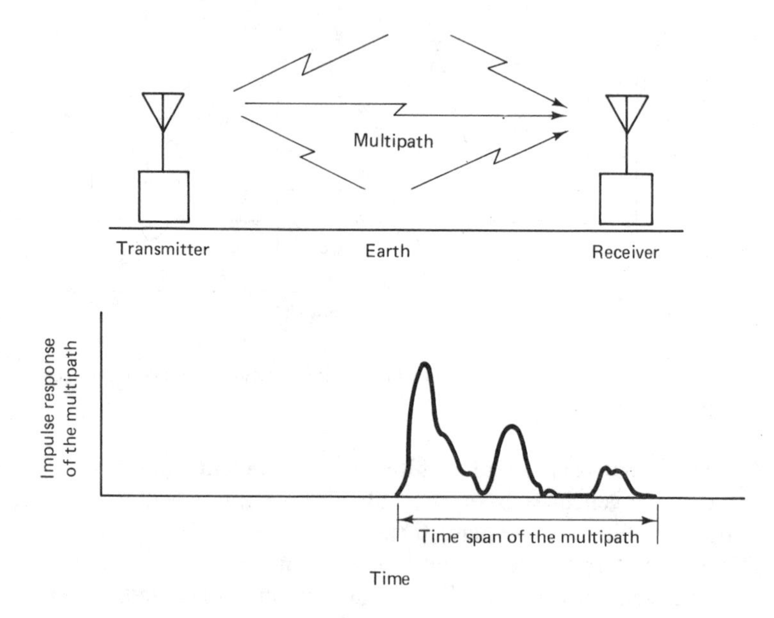

Figure 9.5 Typical dispersive channel and its impulse response.

peak at zero lag and become essentially zero for other lags. Maximal-length shift register sequences have these properties and have been used widely in this application [1]. The spectrum of a sequence of such strings tends to be very broad even for a regular pattern of ONEs and ZEROs: hence the term "spread-spectrum communications." Systems of this type are very effective in the presence of strong broadband additive interference.

The effects of multipath, when present, could be disastrous to a spread-spectrum system of the type considered here, however. Bit strings could become scrambled at the receiver as a result of propagation through multipath, that is, parallel propagation through many natural paths from transmitter to receiver, each with its own delay time. The responses of the various paths add linearly at the receiver causing scrambling or smearing. The goal of adaptive techniques used in conjunction with the spread-spectrum techniques described above is to "unravel" the multipath and to permit communication essentially as if multipath were not present.

Figure 9.6 shows a block diagram of a spread-spectrum communication system working in this illustrative case with a simple nondispersive, no-multipath channel. Channel noise is assumed to be present. At the transmitter, a key is switched to generate the data waveform, turning on a ONE or ZERO pseudorandom sequence as required to represent the information to be transmitted. Both the ONE and ZERO sequences are initiated at the same instant, triggered by a clock, and are repeated in accordance with the information pattern. The key must remain in the same state throughout the transmission of the entire ONE or ZERO sequence, whichever is selected. The key can then remain in the same state or switch to the opposite state, depending on the next information bit to be transmitted. The transmitted waveform, a sequence of concatenated ONE and ZERO sequences, propagates through the channel. Except for time delay, the receiver sees the same waveform plus additive channel noise. The receiver has a clock that also triggers the known ONE and ZERO sequences. The receiver clock generates triggers at exactly the same rate as the clock at the transmitter, but its pulses are phased relative to the transmitter clock taking into account the channel delay. The receiver ONE and ZERO sequences are cross-correlated against the noisy signal received, and if the local clock is phased properly, one of the cross-correlator outputs will build up to a peak value just before the clock triggers the repeat of the local sequences. Since the channel delay would generally be unknown, the phase of the local clock could be gradually slewed to maximize the peak outputs of the cross-correlators. A comparator produces the data sequence at the receiver output by periodically deciding (just before the local clock pulses occur) which cross-correlator has the largest output. If the ONE cross-correlator has the largest output at the proper sensing time, the system output will be ONE, and so on.

With a noise-free channel, the output of only one of the cross-correlators will peak at the proper sensing time; the other cross-correlator output will remain at a very low level. Noise confuses the outputs of the two cross-correlators, however, and forces decision making on the basis of taking the largest output. In designing such a system, some prior knowledge of channel signal-to-noise ratio is generally useful.

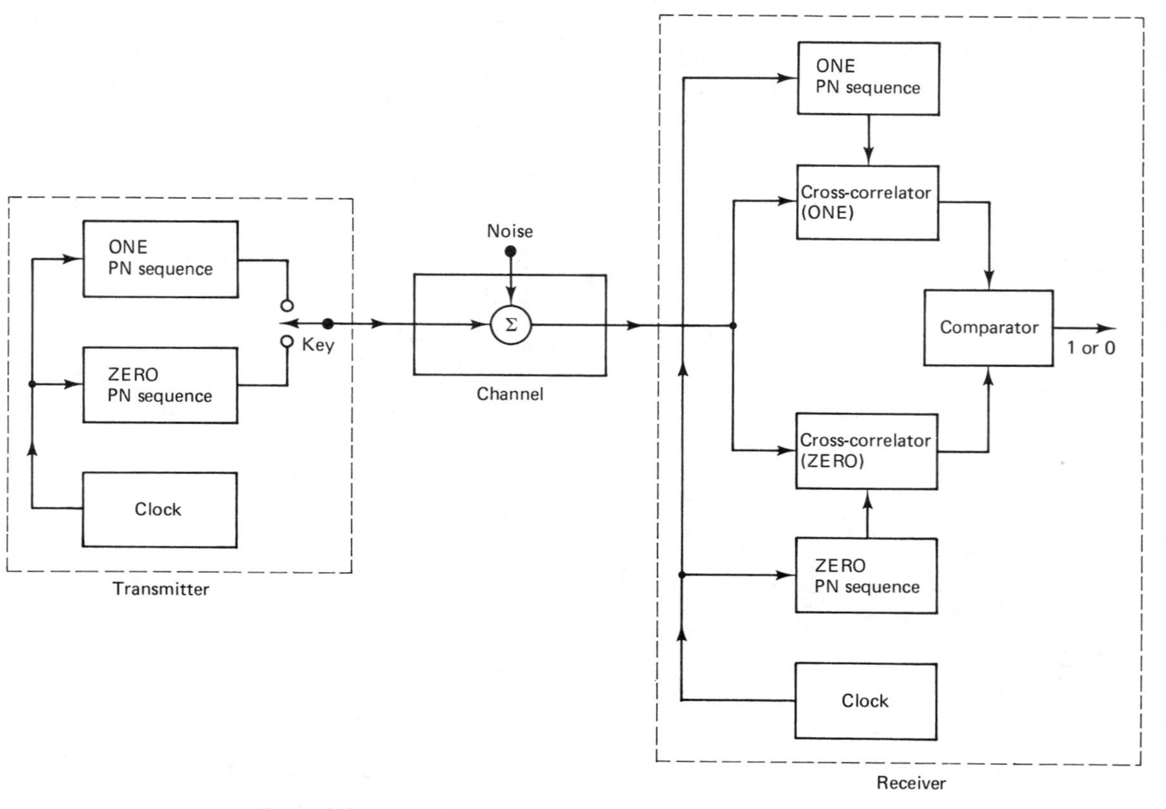

Figure 9.6 Block diagram of a spread-spectrum transmitter and receiver.

The poorer the signal-to-noise ratio, the longer one would make the ONE/ZERO code strings. Averaging in the cross-correlation process reduces the effects of random channel noise.

This form of spread-spectrum system provides resistance to background noise, to jamming, and to other forms of interference. A measure of privacy in communication is also afforded by this type of system, since the ONE and ZERO codes could be made unknown except to the intended recipient.

We consider next the case where the channel is not only noisy but afflicted with multipath. Suppose, for example, that the impulse response of the channel consists of a pure delay followed by an impulse response that is smeared out over a finite time span, as sketched in Figure 9.5. Suppose that the time duration of the ONE or ZERO sequence is comparable to the time span of the multipath. Then, when the transmitted signal is convolved with the channel impulse response, severe interference will be experienced at the receiver—interference within the bit code and from one bit code sequence to the next. The phenomenon is known as "intersymbol interference." Our goal is to solve this problem with adaptive filtering, by modeling the multipath characteristics of the channel.

Modeling an unknown channel to obtain a close approximation to its impulse response could be done as shown in Figure 9.7. ONEs and ZEROs are not transmitted in this case. Instead, a single known pseudorandom sequence is transmitted cyclically into the channel. The channel output is observed at the receiving site. The adaptive filter output is compared with the channel output, which in this case is the desired response. The filter is adapted to minimize mean-square error, where the error is the difference between the channel output and the adaptive model output. Problems of synchronization introduced by the unknown bulk delay of the channel are avoided by the cyclic repetition of the pseudorandom sequence. It is necessary, however, that the transmitter and receiver clocks run at the same rate in order for the adaptive filter to converge to a model of the multipath channel. The time duration of the pseudorandom sequence should be longer than the time span of the multipath (the duration of the channel impulse response, excluding the bulk time delay). The memory time of the adaptive filter must be at least as great as the time span of the multipath. Notice that, for the system configuration of Figure 9.7, channel noise has no effect on the converged least-squares solution for the channel model. The cross-correlation in Figure 9.7 is used to measure the effectiveness of the adaptive model, to slew the receiver clock, and so on.

A digital communication system using fixed nonadaptive channel models is diagrammed in Figure 9.8. An unrealistic assumption is made here for sake of argument—that an exact model of the channel is available at the receiver. As with the system of Figure 9.6, the two pseudorandom data bit sequences representing ONE and ZERO are selected to correspond at both transmitter and receiver sites. Once again, the data bit train is encoded by setting the key to select the proper code sequence at the transmitter. At the receiver, identical versions of the channel model are driven by the ONE and ZERO sequences, which are repeated as described previously. The model outputs are cross-correlated against the received channel

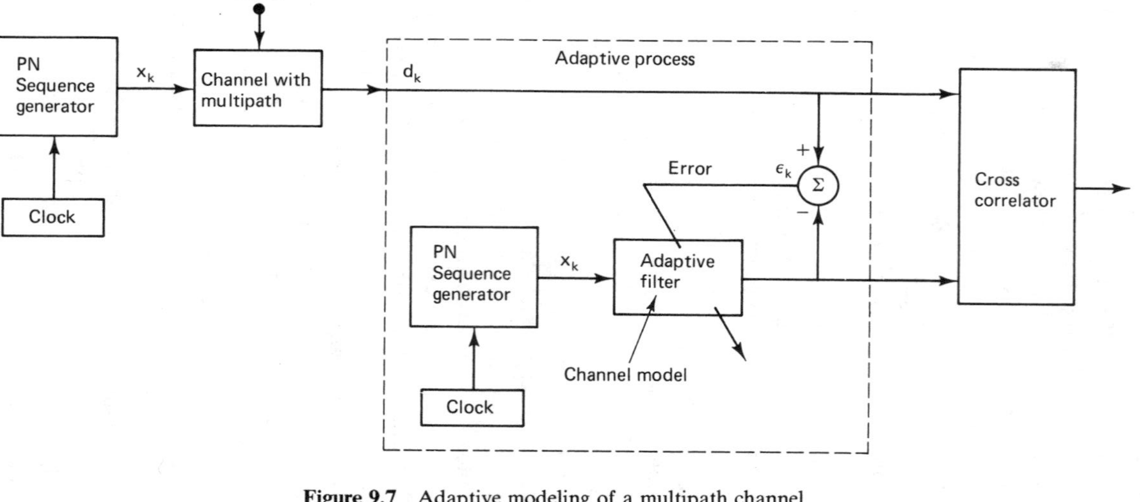

Figure 9.7 Adaptive modeling of a multipath channel.

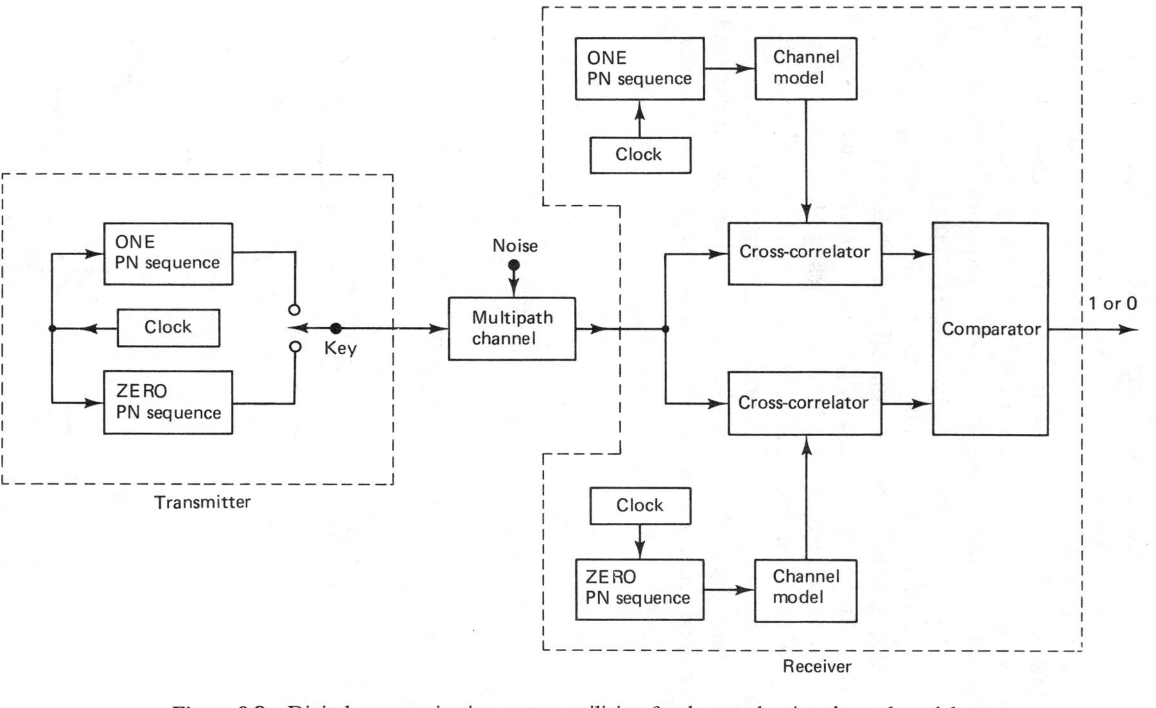

Figure 9.8 Digital communication system utilizing fixed, nonadaptive channel models.

output in a continuous data-flow process. The comparator makes ONE or ZERO choices as one or the other cross-correlator output peaks up at the proper time instants, spaced corresponding to the transmission rate of the information bits. The two receiver clocks are synchronized, and their phasing would need to be slewed to maximize the correlator output peaks.

In a practical system, some method must exist at the receiver to find the channel model. The method for obtaining the channel model by the adaptive process of Figure 9.7 is a workable one, except that transmission of real data is precluded since only one pseudorandom sequence is transmitted and repeated over and over again. A more comprehensive approach for channel modeling during actual data transmission is suggested by the scheme of Figure 9.9. The transmitted signal is generated as before, consisting of ONE and ZERO sequences representing the data. At the receiver, locally generated synchronized ONE and ZERO sequences are summed and fed to the adaptive modeling filter, whose output is compared with the signal emerging from the multipath channel. The filter is adapted to make a best least-squares match to the channel output. Since the adaptive filter input contains a sum of both the ONE and ZERO sequences, the received data will be correlated with one of these sequences or the other, roughly depending on whether a ONE or ZERO

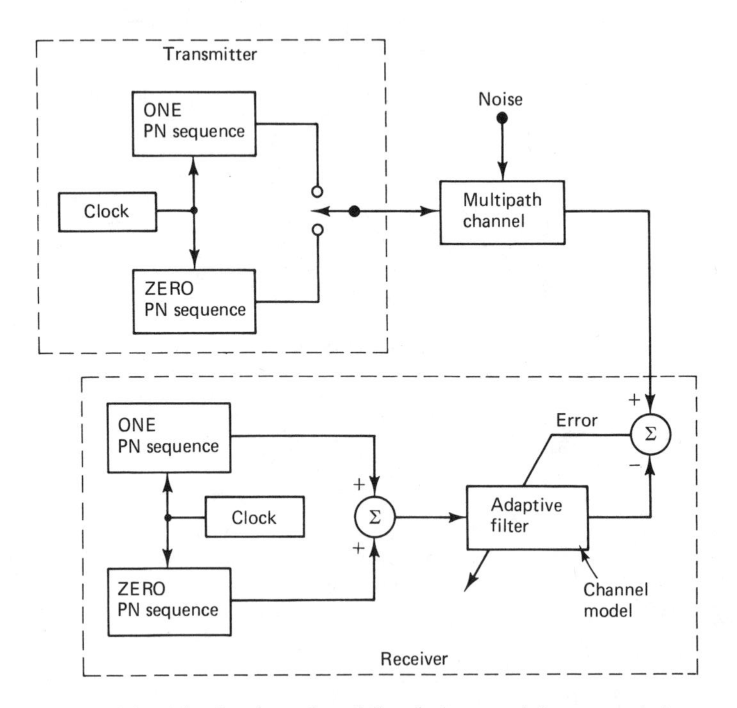

Figure 9.9 Adaptive channel modeling during actual data transmission.

is being received at the moment. The adaptive filter of Figure 9.9 will converge to a least-squares solution which will be identical to the solution resulting from the scheme of Figure 9.7, except for a scale factor. As in Chapter 2 [Equation (2.17), etc.] the least-squares solution is $\mathbf{R}^{-1}\mathbf{P}$. Comparing the two modeling schemes, the R-matrices at the inputs to the adaptive filters will be identical except for a factor of 2, but the P-vectors will be the same.

Consider the following argument. Let the pseudorandom (PN) sequence in Figure 9.7 be, for sake of argument, the same as the ONE sequence in Figure 9.9. This sequence repeated over and over again will have an associated R-matrix. The components of the P-vector for the system of Figure 9.7 will be equal to the cross-correlation function between the repeated PN sequence and the multipath channel output. Referring now to Figure 9.9, the P-vector will be the same here as above regardless of the keying activity at the transmitter since the channel output will be correlated with the adaptive filter input whether the received signal is derived from transmission of either the ONE sequence or the ZERO sequence. (Recall that the ONE sequence is designed to be essentially uncorrelated with the ZERO sequence.) The R-matrix, on the other hand, will be twice as large for the adaptive filter of Figure 9.9 as for that of Figure 9.7. The reason is that in Figure 9.9, the R-matrix equals the sum of the R-matrix due to the repeated ONE sequence alone and that due to the repeated ZERO sequence alone. (Recall that the two sequences are designed to have identical autocorrelation properties.) The result is that the converged weight vector in Figure 9.9 will be half that of Figure 9.7. Since scaling in the channel model is unimportant when applied as in Figure 9.8 (where the final output decisions are made by a comparator), the scheme of Figure 9.9 will be as effective as that of Figure 9.7, yet it will permit actual data transmission. This scheme for adaptation during data transmission was invented by Michael J. Ball.†

It should be noted once again that noise in the channel will not affect the form of the least-squares solution but will cause noise in the adaptive weights. Good results with high channel noise will necessitate slow adaptation (keeping misadjustment low). This would only be workable when the channel impulse response is stationary, or if nonstationary, it must be slowly varying. Rapid variation in the multipath characteristics accompanied by high channel noise will cause the scheme to fail. Good success will result with either low noise or slow channel variation (most physical channels vary over time) or both. A system utilizing adaptive channel models and incorporating Ball's method of adapting the channel model while transmitting and receiving actual data is shown in Figure 9.10. The cross-correlators are represented by multipliers followed by "leaky" (single-pole) integrators.

Slewing the phase of the receiver clock is not necessary since the impulse response of the adaptive filter would advance or retard itself as required to make its output achieve a best fit to the actual channel output. This would automatically cause the correlator output peaks to be maximized. The system of Figure 9.10 has been tested for acoustic communication and has been shown to be workable in the presence of multipath, channel noise, and slow channel nonstationarity.

† Private communication.

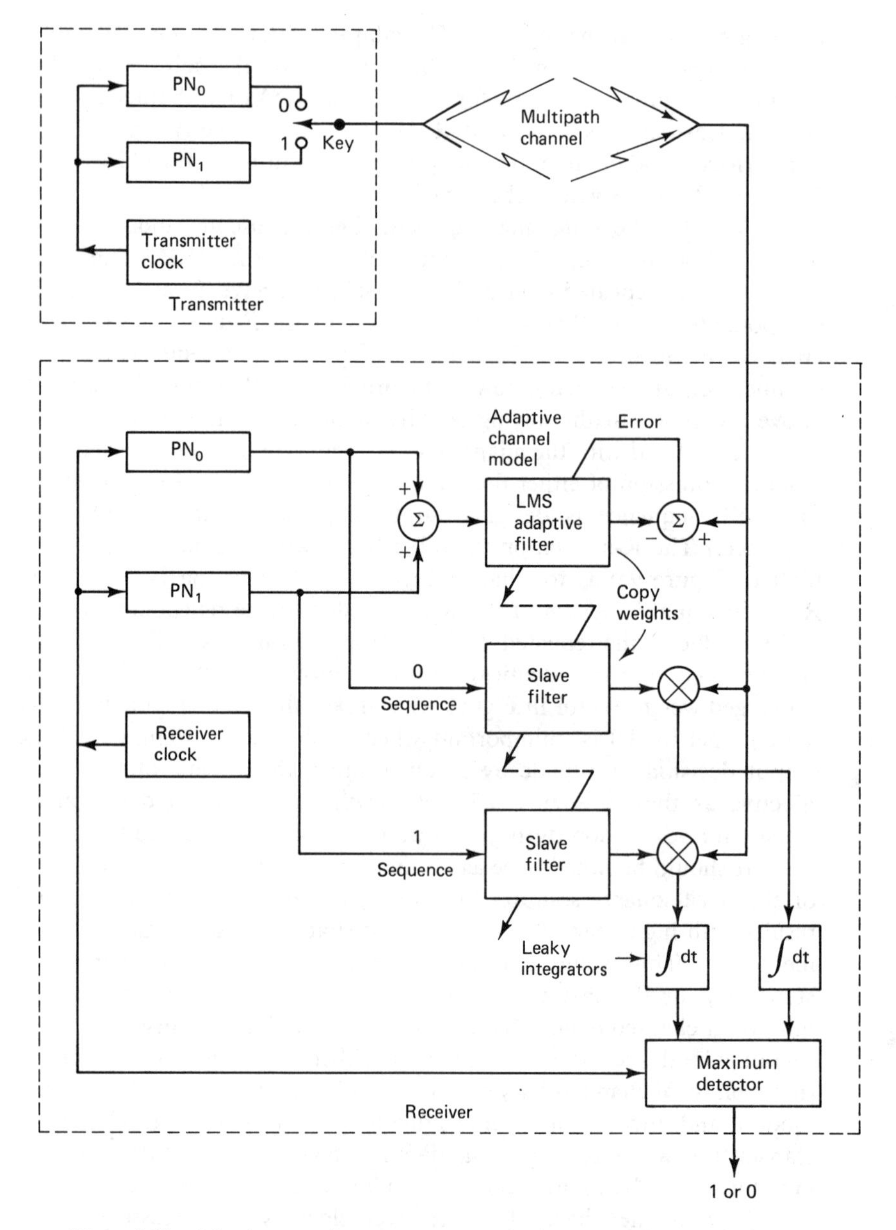

Figure 9.10 Adaptive spread-spectrum communication system for multipath channel. Note that the 1 and 0 pseudorandom sequences are known at both transmitter and receiver and could serve as a cipher. Transmitter and receiver clocks are assumed to be synchronized.

ADAPTIVE MODELING IN GEOPHYSICAL EXPLORATION

Seismic reflection technology is the primary tool used in exploration for oil and gas. Geologic formations have been laid down over tens of millions of years. These are layers of material such as sand, sandstone, silt, shale, clay, and so on. In order that hydrocarbons such as oil and gas be able to accumulate, there must be a suitable reservoir. This generally consists of a porous rock, for example a layer of sand, and an impervious rock to cap and seal the reservoir to prevent the escape of the hydrocarbon. Oil and gas are generally lighter than surrounding fluids and therefore tend to float upward. A typical material that would cap a reservoir might be clay or shale. Natural layerings and faultings in the earth create traps or potential reservoirs. Most traps contain salt water rather than hydrocarbons. Thus the existence of a trap is a necessary condition for the accumulation of oil or gas, but unfortunately not a sufficient condition. When exploring in a new area, one uses seismic reflection techniques to identify traps or to identify regions where traps are likely to exist. In most cases, this provides the best and most useful information for exploration purposes.

In seismic exploration, a source of seismic energy is typically activated on the surface of the earth. On land, dynamite may be exploded to apply a seismic impulse to the earth. Or a source of vibration could be used which could be pulsed or continuous. On water, a spark source or an air gun could be used to impart underwater mechanical impulses which propagate through the water to the sea bottom and into the earth below. The earth behaves like an elastic medium, approximately linear. As the seismic waves (which are like acoustic waves except that the earth is able to propagate shear waves as well as compressional waves) propagate from one geologic layer into another, change of material often is accompanied by change in seismic impedance, which causes reflections. Detection of reflections allows one to detect changes in material and thus detect layering in the subsurface. Formation layers frequently extend laterally over distances of many kilometers. By measuring the time delay along a reflection path from the surface down to a reflection plane and back up to the surface and invoking some knowledge of the speed of propagation, one may estimate the depth of the reflection plane. If this plane is sloped ("dipped") or if it is curved rather than flat, the nonflatness can be determined by measuring two-way travel time from many points on the earth's surface down to the reflection plane. In this way, layerings and potential traps and locales where traps might exist can be identified.

Adaptive modeling techniques can be used to measure two-way travel times from the surface down to multiple-layered reflection planes. Described here is a method for doing this based on U.S. patent 2,275,735 by R. T. Cloud, filed on June 23, 1939, and issued as a patent on March 10, 1942. We shall take some liberties with Cloud's invention, modernizing it by incorporating adaptive filtering techniques.

Figure 9.11 is based on Cloud's patent. Signals are received by geophones 13 and 14. These are similar to low-frequency microphones which are coupled to the

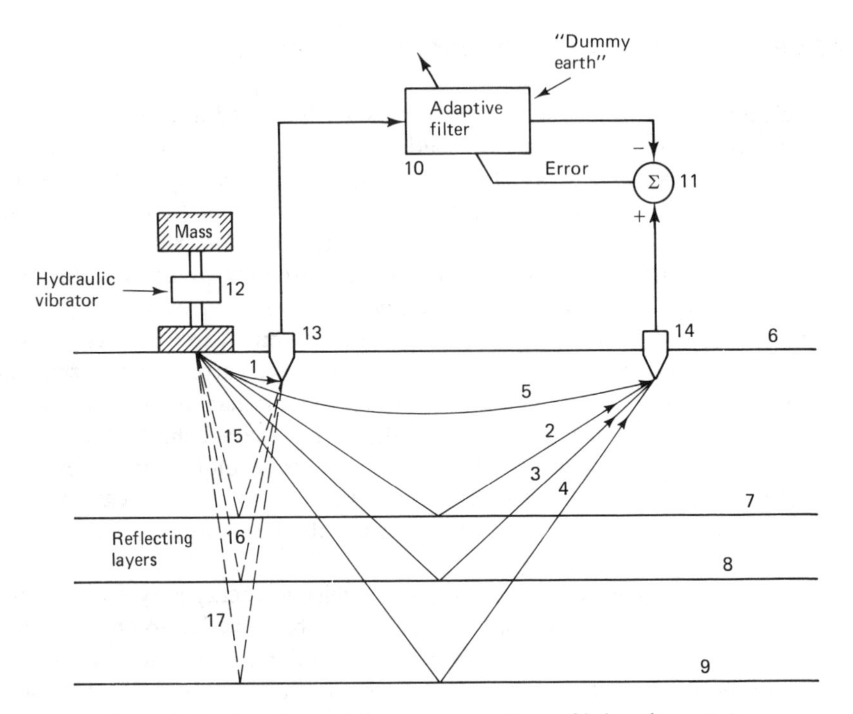

Figure 9.11 Adaptive modeling to measure the earth's impulse response.

earth. They produce output voltage waveforms corresponding to the vertical velocity of motion of the earth's surface in response to seismic waves. These waves propagate along ray paths as indicated in Figure 9.11. Geophone 13 is placed very close to the seismic source. It receives radiation from the source via a direct path 1. We will assume that this path is so short that its propagation delay is negligible. Geophone 13 also receives reflections from layers 7, 8, and 9 via the dashed ray paths 15, 16, and 17. We will assume that the signals of geophone 13 resulting from the dashed paths are negligible in amplitude compared to the strong signals from the nearby seismic source arriving via direct path 1. Thus the output of geophone 13 is synchronized with and corresponds to the earth motion imparted by the seismic source.

The seismic vibrations received by geophone 14 via paths 2, 3, and 4 are the "layer-identifying" signals. Path 5 is a surface duct and is not of geologic importance. If the seismic source in Figure 9.11 were to impart an ideal (perfect) impulse to the earth, the electrical impulse response at geophone 14 might appear as sketched in Figure 9.12. The adaptive filter of Figure 9.11 should develop an impulse response, after convergence, that matches this ideal impulse response. To achieve this result, the seismic source would need to deliver a broadband signal to the earth.

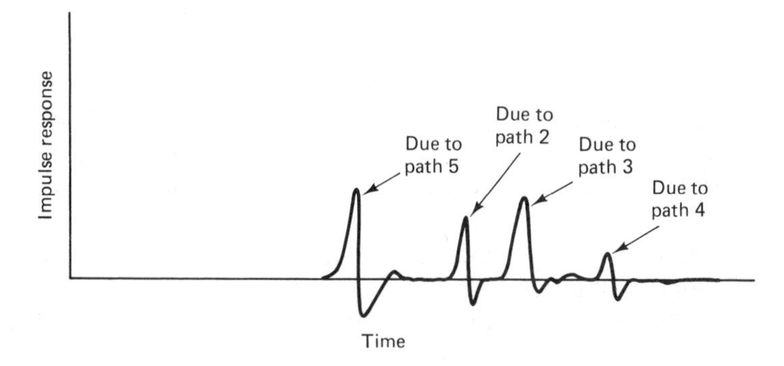

Figure 9.12 Seismic impulse response from source 12 to geophone 14.

This could be a sinusoidal signal of continuously varying frequency as suggested in Cloud's patent, or it could be broadband random noise. The adaptive filter would become, upon convergence, a "dummy earth," quoting Cloud. The time delays of the various reflecting paths would be apparent from the converged adaptive filter impulse response. These delays have important significance in geological studies.

The Cloud patent did not contemplate a dummy earth with a very large number of delay-line taps. The number of taps was chosen to be equal to the presumed number of discrete paths from the vibration source to geophone 14. The tap spacings and tap weights were adjusted manually to make the error as small as possible as seen on an oscilloscope. It is interesting to note that Wiener had not yet developed his filter-optimizing concepts and Cloud was already developing a manually adapted adaptive filter. Minimizing the mean-square error was apparently not a part of Cloud's idea, however. Nor were adaptive techniques available for adapting a large number of weights automatically. Adjusting the delays automatically is still a problem at the time of this writing.

A transversal filter with a very large number of weights would converge to a solution having most of the weights near zero, with a few having large values as required to match the actual character of the geologic multipath. A more efficient modeling process would use weights and taps sparingly as Cloud suggests. To do this with an automatic adaptive filter, it would be necessary to adapt both weight values and delay times. For fixed delays, whether uniformly spaced or not, the mean-square error would be a quadratic function of the weights, with a stationary seismic source. On the other hand, the mean-square error would not be a quadratic function of the tap delay spacing. Methods of adapting the tap spacings are under development [3].

Somehow Cloud's method was lost in the labyrinth of technology. But with the advent of adaptive filtering techniques and digital software and hardware for their implementation, Cloud's method is now a feasible tool in geological exploration. A new patent application modernizing this method has been filed by B. Widrow.

ADAPTIVE MODELING IN FIR DIGITAL FILTER SYNTHESIS

The synthesis of digital filters can be accomplished by making use of adaptive modeling techniques. The basic idea for the synthesis of FIR digital filters is represented in the block diagram of Figure 9.13. The adaptive filter, upon convergence of the adaptive process, will assume an impulse response that best satisfies a set of design specifications. These specifications are embodied in the box labeled "pseudofilter." The pseudofilter does not exist because a filter to meet the exact specifications will generally not be physically realizable. The pseudofilter is conceptual only. Its purpose is to tie the filter synthesis problem to the plant modeling

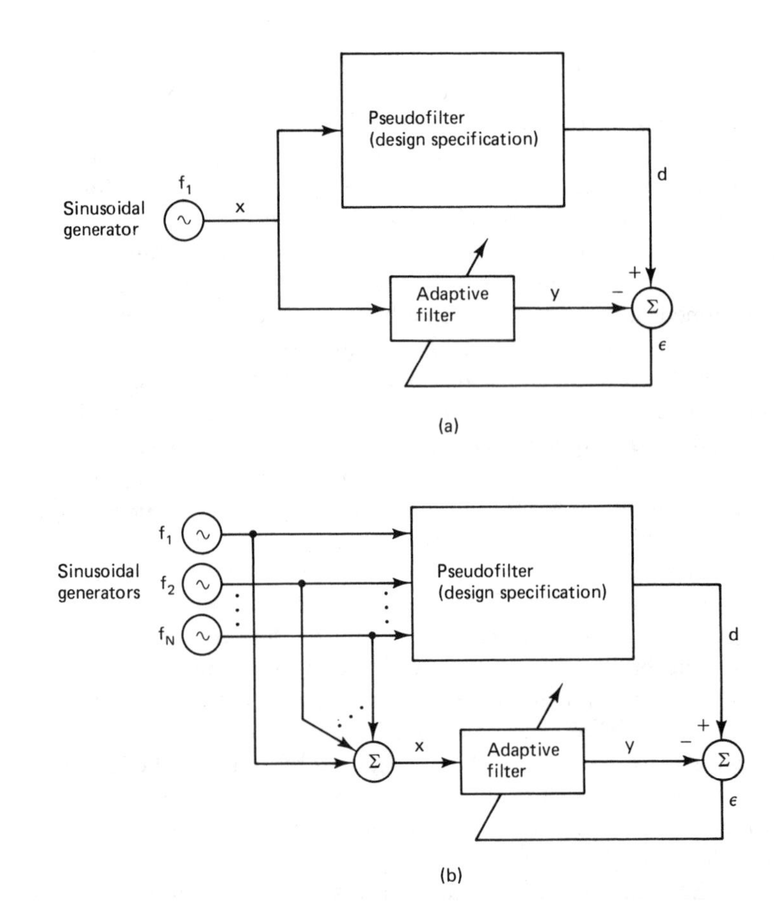

Figure 9.13 Scheme for the adaptive synthesis of a specified filter: (a) excitation of a single frequency, f_1; (b) excitation at N different frequencies.

problem. It need not exist physically in order for filter synthesis to be performed, as we shall see.

Let us assume that the filter specifications are given in the form of a frequency response, as discussed in Chapter 7, that is, a set of requirements that the filter have prescribed gain magnitude and phase characteristics at the discrete frequencies $f_1, f_2, \ldots, f_N$, measured in hertz. Generally, the number of weights to be used in the digital filter will be specified, thus defining L, the size of the adaptive filter. The adaptive process finds a design solution that is a best fit (in the minimum mean-square-error sense) to the specifications.

Referring to Figure 9.13, the adaptive filter models the pseudofilter derived from the design specifications. These specifications cannot in most cases be met perfectly in their entirety. However, we imagine the existence of the pseudofilter having a frequency response, magnitude, and phase perfectly meeting the design specifications.

In Figure 9.13(a), an input sinusoid of the form

$$x(t) = \sin 2\pi f_1 t \tag{9.5}$$

is applied to both the pseudofilter and the adaptive filter. The frequency f_1 is one of the specification frequencies. The output of the pseudofilter, which is assumed to be linear, is

$$d(t) = a_1 \sin(2\pi f_1 t + \theta_1) \tag{9.6}$$

The adaptive filter has this output as its desired response. Notice that it is not necessary for the pseudofilter to exist. Only its output signal is needed for the adaptive process, and this output can be readily constructed knowing the design specifications. The coefficient a_1 is the design response magnitude at frequency f_1, and the angle θ_1 is the design phase shift at frequency f_1.

In order for the specifications to be met (or at least closely approximated) at many frequencies simultaneously, an input comprising a sum of sinusoids, one for each of N specification frequencies, is applied to both the pseudofilter and the adaptive filter in Figure 9.13(b). This input is

$$x(t) = \sum_{l=1}^{N} \sin 2\pi f_l t \tag{9.7}$$

The output of the pseudofilter, the desired response for the adaptive filter, is accordingly

$$d(t) = \sum_{l=1}^{N} a_l \sin(2\pi f_l t + \theta_l) \tag{9.8}$$

When specifications cannot be perfectly met at all frequencies, it is sometimes desirable to have specifications more tightly met at certain frequencies than at others. Certain parts of the design frequency response may be more critical than others. This can be accomplished easily by having input sinusoids with various individual amplitudes, rather than all unit amplitudes, and scaling the components of $d(t)$ accordingly. The larger the amplitude of the input sinusoid, the more tightly will the specification be held at its frequency. When practicing this technique, the lth individual input sinusoid is scaled by c_l and is given by

$$c_l \sin 2\pi f_l t \tag{9.9}$$

with c_l being a positive constant "cost function" for all l. The input signal, again a sum of sinusoids, is now

$$x(t) = \sum_{l=1}^{N} c_l \sin 2\pi f_l t \tag{9.10}$$

The desired response, the output of the pseudofilter, is

$$d(t) = \sum_{l=1}^{N} a_l c_l \sin(2\pi f_l t + \theta_l) \tag{9.11}$$

Once again, the design specification magnitude is a_l and the phase shift is θ_l at frequency f_l.

The adaptive filter converges to a least-squares solution which provides a best least-squares fit to the design specifications. The form of this solution is of interest. The least-squares solution has been discussed in previous chapters, particularly Chapters 2 and 7. The algebraic form of the least-squares solution in the case of the adaptive linear combiner was given first in (2.17). Using the correlation notation in (7.62) and (7.63), the solution is

$$\mathbf{W}^* = \mathbf{R}^{-1}\mathbf{P} = \begin{bmatrix} \phi_{xx}(0) & \cdots & \phi_{xx}(L) \\ \vdots & & \vdots \\ \phi_{xx}(L) & \cdots & \phi_{xx}(0) \end{bmatrix}^{-1} \begin{bmatrix} \phi_{dx}(0) \\ \vdots \\ \phi_{dx}(L) \end{bmatrix} \tag{9.12}$$

As before, $L + 1$ is the number of weights in the adaptive linear combiner. Since we know the signals d and x, we can compute the correlation functions in (9.12). As in Chapter 7, let us define

$$T \triangleq \text{time step between samples} \tag{9.13}$$

Then, from (7.38) and (9.10) we have

$$\begin{aligned} \phi_{xx}(n) &= E[x(t - nT)x(t)] \\ &= E\left[\sum_{l=1}^{N} c_l \sin 2\pi f_l (t - nT) \sum_{m=1}^{N} c_m \sin 2\pi f_m t\right] \end{aligned} \tag{9.14}$$

Since the expected value of the product of two sinusoidal time functions of different frequencies is zero, (9.14) becomes

$$\phi_{xx}(n) = E\left[\sum_{l=1}^{N} c_l^2 \sin 2\pi f_l(t - nT) \sin 2\pi f_l t\right] \tag{9.15}$$

Using the trigonometric identity

$$\sin 2\pi f_l(t - nT) = \sin 2\pi f_l t \cos 2\pi f_l nT - \cos 2\pi f_l t \sin 2\pi f_l nT \tag{9.16}$$

(9.15) becomes

$$\begin{aligned}\phi_{xx}(n) &= E\left[\sum_{l=1}^{N} c_l^2 \sin^2 2\pi f_l t \cos 2\pi f_l nT - c_l^2 \sin 2\pi f_l t \cos 2\pi f_l t \sin 2\pi f_l nT\right] \\ &= E\left[\sum_{l=1}^{N} c_l^2 \sin^2 2\pi f_l t \cos 2\pi f_l nT\right] \\ &= \sum_{l=1}^{N} \frac{1}{2} c_l^2 \cos 2\pi f_l nT \end{aligned} \tag{9.17}$$

In obtaining (9.17), we recall that sine and cosine waves are uncorrelated and that the mean square of a sine wave is half the square of its amplitude. Thus in (9.17) we have all of the elements of the **R**-matrix in (9.12). The elements of the cross-correlation vector **P** may be found in like manner. From (7.37), (7.63), (9.10), and (9.11), we have

$$\begin{aligned}\phi_{dx}(-n) &= E[x(t - nT)d(t)] \\ &= E\left[\sum_{l=1}^{N} c_l \sin 2\pi f_l(t - nT) \sum_{m=1}^{N} a_m c_m \sin(2\pi f_m t + \theta_m)\right] \\ &= E\left[\sum_{l=1}^{N} a_l c_l^2 \sin 2\pi f_l(t - nT) \sin(2\pi f_l t + \theta_l)\right] \\ &= E\left[\sum_{l=1}^{N} a_l c_l^2 \sin(2\pi f_l t - 2\pi f_l nT - \theta_l) \sin(2\pi f_l t)\right] \\ &= E\left[\sum_{l=1}^{N} \frac{1}{2} a_l c_l^2 \cos(2\pi f_l nT + \theta_l)\right] \end{aligned} \tag{9.18}$$

Using (9.17) and (9.18) in (9.12), we can write explicitly the least-squares solution for the adaptive filter weights. The solution is

$$\mathbf{W}^* = \begin{bmatrix} \sum_{l=1}^{N} c_l^2 & \sum_{l=1}^{N} c_l^2 \cos 2\pi f_l T & \cdots & \sum_{l=1}^{N} c_l^2 \cos 2L\pi f_l T \\ \sum_{l=1}^{N} c_l^2 \cos 2\pi f_l T & \sum_{l=1}^{N} c_l^2 & & \\ \vdots & & \ddots & \\ \sum_{l=1}^{N} c_l^2 \cos 2L\pi f_l T & & & \sum_{l=1}^{N} c_l^2 \end{bmatrix}^{-1}$$

$$\times \begin{bmatrix} \sum_{l=1}^{N} a_l c_l^2 \cos(\theta_l) \\ \sum_{l=1}^{N} a_l c_l^2 \cos(2\pi f_l T + \theta_l) \\ \vdots \\ \sum_{l=1}^{N} a_l c_l^2 \cos(2L\pi f_l T + \theta_l) \end{bmatrix} \tag{9.19}$$

Some simplifications in this formulation result when the various specification frequencies are uniformly spaced and when the input sine waves are all of the same amplitude. Equation (9.19) provides a general form of the solution.

The least-squares solution in (9.19) does not in itself give much insight into the filter design. In some ways the adaptive process in Figure 9.13 is much more appealing. The least-squares solution is often valuable in the computer implementation of the design process, however, because the **R**-matrix and the **P**-vector in (9.19) can be computed directly from the design specifications. If the adaptive weights are fixed at $\mathbf{W}^*$, the least-squares solution, the mean-square error is

$$\xi_{\min} = \sum_{l=1}^{N} \frac{1}{2} c_l^2 |S_l - H_l^*|^2 \tag{9.20}$$

where S_l is the complex transfer function of the specified pseudofilter at frequency f_l, that is,

$$S_l \triangleq a_l e^{j\theta_l} \tag{9.21}$$

and H_l^* is the complex transfer function of the optimal adaptive linear combiner with weights $\mathbf{W}^*$ at frequency f_l. In Chapter 7 we saw that the transfer function for the adaptive transversal filter is

$$H^*(z) = \sum_{n=0}^{L} w_n^* z^{-n} \tag{9.22}$$

and we also saw that $e^{j\omega} = e^{j\Omega T}$ is substituted for z to obtain the frequency response. The normalized angular frequency corresponding to f_l is, of course,

$$\omega_l = 2\pi f_l T \tag{9.23}$$

so H_l^* in (9.20) becomes

$$H_l^* = \sum_{n=0}^{L} w_n^* e^{-j2\pi n f_l T} \tag{9.24}$$

In obtaining (9.20) we have used the idea that the error is a sum of sinusoids and that the power of the sum is the sum of powers when the sinusoids differ in frequency.

The cost weighting can be determined by "cut and try." No general analytical procedure for its determination has been devised so far. When designing a filter, one might make all the c's equal at the outset, then examine the resulting filter frequency response, then increase the values of the c's at frequencies where one wishes closer adherence to specifications, and so on. Experience has shown that this is not generally a difficult process. Although one would really like to have an analytical procedure to determine the c's, the cut-and-try method seems to be quite efficient.

Without using the least-squares solution (9.19), one can adjust the adaptive filter using the LMS algorithm and thereby closely approximate the optimal solution. The procedure is simple and involves generating the sums of sinusoids in (9.10) and (9.11) to provide the proper training inputs for the adaptive filter, as illustrated in Figure 9.13. The adaptive method is often used when the filter to be designed requires a large number of weights. Using the formula in (9.19) would require the solution of a large number of simultaneous linear equations. If the number of weights is 256, for example, one would need to solve 256 linear equations for 256 unknowns. Depending on computer characteristics, one might find it more convenient to use the LMS algorithm for filter design rather than to use the analytical approach.

Because of the flexibility of this synthesis approach, unusual filter designs can be realized in addition to the more classic low-pass, high-pass, and bandpass designs. Figure 9.14 shows the results of an attempt to design a 50-weight filter by specifying its frequency response at 100 frequencies spaced evenly over the entire Nyquist interval, from zero to one-half the sampling frequency. The goal was to design a filter whose amplitude response was specified to vary linearly on a dB scale from -50 dB at zero frequency to 0 dB at one-fourth the sampling frequency, and to have a constant gain of -60 dB from one-fourth to one-half the sampling frequency. The specification is indicated by small crosses on the phase and magnitude plots. The

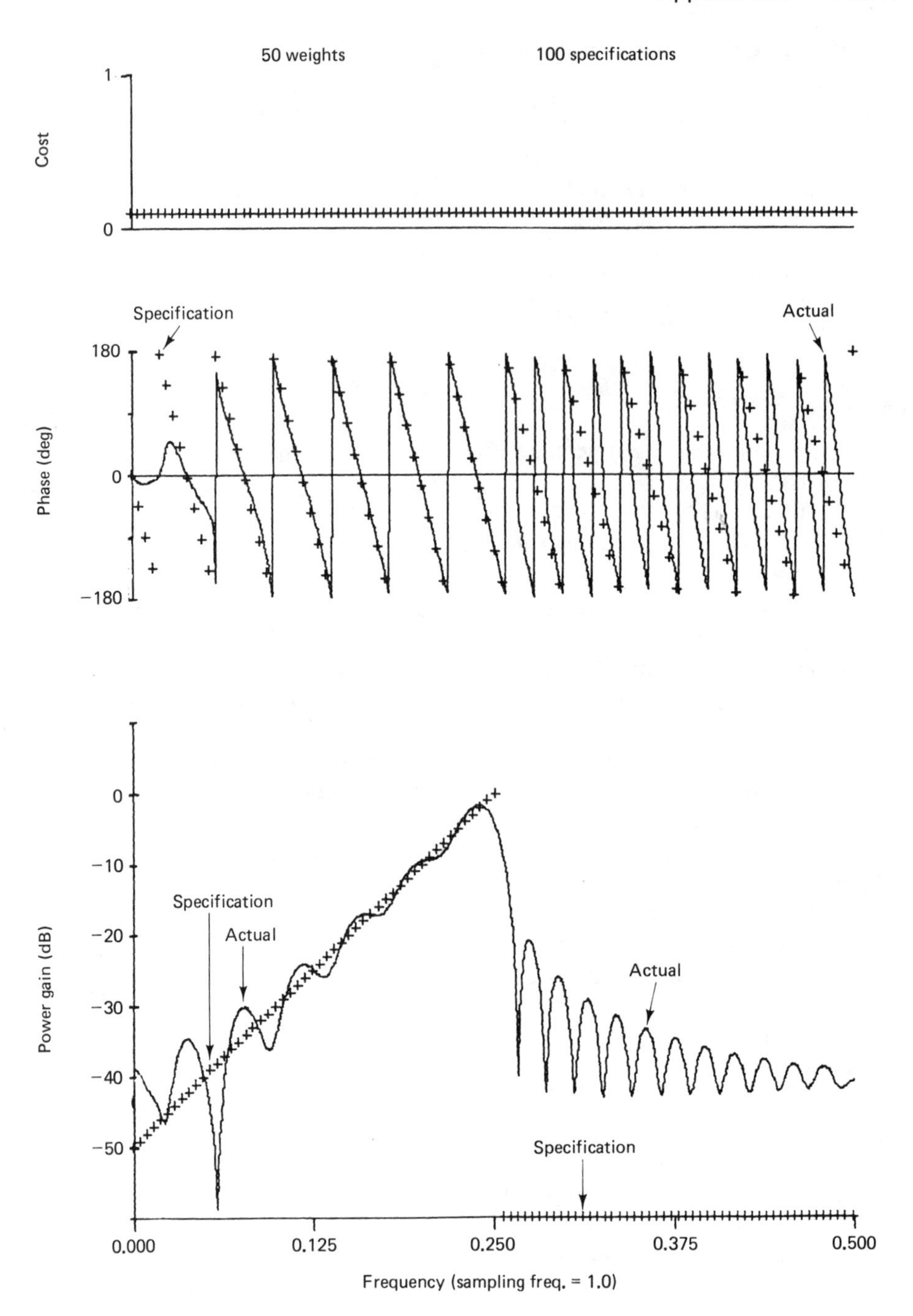

Figure 9.14 Adaptive synthesis of an unusual digital filter specification using a uniform cost function.

least-squares solution is superimposed on these plots. The specification required a linear phase characteristic providing a phase shift of about -13π radians from zero frequency to one-fourth the sampling frequency. Note that the phase and amplitude specification were more closely met where the filter gain magnitude was high (see Exercise 24).

In Figure 9.14, the c's, comprising the cost function, were chosen equal, as shown at the top of the figure. The same design was attempted again, but this time the c's were not all chosen to be equal. The results are shown in Figure 9.15. By increasing the values of the c's at some frequencies and reducing the values at other frequencies, an improved frequency response characteristic resulted. The design results were displayed using an interactive computer terminal. The effects of changes in the c's were immediately visible. Raising the c's of course causes tighter adherence to the specifications at the corresponding frequencies and reducing the c's has the opposite effect. One must closely compare the results shown in Figures 9.14 and 9.15 to see how the varying cost function was used. In some cases, a single isolated c-value was increased to suppress an individual undesired lobe of the frequency response.

In many cases, a linear-phase characteristic or a zero-phase-shift characteristic is desired. As we saw in Chapter 7, linear phase corresponds to a time delay while zero phase corresponds to no time delay. Figure 9.16 shows first a noncausal filter that has an exact phase shift of zero at all frequencies, then a causal version that is the same except for a time delay.

The tap weights are chosen in a symmetrical fashion about the central weight. Using the relationships in Chapter 7, we can easily show the linear phase characteristics of these filters. For the upper filter we have

$$H(z) = w_0 + \sum_{l=1}^{L} w_l(z^l + z^{-l})$$

$$H(e^{j\omega}) = w_0 + 2\sum_{l=1}^{L} w_l \cos \omega \tag{9.25}$$

Since the transfer function is real, the phase must be zero. For the lower filter we have the same transfer function multiplied by z^{-L} (i.e., by $e^{-jL\omega}$): hence the linear-phase characteristic. In the event that there is an even number of weights in the linear-phase filter, the two central weights must be made equal and the other weights chosen symmetrically about the central weights.

An adaptive process to adjust the linear-phase weights to minimize mean-square error while maintaining the symmetry constraint can be obtained by modifying the LMS algorithm. Referring to the zero-phase diagram in Figure 9.16(a), one adapts the central weight w_0 in accord with (6.3):

$$w_{0,k+1} = w_{0k} + 2\mu\varepsilon_k x_k \tag{9.26}$$

The weights surrounding w_0 must then be initialized in pairs to equal values and

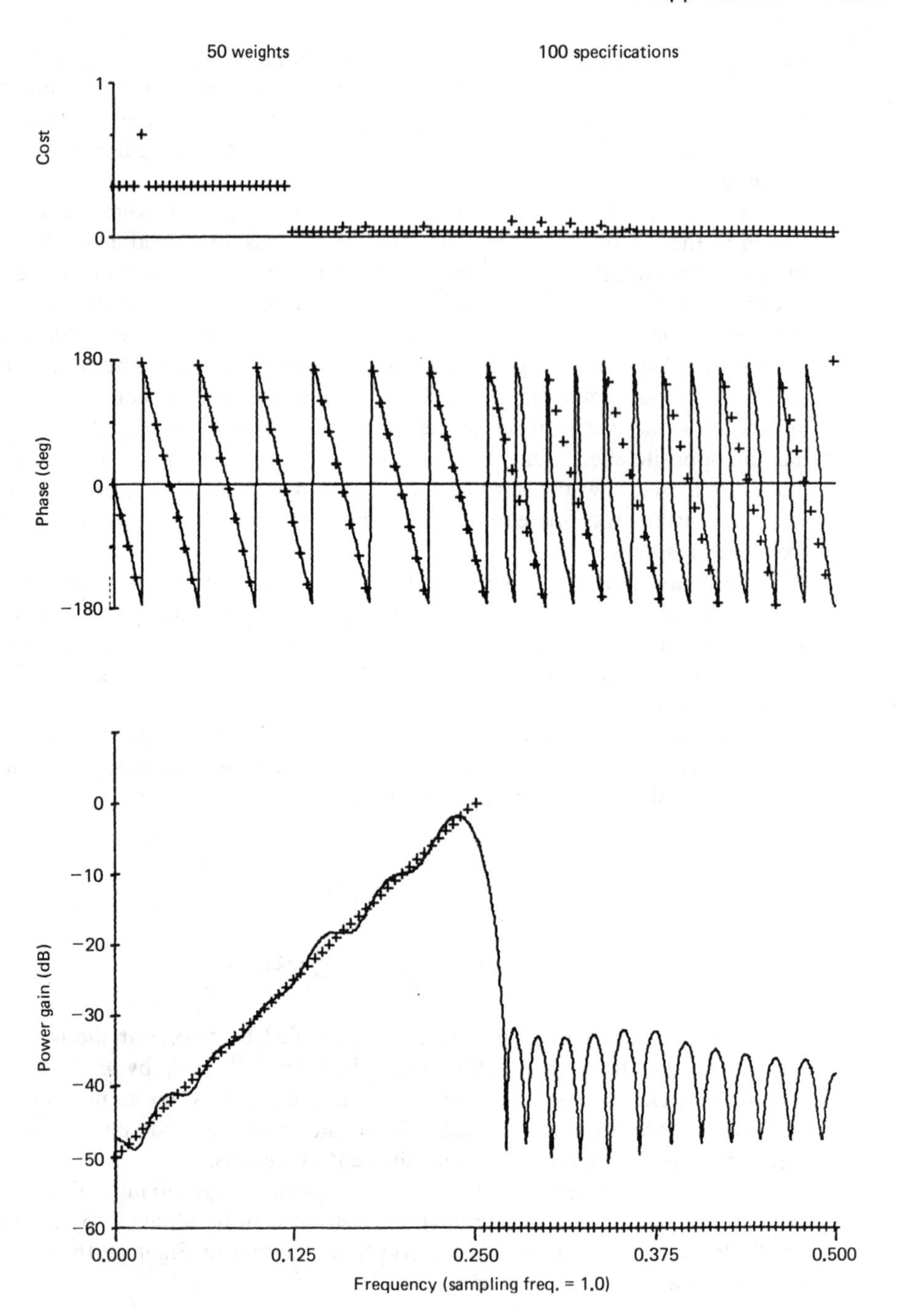

Figure 9.15 Adaptive filter synthesis as in Figure 9.14, but with a nonuniform cost function.

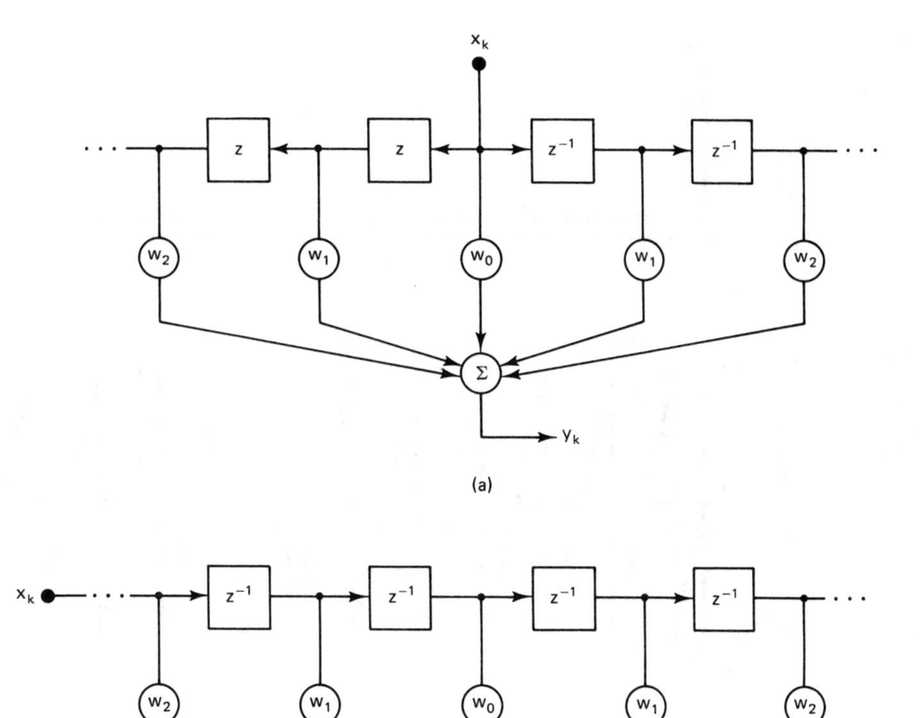

Figure 9.16 (a) Noncausal zero-phase filter; (b) causal linear-phase filter. The two versions are identical except for a time delay.

then adapted identically in pairs according to

$$w_{-l,k+1} = w_{l,k+1} = w_{lk} + \mu\varepsilon_k(x_{k+l} + x_{k-l}) \qquad 1 \le l \le L \tag{9.27}$$

We can see that this algorithm will satisfy the symmetry constraint, ensuring a linear phase characteristic. It is also clear that with the weights locked together in symmetric pairs, the mean-square error is a quadratic function of the weights (although the number of degrees of freedom is roughly equal to half the number of weights). The gradient estimate is unbiased, so the algorithm is again a steepest-descent algorithm.

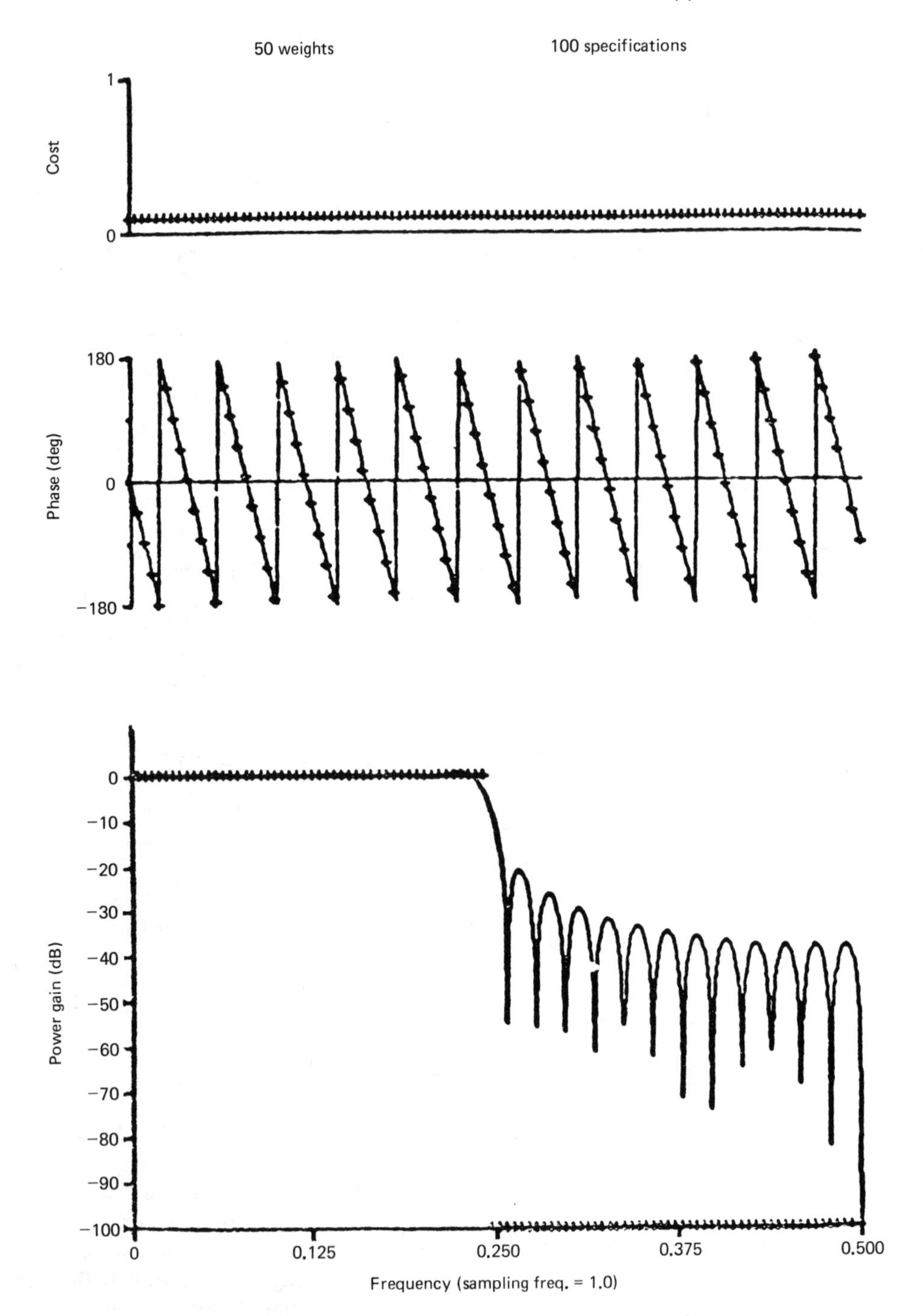

Figure 9.17 Adaptive design of a low-pass filter with linear phase, using a uniform cost function.

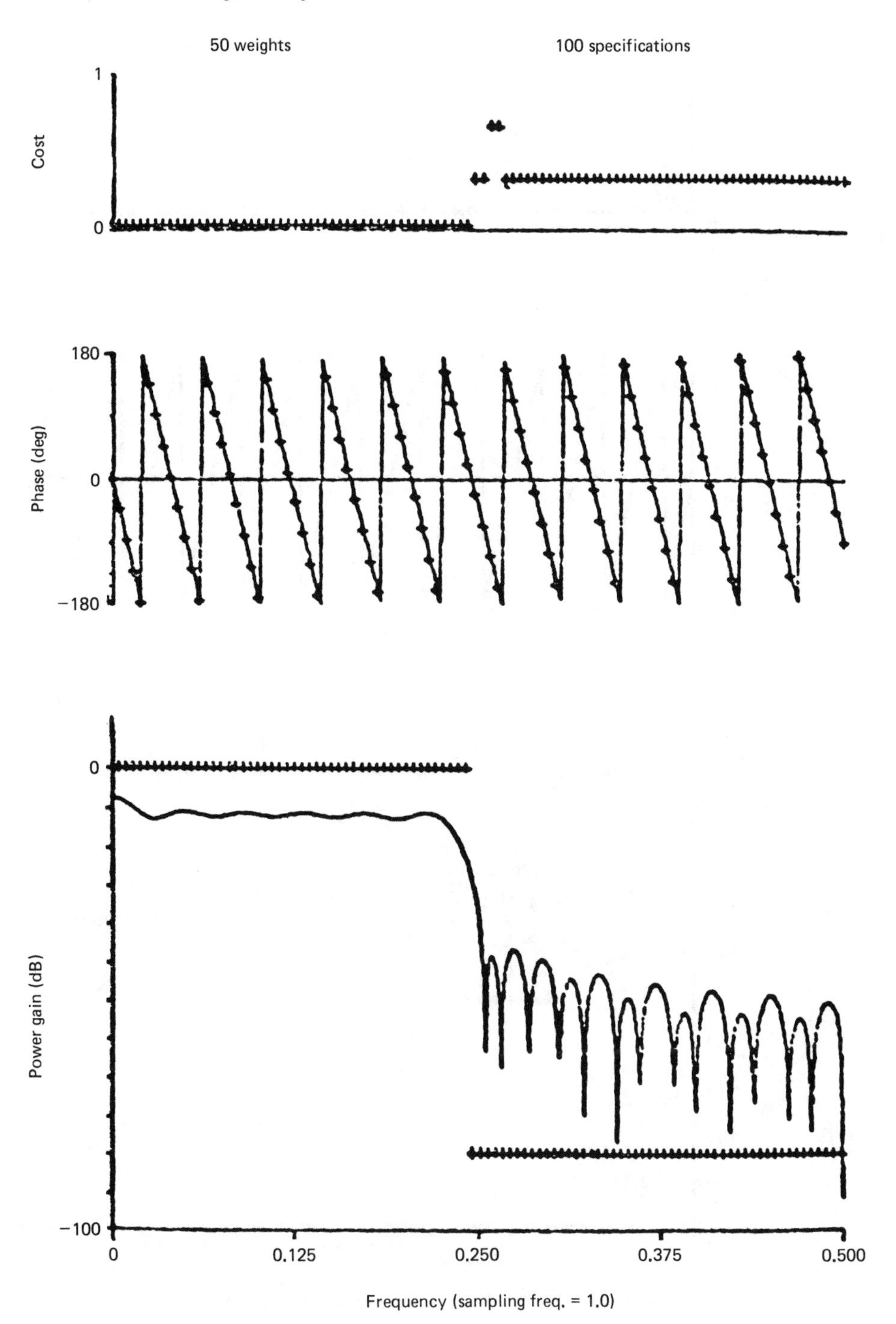

Figure 9.18 Adaptive design similar to Figure 9.17 but with a nonuniform cost function.

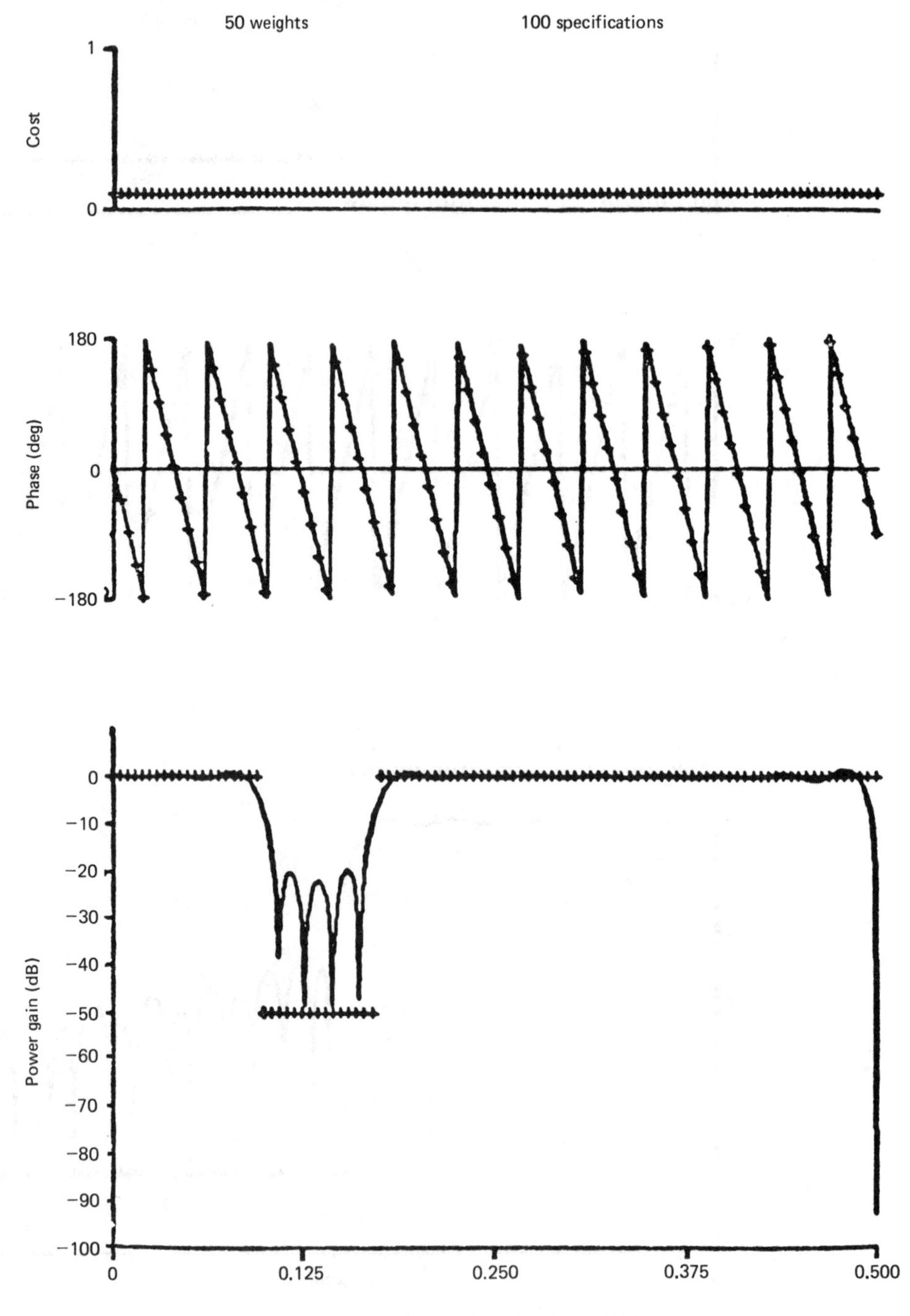

Figure 9.19 Adaptive design of a linear-phase notch filter.

Thus the algorithm in (9.26) and (9.27) will always give a linear-phase characteristic, but will otherwise permit a wide variation in frequency response. Let us now consider an example.

Figure 9.17 shows the results of an attempt to synthesize a 50-weight low-pass filter with linear phase. As expected, the phase response meets the specification, as shown in the figure. The amplitude specification is almost perfectly met between zero frequency and one-fourth the sampling rate. Between one-fourth and one-half the sampling rate, the specification calls for a gain magnitude of -100 dB, as shown in the lower plot. This specification was not achieved, but the gain everywhere in that range was lower than -20 dB and in most places was lower than -30 dB. This form of the filter, although suboptimal, would be useful for some applications.

Better results could have been obtained by using more weights, because the 50-weight filter could control the complex gain only at 25 frequencies. On the other hand, better results may also be obtained by adjusting the c's. For example, the cost function was adjusted as shown in Figure 9.18. Again the phase response meets the specification exactly as expected. To obtain the amplitude response, the cost function was reduced in the passband and increased in the stop band. The result was a 10-dB loss of gain in the passband, but the stop-band gain was substantially reduced to below -40 dB. This makes the stop-band gain at least 30 dB below the passband gain, thus producing an improved design. The cost function was made especially high at a few frequency points to suppress the strongest unwanted lobes in the stop band, near the transition between passband and stop band.

As a final example, a notch filter with a broadband notch and a linear-phase characteristic was designed using these methods. The results are shown in Figure 9.19. A flat response was obtained in the passbands. The stop-band gain specification of -50 dB was not achieved perfectly, but at least -20 dB was obtained. Other designs may be produced similarly, and the method may be especially useful in cases where the specifications are unusual and analytical "cookbook" methods therefore do not exist.

EXERCISES

1. The modeling problem shown below is similar to the one discussed at the beginning of the chapter. Assume that x_k is given by

$$x_k = \text{RANDOM}(1.) - 0.5 \qquad 0 \leq k \leq 500$$

with the random numbers coming from the routine in Appendix A, initialized with 12357 as shown in the appendix. For this exercise assume that there is no noise, let $L = 1$ and $\mu = 0.1\mu_{\text{max}}$, and make plots of w_{0k} and w_{1k} versus k over a range of k long enough to show convergence.

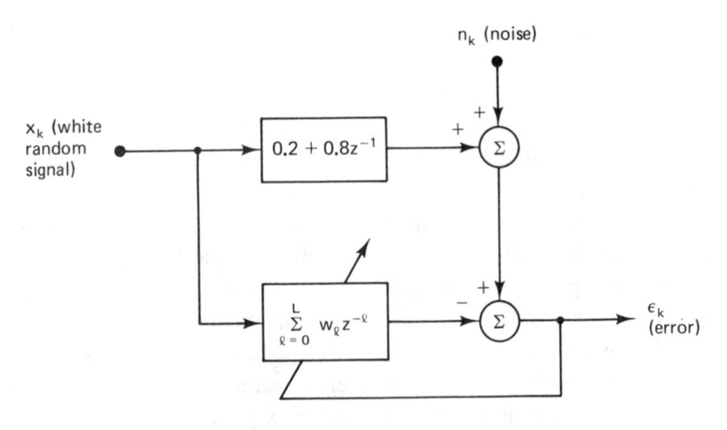

2. What is the theoretical learning-curve time constant in Exercise 1?
3. Check the answer to Exercise 2 by running Exercise 1 100 times (do not reinitialize the random routine) and plotting a curve of $\hat{E}[\varepsilon_k^2]$ versus k over a range large enough to show convergence.
4. Repeat Exercise 3 with $\mu = 0.05\mu_{\max}$ and explain the different learning curve.
5. Repeat Exercise 1 with $L = 3$. Plot all four weights on the same graph and explain the behavior of the weights versus k.
6. Repeat Exercise 3, this time using independent white noise, $[n_k]$, with $E[n_k^2] = 0.00833$. Explain the different learning curve. Comment on the learning-curve time constant.
7. For the system shown below, let x_k again be generated as in Exercise 1, let $\mu = 0.2\mu_{\max}$, let $L = 2$, and assume no noise. Make a theoretical plot of $E[\varepsilon_k^2]$ and an experimental plot of ε_k^2 versus k on the same graph.

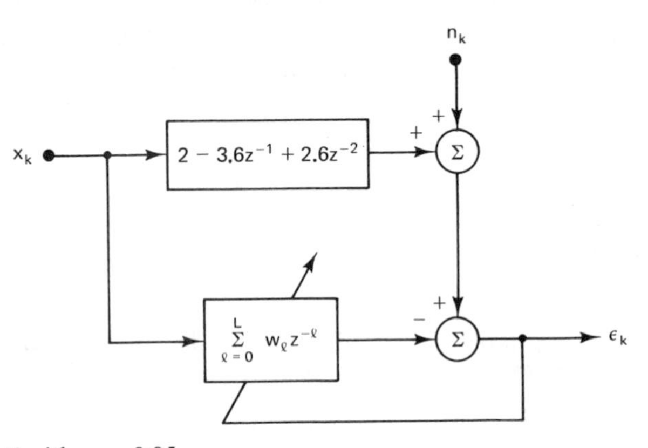

8. Do Exercise 7 with $\mu = 0.05\mu_{\max}$.
9. Do Exercise 7 with $[n_k]$ a white-noise sequence having power $E[n_k^2] = 0.01$.
10. Assume the conditions of Exercise 7, but let $L = 1$. Without reinitializing the random routine, make 100 adaptive runs out to $k = 200$ iterations, and plot an experimental curve of $E[\varepsilon_k^2]$ versus k. Comment on the excess MSE and the learning time constant.

11. Assume the conditions of Exercise 7, but this time with $L = 3$. Make three adaptive runs without reinitializing the random routine, and plot the weights versus k for each run. Compare the curves and comment.
12. Assume again the conditions of Exercise 7, but for this exercise let x_k be a unit sinusoid,

$$x_k = \sin\frac{2\pi k}{15}$$

and let $L = 1$. Make an experimental plot of ε_k^2 as well as the two weights versus k, and discuss the results.
13. In the modeling situation shown below, describe the conditions under which ε_k can be driven to zero. Assume that x_k is as described in Exercise 1.

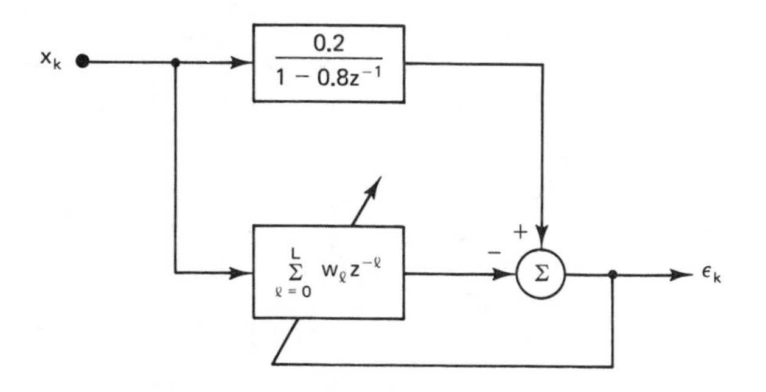

14. For the conditions in Exercise 13, plot a theoretical curve of $E[\varepsilon_k^2]_{\min}$ versus L.
15. In addition to the conditions in Exercise 13, let $L = 8$ and $\mu = 0.2\mu_{\max}$. Plot a curve of ε_k^2 versus k over a range of k long enough to show convergence.
16. Make an adaptive run with the multiple-signal configuration in Figure 9.4 using the following plant:

$$H_{11}(z) = 1 - z^{-1}$$

$$H_{12}(z) = H_{21}(z) = 0.4z^{-1}$$

$$H_{22}(z) = 1 + 0.8z^{-1} + 0.8z^{-2}$$

Let $[x_{1k}]$ and $[x_{2k}]$ be white random sequences generated by taking *alternate* samples from $[x_k]$ in Exercise 1, starting with x_{10}, then x_{20}, then x_{11}, and so on. Let each adaptive model be of the form

$$G(z) = w_0 + w_1 z^{-1} + w_2 z^{-1}$$

and let $\mu = 2$. Make the run long enough to show convergence and plot ε_{1k}^2 and ε_{2k}^2 versus k to illustrate convergence.
17. Explain in your own words why the term "spread spectrum" is used when information bits are represented by pseudorandom sequences.

18. To provide an exercise that involves adaptive modeling of a multipath channel, we modify Figure 9.7 as shown below, with the following specifications:

PN sequence: 11101000
Impulse response of
multipath channel: $H(z) = 1 - 0.5z^{-1} + 0.25z^{-2} + 0.4z^{-7} - 0.2z^{-8} + 0.12z^{-9}$
Adaptive filter length: $L + 1 = 12$
$\mu = 0.4$

Using the repeated PN sequence, run the adaptive process to convergence. Plot ε_k^2 versus k. Plot and compare the channel and adaptive impulse responses.

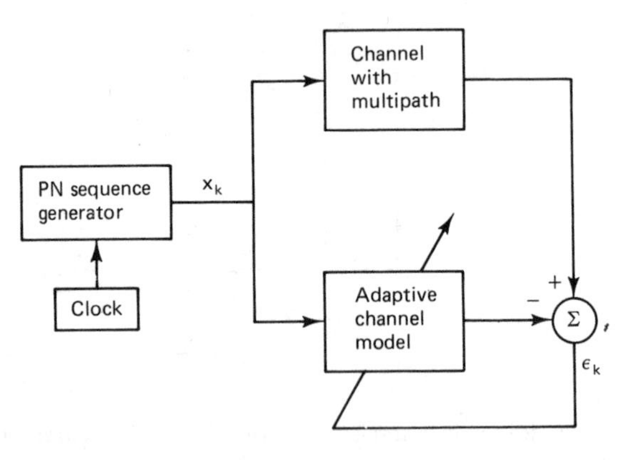

19. Do Exercise 18 using the PN sequence 1111000011010010. Comment on any overall changes in performance.

20. The diagram below represents a situation similar to Figure 9.11, where we are attempting to measure the earth's impulse response. Let the earth response be

$$H(z) = z^{-102} + z^{-112} + z^{-119}$$

which represents three earth paths (delays) from source to geophone. The delay, z^{-100}, is

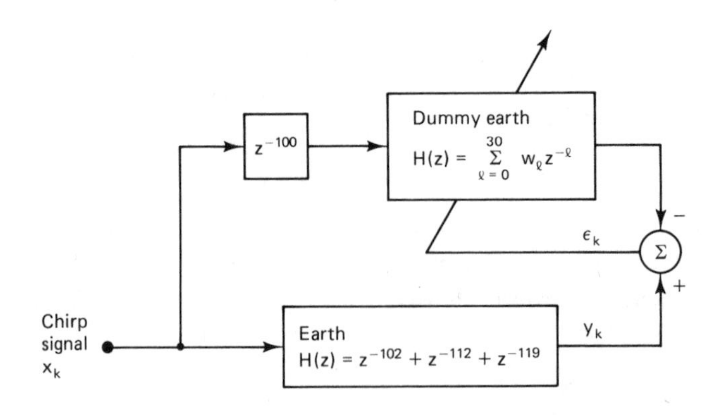

to compensate for most of the path delay. Let the signal x_k consist of chirps of the form

$$\sin\frac{20\pi k}{220 - k}$$

repeated at $k = 0, 200, 400, \ldots$. First plot the earth response, y_k. Then select an appropriate value for μ and plot ε_k^2 versus k to convergence. Discuss the form of the converged weight vector.

21. In the adaptive filter synthesis application, suppose that the input, x_k, consists of N unit sine waves spaced evenly at N frequencies from zero up to (but not including) one-half the sampling rate. Provide a simple formula for x_k as a function of k.

22. Do Exercise 21 using cosines in place of sines.

23. Suppose that a filter is to be synthesized adaptively as in Figure 9.13(b), using equal excitation of 16 frequencies spaced evenly as described in Exercise 21. The pseudofilter has unit gain at all frequencies and phase shift, θ, shown below. Assume a uniform cost function and 12 adaptive weights. Select an appropriate μ and run the LMS algorithm to convergence. Then plot the gain and phase shift of the adaptive filter. Suggest possible subsequent modifications to the cost function.

24. When a filter is synthesized as in Figure 9.13(b) and the pseudofilter has varying amplitude gain, explain why the synthesis tends to be more accurate at frequencies where the pseudofilter gain is high.

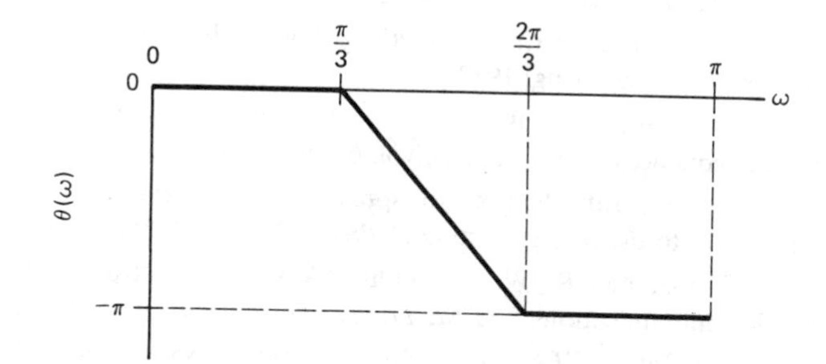

25. A digital signal s_k is transmitted through a linear transmission channel that introduces both distortion and additive noise. The transfer function of the channel is

$$H_c(z) = \frac{1}{1 - 0.3z^{-1}}$$

The channel noise (referred to the output) is uncorrelated with the signal and has autocorrelation

$$\phi_{nn}(n) = a\delta(n)$$

At the receiving (output) end, a Wiener filter $H_o(z)$ (which could be causal or noncausal) is to be used to unravel the effects of the channel while minimizing noise. The desired receiver output is the signal s_k itself. Its autocorrelation is

$$\phi_{ss}(n) = \delta(n)$$

(a) Find an expression for $H_o(z)$.
(b) Let $a = 0$. Find $H_o(z)$. Explain this result.
(c) Let $a = 1$. Find $H_o(z)$, and find h_{ok}, the optimal impulse response.

ANSWERS TO SELECTED EXERCISES

2. $\tau_{mse} = 5$ iterations.

6. τ_{mse} is not affected.

REFERENCES AND ADDITIONAL READINGS

1. S. W. Golomb (Ed.), *Digital Communications with Space Applications*. Englewood Cliffs, N.J.: Prentice-Hall, 1964.
2. R. C. Dixon (Ed.), *Spread-Spectrum Techniques*. New York: IEEE Press, 1976.
3. D. M. Etter and S. D. Stearns, "Adaptive estimation of time delays in sampled data systems," *IEEE Trans. Acoust. Speech Signal Process.*, vol. ASSP-29, pp. 582–587, June 1981.
4. B. Widrow, P. F. Titchener, and R. F. Gooch, "Adaptive design of digital filters," *Proc. ICASSP-81*, pp. 243–246.
5. D. L. Soldan, "A comparison of adaptive algorithms used for designing phase compensation filters," *Proc. ICASSP-81*, pp. 542–545.
6. *IEEE Transactions on Communications*, Special Issue on Spread-Spectrum Communications, pp. 745–865, Aug. 1977.
7. D. V. Sarwate and M. B. Pursley, "Crosscorrelation properties of pseudorandom and related sequences," *Proc. IEEE*, vol. 68, pp. 593–619, May 1980.
8. G. L. Turin, "Introduction to spread-spectrum antimultipath techniques and their application to urban digital radio," *Proc. IEEE*, vol. 68, pp. 328–353, Mar. 1980.
9. G. R. Cooper and R. W. Nettleton, "A spread-spectrum technique for high-capacity mobile communications," *IEEE Trans. Veh. Technol.*, pp. 264–275, 1978.
10. T. Kailath (Ed.), *IEEE Trans. Autom. Control*, Special Issue on System Identification and Time Series Analysis, vol. AC-19, no. 6, Dec. 1974.
11. D. Graupe, *Identification of Systems*, New York: Van Nostrand Reinhold, 1972.
12. P. Eykhoff, *System Identification*. New York: Wiley, 1974.
13. J. G. Proakis, *Digital Communications*. New York: McGraw-Hill, 1983, Chap. 8.

chapter 10

Inverse Adaptive Modeling, Equalization, and Deconvolution

In Chapter 9 the fundamental subject of direct modeling and adaptive system identification was discussed together with several applications of the technique to practical problems. In this chapter we consider another form of modeling, inverse modeling, again with examples that show how the method is used in solving practical problems. The inverse model of a system having an unknown transfer function is itself a system having a transfer function which is in some sense a best fit to the reciprocal of the unknown transfer function. Sometimes the inverse model response contains a delay which is deliberately incorporated to improve the quality of the fit.

The best way to appreciate the importance of inverse modeling is to visualize useful applications. One such application to speech processing was mentioned at the end of Chapter 8. Applications to communications systems and to digital filter synthesis are described in this chapter. Applications to control systems are also possible, in which an inverse model of an unknown plant is used to generate control command signals to the plant. Adaptivity is required when the plant characteristics are unknown or slowly varying. These applications to control systems will be described more extensively in Chapter 11. Other applications (e.g., to the deconvolution of seismic data) are described in the literature [18, 31].

In the communication applications in this chapter, we discuss dispersive channels such as telephone and radio channels. A dispersive channel is one in which signals at different frequencies travel with different velocities, or with different group delays. To understand these terms, suppose that we have a channel whose transfer function (as described in Chapter 7) is $H(\omega)$, where ω is the normalized frequency. If we multiply $H(\omega)$ by $\exp(-jn\omega)$, we introduce a *delay* of n time steps into the channel. More generally, if we multiply $H(\omega)$ by $\exp(-j\Theta)$, where Θ is a function of ω, we introduce a *group delay* of $d\Theta/d\omega$ time steps into the channel [1]. In a physical channel such as an acoustic path or a radio link, the overall or total group delay is equal to the distance in the channel divided by the propagation velocity of the signal. The channel is said to be *dispersive* [2] if the group delay is a nonconstant function of frequency (i.e., if $d^2\Theta/d\omega^2 \neq 0$).

In a dispersive channel, an adaptive filter can be placed at the receiving end to "equalize" the channel, that is, to provide a frequency and phase response in the signal pass band which is the inverse or reciprocal of that of the channel itself, and thus compensate for the dispersion. The waveform of the transmitted signal arrives at the receiver convolved with the impulse response of the channel. The equalization filter at the receiver input "deconvolves" the channel characteristic and restores the original waveform. Once again, adaptivity is required when the channel characteristics are unknown or slowly time varying.

GENERAL DESCRIPTION OF INVERSE MODELING

One way to perform inverse modeling is shown in Figure 10.1(a). The unknown transfer function to be inverse modeled, called the plant in control systems terminology, is seen with input s_k. Noise n_k is added to its output to represent internal plant noise. The noisy plant output x_k is available as an input to the adaptive filter. Upon convergence, the adaptive filter output is a best least-squares match to the plant input. As in previous chapters, the minimum mean-square error generally depends on the number of weights in the inverse model, and a better fit giving lower mean-square error is sometimes possible if one is willing to add more weights to the adaptive system.

Assume for the moment that the noise n_k is zero and assume that upon convergence, the error ε_k is very small. Let the value of μ in the LMS algorithm (Chapter 6) be small so that convergence is slow with negligible misadjustment. The adaptive filter impulse response then is essentially static. Its transfer function will be the reciprocal of the transfer function of the plant. The transfer function of the plant cascaded with the converged adaptive inverse will be unity, and the impulse response of the combination will be a unit impulse with no delay.

The ability to form an inverse with low mean-square error will generally be limited by three factors:

1. The presence of plant noise n_k causes the output of the adaptive inverse filter to be noisy, raising the mean-square error. Also, the noise n_k affects the converged impulse response of the adaptive inverse filter. This impulse response is a least-squares solution to the problem of simultaneously rejecting the noise n_k and providing an inverse to the plant. In the presence of noise, the transform of the converged adaptive impulse response, a least-squares solution, will not generally be the reciprocal of that of the plant.

2. The plant is generally a causal system, and generally the signal s_k will be delayed as it goes through a physical plant. Such conditions would require the inverse to be a predictor, a task that can only be performed approximately by a causal adaptive filter in a statistical sense. In many applications a delayed inverse is acceptable, alleviating the need for prediction in the adaptive filter. The scheme of

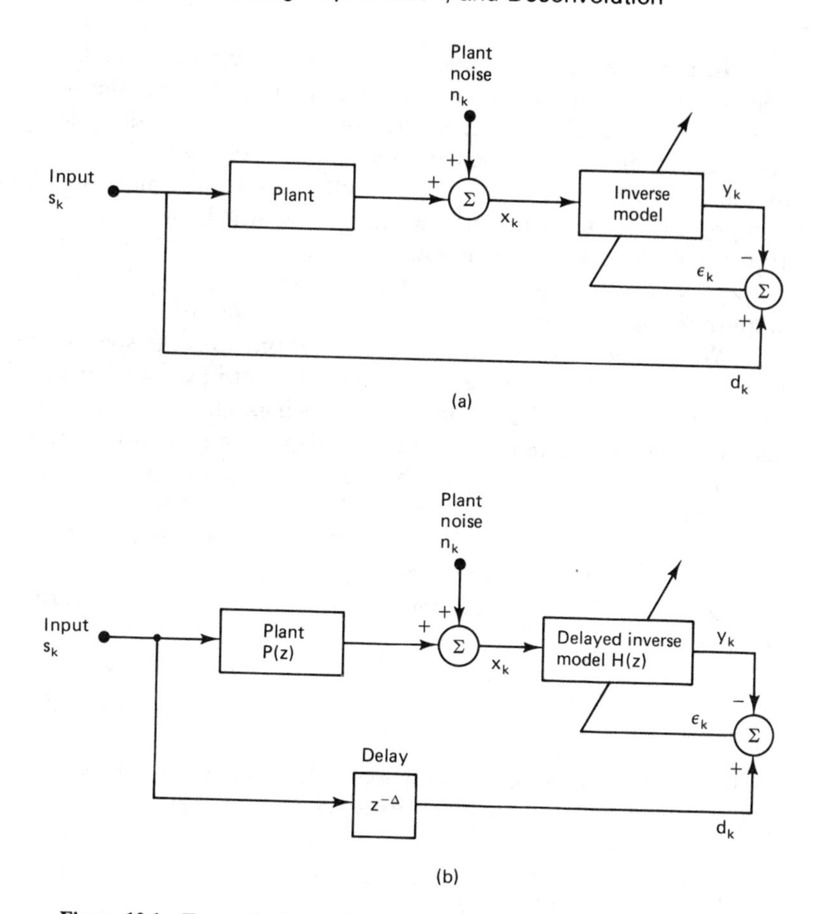

Figure 10.1 Types of adaptive inverse modeling: (a) undelayed; (b) delayed.

Figure 10.1(b) produces such an inverse. Inclusion of the delay of Δ samples generally permits much lower values of minimum mean-square error and causes the converged adaptive impulse response, when convolved with that of the plant, to approximate an impulse with a delay of Δ. A delayed inverse is also advantageous whenever the plant is non-minimum-phase, that is, when the plant has zeros outside the unit circle in the z-plane. A reciprocal transfer function would then have poles outside the unit circle. In order for such an inverse to be stable, the impulse response would need to be left-handed in time (i.e., noncausal). A *delayed* non-causal impulse response could be approximated, however, by a causal impulse response truncated in time.

3. The adaptive filter, when realized as an adaptive transversal filter, has a finite impulse response. Such an impulse response can only approximate an infinite impulse response when the latter is required to realize the optimal inverse.

In the absence of the noise n_k, excellent inverses are realizable when enough weights are utilized within the adaptive filter, and when the delay Δ is suitably chosen. The latter choice is generally not critical. When a delay in the inverse is not disadvantageous to the particular application, a good rule of thumb is to let Δ be set to half the time length of the adaptive filter. Generally, much smaller delays would also prove to be effective in producing inverses which, in the configuration of Figure 10.1(b), would perform with low mean-square error. If noise n_k is present, its properties may have some effect on the best choice of Δ. Also, if n_k is of significant amplitude, distortion of the inverse solution will result.

We will now derive least-squares solutions to some specific inverse modeling problems. These least-squares solutions represent ideal performance in that misadjustment is not considered for the time being. This level of performance can be achieved only in the limit as the rate of adaptation is reduced toward zero.

Refer to Figure 10.1(b). The transfer function of the plant is $P(z)$. The autocorrelation function of the input, s, is $\phi_{ss}(l)$ and the power spectrum [Equation (7.42)] is $\Phi_{xx}(z)$. The delayed inverse filter has the transfer function $H(z)$ as shown in the figure. Let us assume that $H(z)$ represents a nonrecursive adaptive filter having an infinite two-sided impulse response. Then the performance surface is, from (7.67),

$$\xi = \phi_{dd}(0) + \sum_{l=-\infty}^{\infty} \sum_{m=-\infty}^{\infty} w_l w_m \phi_{xx}(l-m) - 2 \sum_{l=-\infty}^{\infty} w_l \phi_{dx}(-l) \tag{10.1}$$

As before, we obtain the least-squares weight vector, $\mathbf{W}^*$, by setting the derivatives of (10.1) with respect to the weights equal to zero. Thus

$$\frac{\partial \xi}{\partial w_k} = 2 \sum_{l=-\infty}^{\infty} w_l \phi_{xx}(k-l) - 2\phi_{dx}(-k) = 0$$

Therefore,

$$\sum_{l=-\infty}^{\infty} w_l^* \phi_{xx}(k-l) = \phi_{xd}(k) \qquad -\infty \leq k \leq \infty \tag{10.2}$$

We used (7.39) and (7.40) to obtain (10.2) from (10.1), and again the star designates that w_l^* is an optimum weight value.

The solution in (10.2) can be transformed just as (7.3) was transformed (with $b_n = 0$) to obtain (7.8). The convolution on the left in (10.2) becomes a product, so we can write the z-transform of the optimum weight vector as

$$\begin{aligned} H^*(z) &= z\text{-transform of } \left[w_k^* \right] \\ &= \frac{\Phi_{xd}(z)}{\Phi_{xx}(z)} \end{aligned} \tag{10.3}$$

Thus the transform of the optimum weights is the ratio of the cross power spectrum between x and d to the power spectrum of x, the input to the adaptive model.

Let us now examine the spectra in (10.3) for the case of Figure 10.1(b). From (7.48), the input power spectrum is

$$\Phi_{xx}(z) = \Phi_{ss}(z)|P(z)|^2 + \Phi_{nn}(z) \tag{10.4}$$

Here we have assumed that the plant noise is independent of the input. The cross spectrum, $\Phi_{xd}(z)$ in (10.3), may be found as follows. From (7.47) in Chapter 7, we have

$$\Phi_{dx}(z) = G(z)\Phi_{dd}(z) \tag{10.5}$$

where $G(z)$ is the transfer function from d to x. In Figure 10.1(b) we can see that the delay does not alter the signal power, so

$$G(z) = z^{\Delta}P(z) \qquad \text{and} \qquad \Phi_{dd}(z) = \Phi_{ss}(z) \tag{10.6}$$

From (7.44) we know that $\Phi_{xd}(z)$ is found from $\Phi_{dx}(z)$ by replacing z with its reciprocal, z^{-1}. Thus, using (10.6) in (10.5) and applying (7.44), we have

$$\begin{aligned} \Phi_{xd}(z) &= \Phi_{dx}(z^{-1}) \\ &= z^{-\Delta}P(z^{-1})\Phi_{ss}(z) \end{aligned} \tag{10.7}$$

Notice that $\phi_{ss}(l) = \phi_{ss}(-l)$; thus $\Phi_{ss}(z) = \Phi_{ss}(z^{-1})$.

So from (10.3), (10.4), and (10.7), the transform of the optimum two-sided weight vector is

$$H^*(z) = \frac{z^{-\Delta}\Phi_{ss}(z)P(z^{-1})}{\Phi_{ss}(z)|P(z)|^2 + \Phi_{nn}(z)} \tag{10.8}$$

A special case of interest is that of inverse modeling without noise. Let the noise n_k be zero. Then, since $|P(z)|^2 = P(z)P(z^{-1})$,

$$H^*(z) = \frac{z^{-\Delta}}{P(z)} \tag{10.9}$$

This result requires, as one would expect, that the optimal transfer function be the reciprocal of the plant transfer function cascaded with a delay of Δ. In this case, the minimum mean-square error would be zero, corresponding to a perfect fit.

Neither a perfect fit (zero mean-square error) nor a perfect inverse is obtained by the least-squares solution when noise n_k is present, as may be verified from (10.8). For example, let the plant transfer function be $P(z)$ as above, and suppose that the plant input is a white signal with unit power. As in (7.34), we then have

$$\Phi_{ss}(z) = 1 \tag{10.10}$$

Let us assume that the plant has white noise at its input. Then the noise n_k in Figure 10.1(b) would not be white but would have its spectrum shaped by $P(z)$. That is, Φ_{nn} would have the form

$$\Phi_{nn}(z) = pP(z)P(z^{-1}) \tag{10.11}$$

where p is the total power of the white noise at the input (see Exercise 1).

Using (10.10) and (10.11) in (10.8), we now have a more specific expression for the transform of the optimal weights:

$$H^*(z) = \frac{z^{-\Delta}\Phi_{ss}(z)}{\Phi_{ss}(z)P(z) + \Phi_{nn}(z)/P(z^{-1})}$$

$$= \frac{z^{-\Delta}}{(1+p)P(z)} \tag{10.12}$$

Since the true delayed inverse transfer function is the same as (10.12) with $p = 0$, we see that the least-squares solution differs only by a scale factor.

When there is noise and the weights are given by (10.12), we can use (7.65) to show the nonzero minimum mean-square error. Using the optimal $H(z)$ in (7.65), we have

$$\xi_{\min} = \phi_{dd}(0) + \frac{1}{2\pi j}\oint\left[H^*(z^{-1})\Phi_{xx}(z) - 2\Phi_{dx}(z)\right]H^*(z)\frac{dz}{z} \tag{10.13}$$

In this result we substitute H^* in (10.12), Φ_{dx} in (10.7) with z^{-1} replacing z, and Φ_{xx} in (10.4). When we also use (10.10) and (10.11), the final result is

$$\xi_{\min} = \frac{p}{1+p} \tag{10.14}$$

Exercise 4 involves the detailed derivation of (10.14) from (10.13). As indicated in this result, the optimal inverse model is able to produce a zero minimum mean-square error only when the noise power, p, is zero.

If the plant noise were different from (10.11), it is clear that a different $H^*(z)$ would result. In most such cases, $H^*(z)$ would be quite different in form from (10.12). The least-squares solution is quite sensitive to noise and could, in many practical circumstances, be rendered ineffective as a plant inverse by the plant noise.

For the noise-free case in (10.9) we can see that $H^*(z)$ is nonrecursive only if the plant, $P(z)$, has no zeros. If $P(z)$ has zeros, the poles of $H^*(z)$ can only be approximated with a finite-length adaptive transversal filter. A two-sided infinite-impulse-response optimal filter can be approximated with a causal finite-length adaptive filter if the length, L, is made sufficiently long and if the delay, Δ, is chosen appropriately.

To illustrate the foregoing remarks, we will consider some more specific cases of inverse modeling.

SOME THEORETICAL EXAMPLES

We will now consider three examples of adaptive inverse modeling, using specific plants in Figure 10.1. An important purpose of these examples is to illustrate some of the problems that were not found in the direct modeling applications in Chapter 9: causality and the determination of an optimal delay.

For the first example we will assume that the plant noise, n_k, is zero. The plant characteristic is given by

$$\textit{First example:} \quad P(z) = 0.2 - 0.5z^{-1} + 0.2z^{-2} \tag{10.15}$$

Let us first assume the case with zero delay in Figure 10.1(a). With no noise, the ideal inverse of (10.15) is

$$\begin{aligned} H^*(z) &= \frac{1}{0.2 - 0.5z^{-1} + 0.2z^{-2}} \\ &= \frac{20/3}{1 - 2z^{-1}} - \frac{5/3}{1 - 0.5z^{-1}} \end{aligned} \tag{10.16}$$

Since the ideal inverse has a pole inside and a pole outside the unit circle, its impulse

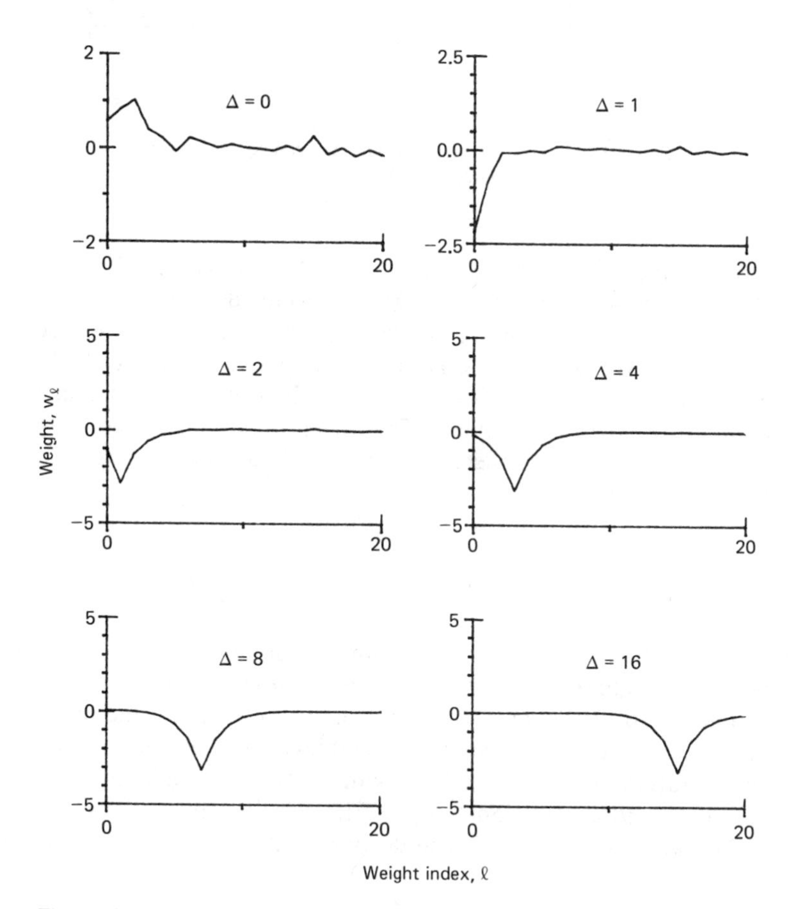

Figure 10.2 (First example) Impulse response of the optimal least-squares adaptive inverse model, using various values of the delay Δ.

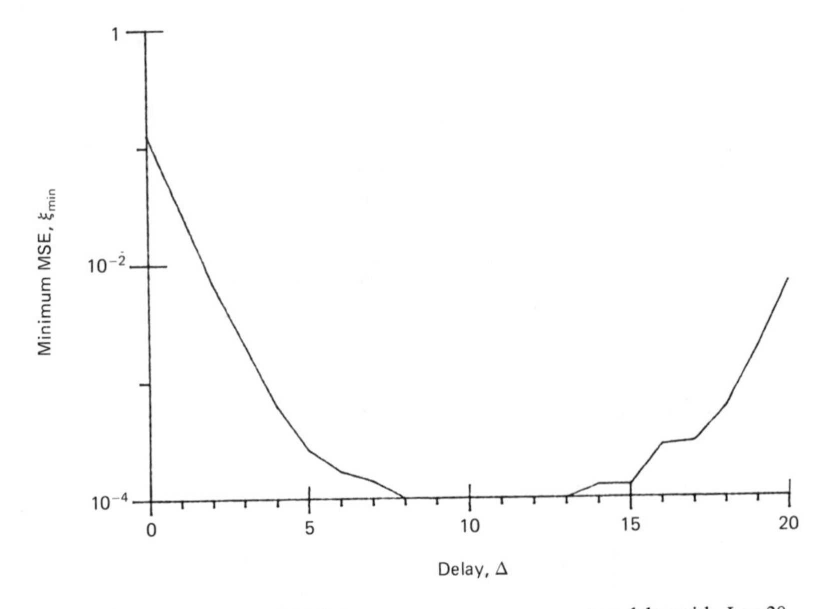

Figure 10.3 (First example) Minimum mean-square error versus delay with $L = 20$. Optimal delay is around $\Delta = 11$.

response must be two-sided to be stable. As discussed in Chapter 7, the term of the partial fraction expansion having a pole outside the unit circle can be made to correspond to a noncausal, stable left-handed impulse response component. See especially Exercise 26 in Chapter 7. Similarly, the term of the partial fraction expansion whose pole is inside the unit circle must correspond to a right-handed impulse response component. The ideal impulse response, h_k^*, the inverse z-transform of $H^*(z)$, is therefore a string of pulses extending infinitely two ways in time, represented by

$$h_k^* = \begin{cases} -\frac{20}{3}(2)^k & k < 0 \\ -\frac{5}{3}(0.5)^k & k \geq 0 \end{cases} \tag{10.17}$$

Equation (10.17) is a least-squares solution whose implementation would require an infinite number of weights as discussed above. An approximate realization of (10.17) by an FIR filter adapted to minimize the mean-square error will produce a solution that generally resembles a truncated version of (10.17). Using the configuration of Figure 10.1(b) with a 21-weight adaptive filter and the LMS algorithm and a white input, some FIR solutions for various values of the delay Δ are shown in Figure 10.2.† The delay provides an additional adjustable parameter.

† The experiments shown in Figures 10.2 through 10.11 were performed by Michael J. Larimore. His assistance is gratefully acknowledged.

Its choice has a substantial effect on the minimum mean-square error achievable and in that sense on the quality of the inverse. The minimum mean-square error versus the delay Δ is shown in Figure 10.3. This error is high for zero delay, decreases to a minimum around $\Delta = 11$, then starts back up again as the delay is increased. The choice of delay is seen to be noncritical, however, and the general rule of thumb, in which we let Δ be half the length of the adaptive filter, applies well here.

The optimal FIR impulse response for $\Delta = 8$, seen in Figure 10.2, is close to the unconstrained two-sided geometric response of (10.17), delayed by 8 units of time. Since the total number of weights used was adequate, the two-sided FIR truncation had very little effect on the optimal impulse response shape and the corresponding performance. Essentially the same shape results, as Δ is varied between 5 and 6; the impulse response is correspondingly delayed. When Δ is less than 5, portions of the impulse response are truncated and the optimal impulse response shape changes. Radical changes take place when the delay Δ goes from 1 to 0. This is accompanied by a sudden increase in the minimum mean-square error. When Δ is too large (e.g., larger than 16) truncation on the right takes place and the minimum mean-square error again goes up. The form of the optimal FIR filters shown in Figure 10.2 for various values of Δ can, of course, be derived analytically starting with (10.13). The algebra is straightforward but not simple. The curve in Figure 10.3 has a slope that is characteristic of many different examples.

Let us now turn to a second example. A more complicated plant impulse response is shown in Figure 10.4. This 41-weight impulse response is a string of numbers having no special mathematical form. An inverse with 21 weights was obtained by LMS adaptive filtering as in example 1. The input signal was white noise. The result is shown in Figure 10.5. The best delay was found to be $\Delta = 7$.

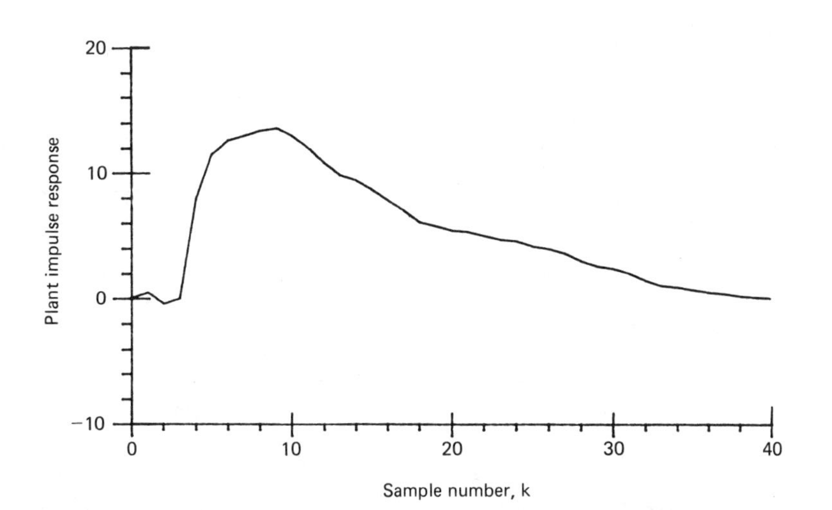

Figure 10.4 (Second example) Plant impulse response.

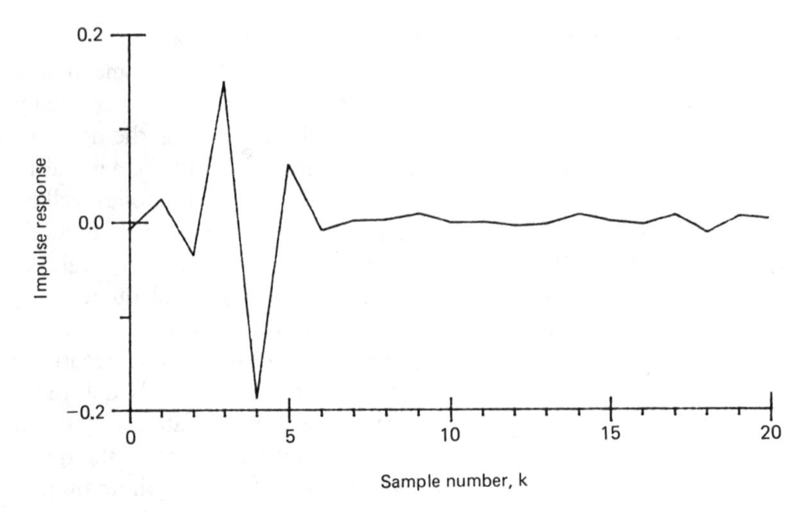

Figure 10.5 (Second example) Impulse response of optimal inverse model with $L = 20$ and $\Delta = 7$.

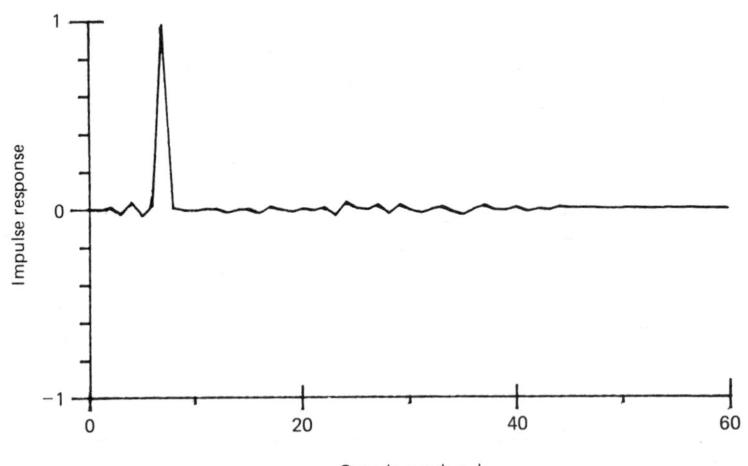

Figure 10.6 (Second example) Impulse response of 41-weight plant cascaded with optimal 21-weight inverse model with delay $\Delta = 7$.

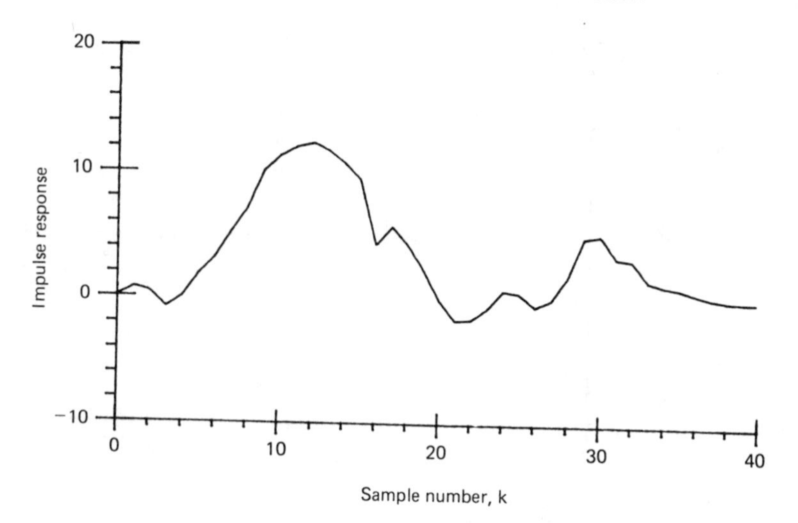

Figure 10.7 (Third example) Plant impulse response.

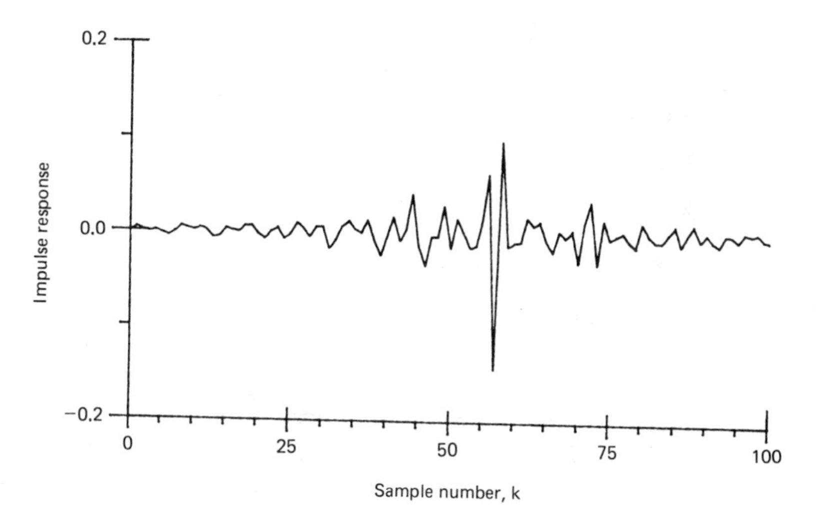

Figure 10.8 (Third example) Impulse response of optimal inverse model with $L = 100$ and $\Delta = 73$.

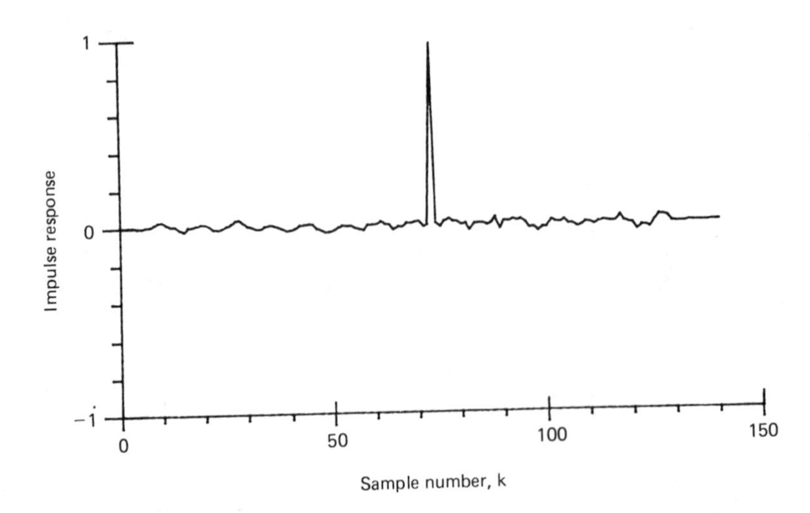

Figure 10.9 (Third example) Impulse response of 41-weight plant cascaded with optimal 101-weight inverse model with $\Delta = 73$.

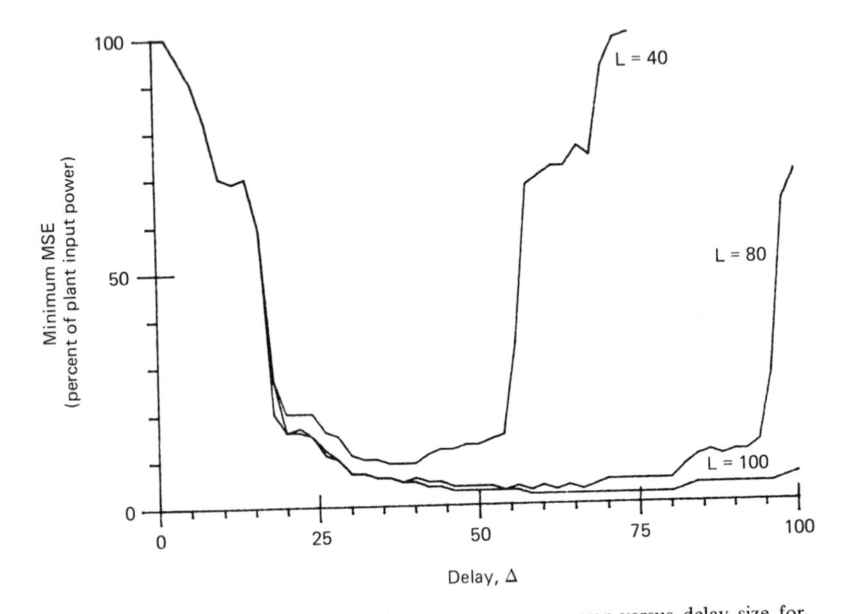

Figure 10.10 (Third example) Minimum mean-square error versus delay size for three adaptive model lengths.

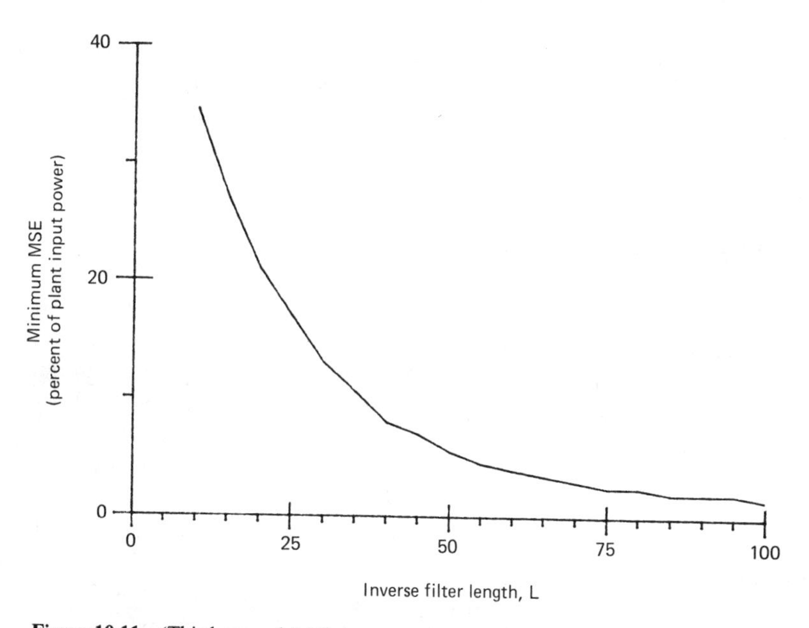

Figure 10.11 (Third example) Minimum mean-square error versus inverse filter size with delay adjusted to near-optimal values.

The smallest value of $\xi_{\min}$ was about 1 percent of the input signal power. Figure 10.6 shows a plot of the 41-weight plant impulse response convolved with that of its 21-weight inverse, with $\Delta = 7$. As expected, the response is essentially a unit impulse located at $k = 7$. Some small sidelobes are present, however.

Finally, for our third example, we choose a 41-weight plant that is more difficult to invert. Its impulse response is shown in Figure 10.7. A 101-weight inverse was tested, and the converged solution with Δ optimized at 73 is shown in Figure 10.8. Convolving this inverse with the original plant gives the cascade impulse response of Figure 10.9, which is approximately a unit impulse at $k = 73$, with small sidelobes. The sidelobes and the minimum mean-square error can be reduced by using more weights in the inverse, if desired. Figure 10.10 shows a plot of the minimum mean-square error versus the delay Δ, as the number of weights is increased from 41 to 81 to 101. In each case, Δ can be optimized, although it is interesting to note that wide ranges of Δ values produce uniformly low minimum mean-square errors. Often the value of Δ for minimizing mean-square error is noncritical. In each of the cases illustrated in Figure 10.10, letting Δ equal half the length of the adaptive filter would not be precisely optimal but would be quite satisfactory. We can see the reason for this in Figure 10.1(b). If the plant itself contains a substantial delay, the optimal value of Δ will exceed $L/2$, but will not be critical if L is large. Finally, with Δ optimized in each case, Figure 10.11 shows how the minimum mean-square error decreases as the number of adaptive weights is increased.

Inverse modeling techniques are very useful and have many important applications. In the following sections we consider applications to communications systems and to digital filter synthesis, and some control systems applications are discussed in Chapter 11.

ADAPTIVE EQUALIZATION OF TELEPHONE CHANNELS

In telephone channels, dispersion causes interference between successive samples (intersymbol interference) and greatly complicates reliable transmission and reception of digital signals. The direct-modeling adaptive communication scheme described in Chapter 9 is useful in cases of severe multipath and poor signal-to-noise ratio, but we recall that the pseudorandom scheme requires many bits to be transmitted per data bit. Here we describe an important application where dispersion is present, but of moderate severity, and where noise levels are low so that pseudorandom coding is not required. The application is the reliable transmission of digital data over conventional telephone circuits, based on an inverse modeling scheme developed by Lucky [11, 12, 16] and others [13, 15, 17, 19, 28, 29]. Instead of a pseudorandom sequence, this method involves the transmission of only one bit for each data bit.

Much work has been done since the mid-1960s on the adaptive equalization of telephone data transmission systems. Typically, in these systems, interference caused by the dispersive characteristics of long-distance telephone circuits necessitates limiting the sampling rate to only a small fraction of the channel bandwidth. With adaptive equalization (now common in commercial digital data transmission systems), it is possible to signal at much larger fractions of the channel bandwidth. The noise in telephone lines is generally low, and typically the main problem is intersymbol interference. To combat this problem, one would ideally require an equalizer in the receiver with a transfer function that is essentially the inverse of the channel transfer function. By approximating the ideal case with adaptive equalization, fivefold increases in data rates through a given channel with the same error probability are not uncommon.

Although systems with multiple carrier frequencies are generally used, we can discuss adaptive equalization by studying a telephone channel with a single carrier being modulated by a digital signal. Accordingly, an ideal channel has the bandpass frequency response characteristic shown in Figure 10.12(a). This response is uniform within the passband and zero outside the passband. The cutoff frequency is at $f_0 \pm f_{co}$ as shown, and the channel bandwidth is $2f_{co}$. The phase characteristic, not shown, is ideally linear, causing only time delay with no dispersion. The channel impulse response, which is the inverse transform of the demodulated channel transfer function, is the sinc function shown in Figure 10.12(b). The group delay of the channel is not included in Figure 10.12(b). Let us assume that this channel is used to carry a single digital data stream, to which the entire bandwidth $2f_{co}$ is dedicated.

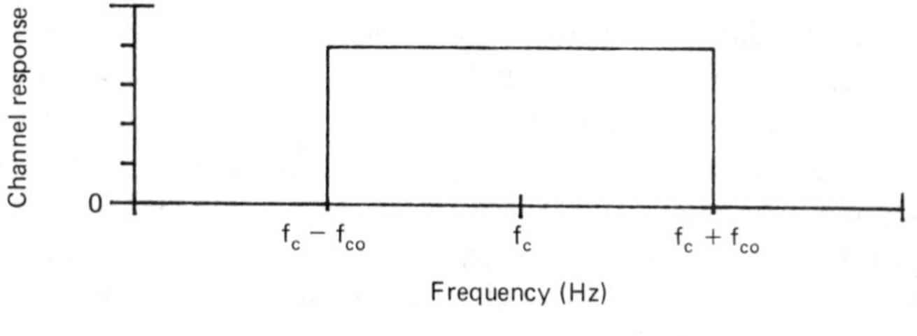

(a)

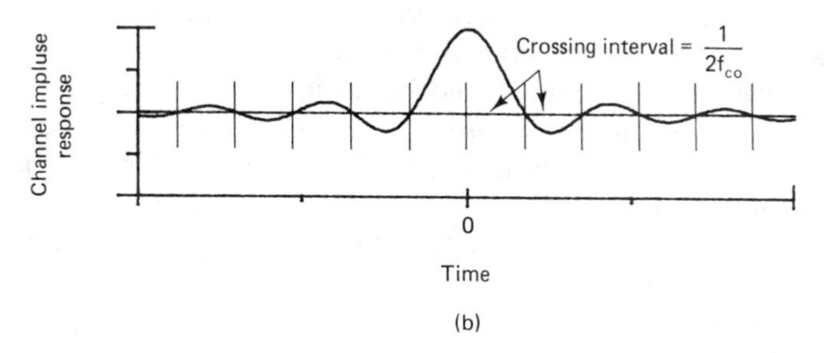

(b)

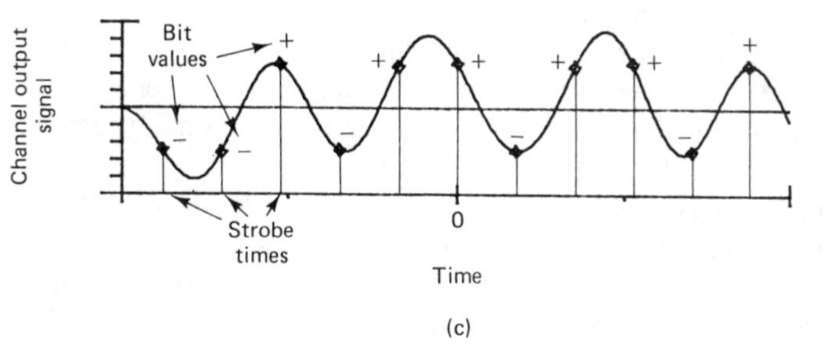

(c)

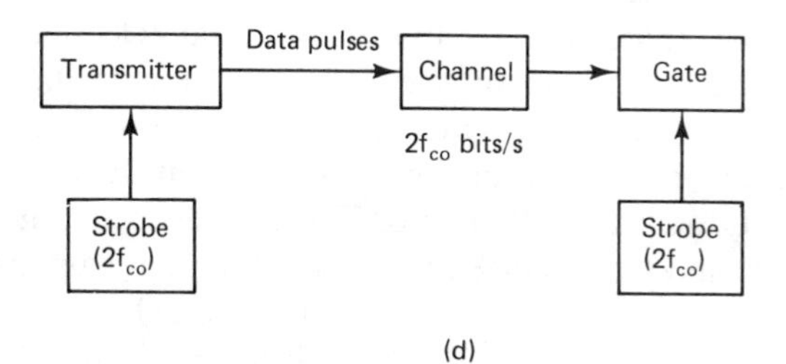

(d)

Figure 10.12 Digital communication in an ideal channel: (a) channel amplitude response; (b) impulse response after demodulation; (c) typical channel output at $2f_{co}$ bits per second (bit value at each strobe time is constant and independent of neighboring bit values); (d) diagram of conventional communication system.

Binary data are generally encoded by using a positive pulse to represent a ONE data bit and a negative pulse to represent a ZERO data bit. Thus a stream of data is transmitted as a string of positive and negative pulses. The channel, being linear, produces at the receiver an analog signal which is the convolution of the data pulses with the channel impulse response. Figure 10.12(c) shows a typical channel output response to a finite burst of data. The binary data are recovered at the receiving site by sampling the channel output. Suppose that the channel has an ideal frequency response and that the original data are transmitted exactly at the Nyquist rate for the channel (i.e., at $2f_{co}$ pulses per second). Then, when one samples the received signal at this rate and precisely phases the receiver sampling strobe pulses so that they are synchronized with the transmitter clock pulses delayed by the channel bulk delay, one obtains with each strobe pulse a response due uniquely to a single data pulse. There should be no interference from neighboring data pulses, because the impulse response of the ideal channel in Figure 10.12(b) has zero crossings which are precisely spaced at intervals of $1/(2f_{co})$. Choosing the right sampling frequency and phase at the receiver allows one to sample the channel output at the peak of each individual sinc pulse, while neighboring sinc pulses are all passing through zero. A communication system based on this principle is diagrammed in Figure 10.12(d).

In reality, telephone channels are nonideal; each channel characteristic differs from all others, and the characteristic of each channel changes slowly but significantly over time. A typical impulse response might resemble somewhat the ideal response of Figure 10.12(b), but the zero crossings would generally not be uniformly spaced over time. Accordingly, no signaling frequency and strobe phase could be chosen such that sampled outputs could be uniquely derived from individual sinc pulses with no interference from neighboring pulses. This type of intersymbol interference could cause errors and could make the communication system especially error prone in the presence of channel noise. Since even nonideal channel impulse responses tend to decay over time in both directions from the response peak as in Figure 10.12(b), one could reduce intersymbol interference by signaling slowly, much slower than the Nyquist rate, leaving adequate spacing betwen impulse responses. This approach was used before the advent of adaptive channel equalizers.

Greater speed and reliability in digital telephone communication has been afforded by the use of adaptive equalization. Figure 10.13(a) is a block diagram of a communication system using channel equalization. The adaptive equalizer forms an inverse transfer function for the channel within the channel passband. Outside this passband, its gain is small or zero. The cascaded gain magnitude versus frequency of the channel and the equalizer, the product of their gain magnitudes, is essentially flat (constant) within the passband, and is essentially zero outside the passband. The equalizer must also correct for phase distortion in the channel. The cascaded phase response (i.e., the sum of the phase responses) must be a linear function of frequency within the passband from zero to f_{co}. If these requirements are met, the combined impulse response is the sinc function described in the preceding section, and

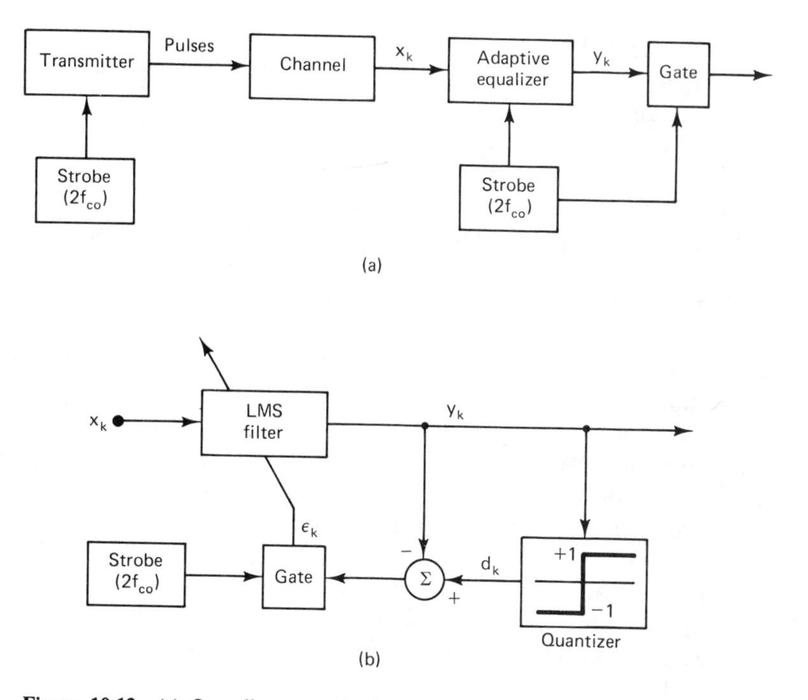

Figure 10.13 (a) Overall communication system with adaptive channel equalization; (b) details of the adaptive equalizer with decision-directed learning.

intersymbol interference can be eliminated. Adaptation is required in the equalizer because of the unknown and time-variable character of the channel itself.

The adaptive process is diagrammed in Figure 10.13(b). An adaptive transversal filter (LMS filter) is used in the inverse modeling configuration described previously. Since the signal bandwidth of the input x_k is limited by the bandwidth of the channel, the adaptive linear combiner is adapted to equalize the channel gain and phase characteristics only over the passband of the channel. In order to adapt, however, a desired response d_k is of course required. If one knew the channel input and accounted for the channel bulk delay, one would have d_k, but usually one does not have this information. If the transmitted signal were available, one would generally not need to communicate at all, although in some systems it is possible to know the transmitted signal part of the time. Periodically, the transmission of information could be interrupted so that a known code sequence could be transmitted, thus allowing intermittent adaptation.

Another method of providing d_k was devised by R. W. Lucky of the Bell Telephone Laboratories. This method uses the filter's own output to provide d_k, thus obviating the need for any prior knowledge of the transmitted signal. Lucky called his method "decision-directed" learning. More specifically, a "desired" signal, $d_k = \text{sign}(y_k)$, is formed by quantizing the filter output, as shown in Figure 10.13(b). Since the data are binary, the sampled output of the properly equalized channel at the strobe times should, except for channel noise, be either $+1$ or -1. Comparing the filter output with the quantized filter output produces the error signal ε_k. The assumption is that the decision made by the quantizer is correct and is a true indication of the binary signal most of the time. Since the equalized outputs should uniquely represent the individual sinc pulse at the proper strobe times, adaptation is permitted only at the strobe times. This is accomplished by gating the error signal, ε_k, using strobe pulses which are synchronized to the signaling rate, as shown in Figure 10.13(b). On the average, if the quantized desired response is true and correct, adaptation progresses in the right direction.

Lucky's method works when the channel is relatively noise free and has a frequency and phase response not having severe distortion. Kennedy [17] has shown that even if only 25 percent of the quantized decisions are initially incorrect, the adaptive filter will converge to an optimum solution.

The filter weights may be initialized to zero, except that one weight near the center of the adaptive linear combiner should be set to unity. Then the equalizer initially has unit gain, and its impulse response evolves during the adaptation process to have the gain and phase characteristic required to perform channel equalization.

Practical adaptive equalizers for telephone channels are digital. Their internal sampling rates typically are multiples of the basic data rate, usually 4 to 16 times faster, and typical adaptive filter sizes are 32 to 64 weights. Thus the total filter memory time spans several cycles of the sinc data pulse.

The degree of success of adaptive equalization can be viewed in what is known as an "eye pattern." An eye pattern is produced by synchronizing an oscilloscope to the strobe signal and connecting the input to the received signal, so that the received signal cycle is repeated over and over on the screen. The result before equalization is shown in Figure 10.14(a), and after equalization in Figure 10.14(b). Without equalization, the eye pattern of Figure 10.14(a) shows a wide disparity in the positive sinc pulses, and a similar disparity in the negative pulses, and thus indicates the presence of distortion in the sinc pulses which is associated with intersymbol interference. After equalization, the eye pattern of Figure 10.14(b) is open and sharp and clear. The positive sinc pulses are tightly clustered. Intersymbol interference has been greatly diminished. The likelihood of mistaking a positive pulse for a negative one is reduced, and sensitivity to random channel noise is also reduced. The adaptive process has forced all positive pulses to have close to $+1$ amplitudes at their strobe times (at the center of the image) and forced all negative pulses to have -1 amplitudes at their strobe times. By equalizing the channel in this way, the adaptive filter effectively eliminates intersymbol interference. Using this approach, a telephone channel with an unequalized error rate as high as 10^{-1} could typically have an equalized error rate of 10^{-6} or even lower.

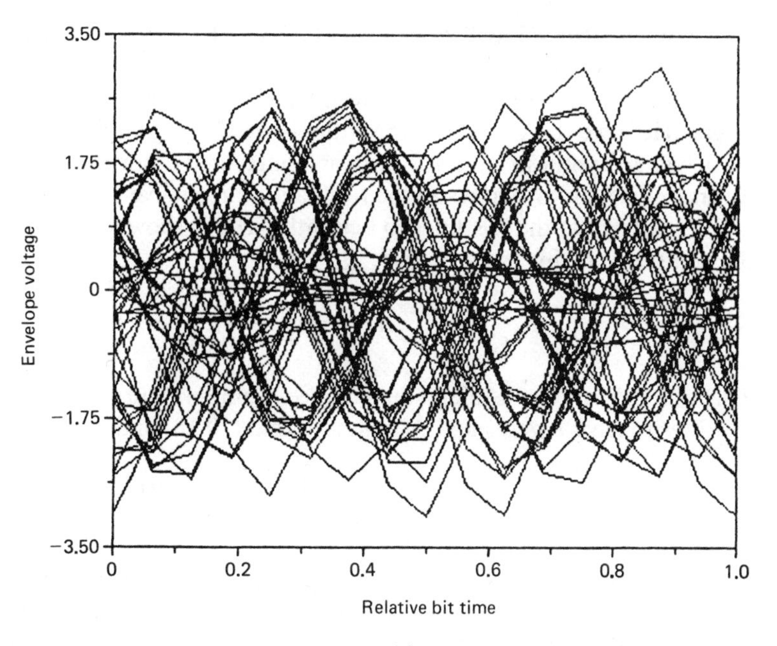

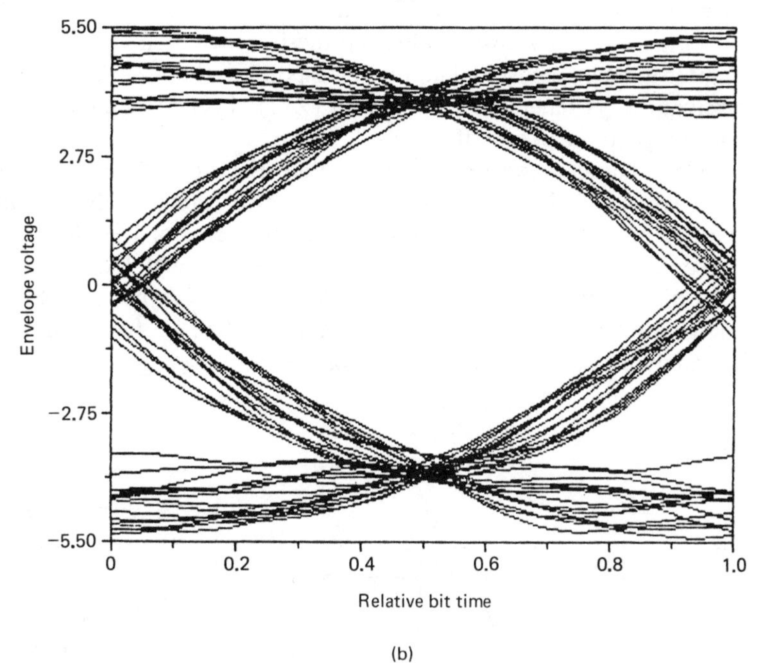

Figure 10.14 Eye patterns produced by overlaying cycles of the received waveform: (a) before equalization; (b) after equalization. (Courtesy of Hughes Aircraft Company.)

ADAPTING POLES AND ZEROS FOR IIR DIGITAL FILTER SYNTHESIS

In Chapter 9 adaptive modeling techniques were used to synthesize FIR digital filters. Here we discuss a similar synthesis of IIR digital filters. We will use direct modeling to synthesize the forward or nonrecursive portion of the IIR filter, and inverse modeling to synthesize the feedback or recursive portion.

The form of the IIR filter to be synthesized is shown in Figure 7.2. As in (7.8), the transfer function is

$$H(z) = \frac{A(z)}{1 - B(z)} \tag{10.18}$$

This transfer function is diagrammed once again in Figure 10.15. Here we will allow the filter to have L zeros and M poles, so that (10.18) in detail is

$$H(z) = \frac{a_0 + a_1 z^{-1} + \cdots + a_L z^{-L}}{1 - b_1 z^{-1} - \cdots - b_M z^{-M}} \tag{10.19}$$

The filter thus has $L + 1$ feedforward weights and M feedback weights. Our goal is to develop an adaptive process that will automatically adjust these weights so that the filter transfer function is a best fit to a set of design specifications. As in Chapter 9, the design specifications are represented by a "pseudofilter."

Figure 10.16 shows a filter synthesis scheme similar to Figure 9.13. As in Chapter 9, the N sinusoidal inputs correspond to the N specification frequencies and the signal d_k is generated as in (9.8) or (9.11). In the IIR filter, both $A(z)$ and $B(z)$ must adapt to the pseudofilter.

In Chapter 8 we examined adaptive IIR algorithms and saw that they had some associated difficulties. The error surfaces are not always unimodal and the adaptive filter can become unstable unless the zeros of $1 - B(z)$ are held inside the unit circle. To get around these difficulties, we will look at a scheme for IIR adaptation that is especially suited to the filter synthesis problem and does not require an IIR adaptive algorithm. In this scheme, $A(z)$ and $B(z)$ are adjusted separately as adaptive transversal filters.

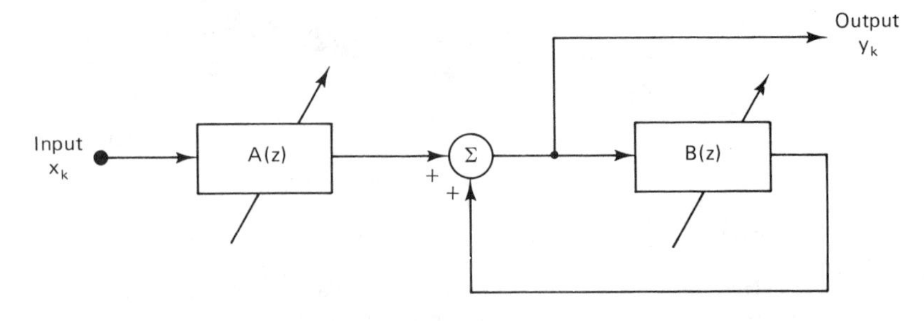

Figure 10.15 Block diagram of the adaptive IIR filter.

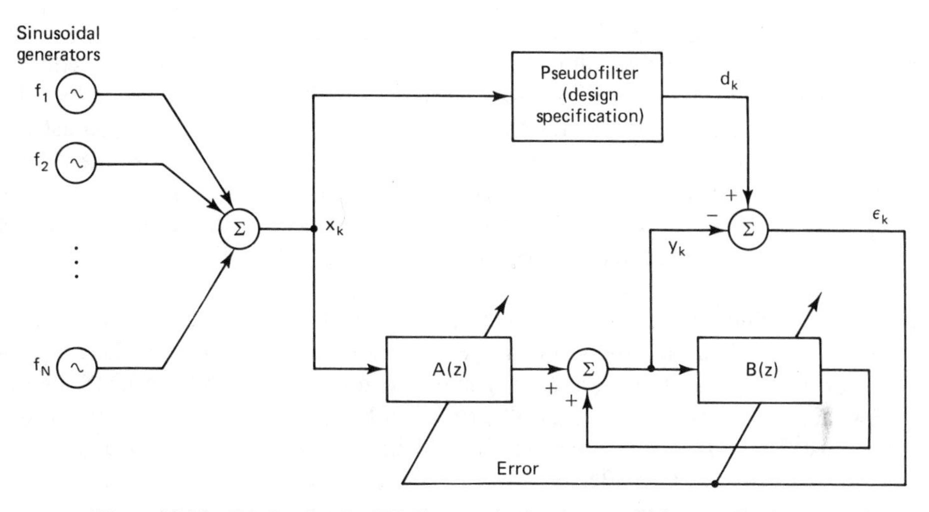

Figure 10.16 Simple adaptive IIR filter synthesis scheme, which generally does not work.

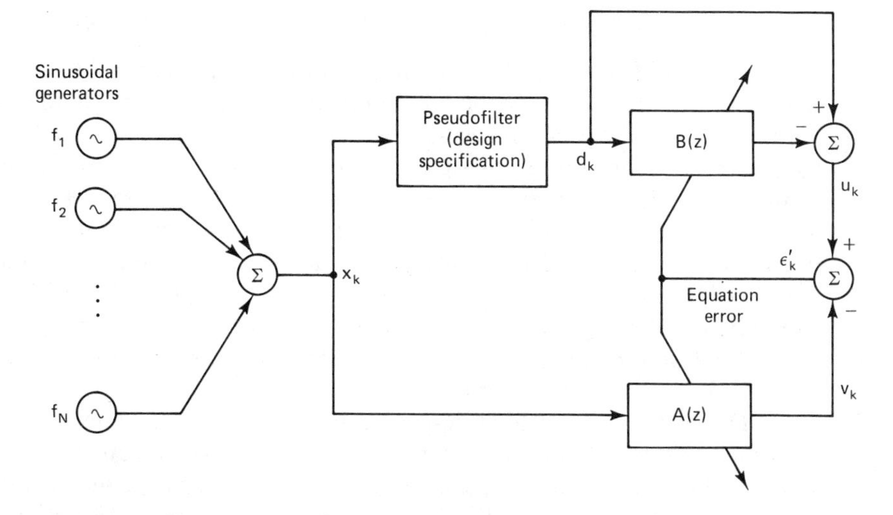

Figure 10.17 IIR filter synthesis by the equation error (simultaneous direct and inverse modeling) method.

The approach that we shall take does not quite minimize the expected value of ε^2. It is called "simultaneous direct and inverse modeling" or "equation error" modeling [3] and is illustrated in Figure 10.17. The error is ε', which is different from ε in Figure 10.16. In many cases, however, we shall see that adjusting A and B to minimize the mean square of ε' provides values of A and B that, if used in the system of Figure 10.16, would come very close to minimizing the mean square of ε. Accordingly, once A and B are found using the configuration of Figure 10.17, the IIR filter is constructed by copying the values of A and B into the system of Figure 10.15.

The roundabout approach in Figure 10.17 is used because the mean square of ε' is a quadratic function of the coefficients of both A and B. The error surface in this case is unimodal, and so A and B can be adapted by the LMS algorithm. The adaptive process is like that illustrated in Figure 9.4, where a common error signal is used to adapt two adaptive filters whose outputs are summed. Note that $B(z)$ is adjusted so that, ideally, $1 - B(z)$ cancels the poles of the pseudofilter, while $A(z)$ is adjusted in the "direct modeling" mode to cancel the zeros of the pseudofilter. Further details of the adaptive process are given by Gooch [32].

The relationship between ε and ε' is interesting. Let the transfer function of the pseudofilter be designated by PS(z). Let the sum of the driving sinusoidal signals be designated by $F(z)$. With $A(z)$ and $B(z)$ fixed, the z-transform of the error ε in Figure 10.16 is

$$E(z) = F(z)\left[\text{PS}(z) - \frac{A(z)}{1 - B(z)}\right] \tag{10.20}$$

Referring now to Figure 10.17, let $A(z)$ and $B(z)$ again be fixed. The z-transform of the error ε' is

$$E'(z) = F(z)[\text{PS}(z)(1 - B(z)) - A(z)] = E(z)(1 - B(z)) \tag{10.21}$$

It is clear that $E'(z)$ and $E(z)$ are in proportion to the factor $1 - B(z)$, which is not fixed but varies as B adapts. Accordingly, adjusting A and B to minimize the mean square of ε' would not necessarily cause the mean square of ε to be minimized. However, if A and B are designed with adequate numbers of weights so that adjustments of A and B exist that would bring the mean square of ε of Figure 10.16 to zero, these same adjustments would obviously bring the mean square of ε' of Figure 10.17 to zero. Adjustments for A and B can therefore be found with the system of Figure 10.17 simply by minimizing the mean square of ε', and will produce the desired IIR design when an adequate number of degrees of freedom is allowed for both A and B. More weights than needed in either or both A or B would be wasteful but not otherwise harmful.

This method of IIR filter design has been tested by computer implementation. Although one can obtain fine control of phase with FIR designs (such as perfectly linear phase; see Chapter 9), the control of frequency response that can be achieved with IIR designs is often much better than that obtained with FIR designs having an equivalent number of weights and delays. We will now discuss some examples of IIR filter design using the LMS algorithm. The examples are similar to those given in Chapter 9, with similar values of μ and similar convergence properties.

The results of a low-pass IIR design are shown in Figure 10.18. Notice the flatness of response in the passband and the sharpness of the dropoff into the stop band. The phase, intended to be zero at all frequencies, was achieved quite well except for the transition region between passband and stop band. Figure 10.19 shows the results of another attempt at the same design problem. In this case, nonuniform cost weighting was used to obtain a sharper transition and deeper cutoff in the stop band. Some uniformity in the passband was sacrificed to achieve this result. In both attempts, the IIR filters had ten feedforward weights and nine feedback weights, and $A(z)$ was noncausal. There were 25 equally spaced specification frequencies.

Figure 10.20 shows the results of the adaptive design of another IIR low-pass filter, but with linear phase specifications in both the passband and the stop band. Once again, there were 10 feedforward weights and nine feedback weights, and $A(z)$ was causal in this case. Nonuniform cost weighting was used, as indicated in the figure. As in Figures 10.18 and 10.19, the magnitude and phase response plots were made from the converged solution by computing magnitude and phase values not only at the specification frequencies but at a large number of frequencies in between. The linear phase requirement was met in the passband, but only poorly realized in the stop band. The magnitude response is seen to be quite good.

An example of the design of an unusual filter is shown in Figure 10.21. The specification required linear phase with a magnitude response in dB that is a sawtoothed function of frequency. The achievement of such a design is difficult without using an adaptive synthesis procedure. Note that responses even closer to the sawtooth and linear phase specifications could be obtained by incorporating more feedforward and feedback weights and by specifying these responses at more frequency points. At present, there is no simple way to predict closeness of fit as a function of the number of weights and the number of specification points. However, one may use a cut-and-try approach with computer implementation. This kind of interactive process can easily be done with a personal computer, or with a time-sharing terminal.

We must now discuss a problem that often arises when designing IIR filters by means of the procedure described above. Refer to Figure 10.17. The transversal filters $A(z)$ and $B(z)$ are easily obtained from convergence of the LMS algorithm or any other least-squares procedure, such as the nonrecursive SER algorithm in Chapter 8. The problem is that when $B(z)$ is used in the synthesized feedback filter of Figure 10.15, the feedback loop sometimes turns out to be unstable. The transfer function of the synthesized filter is

$$\frac{A(z)}{1 - B(z)} \tag{10.22}$$

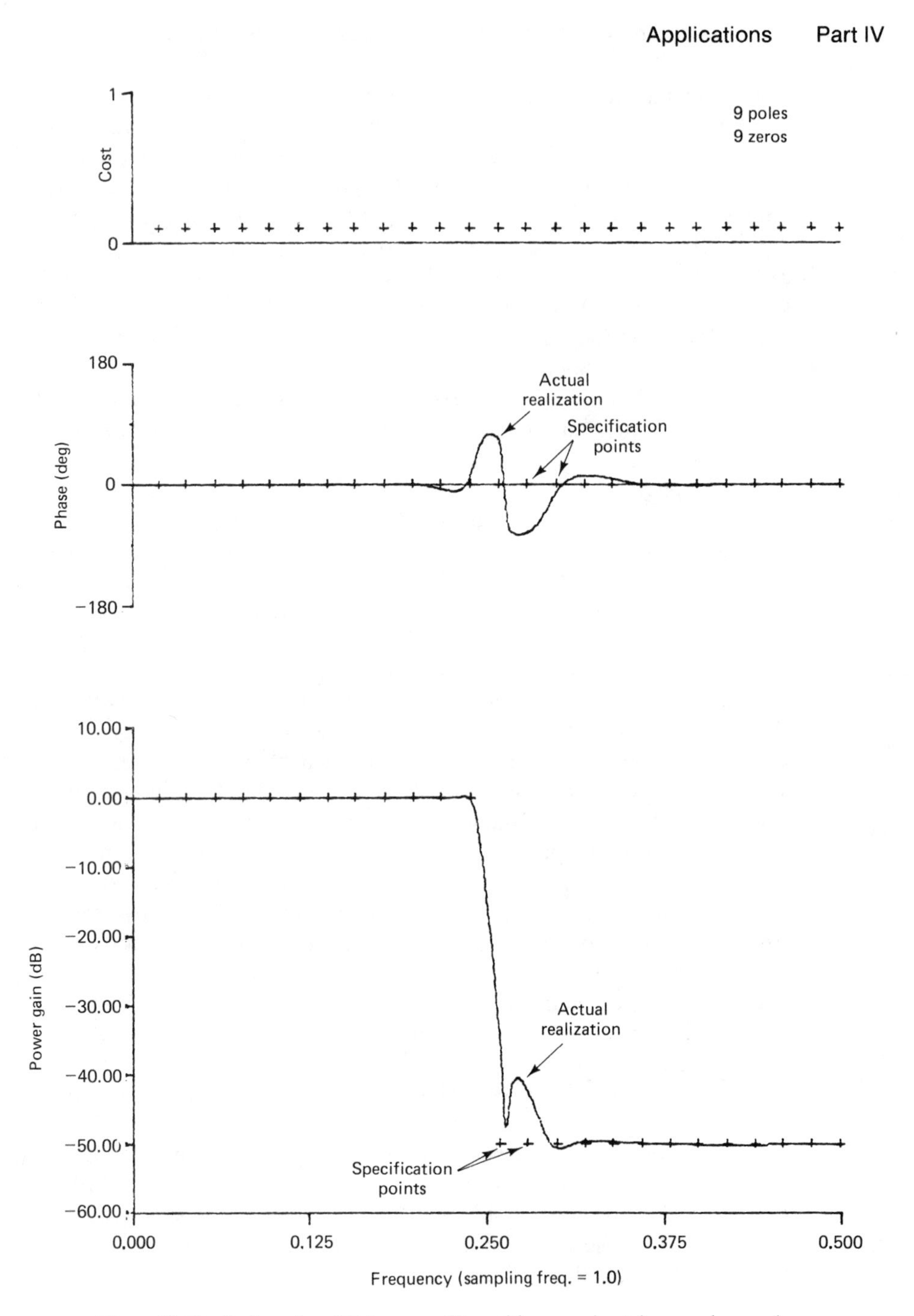

Figure 10.18 Design of an IIR low-pass filter with approximately zero phase, using a least-squares algorithm to determine $A(z)$ and $B(z)$ in Figure 10.17.

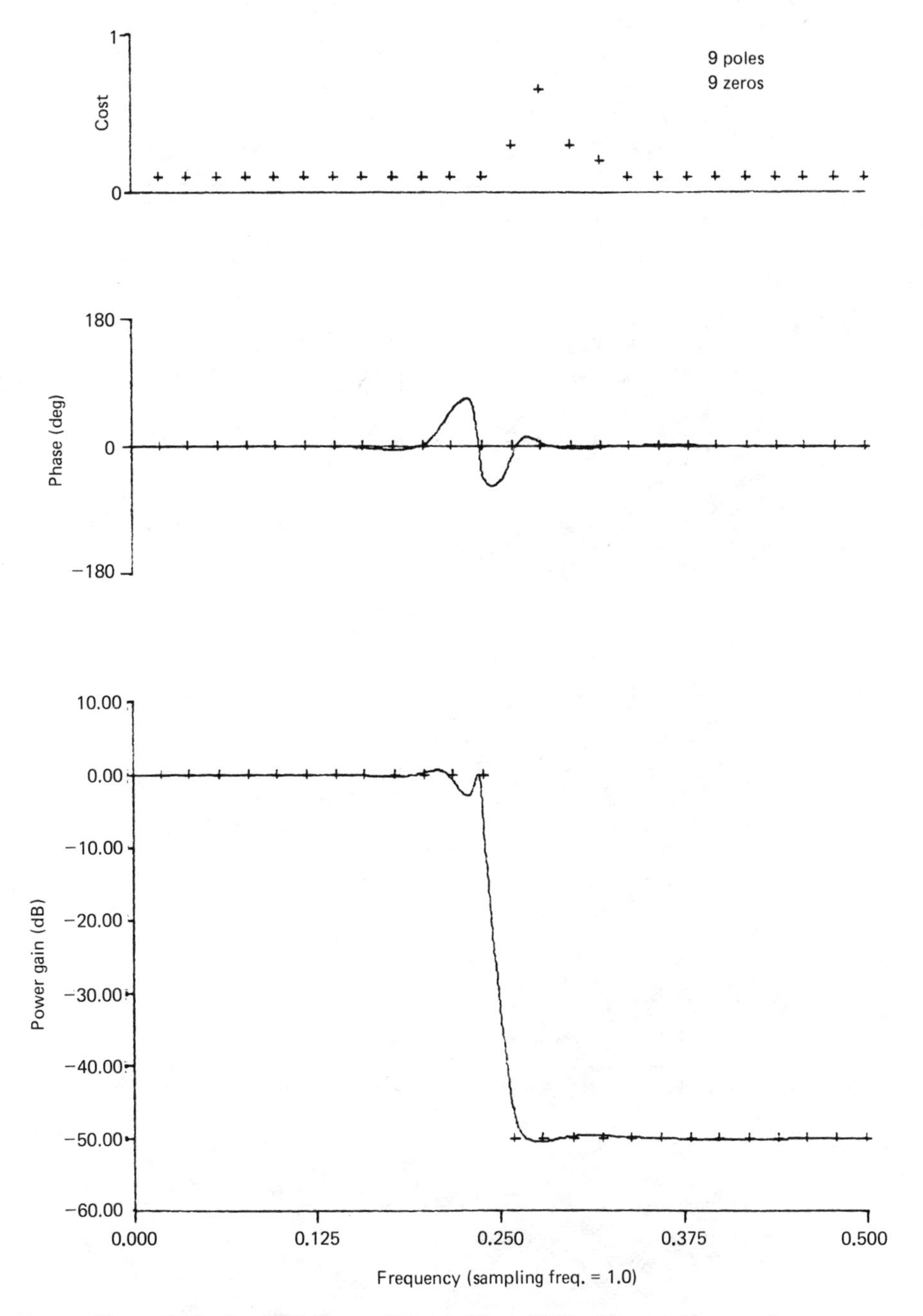

Figure 10.19 Same IIR filter as shown in Figure 10.18 with nonuniform costing to improve the transition band performance.

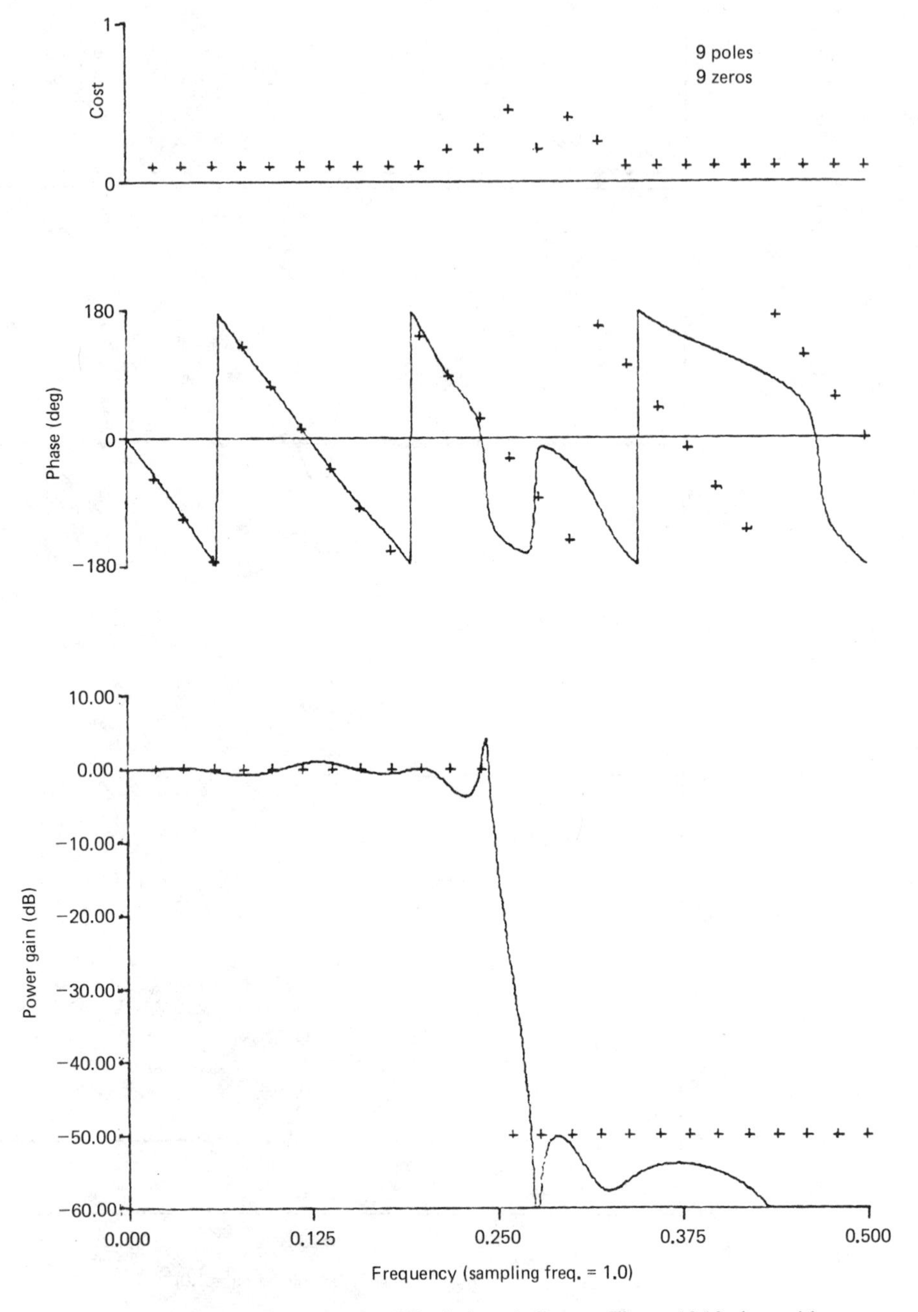

Figure 10.20 Another low-pass IIR design similar to Figure 10.19, but with linear-phase specifications.

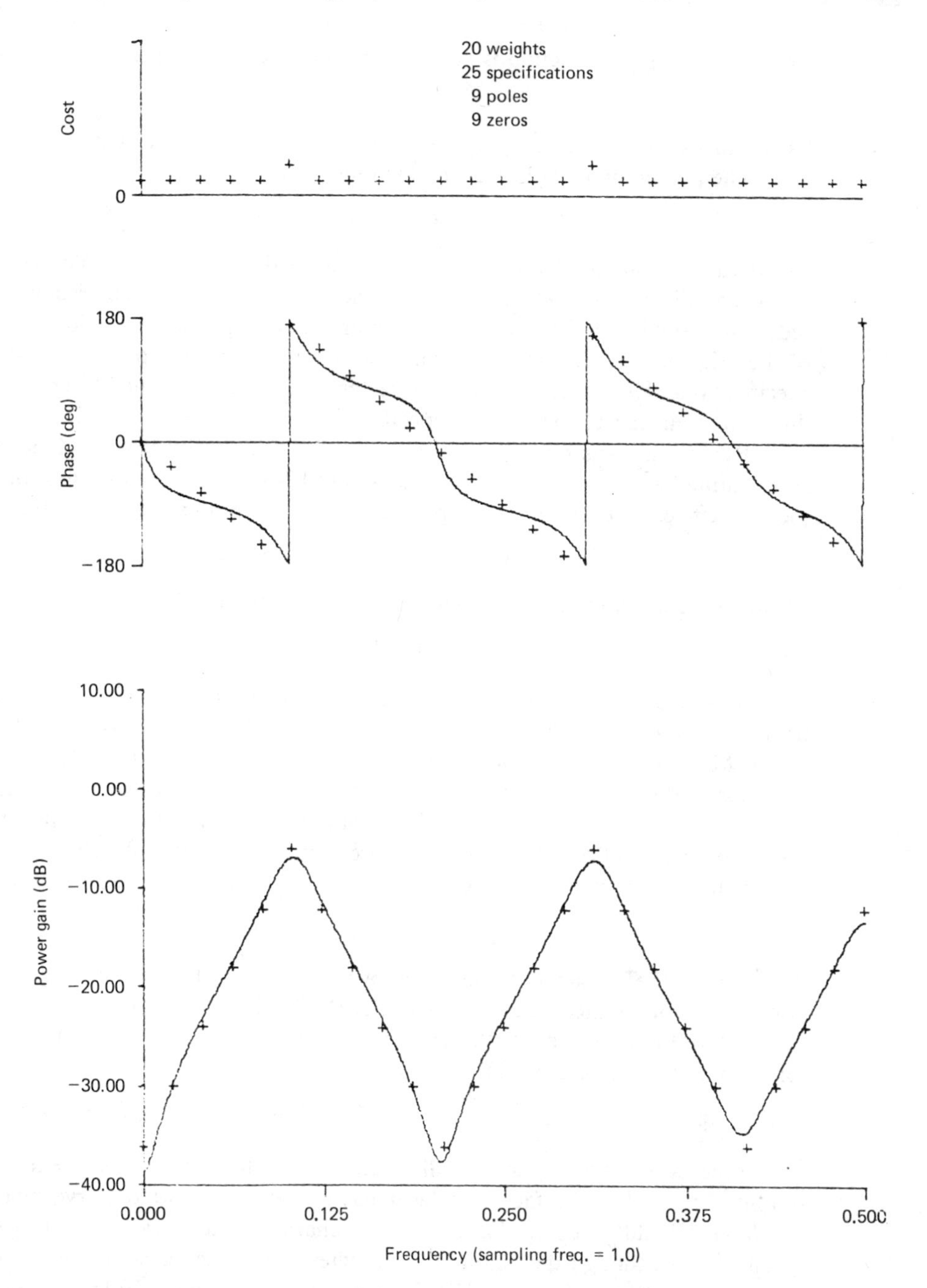

Figure 10.21 Adaptive design of an IIR filter having linear phase and sawtooth logarithmic gain.

This causal transfer function is stable only when the polynomial

$$1 - B(z) \tag{10.23}$$

has all of its roots $z_1, z_2, \ldots, z_M$ inside the unit circle in the z-plane.

The polynomial (10.23) can be factored as

$$1 - B(z) = z^{-M}(z - z_1)(z - z_2)\cdots(z - z_M) \tag{10.24}$$

In all cases, a stable IIR filter can be redesigned from an unstable IIR filter by replacing all roots of (10.24) outside the unit circle by their reciprocals. The redesigned stable filter will then have a magnitude response that is identical to that of the original unstable IIR filter, which, of course, in turn approximates the design specifications. However, the phase characteristic of the redesigned filter may be very different from that of the original specification.

Replacing roots outside the unit circle by their reciprocals is done easily by substituting z^{-1} for z in all of the factors of (10.24) that correspond to roots outside the unit circle. In other words, suppose that the polynomial factor in (10.24) is

$$(z - z_p) \qquad \text{where } |z_p| > 1 \tag{10.25}$$

This factor should then be replaced in the polynomial by the factor

$$\left(z^{-1} - z_p\right) \tag{10.26}$$

The remaining factors corresponding to roots inside the unit circle should appear unchanged in (10.24). The result is a new polynomial having all roots inside the unit circle. Multiplying the factors together to find a new $B(z)$ as in (10.24), one can incorporate the new $B(z)$ into the system of Figure 10.15 to form a stable IIR filter.

The factor replacement to achieve stability has no effect on the amplitude response for the following reason. Suppose that z_p above is real. Then, as in (7.19), one of the factors in the amplitude gain is $|z - z_p|$ with $z = e^{j\omega}$, and clearly

$$|z - z_p| = |z^{-1} - z_p| \qquad \text{for all } z = e^{j\omega} \tag{10.27}$$

Suppose, on the other hand, that z_p above is complex. Then its conjugate $\bar{z}_p$ must also be a root of the denominator of $H(z)$. We isolate the factors $(z - z_p)$ and $(z - \bar{z}_p)$ and replace them with the factors $(z^{-1} - z_p)$ and $(z^{-1} - \bar{z}_p)$. Once again, for a general value of z on the unit circle,

$$|(z - z_p)(z - \bar{z}_p)| = \left|\left(z^{-1} - z_p\right)\left(z^{-1} - \bar{z}_p\right)\right| \qquad \text{for all } z = e^{j\omega} \tag{10.28}$$

The same reasoning applies to all other factors needing replacement to ensure stability. Note that the factor replacement process does not preserve phase. Thus the final IIR filter could have a phase characteristic very different from that specified, even though the amplitude characteristic is very close to the specification.

As a matter of fact, the IIR filter described in Figure 10.18 has the type of instability just described. If we replace the poles as in (10.26), we obtain the stable filter described in Figure 10.22. Just as in Figure 10.18, 10 feedforward weights and nine feedback weights were used. Note the close adherence to the specifications of

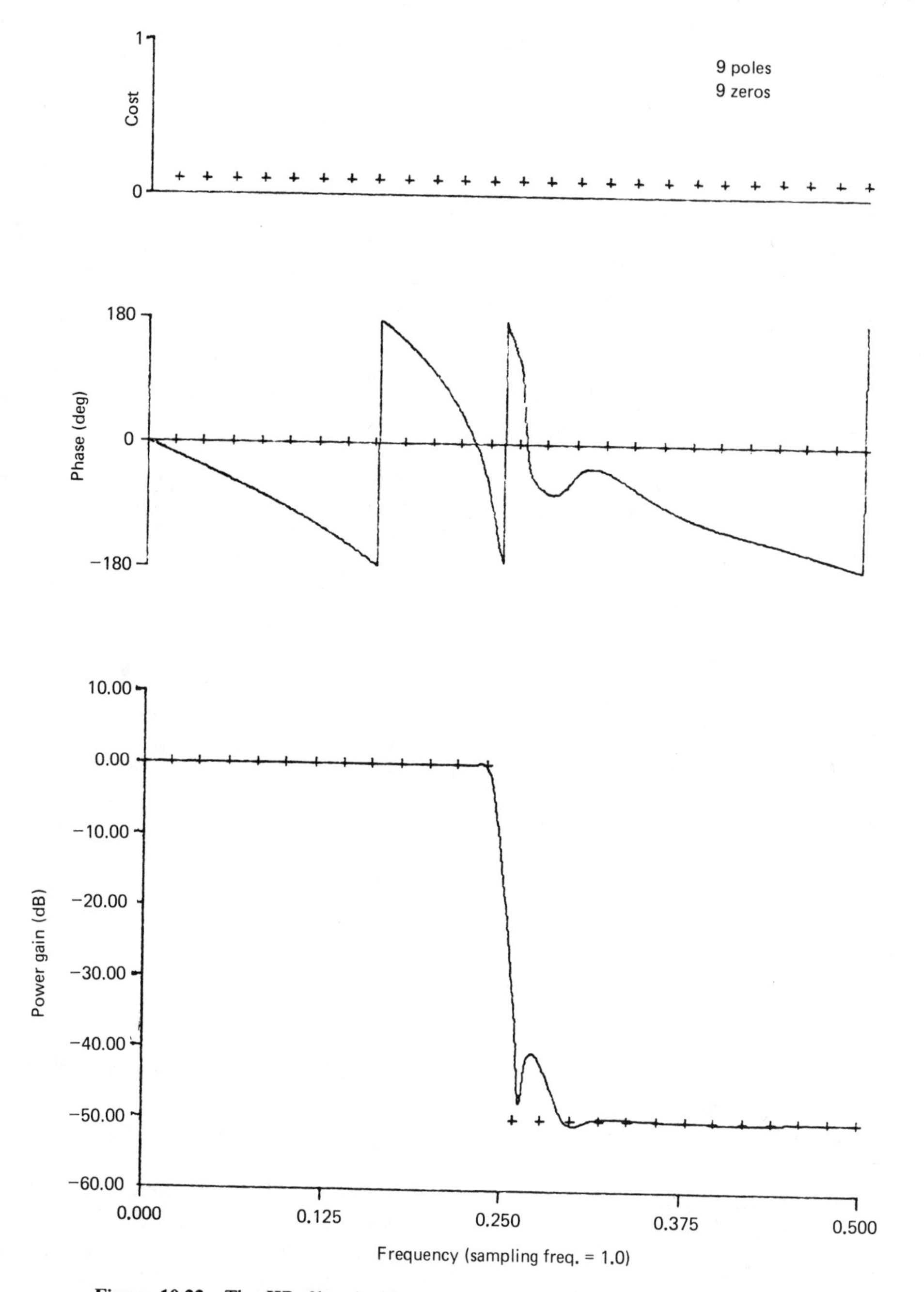

Figure 10.22 The IIR filter in Figure 10.18 with unstable poles reciprocated to achieve a stable filter.

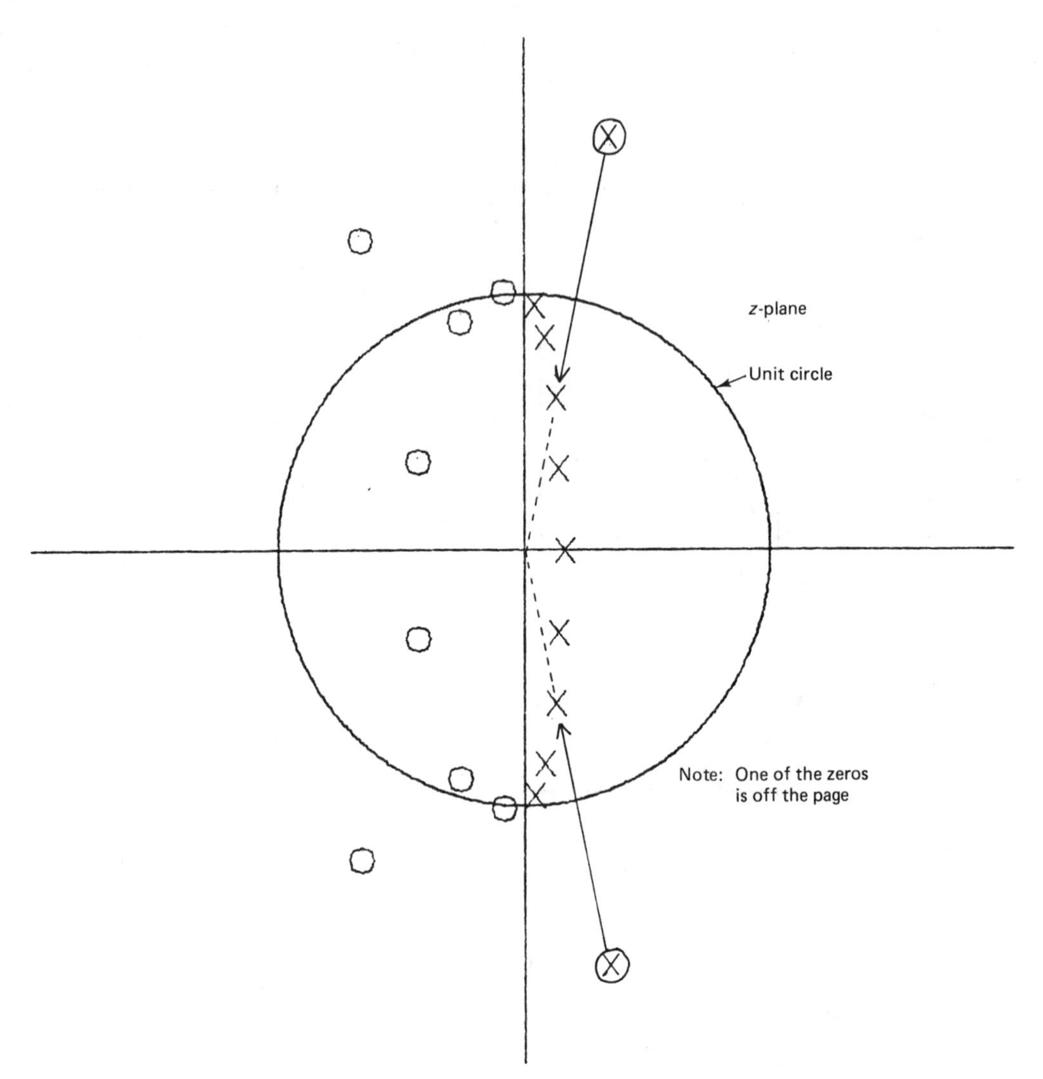

Figure 10.23 Pole–zero plot for the IIR filter of Figures 10.18 and 10.22, showing the reciprocation of poles to achieve stability.

the amplitude characteristic, and the discrepancy between the realized and specified phase characteristic. Figure 10.23 shows a pole–zero plot for the filter transfer functions of Figures 10.18 and 10.22 and indicates how the unstable conjugate poles were reciprocated as in (10.26) to create a stable filter.

If our goal is to match both amplitude and phase specifications while maintaining a stable IIR adaptive filter, at least two general approaches are available. We will describe each of these briefly in general terms.

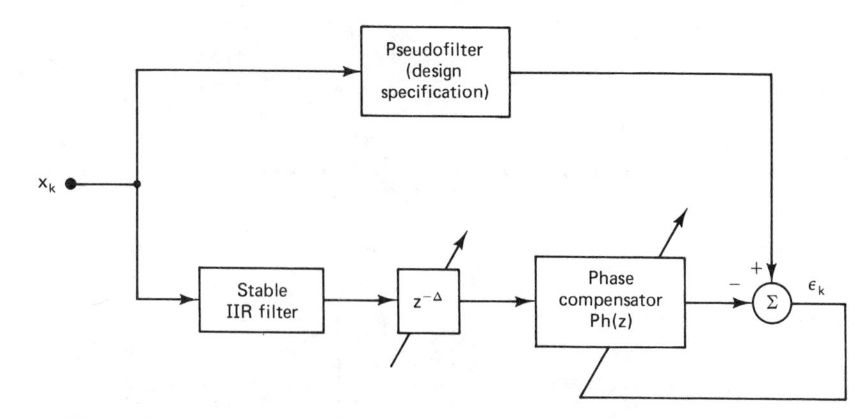

Figure 10.24 Using a phase compensator to correct the phase error caused by reciprocation of poles.

The first approach is illustrated in Figure 10.24. Here the stable IIR filter is designed as already described, with pole reciprocation to achieve stability. The stable IIR filter is then frozen and placed in cascade with a variable delay and a phase compensator. These are adjusted to minimize the mean-square error while the stable IIR filter is held constant. The phase compensator is a filter with unit amplitude gain and variable phase. It consists of one or more cascaded sections of the form

$$\text{Ph}(z) = \frac{b_2 + b_1 z^{-1} + z^{-2}}{1 + b_1 z^{-1} + b_2 z^{-2}} \tag{10.29}$$

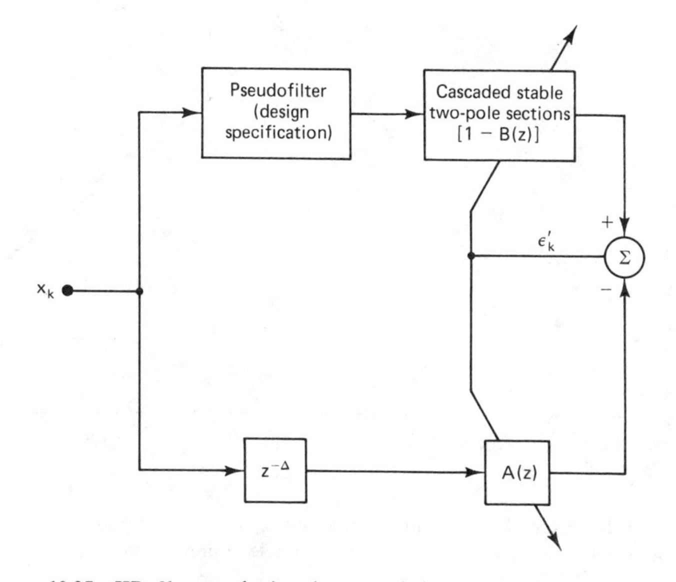

Figure 10.25 IIR filter synthesis using cascaded two-pole sections to implement $1 - B(z)$.

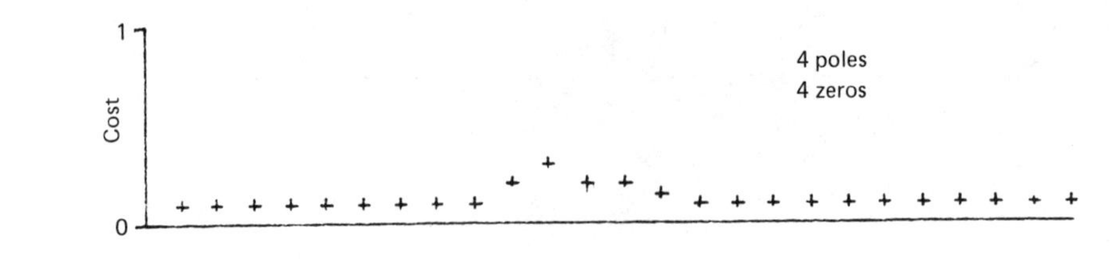

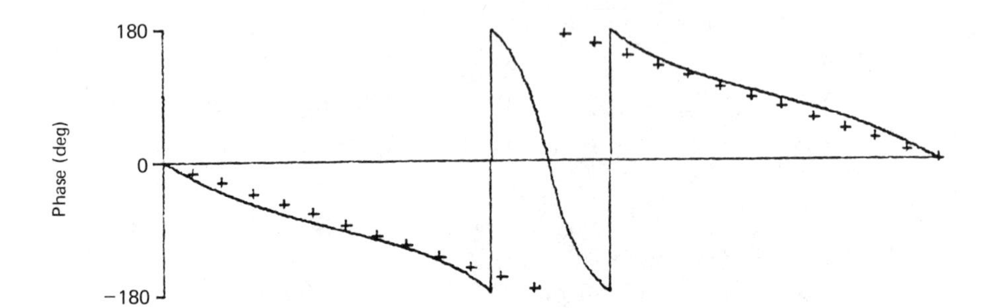

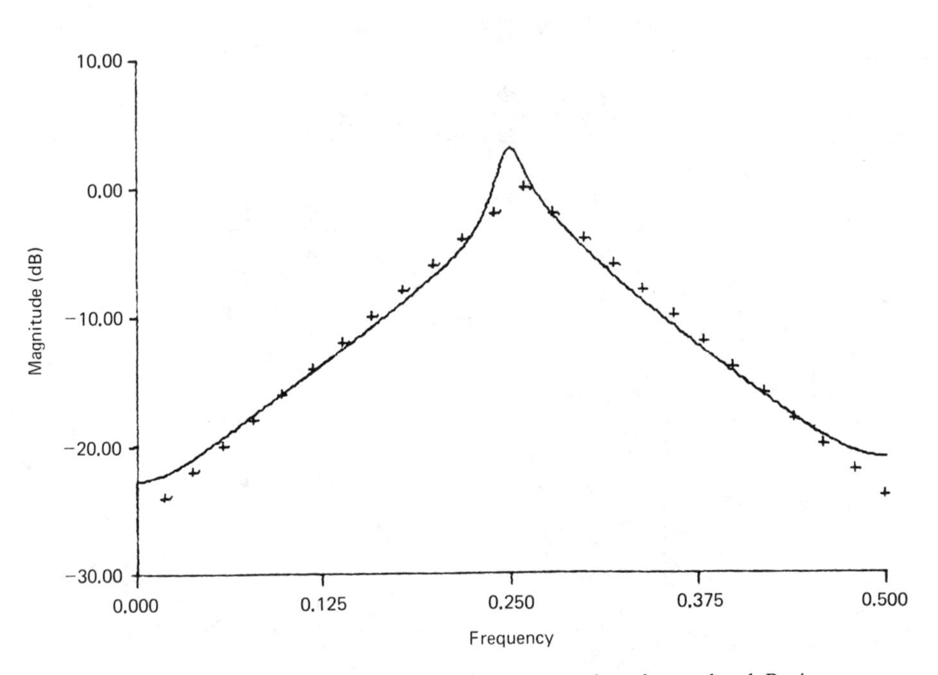

Figure 10.26 Example of IIR filter synthesis using a triangular passband. Reciprocation of poles was required, so only the amplitude response is matched.

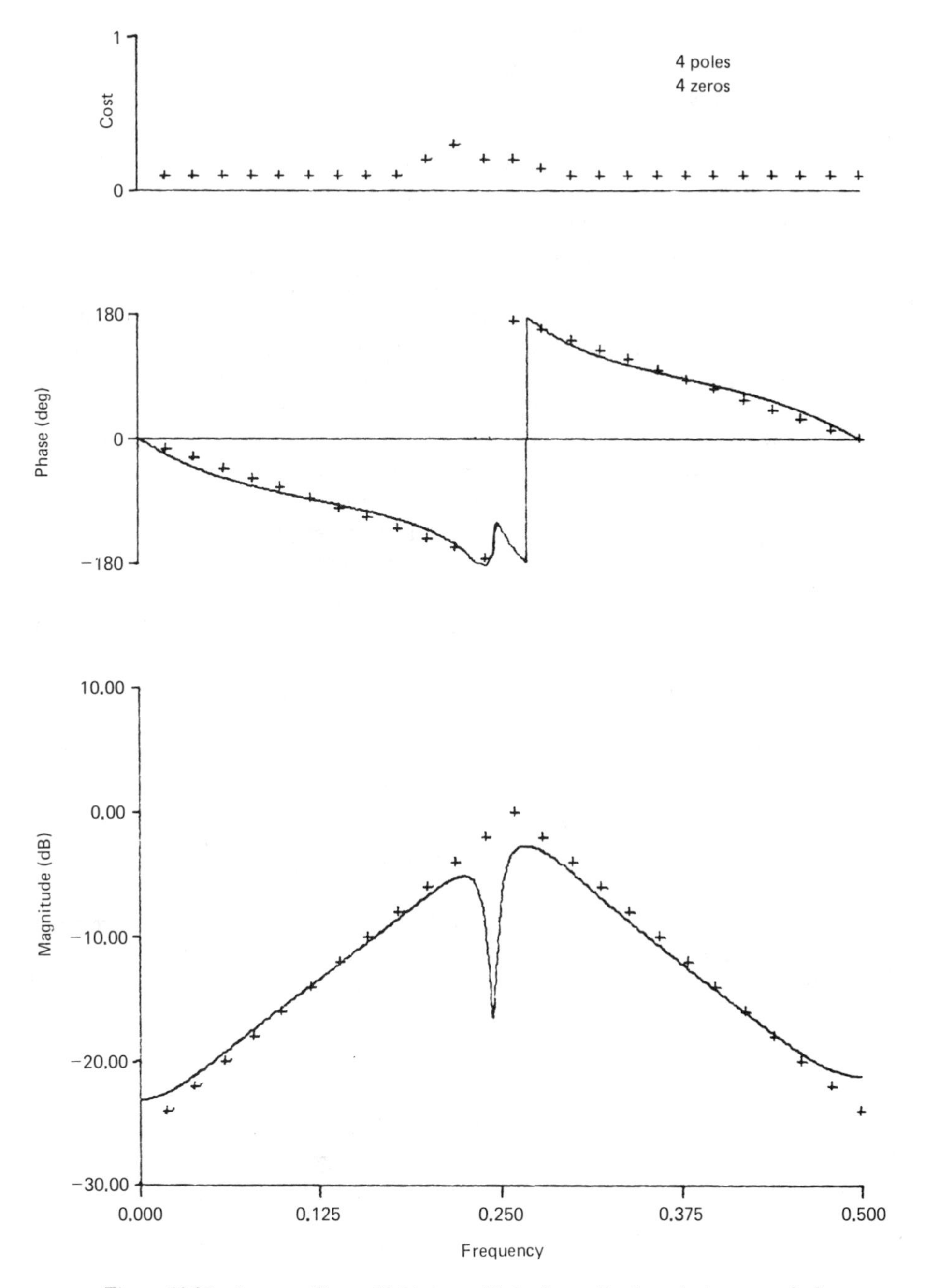

Figure 10.27 Same as Figure 10.26, but with both amplitude and phase matched using the method in Figure 10.25.

As in Figure 8.9, the weights b_1 and b_2 must be adjusted such that the poles of Ph(z) remain inside the unit circle. The final IIR filter design thus consists of the stable IIR filter in Figure 10.24 in cascade with the delay and the phase compensator.

The second method for matching both the amplitude and the phase of the pseudofilter is illustrated in Figure 10.25. Here the feedback polynomial, $1 - B(z)$, is composed of cascaded second-order sections. Then the poles of $1 - B(z)$ can be kept inside the unit circle as in Figure 8.9, and the resulting IIR filter will again be stable. The delay, $z^{-\Delta}$, may be required as before if the pseudofilter contains a significant group delay. We will not pursue the method of Figure 10.25 here, although the adaptive operation of cascade structures has been investigated by David [4].

Of the two methods just described, the second is considered at present to be the most straightforward. An experiment was performed to test this method, and the results are shown in Figures 10.26 and 10.27. First, Figure 10.26 shows the "magnitude only" type of synthesis described previously, and we can see that the phase characteristic has been badly warped by reflecting zeros inside the unit circle in $1 - B(z)$. Then Figure 10.27 shows the synthesis using the adaptive method in Figure 10.25, in which both magnitude and phase specifications are matched. Note that the magnitude response is better in Figure 10.26, where we are not trying to match the phase as well. Both designs could be improved by increasing the total number of poles and zeros.

EXERCISES

1. Show that the plant noise is given by Equation (10.11) when the white-noise power is p at the "front end" of the plant.
2. In the inverse modeling situation shown below, suppose that s_k is independent of s_{k+i} for all $i \neq k$ and that $\phi_{ss}(0) = 1$. Also, suppose that n_k is white noise independent of s_k, with $\phi_{nn}(0) = p$. Derive expressions for the power spectra $\Phi_{ss}(z)$, $\Phi_{xx}(z)$, $\Phi_{dd}(z)$, and $\Phi_{dx}(z)$.

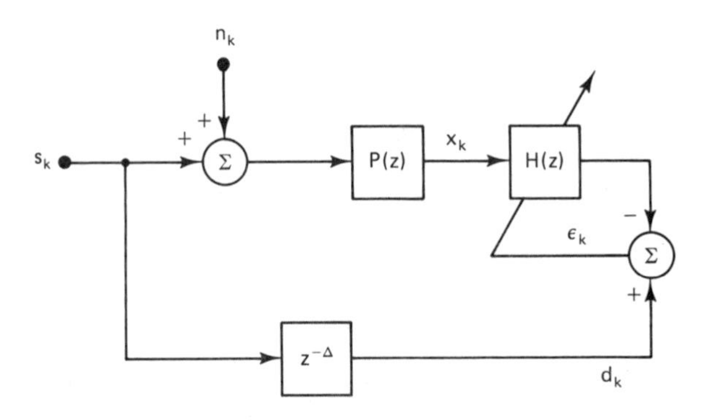

3. Derive an expression for the optimum inverse model, $H^*(z)$, for the situation above, beginning with the general expression in Equation (10.8).
4. Using the answer to Exercise 3 in Chapter 7, and Equation (7.65), derive a simple expression for the minimum mean-square error in the system above.
5. What are the minimum mean-square error and optimum weight values for the modeling situation shown below?

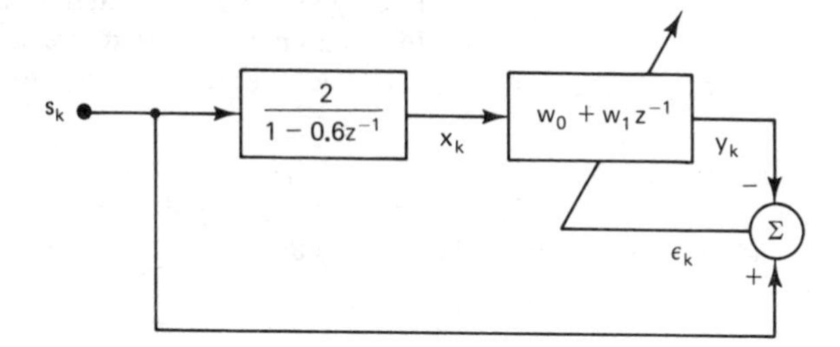

6. With w_0 and w_1 optimized, use the filtering algorithms in the system shown in Exercise 5 to show that ε_k is exactly zero for $k \geq 0$.
7. Demonstrate the convergence of the LMS algorithm in Exercise 5. Let $s_k =$ RANDOM(1.) − .5, as in Appendix A. Use $\mu = 0.08$ and let $w_0 = w_1 = 0$ initially. Plot ε_k^2 over a range of k long enough to show convergence.
8. Demonstrate the ability of the LMS algorithm to track a time-varying plant by plotting the weights w_0 and w_1 for the system below for $0 \leq k \leq 1500$. Use s_k and μ as in Exercise 7. Assume initially that $w_0 = 0.5$ and $w_1 = -0.3$.

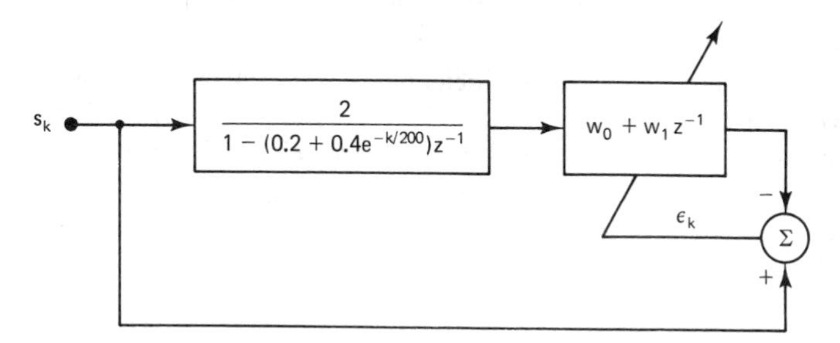

9. Write an expression for the error, ε_k, in terms of the input signal, s_k, for the inverse modeling system shown here:

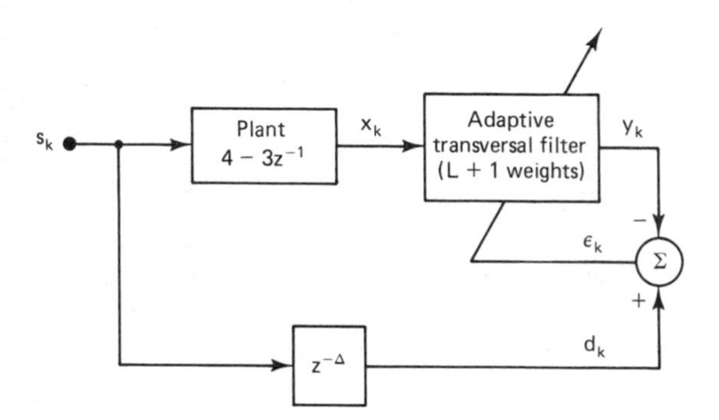

10. For the system in Exercise 9, assume that s_k is a white signal with unit power. Write an expression for the performance function, $\xi = E[\epsilon_k^2]$, in terms of the adaptive weights, w_0 through w_L.

11. For the system in Exercise 9, let s_k be a white signal with unit power, and assume that $L = \Delta = 1$. Using the answer to Exercise 10, solve analytically for the optimum weight values.

12. For the system in Exercise 9, let s_k be a white signal, let $L = 15$, and let $\Delta = 8$. Program the LMS algorithm to converge to $\xi_{\min}$. Plot the final impulse response of the adaptive linear combiner. Then plot the convolution of the latter with the impulse response of the plant.

13. In the inverse modeling situation shown below, let s_k be a white signal and let $n_k = 0$. Let the LMS algorithm converge and then plot the impulse response of the optimal inverse model, and the convolution of the two impulse responses.

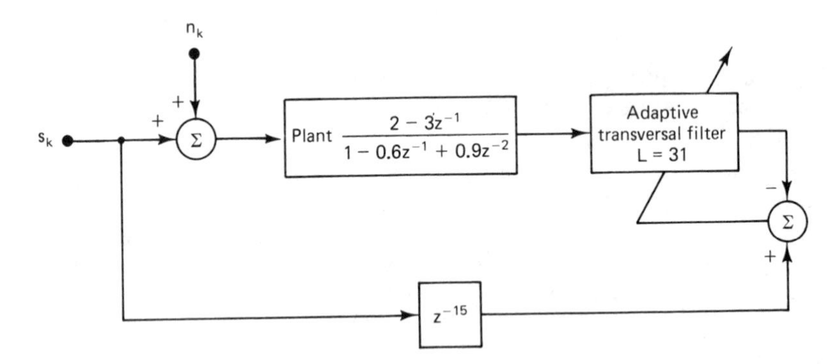

14. In the system shown in Exercise 13, determine experimentally the minimum mean-square error when the noise power, $\phi_{nn}(0)$, is 0.00, 0.01, and 0.10 times the signal power, $\phi_{ss}(0)$. Assume that s_k and n_k are both white but independent of each other.

15. In the system shown below, plot an LMS learning curve by averaging ϵ_k^2 over 100 adaptive runs. Use $\Delta = 15$, let s_k be white with unit power, and let μ be 10% of the total power in x_k.

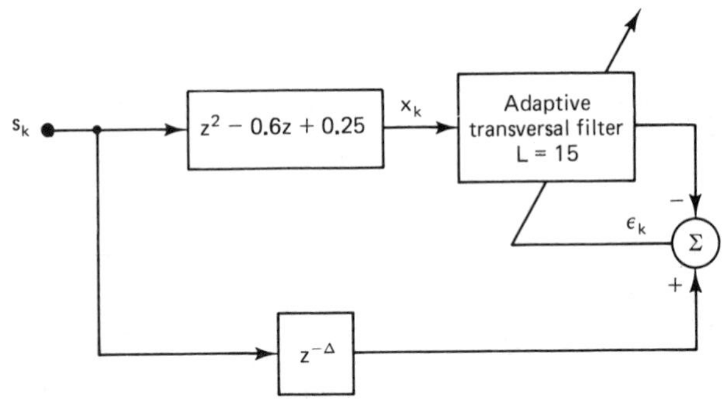

16. Run the system in Exercise 15 to convergence using the LMS algorithm with $\mu = 0.1$, and plot $\xi_{\min}$ versus Δ for $\Delta = 0$ through 15. To estimate $\xi_{\min}$, take an average of 100 values of ϵ_k^2 after convergence at each value of Δ.

17. Explain in your own words the operation of the adaptive equalizer in Figure 10.13(b).
18. In the "equation error" IIR synthesis model in Figure 10.17, show that the error surface is quadratic by deriving an expression similar to Equation (7.67).
19. Assume that $L = 0$ and $\mu = 1$, that is, that $A(z)$ and $B(z)$ both have a single weight in Figure 10.17. Express the values of these weights that minimize the mean-square error in Figures 10.16 and 10.17. Are these values the same?
20. Using the LMS algorithm, run the IIR design experiment in Figure 10.18. Use $L = M = 9$ in (10.19) and use a value of μ that will cause the equation error model to converge within a few thousand iterations. Make a plot of $\varepsilon_k'^2$ versus k. After convergence, plot a smooth magnitude response curve as in Figure 10.18.
21. Show the location of the poles of the IIR filter in Exercise 20. Compare your result with Figure 10.23. If necessary, reflect all unstable poles to the interior of the unit circle and give the new (stable) version of the IIR transfer function.
22. Show that the gain of the phase compensator in Equation (10.29) is the same at all frequencies. Find an expression for the phase, $\theta(\omega)$, where ω is the normalized frequency [Equation (7.16)].
23. Using the method of Figure 10.25, try to produce the IIR filter synthesis in Figure 10.27. Use $L = M = 4$, and vary the cost function approximately as shown in the figure, so as to produce results similar to those shown.

ANSWERS TO SELECTED EXERCISES

1. See Equations (7.73), (7.74), and (7.57).
2. $\Phi_{ss}(z) = \Phi_{dd}(z) = 1,\ \Phi_{xx}(z) = (1 + p)|P(z)|^2,\ \Phi_{dx}(z) = z^{\Delta}P(z)$
3. $H^*(z) = z^{-\Delta}/[(1 + p)P(z)]$
4. $\xi_{\min} = p/(1 + p)$
9. $\varepsilon_k = s_{k-\Delta} - \sum_{l=0}^{L} w_l(4s_{k-l} - 3s_{k-l-1})$
10. $\xi = 1 + 25\sum_{l=0}^{L} w_l^2 - 24\sum_{l=1}^{L} w_l w_{l-1} - 8w_\Delta + 6w_{\Delta-1}$
11. $w_0, w_1 = -0.0561;\ 0.1331$
22. $\tan\theta(\omega) = \dfrac{2b_1(b_2 - 1)\sin\omega + (b_2^2 - 1)\sin 2\omega}{2b_2 + b_1^2 + 2b_1(b_2 + 1)\cos\omega + (b_2^2 + 1)\cos 2\omega}$

REFERENCES AND ADDITIONAL READINGS

1. A. Papoulis, *The Fourier Integral and Its Applications*. New York: McGraw-Hill, 1962, Sec. 7.5.
2. P. O. Bhatnayar, *Nonlinear Waves in One-Dimensional Dispersive Systems*. Oxford: Clarendon Press, 1979, Chap. 1.

3. B. Widrow, P. F. Titchener, and R. P. Gooch, "Adaptive Design of Digital Filters," *Proc. ICASSP-81*, pp. 243–246, Mar. 1981.
4. R. A. David, "IIR adaptive algorithms based on gradient search techniques," Stanford Univ., Stanford, Calif., Aug. 1981, Chap. 3 (Ph.D. dissertation).
5. J. Makhoul, "Linear prediction: a tutorial review," *Proc. IEEE*, vol. 63, pp. 561–580, Apr. 1975.
6. J. Makhoul, "Spectral linear prediction: properties and applications," *IEEE Trans. Acoust. Speech Signal Process.*, vol. ASSP-23, pp. 283–296, June 1975.
7. J. D. Markel, "Formant trajectory estimation from a linear least-squares inverse filter formulation," Speech Communications Research Lab, Inc., Santa Barbara, Calif., AD734679, Oct. 1971.
8. J. D. Markel, "Digital inverse filtering—a new tool for formant trajectory estimation," *IEEE Trans. Audio Electroacoust.*, vol. AU-20, pp. 129–137, June 1972.
9. B. S. Atal and S. L. Hanauer, "Speech analysis and synthesis by linear prediction of the speech wave," *J. Acoust. Soc.*, vol. 50 (part 2), pp. 637–655, 1971.
10. C. Klayman et al., "Real-time implementation of a linear predictive coding system," in *Proc. Net. Commun. Conf.*, Paper 29E, Nov. 1973.
11. R. W. Lucky, "Automatic equalization for digital communication," *Bell Syst. Tech. J.*, vol. 44, pp. 547–588, Apr. 1965.
12. R. W. Lucky, "Techniques for adaptive equalization of digital communication systems," *Bell Syst. Tech. J.*, vol. 45, pp. 255–286, Feb. 1966.
13. A. Gersho, "Adaptive equalization of highly dispersive channels for data transmission," *Bell Syst. Tech. J.*, vol. 48, pp. 55–70, Jan. 1969.
14. M. M. Sondhi, "An adaptive echo canceller," *Bell Syst. Tech. J.*, vol. 46, pp. 497–511, Mar. 1967.
15. J. G. Proakis, *Digital Communications*. New York: McGraw-Hill, 1983.
16. R. W. Lucky, J. Salz, and E. J. Weldon, Jr., *Principles of Data Communication*. New York: McGraw-Hill, 1968.
17. J. C. Kennedy, "Equalization of digital communication channels using bootstrap mean square error criterion," Stanford Univ., Stanford, Calif., May 1971 (Ph.D. thesis).
18. L. J. Griffiths, F. R. Smolka, and L. D. Trembly, "Adaptive deconvolution: a new technique for processing time-varying seismic data," *Geophysics*, June 1977.
19. S. Qureshi, "Adaptive equalization," *IEEE Commun. Mag.*, p. 9, Mar. 1982.
20. M. M. Sondhi and D. A. Berkley, "Silencing echoes on the telephone network," *Proc. IEEE*, vol. 68, p. 948, Aug. 1980.
21. F. K. Becker and H. R. Rudin, "Application of automatic transversal filters to the problem of echo suppression," *Bell Syst. Tech. J.*, vol. 45, pp. 1847–1850, Dec. 1966.
22. M. M. Sondhi and A. J. Presti, "A self-adaptive echo canceller," *Bell Syst. Tech. J.*, vol. 45, pp. 1851–1854, Dec. 1966.
23. M. M. Sondhi, "An adaptive echo canceller," *Bell Syst. Tech. J.*, vol. 46, pp. 497–511, Mar. 1967.
24. N. Demytko and L. K. Mackechnie, "A high speed digital adaptive echo canceller," *Aust. Telecommun. Rev.*, vol. 7, pp. 20–27, 1973.

25. S. J. Campanella, H. G. Suyderhoud, and M. Onufry, "Analysis of an adaptive impulse response echo canceller," *Comsat Tech. Rev.*, vol. 2, no. 1, pp. 1–37, Spring 1972.

26. D. L. Duttweiler, "A twelve-channel digital echo canceller," *IEEE Trans. Commun.*, vol. COM-26, pp. 647–653, May 1978.

27. D. L. Duttweiler and Y. S. Chen, "Performance and features of a single chip FLSI echo canceller," *Proc. NTC 79*, Washington, D.C., Nov. 17–29, 1979.

28. J. G. Proakis, "Adaptive digital filters for equalization of telephone channels," *IEEE Trans. Audio Electroacoust.*, vol. AU-18, no. 2, pp. 484–497, June 1970.

29. J. G. Proakis and J. H. Miller, "An adaptive receiver for digital signaling through channels with intersymbol interference," *IEEE Trans. Inf. Theory*, vol. IT-15, no. 4, pp. 484–497, July 1969.

30. R. A. David, "A cascade structure for equation error minimization," *Proc. 16th Asilomar Conf. Circuits, Syst., Comput.*, p. 182, Nov. 8, 1982.

31. J. J. Kormylo and J. M. Mendel, "Maximum-likelihood seismic deconvolution," *IEEE Trans. Geoscience and Remote Sensing*, vol. GE-21, pp. 72–82, Jan. 1983.

32. R. P. Gooch, "Adaptive pole-zero filtering: the equation error approach," Stanford Univ., Stanford, Calif., June 1983, Chap. 5 (Ph.D thesis).

chapter 11

Adaptive Control Systems

In this chapter we present some adaptive control applications of the direct and inverse adaptive modeling procedures in Chapters 9 and 10. Our purpose is to illustrate the use of adaptive modeling in adaptive control, but not to provide a complete discussion of the subject. There is already a large amount of literature on adaptive control, an example being the text by Landau [1].

In control theory, the "plant" is the physical system to be controlled. It produces an output response to an input control signal. An example of a plant might be the control surfaces and the airframe of an aircraft that respond to an input command signal, or the device that delivers fuel and its associated combustion system that responds to a demand for heat or power.

In addition to the plant there is also the "controller" which, as shown in Figure 11.1, provides an input signal to the plant. The controller receives information from the external world and also, in the case of feedback control systems, from the plant output. The controller may be a fixed linear or nonlinear system, or it may be time varying so that it can adapt to changing plant or environmental conditions. In this chapter we will be concerned with adaptive linear control systems, that is, with adaptive control systems that become linear once their internal adaptive components converge.

An example of such a system is shown in Figure 11.2. It is called a "unity feedback" system. The output, $c(t)$, is subtracted from the input, $r(t)$, to form an error signal, $e(t)$. The latter then drives the controller, which tries continuously to drive the plant to cause $e(t)$ to be driven toward zero. The intent is that $c(t)$ should remain reasonably close to $r(t)$, and of course the overall transfer function from $r(t)$ to $c(t)$ must be stable.

Although this system is simple in conception, it is rather inefficient and difficult to deal with from the viewpoint of adaptation. The controller could be of the form of a transversal filter, but to adapt it, one would need in real time its desired output corresponding to the required real-time plant input signal. This "desired" signal is very difficult to obtain when the plant is unknown, however. If it were available, so

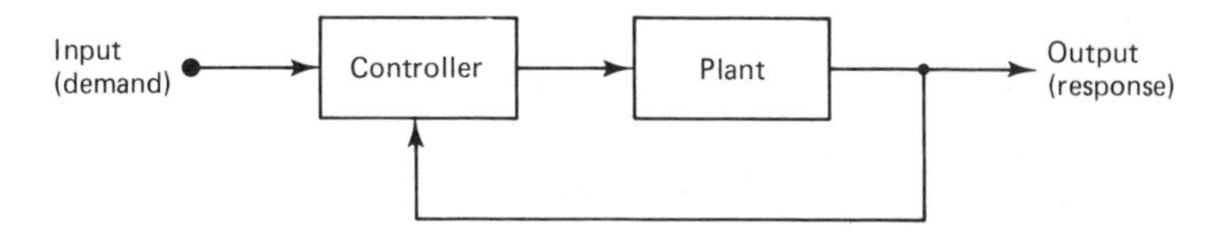

Figure 11.1 Plant with feedback control.

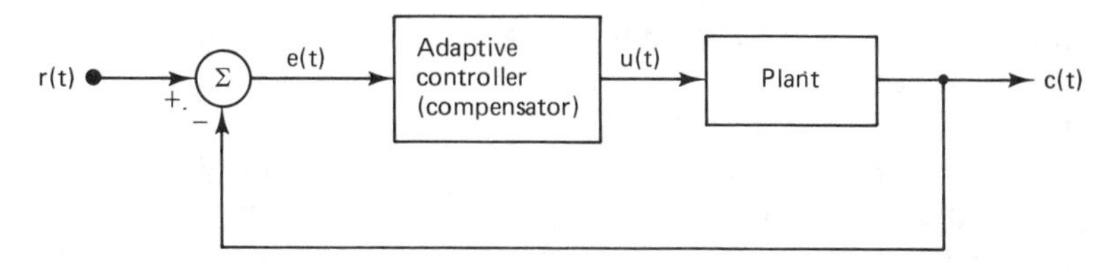

Figure 11.2 Unity feedback control system with an adaptive series compensator.

much would be known that the compensator and the feedback loop would be unnecessary.

Another approach to the adaptation of the system of Figure 11.2 is the following. Suppose that the purpose of adaptation is the minimization of the servo error, $e(t)$, in the mean-square sense. Gradient components could be measured by perturbing the compensator weights. The mean-square error could be minimized by using a gradient method such as the differential steepest-descent method of Chapter 5. There are two difficulties here that limit this approach. Regardless of the method used in perturbing the weights, whether one at a time or all at once, the system transients must be allowed to die out before measurements can be taken each time the compensator adjustments change or the plant parameters change. Furthermore, assuming that the gradient can be measured successfully, the mean-square-error performance function will not be unimodal, as described in Chapter 7. Instability could result from improper controller adjustment. One must be concerned simultaneously with stability of the control system and stability of its adaptive algorithm.

In this chapter we illustrate two approaches to adaptive control that circumvent the difficulties just described in connection with Figure 11.2. The first approach is based on direct modeling of the plant as in Chapter 9, and the second on inverse modeling as in Chapter 10.

ADAPTIVE MODEL CONTROL

The first approach to adaptive control, called adaptive model control (AMC), operates on the following principle. We form a model of the plant using direct modeling, and use the model to determine control inputs to the plant which will cause desired plant outputs. We then apply these same control inputs to the actual plant, and the result is plant outputs that closely match the desired outputs. This type of control is in a sense open loop, but in fact the loop is closed through the adaptive process.

To illustrate the AMC idea we will use the blood pressure regulating system shown in Figure 11.3. The experimental work on this system was performed by students at Stanford University and supervised by Mark Yelderman, William New, Noel Thompson, and Cristy Schade, whose assistance is gratefully acknowledged. The goal is to develop a closed-loop control system for regulating a patient's blood pressure. Thus the plant input in this case is the drug flow in Figure 11.3, and the plant output is the blood pressure. Experiments were performed with canine subjects, who survived the procedure in good condition.

To control the blood pressure, a powerful drug called Arfonad is injected into the animal. This drug has the effect of disabling the natural blood-pressure regulating system of the animal and inducing a prolonged state similar to a state of shock. If left alone, the blood pressure could drop close to zero and irreversible damage could be done to the animal. To prevent this, a vasoconstrictor drug called norepinephrine is infused slowly over many hours to compensate by raising the blood pressure. A computer continually monitors blood pressure and regulates the rate of infusion of the vasoconstrictor. The ultimate purposes of this work are to develop computer controls for human intensive care systems, and to further the development of adaptive control systems.

Typical dynamic responses of the mean animal blood pressure to step changes in the rate of infusion of the vasoconstrictor drug are sketched in Figure 11.4. The shape of the response curve depends on the size, type, and especially on the condition of the animal. An animal in good health will respond to small increases in vasoconstrictor drug flow by eventually settling the blood pressure back to the

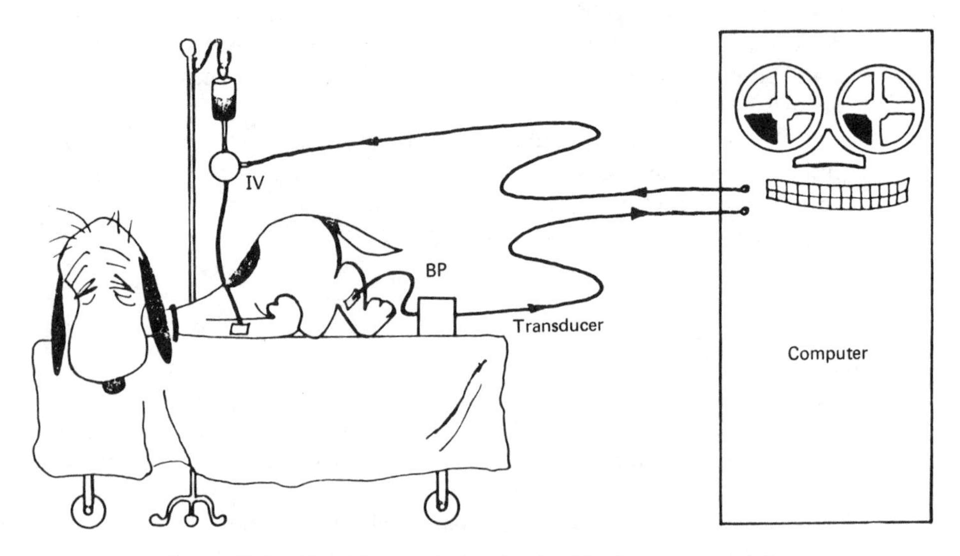

Figure 11.3 Closed-loop control system for blood pressure regulation.

original set point level. A sick animal will not be able to compensate for even moderate increases in vasoconstrictor inputs and hence the blood pressure will increase in a predictable manner and then settle at a higher level. Large variations in animal responses to the vasoconstrictor have been observed. Typically, there is a transport lag of 10 to 20 seconds before the animal responds, and settling times are usually from 50 to 100 seconds.

The system illustrated in Figure 11.3 gives the appearance of being an ordinary feedback control system, but it is not. The dynamic response of an animal (including the transport lag) is often too variable to be managed by conventional feedback control.

A block diagram of the adaptive control system is shown in Figure 11.5. The functions labeled "forward-time calculation" and "adaptive model," which are described below, as well as many data logging and data display functions which are not shown but are necessary in the laboratory setup, are implemented using a minicomputer. The "zero-order hold" holds each sample value, x_k, during the

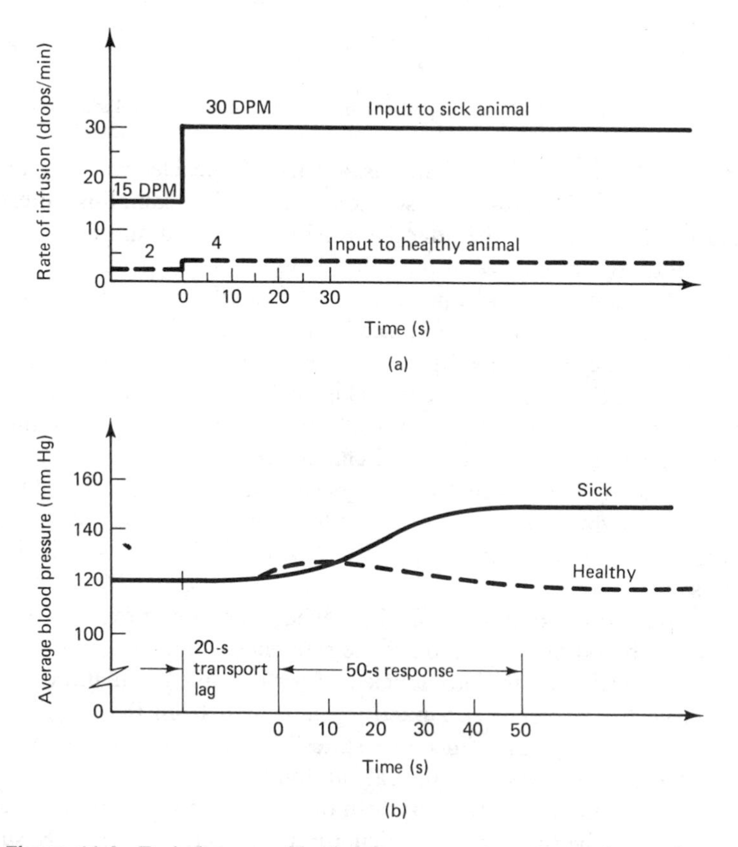

Figure 11.4 Typical average blood pressure responses to step changes in vasoconstrictor infusion rates.

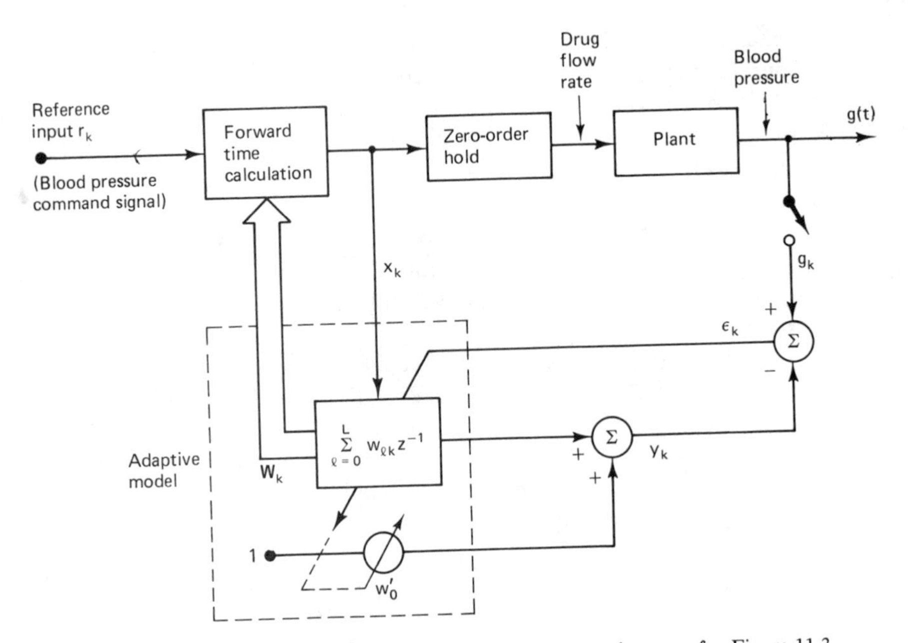

Figure 11.5 Block diagram of the adaptive model control system for Figure 11.3.

interval between samples, and is a part of the electronic system interfacing the computer to the drug-flow solenoid valve. The sampling interval is 5 seconds. During each interval the adaptive model is updated and a new drug rate (in drops per minute) is calculated as described below.

The adaptive model in Figure 11.5 is a 20-weight FIR filter ($L = 19$) covering a total real-time window of 95 seconds. A bias weight (w'_0) is included to represent the ambient average blood pressure when the input drug rate is zero. One can see in Figure 11.5 that the adaptive model is an example of the "direct model" discussed in Chapter 9. Instead of having a fixed linear compensator where the weight values are not functions of the input-signal characteristics, the adaptive process automatically adjusts the weights so that for the given input-signal statistics, the model provides a best minimum-mean-square-error fit to a sampled version of the combination of the zero-order hold and the plant. The LMS algorithm was used in all experimental runs.

Referring again to Figure 11.5, when operating properly, the system causes the animal's blood pressure to track the reference command input r_k. The plant control x_k is derived from the box labeled "forward-time calculation." This box generates x_k from the reference command input r_k and from the weight vector $\mathbf{W}_k$ and the input vector $\mathbf{X}_k$ (the "memory states" of the model). Let us now consider the operation of the forward-time calculation box.

Assuming that $E[\epsilon_k^2]$ has been driven close to zero by the adaptive process, the objective of the forward time calculation is to derive x_k from r_k such that r_k and y_k are equal. If r_k and y_k are equal, then (with ϵ_k^2 small) the plant output, g_k, will be close to r_k, which is our ultimate goal. Thus the forward-time calculation must become essentially an inverse of the adaptive model in Figure 11.5. Since the plant involves a delay, the inverse must therefore look forward in time.

We form the inverse model in the following way. The entire weight vector $(w'_{0k}, \mathbf{W}_k)$ is updated at each iteration in accordance with the LMS algorithm. In the adaptive model, if we assume that y_k and r_k are equal as discussed above, we have for the kth iteration

$$y_k = w'_{0k} + \sum_{n=0}^{L} w_{nk} x_{k-n} \qquad \text{(adaptive model)} \tag{11.1}$$

Therefore,

$$x_k = \frac{1}{w_{0k}} \left[r_k - w'_{0k} - \sum_{n=1}^{L} w_{nk} x_{k-n} \right] \qquad \begin{pmatrix} \text{forward-} \\ \text{time} \\ \text{calculation} \end{pmatrix} \tag{11.2}$$

For this kind of "inversion" to work, we have to assume that x_k excites the plant at an adequate level so that adaptive modeling can take place. If not, a small "dither signal" can be added to the plant input.

We also have to assume that w_{0k} in (11.1) does not get close to zero, and of course there is no guarantee of this in LMS adaptation. In fact, when there are delays in the plant such as the transport lag in Figure 11.4, w_0 will tend to be small and noisy, and the values of x_k computed by formula (11.2) may be very large and erratic, since division by w_0 is required. Therefore, in the blood-pressure control system where massive doses of drug are undesirable and negative doses are generally impossible, we modify the adaptive portion to account for the transport lag.

The modification consists of constraining the first several weights of the adaptive model to zero. The number of zero-constrained weights corresponds to the transport lag of the plant, which would be obtained from a priori knowledge. Suppose, for example, that the first two weights w_{0k} and w_{1k} are constrained to zero. Then the current and previous inputs to the adaptive model, x_k and x_{k-1}, have no effect on its output, but $x_{k-2}, x_{k-3}, \ldots, x_{k-L}$ do have an effect. Choosing inputs to cause the current model output to be equal to r_k, we now have

$$y_k = w'_{0k} + \sum_{n=2}^{L} w_{nk} x_{k-n} \tag{11.3}$$

This result would allow us to calculate x_{k-2} as we did in (11.2), but we really wish to know x_k, not x_{k-2}. Therefore, let us now shift (11.3) two units of time into the future:

$$y_{k+2} = w'_{0,k+2} + \sum_{n=2}^{L} w_{n,k+2} x_{k+2-n} \tag{11.4}$$

Now let us assume that the weights change slowly so that we may use current weight values in place of future values. Then, with y_k equal to r_k again, we have

$$x_k = \frac{1}{w_{2k}}\left(r_{k+2} - w'_{0k} - \sum_{n=3}^{L} w_{nk}x_{k+2-n}\right) \tag{11.5}$$

This result requires knowledge of the reference command input two units of time into the future. Sometimes future values of this input are available, and (11.5) can be used to achieve the control objective. When only r_k, the current reference command input is available, (11.5) can be modified to use the available knowledge, as follows:

$$x_k = \frac{1}{w_{2k}}\left(r_k - w'_{0k} - \sum_{n=3}^{L} w_{nk}x_{k+2-n}\right) \tag{11.6}$$

Using (11.6) will cause the model output to track the reference command delayed by two time units. This delay in the response is thus an inevitable consequence of the plant transport delay.

The AMC techniques in Figure 11.5 have been used a number of times in experiments to regulate and control the average blood pressure in animals. In these experiments, the standard deviation of the noise in the blood-pressure sensing instrumentation has been about 5 to 10 millimeters of mercury (mm Hg). The mean

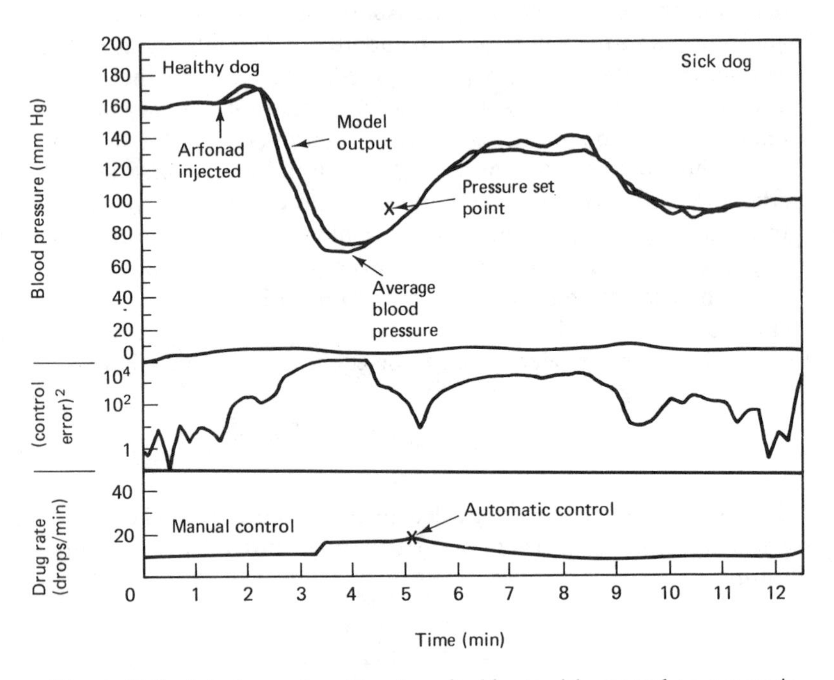

Figure 11.6 Actual run: transition from healthy to sick, manual to automatic control.

blood pressure is typically regulated to within about 2 to 4 mm Hg in steady state and could be off about 5 to 10 mm Hg temporarily under extreme transient conditions. Typical startup settling times were of the order of 2 minutes, somewhat longer than the total time window spanned by the adaptive plant model. To start the system as quickly as possible, the initial weight values in the modeling process are usually taken from a previous run. Initial values are not critical, but if they are close to correct, there will be only a small startup transient.

Figures 11.6 through 11.9 present in sequence results developed during an experimental run while controlling an animal's average blood pressure. The dog was normal until the Arfonad was injected, whereupon the blood pressure plummeted, as seen in Figure 11.6.

In this run the adaptive-model weights began to form, starting from initial settings, at the very outset before the Arfonad was injected. The two upper tracings show the actual average blood pressure of the animal and the output of the model, respectively. Note how they stay closely together. They stay moderately close together even in periods of great stress such as just after the Arfonad was injected.

At the beginning of the run, the flow rate of the vasoconstrictor (the "drug rate" in the lower plot in Figure 11.6) was manually set at 10 drops per minute. This was manually raised to 20 drops per minute after the Arfonad took hold. Raising the drug rate checked the blood pressure decline. Soon thereafter, as indicated by the cross on the drug-rate tracing, control of the drug rate was turned over to the

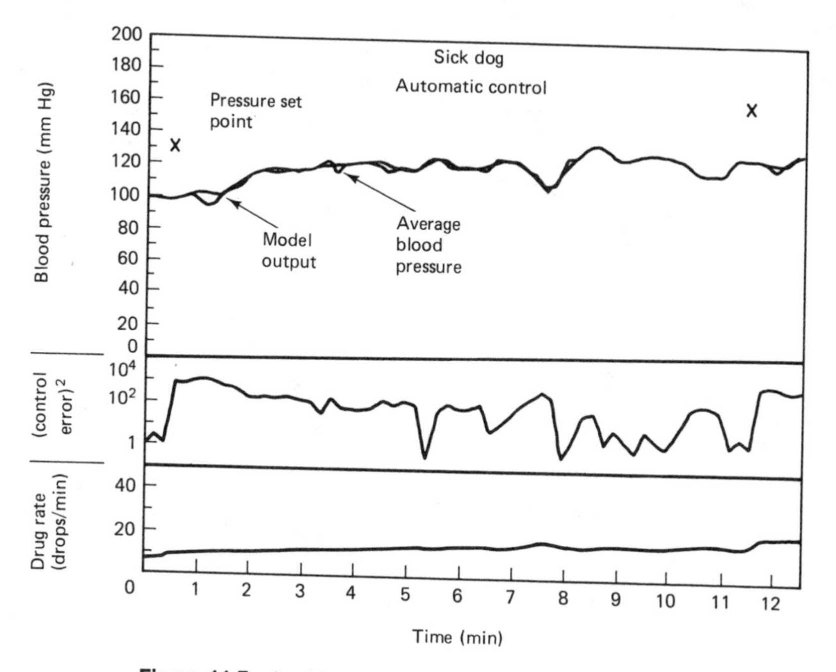

Figure 11.7 Actual run: control of sick dog's blood pressure.

automatic system and remained automatic thereafter. A pressure set point was entered through the computer keyboard, and this level was indicated by the cross near the upper two tracings. The control system then had the job of raising the animal blood pressure to the set point and controlling it in spite of natural disturbances. The middle curve shows a running average mean-square error (on a log scale) between the plant and the adaptive model.

The total time span of the adaptive model was 95 seconds. The model contained 20 taps with 5-second delays between taps of the adaptive transversal filter. Once automatic control was established, the blood-pressure settling time was about 5 minutes. Thus the system settling time was about three times as long as the memory span of the model, which is a short settling time for an adaptive control system.

Figure 11.6 actually represents a portion of a long run in which the computer controlled the blood pressure for several hours with the animal under different degrees of influence of Arfonad. The results in terms of control were uniformly good, and the response data of Figures 11.7 and 11.8 are typical. The data records of Figures 11.6 through 11.8 are contiguous with slight time overlaps, and represent responses to changing set-point values. In each case, the settling time was approximately 5 minutes.

The weight values of the FIR model were recorded at several times during the run and are plotted in Figure 11.9. The weight values are arranged chronologically

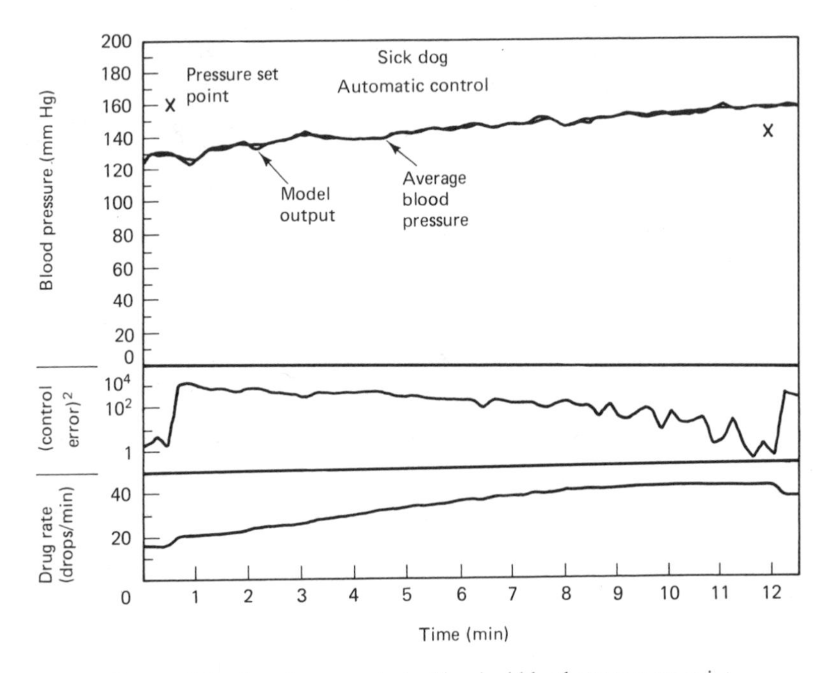

Figure 11.8 Actual run: control with raised blood-pressure set point.

along the line so that they represent the model's view of the animal's impulse response. The twenty-first weight is the bias weight, w_0' in Figure 11.5. The impulse response in the top frame was taken before the Arfonad was injected and here the animal was very sensitive to the vasoconstrictor drug. The next frame was taken after the Arfonad was injected and had taken effect, just before the automatic control was turned on. The shape of the response was changed somewhat, and the sensitivity level changed greatly. As time progressed there were no further drastic changes in the animal impulse response. This lack of further change was also considered a significant result.

In summary, we have described a real-time computer control system for regulating the blood pressure of a subject in a prolonged state of shock. The system controls the infusion rate of a vasoconstrictor drug and monitors the blood pressure. An adaptive model of the subject's drug response in terms of blood pressure is used to derive the required input for control of future blood pressure values. An adaptive linear combiner is used for the model, and control is derived by forward time calculation based on the model impulse response. This is a control technique that we have called AMC. It is based on direct adaptive modeling of the unknown plant.

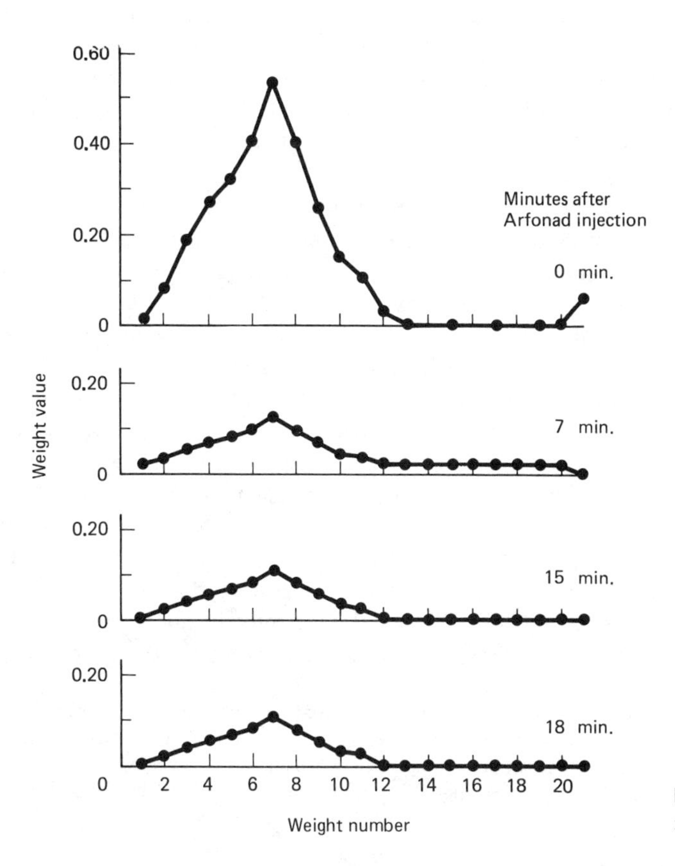

Figure 11.9 Impulse response of the model in Figure 11.5 at various times.

ADAPTIVE INVERSE CONTROL

Adaptive inverse control is another approach to the adaptive control problem, based on the inverse modeling techniques described in Chapter 10. It was developed primarily to accommodate situations where the plant might be "non-minimum-phase," that is, might have transfer function zeros in the right half of the s-plane or, in terms of a discrete (digital) model, might have zeros outside the unit circle in the z-plane.† Controlling a non-minimum-phase plant by the AMC method could lead to difficulties, because the output signal of the forward-time calculation must have a transform essentially equal to the product of the transform of the reference command input and the reciprocal of the plant transfer function (refer to Figure 11.5). If the plant model has zeros outside the unit circle, the control signal x_k will have a transform with poles outside the unit circle. In this case the control signal will be unstable. In the physical world, this signal will increase in magnitude until some part of the system is driven into saturation, causing loss of control.

In other words, an unknown plant can be made to track an input command signal when this signal is applied to a controller whose transfer function approximates the inverse of the plant's transfer function. The controller output becomes the driving signal for the plant. In the adaptive inverse control method, the parameters of the controller are obtained by an adaptive inverse modeling process applied to the plant. If the controller is realized as an adaptive transversal filter whose weights are adapted by a least-squares algorithm such as the LMS algorithm, we will show that the stability of the controller can be assured whether or not the plant is minimum phase.

The inverse model of the unknown plant can be formed as shown in Figure 11.10. The adaptive filter input is the plant output and the filter is adapted to cause its output to be a best least-squares fit to the plant input. A close fit implies that the cascade of the unknown plant and the LMS filter have a transfer function of essentially unit value, at least within the frequency band of the plant input signal. Close fits can be achieved in general with adaptive transversal filters of sufficient length, even when the unknown plant has many poles and zeros.

If a continuous plant is stable, all of its poles lie in the left half of the s-plane. Some of its zeros could lie in the right half-plane, however, and then the plant would be non-minimum phase. The inverse of a minimum-phase plant has all of its poles in the left half-plane, and its inverse is therefore stable. But in the case of a non-minimum-phase plant, the inverse model in Figure 11.10(a) may be unstable. This basic type of instability may be removed by including the inverse modeling delay, $z^{-\Delta}$, shown in Figure 11.10(b). This delay effectively allows the adaptive model to have a two-sided impulse response as described previously in connection

† In the real world, plants to be controlled are continuous and are represented as having poles and zeros in the s-plane. Digital controllers work with samples of the continuous plant inputs and outputs and therefore "see" the plant as if it were discrete. The discussions, analyses, and computer simulations presented here are taken as if the plant were digital. A more complete discussion of this subject is contained in Franklin and Powell [2].

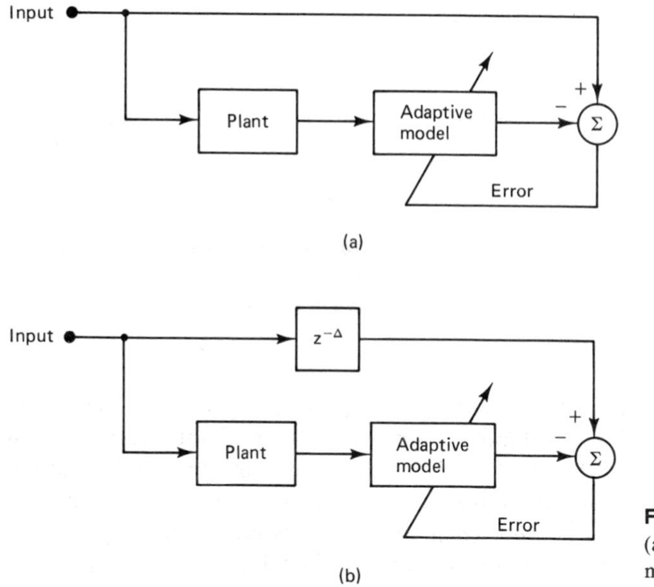

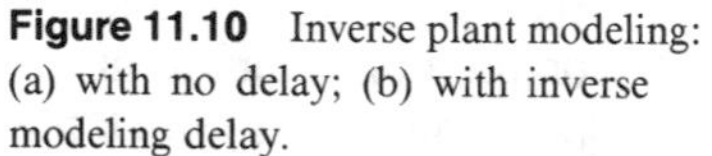
Figure 11.10 Inverse plant modeling: (a) with no delay; (b) with inverse modeling delay.

with Figure 10.1. Thus, by allowing the delay in Figure 11.10(b), we can obtain approximate delayed inverse models to both minimum-phase and non-minimum-phase plants without knowing a priori whether or not the plant is minimum phase. However, as discussed in Chapter 10, some knowledge of plant characteristics will always be helpful when choosing the delay Δ and the length of the transversal filter used for inverse modeling.

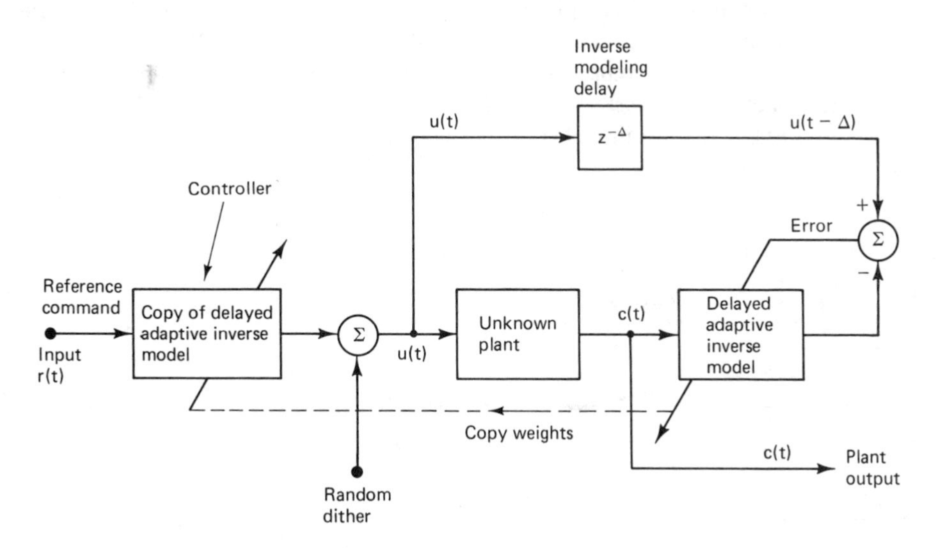

Figure 11.11 Adaptive inverse model control system.

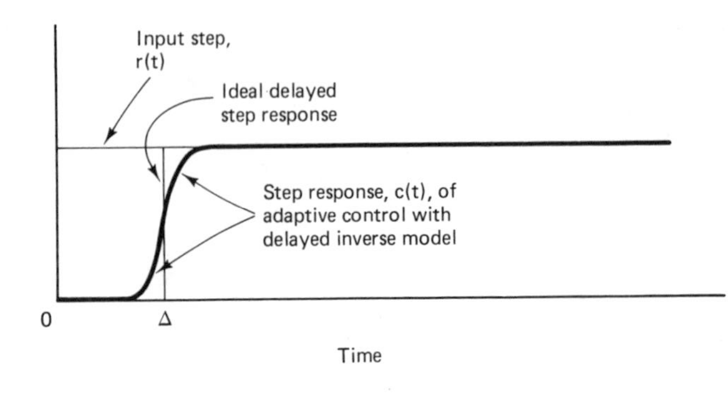

Figure 11.12 Comparison of ideal step response versus adaptive inverse control system step response.

The adaptive inverse control method is illustrated in Figure 11.11. Here the delayed adaptive inverse, being an adaptive transversal filter without feedback, is an approximate stable inverse of the plant. The concept, including the random dither signal, is similar to the AMC concept above. The controller in Figure 11.11 is a copy of the approximate inverse model, and when the system is operating properly, the plant output will track the reference command signal, which is applied as an input to the controller. The controller output is the driving function for the plant. If the controller were an exact delayed plant inverse, the plant output, assuming no noise, would be an exact copy of the input reference command, but delayed, that is,

$$c(t) = r(t - \Delta) \tag{11.7}$$

A step change in the reference command would cause a step change in the plant output after a delay of Δ seconds. If the inverse model is imperfect but a good approximation to the plant inverse, the step response $c(t)$ might appear as illustrated in Figure 11.12. The ideal delayed step response is superimposed in the same figure. For comparison, Figure 11.13 shows a typical step response of a simple feedback control system. It is often interesting to compare this type of response with that of the inverse model control scheme.

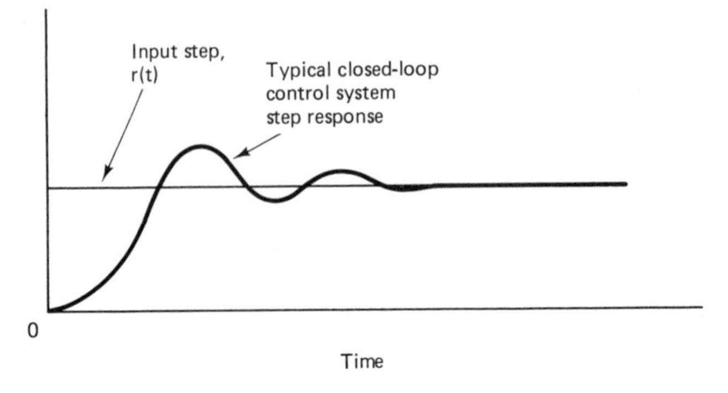

Figure 11.13 Step response of a conventional closed-loop control system.

A problem called "plant drift" is often observed in feedback control systems that have at least one stage of integration within the control loop. Plant drift is observed at the plant output as a low-frequency random component superimposed on other plant outputs, and it occurs spontaneously, not in response to plant inputs. When this effect is strong enough to require a cure, one can use the method based on a bias weight in the inverse model, which is illustrated in Figure 11.14.

Assume that the plant output has an additive drift, d, and that the adaptive plant inverse model has an adaptive bias weight, w_0'. Referring to Figure 11.14(a),

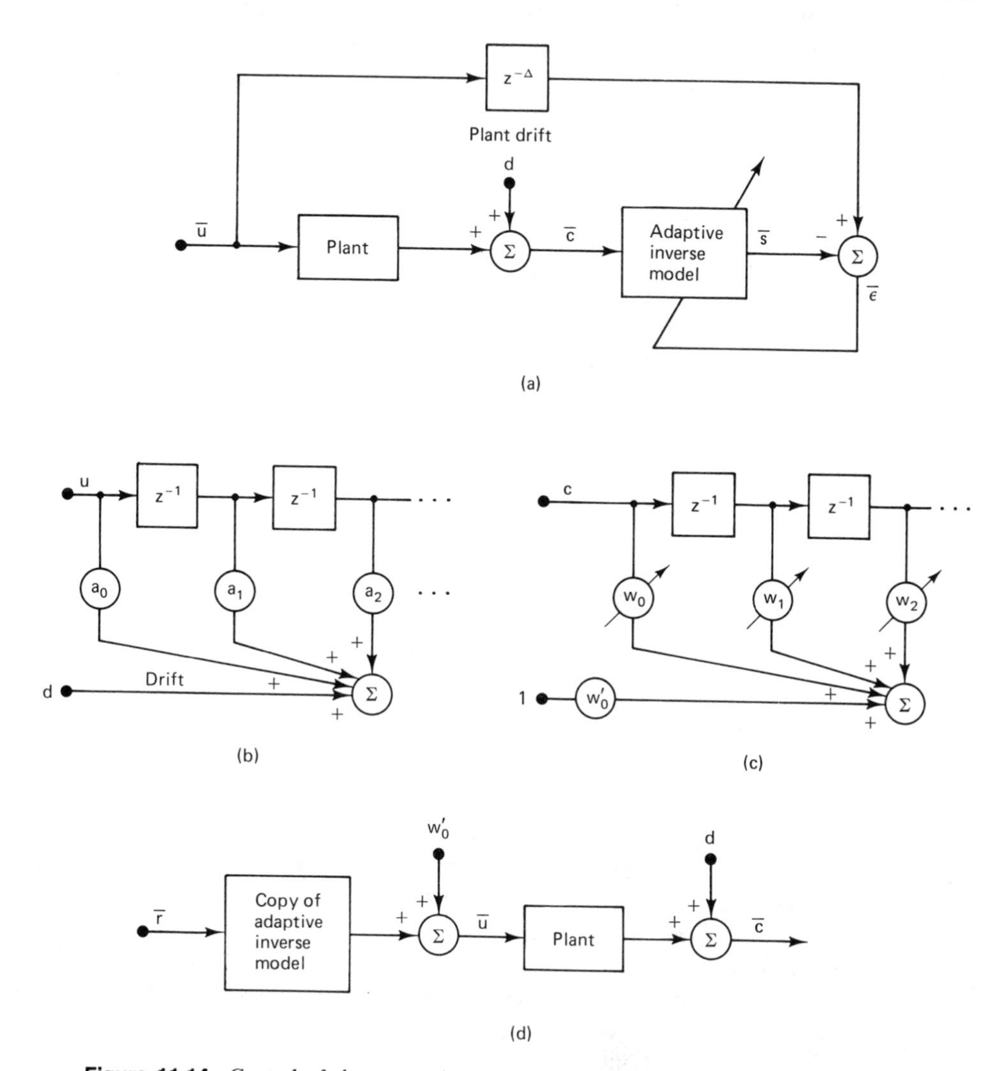

Figure 11.14 Control of the mean plant output in the presence of unknown plant drift: (a) inverse modeling process; (b) representation of plant with drift; (c) adaptive inverse model, including bias weight; (d) control process.

the mean value of the plant input is defined as $\bar{u}$, the mean of the plant output is $\bar{c}$, and the mean of the inverse model output is $\bar{s}$. For the sake of this discussion, the plant is represented in Figure 11.14(b) as an FIR filter having the discrete impulse response $[a_0, a_1, a_2, \ldots, a_M]$. Using these definitions, the mean of the plant output $\bar{c}$ can be expressed as a function of the plant drift and the mean of the plant input $\bar{u}$:

$$\bar{c} = d + \bar{u} \sum_{m=0}^{M} a_m \tag{11.8}$$

The adaptive inverse model, which is an FIR filter with a bias weight, is shown in Figure 11.14(c). The mean value of its output is

$$\bar{s} = w_0' + \bar{c} \sum_{l=0}^{L} w_l \tag{11.9}$$

Since the adaptive process minimizes mean-square error, the bias weight w_0' will be adapted so that the error will be unbiased (any bias will needlessly increase the MSE). Referring to Figure 11.14(a), we may conclude that

$$\bar{s} = \bar{u} \tag{11.10}$$

Referring now to Figure 11.14(d), we recall that the reference command input for the entire system with mean $\bar{r}$ is applied to a copy of the plant inverse.

Since this copy of the plant inverse drives the plant with a signal whose mean has been defined to be $\bar{u}$, driving the copy with an input whose mean is $\bar{r}$ causes its output mean to be

$$\bar{u} = w_0' + \bar{r} \sum_{l=0}^{L} w_l \tag{11.11}$$

Combining (11.10) and (11.11) yields

$$\bar{s} = w_0' + \bar{r} \sum_{l=0}^{L} w_l \tag{11.12}$$

Comparing (11.12) with (11.9) gives the desired result,

$$\bar{c} = \bar{r} \tag{11.13}$$

This result implies that regardless of the amount of plant drift, d, the mean of the plant output, $\bar{c}$, is equal to the mean of the reference command input $\bar{r}$. In effect, the feedback in the adaptive process compensates for the plant drift.

Since this analysis of compensation for plant drift applies to zero-frequency (dc) drift, it leaves some unanswered questions. For example, how fast can the drift, d, change without causing an output bias? How fast can the mean of the input reference command change without producing a significant error in the mean plant output? What are the effects of dynamic inverse modeling errors on system performance? These questions are topics of current research.

Another subject of importance in adaptive inverse modeling is the use of a dither signal to promote smooth and constant adaptation. Refer again to Figure

11.11. A random dither signal is injected to sustain the adaptive process when sufficient ambient signal activity is lacking. The dither is beneficial to the adaptive process but causes noise at the plant output, and should thus be kept small in amplitude so as not to unduly disturb the plant, but be strong enough to sustain the adaptive process. The design of adequate dither is another subject of current research.

EXAMPLES OF ADAPTIVE INVERSE CONTROL

Several examples of the adaptive inverse control system in Figure 11.11 have been simulated. Both the plant and the controller were simulated as discrete systems, and different combinations of plant pole and zero locations were tried. The following results are typical. We consider first a plant having two poles and no zeros. The pole locations of this plant (1) are shown in the z-plane in Figure 11.15, and its step response is shown in Figure 11.16. The inverse of this plant, with two zeros, was easily realized by the adaptive FIR filter which had available 32 weights in the simulation. The impulse response of the adaptive inverse after convergence is plotted in Figure 11.17. Using this inverse as a controller as in Figure 11.11, the step response of the entire system is exact, as shown in Figure 11.18. Note the effect on this step response of the inverse modeling delay Δ.

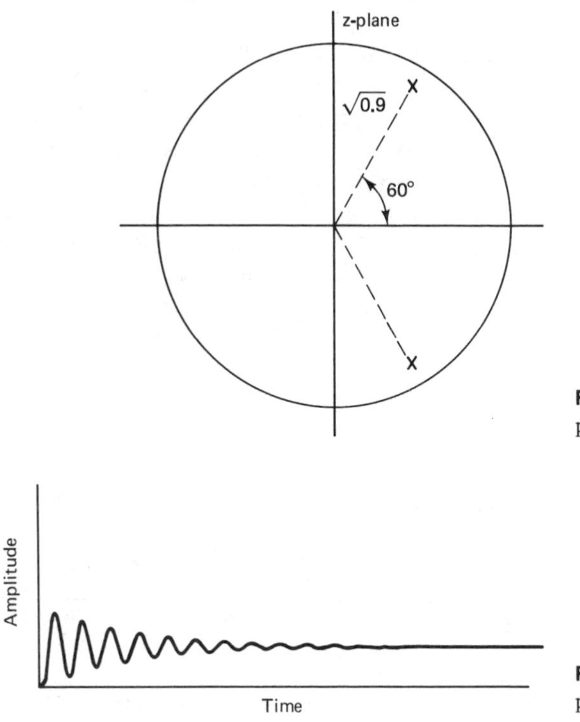

Figure 11.15 Locations of poles of plant 1.

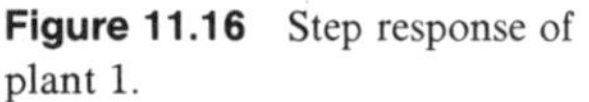

Figure 11.16 Step response of plant 1.

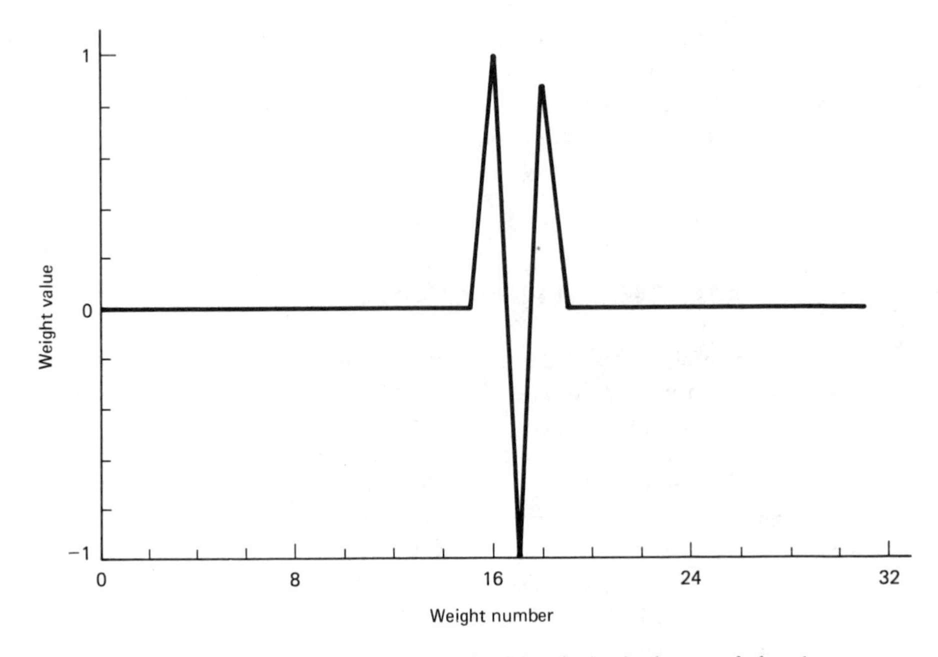

Figure 11.17 Impulse response of the delayed adaptive inverse of plant 1.

A potentially more difficult case is that of controlling plant 2, which has a pole at the origin and a single zero located as shown in Figure 11.19, and is thus a non-minimum-phase plant. Again, the delayed adaptive inverse model was formed as in Figure 11.11 and used as a controller. The impulse response of the adaptive inverse with $L = 32$ and $\Delta = 16$ is shown in Figure 11.20 and the step response of the resulting control system is shown in Figure 11.21. In this case, using the delay, there was no real difficulty in controlling the non-minimum-phase plant. The step response is perfect except for a small amount of ringing.

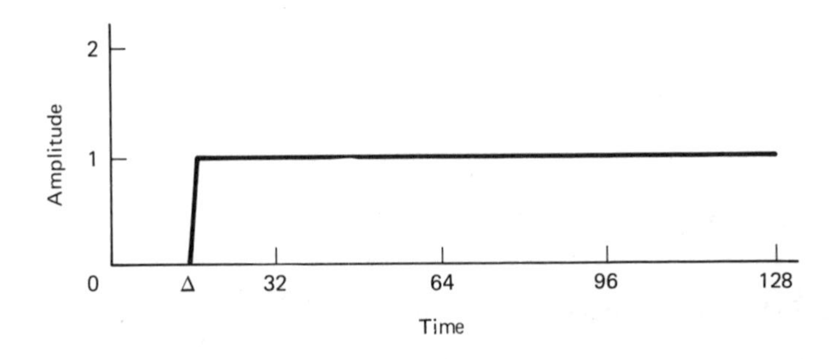

Figure 11.18 Step response of adaptive inverse control system with plant 1.

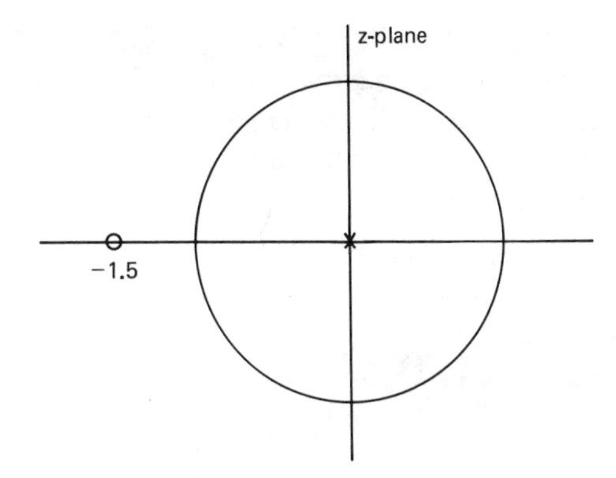

Figure 11.19 Pole and zero of plant 2 at 0.0 and −1.5, respectively.

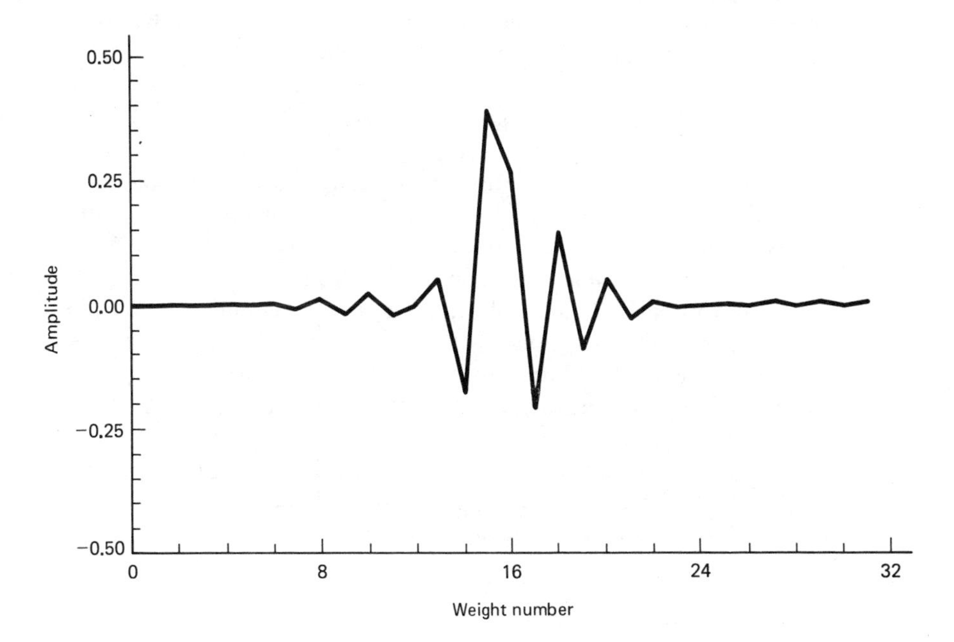

Figure 11.20 Impulse response of the delayed adaptive inverse of plant 2 with $L = 32$ and $\Delta = 16$.

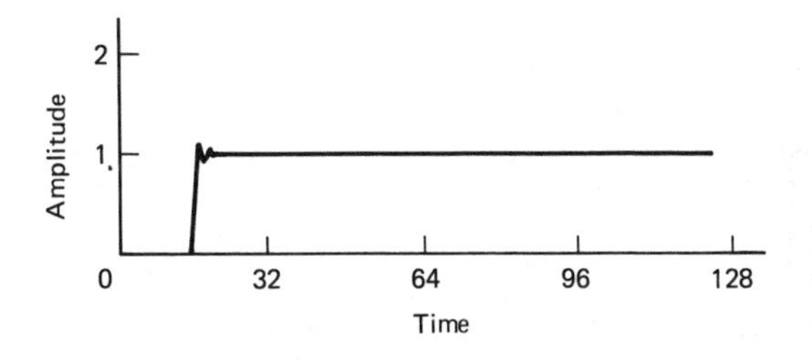

Figure 11.21 Overall step response of the adaptive inverse control system for plant 2.

As seen in the preceding examples, the tightness of control depends on the closeness of fit between the true inverse and the inverse realized by the adaptive filter. Also, the characteristics of the dither signal and the number of weights $(L + 1)$ in the adaptive inverse affect the overall control in ways that depend on the signal characteristics as well as the plant.

PLANT NOISE AND THE FILTERED-*x* LMS ALGORITHM

Let us now examine the general problem of plant noise in a control system. We have seen how the inclusion of a bias weight in the adaptive inverse model alleviates the problem of low-frequency plant drift. However, there is a difficulty in using this same remedy against plant noise at higher frequencies, a little above zero and beyond. The nature of the difficulty can be visualized from inspection of Figure 11.22. In most physical situations, plant noise can be represented as additive noise, usually nonwhite, at the plant output, regardless of its source within the plant itself. The effects of such noise on the inverse modeling process are apparent. This noise is an additive input to the adaptive filter and is not correlated with the desired-response input to the control system. As the adaptive process converges, the inverse model approaches the Wiener solution which, from (2.17), is of the form $\mathbf{R}^{-1}\mathbf{P}$. The plant noise will have no effect on $\mathbf{P}$, but will have an effect on $\mathbf{R}$ and of course on $\mathbf{R}^{-1}$. The result of plant noise is therefore to cause the adaptive solution to be generally different from that of a close approximation to the delayed inverse. Thus, if the noise is significant, the adaptive inverse control approach as we have seen it expressed above may not be satisfactory.

The problem of plant noise has motivated the development of a new algorithm, the "filtered-x" LMS algorithm, which allows adaptation of the inverse filter placed forward of the plant in the cascade sequence. With this approach, partially il-

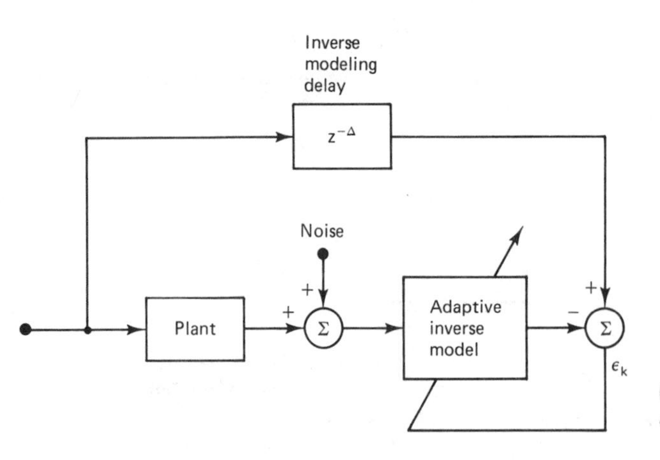

Figure 11.22 Adaptive inverse modeling of a noisy plant.

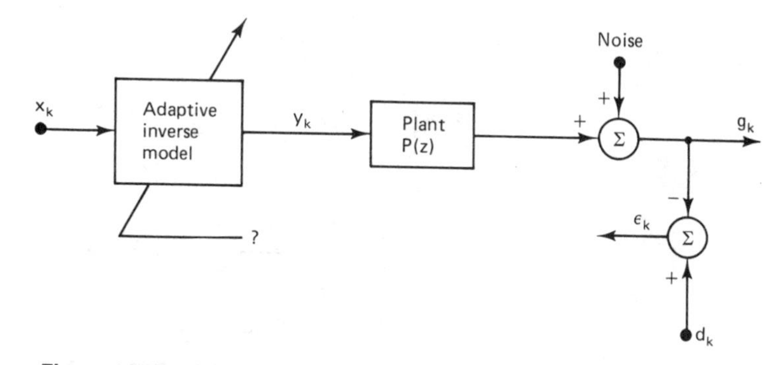

Figure 11.23 Adaptive inverse placed forward of plant with additive noise.

lustrated in Figure 11.23, plant noise does not appear in the adaptive filter input. Even though the plant noise clearly is a component of the error ε_k, as seen in Figure 11.23, this noise will have no effect on the converged solution for the inverse filter provided that the adapting input can be derived properly.

If we assume an FIR model, the mean square of error ε_k in Figure 11.23 is a quadratic function of the adaptive filter weights. Therefore, adaptation has the potential for smooth convergence with a unimodal performance surface. But if the LMS algorithm is to be used to adapt the inverse filter, as suggested in Figure 11.23, we must derive an appropriate "desired" signal to compare with the model output, y_k. The signals d_k and ε_k will not suffice, because ε_k is the error at the plant output, not at the adaptive filter output. If ε_k is used directly with the LMS algorithm to adapt the inverse filter, the adaptive process is almost guaranteed to be unstable, or if not, to find an irrelevant solution. If we are to use ε_k, we must make a fundamental change in the adaptive algorithm to permit its use in this situation. This change results in the filtered-x LMS algorithm.

To make such a change, we must examine the structure of the LMS algorithm as applied to the adaptation of an FIR filter. A detailed schematic is shown in Figure 11.24. First, Figure 11.24(a) shows the overall adaptive filter and signals, similar to Figure 6.1. Then Figure 11.24(b) shows the details of the LMS algorithm in (6.3). The two diagrams represent exactly the same system. The more detailed representation is necessary for the construction of the filtered-x LMS algorithm.

Returning to the problem of Figure 11.23, that is, how to adapt the adaptive filter, we neglect the plant noise for the moment and commute the plant with the LMS filter as shown in Figure 11.25(a), whereupon the LMS algorithm is applied directly. Adjusting the weights to minimize the mean square of ε_k will yield the correct set of weights to minimize the mean square of ε_k in Figure 11.23 with plant noise neglected. Next regard the system diagrammed in Figure 11.25(b). Here the adaptive filter and the plant $P(z)$ are cascaded as originally in Figure 11.23. If the adaptive process shown in Figure 11.25(c) turns out to produce the same set of weights as that of Figure 11.25(a) and (b), it (the filtered-x LMS algorithm) is a viable solution to the adaptation problem of Figure 11.23.

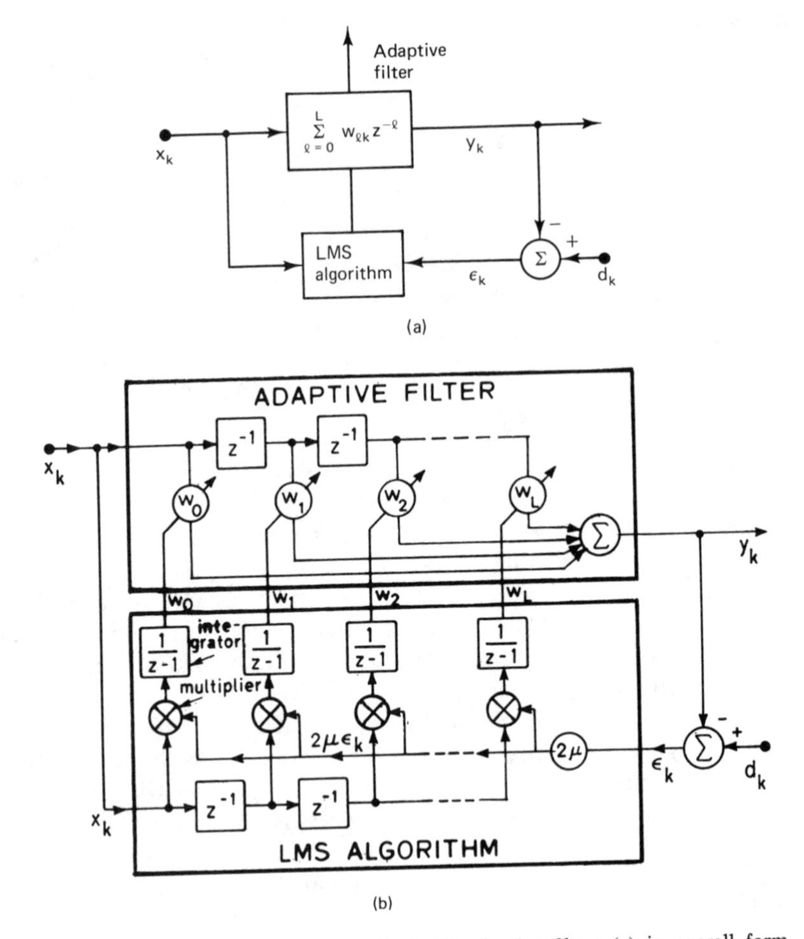

Figure 11.24 Block diagrams of the LMS adaptive filter: (a) in overall form similar to Figure 6.1; (b) showing the details of the LMS filter of (6.3).

Comparing the systems of Figure 11.25(a) and (b), it is clear that the input signal vectors available to the LMS algorithms in both systems are identical at all times. The errors, the ε_k's, are not necessarily identical to each other at all times, however. They would be identical if the adaptive FIR filter weight vectors were identical to each other at all times and if the plant and the adaptive filter were commutable.

For the same input, the same output results when the positions of two cascaded filters are commuted provided that the filters are linear and time invariant. The adaptive filter, however, is neither linear nor time invariant, as we see in Figure 11.24(b). On the other hand, the adaptive filter and the plant would, to a good approximation, be commutable if the plant were linear and if the time variations of the impulse responses of both the plant and the adaptive filter took place with time

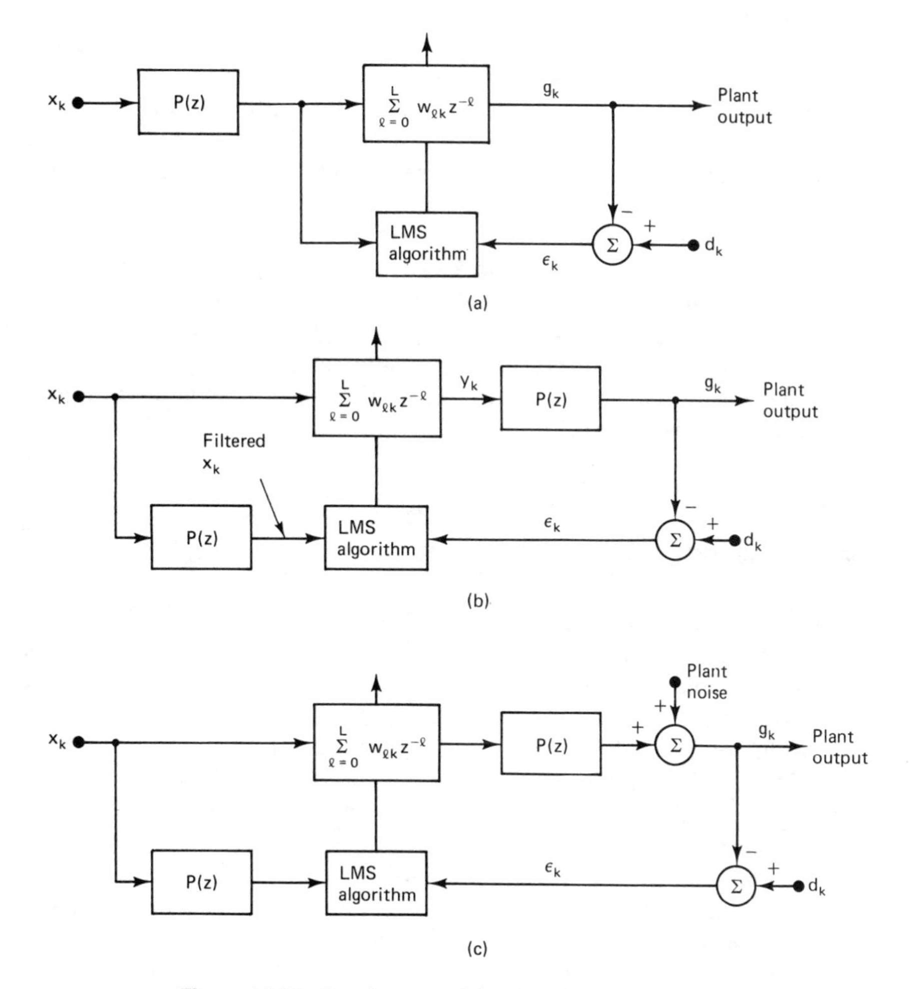

Figure 11.25 Development of the filtered-X LMS algorithm.

constants long compared to the combined memory times or time constants of the adaptive filter and the plant. Thus, with slow adaptation, the adaptive filter may be considered linear and commutable with $P(z)$.

Assuming commutability of the plant and the adaptive filter and assuming that the adaptive weight vectors of the systems of Figure 11.25(a) and (b) are initialized identically, these weight vectors would then undergo identical changes and would therefore track each other. As such, the adaptive process of Figure 11.25(b) would be suitable for the adaptation problem of Figure 11.23.

Convergence of the filtered-x LMS algorithm as defined by Figure 11.25(b) has been demonstrated in a variety of situations. Although the arguments above indicate that adaptation must take place very slowly, rapid adaptation has been achieved in most cases with no particular difficulty. Other derivations of the filtered-x LMS algorithm are currently being examined in an attempt to account for its robustness. Physically, it appears to work as well as the LMS algorithm itself. The choice of initial conditions for the filtered-x LMS algorithm is not important. The algorithm is stable and transients die out just as with the conventional LMS algorithm.

With this new algorithm, we must again consider the effects of plant noise on the converged solution of the weight vector. Figure 11.25(c) shows the application of the filtered-x LMS algorithm to a system with a noisy plant. One can show that the expected value of the adaptive weight vector is the same for the system of Figure 11.25(b) (zero plant noise) as for the system of Figure 11.25(c), where there is plant noise. Although the expected value of the adaptive weight vector is unaffected by independent plant noise, plant noise will of course cause an additional misadjustment in this weight vector.

INVERSE CONTROL USING THE FILTERED-*x* LMS ALGORITHM

The plant transfer function, $P(z)$, appears twice in Figure 11.25(b) and (c), so we must consider how to use the filtered-x method in an actual inverse control situation. One possible scheme is shown in Figure 11.26, in which two separate adaptive processes are used. One produces a direct model of the plant, $\hat{P}(z)$, and the other is the filtered-x LMS process that estimates the plant's delayed inverse. The direct plant model, $\hat{P}(z)$, is exactly copied and used in the filtered-x algorithm in place of the unavailable plant, $P(z)$ itself. Although none of the foregoing arguments relating to the derivation of the filtered-x algorithm would lead one to expect that $\hat{P}(z)$ need not be very precise when incorporated into the algorithm, experience has shown this to be the case. The filtered-x algorithm appears to be robust, and current evidence indicates that the most important attribute of $\hat{P}(z)$ is that its impulse response have at least as great a transport delay as that of $P(z)$ itself. The transport delay is defined as the time from the initiating input pulse to the first nonzero output response.

In the system of Figure 11.26, a delayed inverse model is used to provide the control signal, u_k, to the plant. The reference command input, r_k, drives the inverse model. As discussed previously, a small dither signal may be mixed with r_k to sustain the adaptive processes in the event of inactivity on the part of r_k. When the individual adaptive processes converge, the overall system response to a step function input will be a close approximation to a delayed step, delayed by time Δ. Although the two adaptive processes are not independent in a strict sense, they behave with slow adaptation as if they were independent. A precise analysis of the entire system of Figure 11.26 is currently in progress.

The plant noise in Figure 11.26 would contribute to noise in the weights in both adaptive processes, but would have no effect on their converged expected

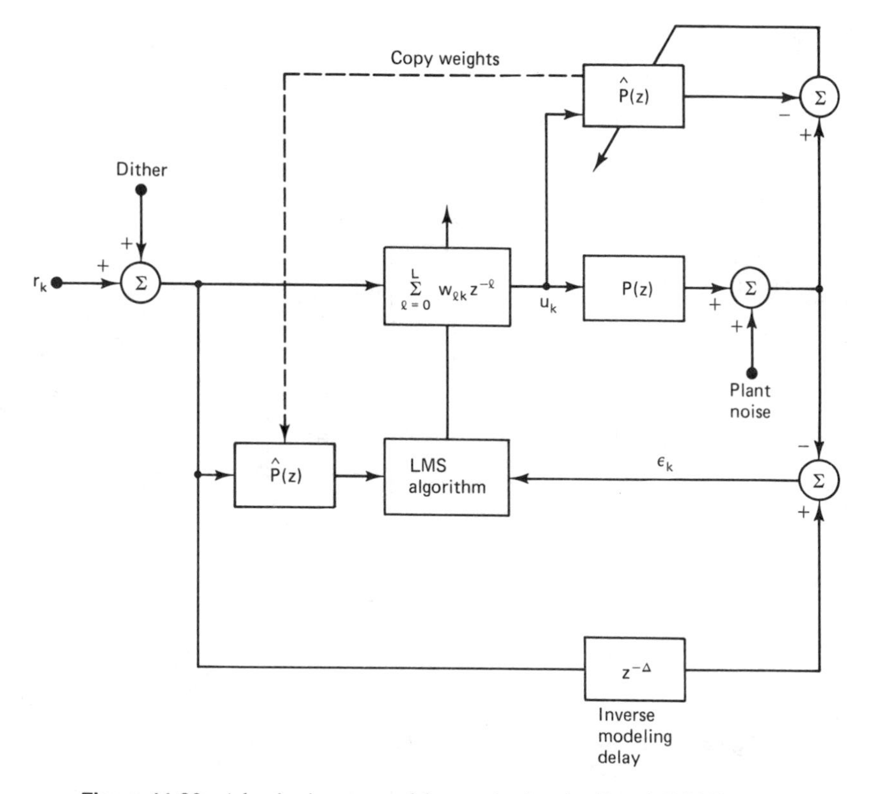

Figure 11.26 Adaptive inverse model control using the filtered-X LMS algorithm.

weight vectors. The plant noise is uncontrolled, and appears at the plant output as if the plant were totally isolated from the rest of the system. In some cases feedback is applied to plants for the purpose of reducing correlated output noise [1–3, 6, 7]. Such feedback could be applied to the plant, $P(z)$, in Figure 11.26 to reduce noise. If this were done, the combination of the plant and its feedback would be regarded by the rest of the system as a new "equivalent" plant which would be modeled and inverse modeled just as $P(z)$ in Figure 11.26.

Although plant noise is not controlled by the system of Figure 11.26, low-frequency plant drift can be easily controlled by adding an adaptive bias weight to the inverse plant model, as illustrated in the system of Figure 11.27. The bias weight w_0' has a fixed unit input in the figure, but it could have any other fixed value. The LMS algorithm is used to adjust w_0'. To have a predetermined convergence rate for this weight, the choice of μ must take into account its fixed input value and the "zero-frequency gain" of the plant $P(z)$, that is, the ratio of the mean of the output of $P(z)$ to the mean of the input. When the bias weight converges, the mean of the error, $E[\varepsilon_k]$, approaches zero and the mean of the plant output, $E[c_k]$, becomes

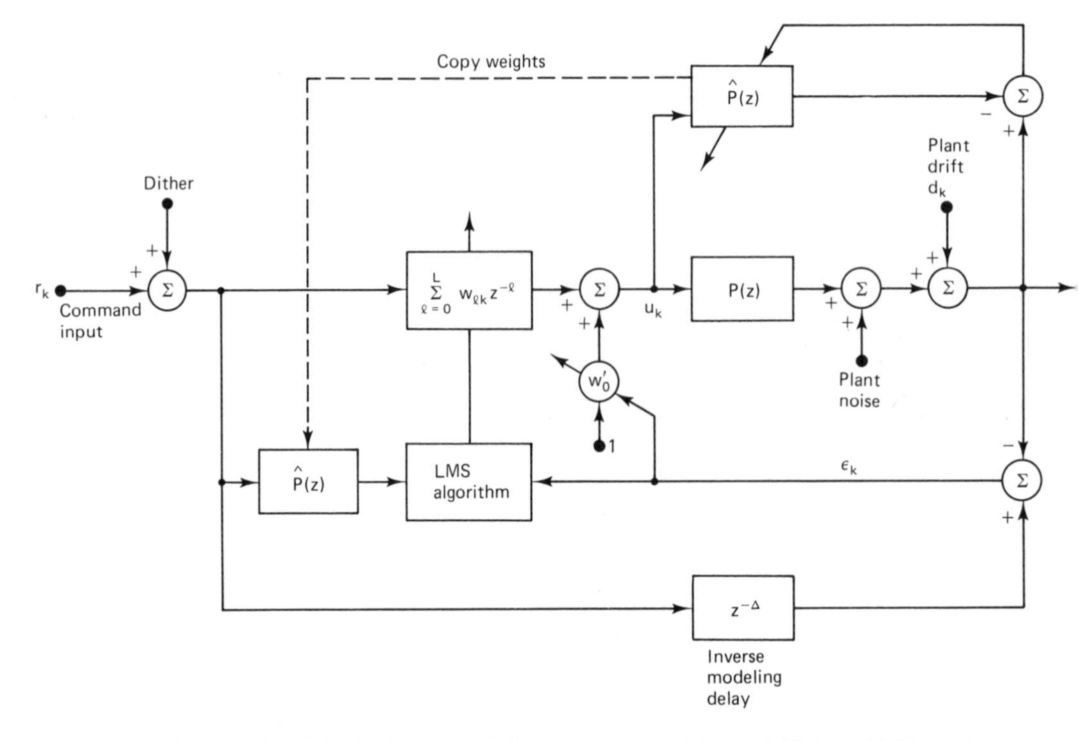

Figure 11.27 Adaptive inverse model control similar to Figure 11.26, but with bias weight added to control plant drift.

equal to the mean of reference command input, $E[r_k]$, regardless of the amount of plant drift (see Exercises 12 through 15).

MODEL REFERENCE CONTROL

With a slight modification of Figure 11.27 we can include the concept of "model reference control," which was introduced by Draper and Li [1–3] in 1961. This concept has influenced much of the control systems literature since that time. The idea is to build, design, or to adapt a system in such a way that its overall input–output response characteristic best matches a reference model response or some form of ideal response. For example, suppose that a high-performance aircraft has a subsonic dynamic response to its control surfaces which is substantially different from its response at supersonic speeds. To provide the pilot with a consistent aircraft response regardless of airspeed, an autopilot accepts the pilot's control inputs and actuates the control-surface servos. The aircraft response to pilot inputs follows the response of a reference model which is chosen by the system designer to give the aircraft a "control feel" that is liked by pilots. Many physical systems are designed to respond like models, and many of these systems are adaptive.

It is easy to modify either Figure 11.27 or Figure 11.11 to include a model-reference feature. We simply replace the inverse modeling delay with the reference model. Then the overall system response will be like that of the reference model rather than a simple delay. The modification is shown in Figure 11.28.

The purpose of the delay ($z^{-\Delta}$) in the systems of Figures 11.11 and 11.27 was to make possible accurate inverse modeling, corresponding to a low mean square of error ε_k. With the delay it is possible to trade away immediate system response for a delayed but more precise response. As discussed above, the delay is needed when the plant, $P(z)$, has transport delay or when the plant is non-minimum phase. When replacing the delay with a reference model, in cases where the delay would be required to form an accurate inverse, generally the reference model should also include such a delay. One must ensure a reference model response that can be accurately realized by the adaptive filter cascaded with the plant $P(z)$, given that the weights of the adaptive filter are set to minimize mean-square error. The concept of Figure 11.28 will perform remarkably well as long as the adaptive system is given a feasible task. One should not expect it to respond faster or more intricately than it is able for the given plant, $P(z)$, and its FIR adaptive controller.

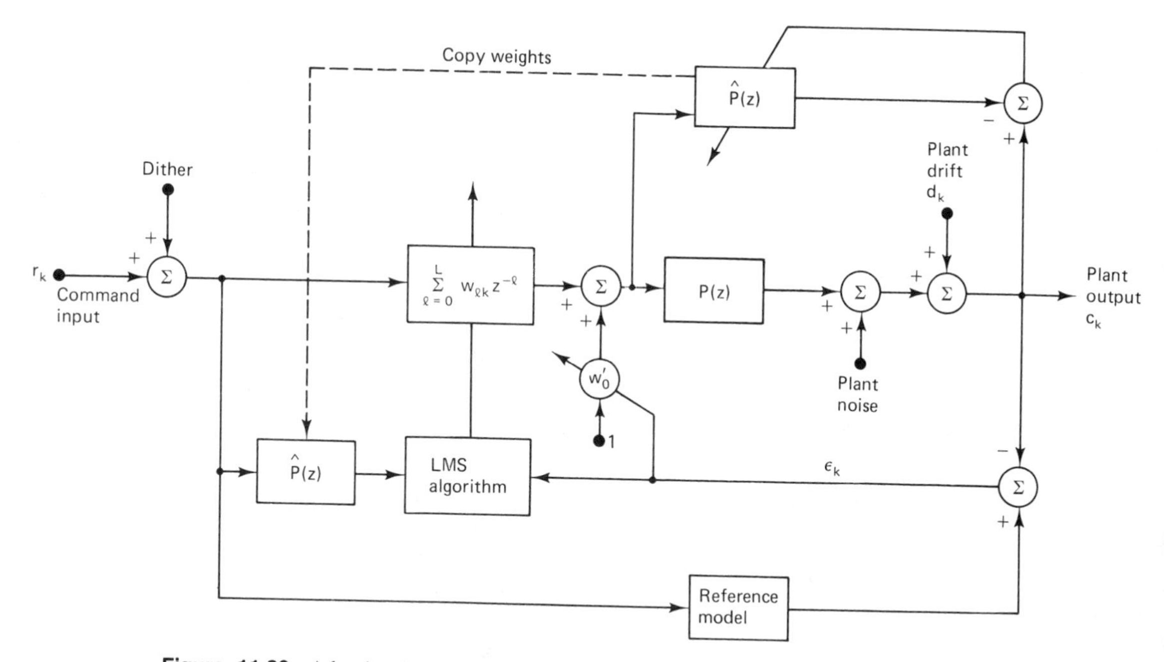

Figure 11.28 Adaptive inverse model control similar to Figure 11.27 but including a reference model.

To provide an example of a model-reference adaptive inverse control system, we consider the following implementation of Figure 11.28:

Plant: $P(z) = \dfrac{2.4z^{-1}(1 - 0.8z^{-1})}{(1 + 0.6z^{-1})(1 - 0.7z^{-1})}$

Reference model: $\dfrac{0.25z^{-1}}{(1 - 0.5z^{-1})^2}$

13 model prefilter weights in $\hat{P}(z)$

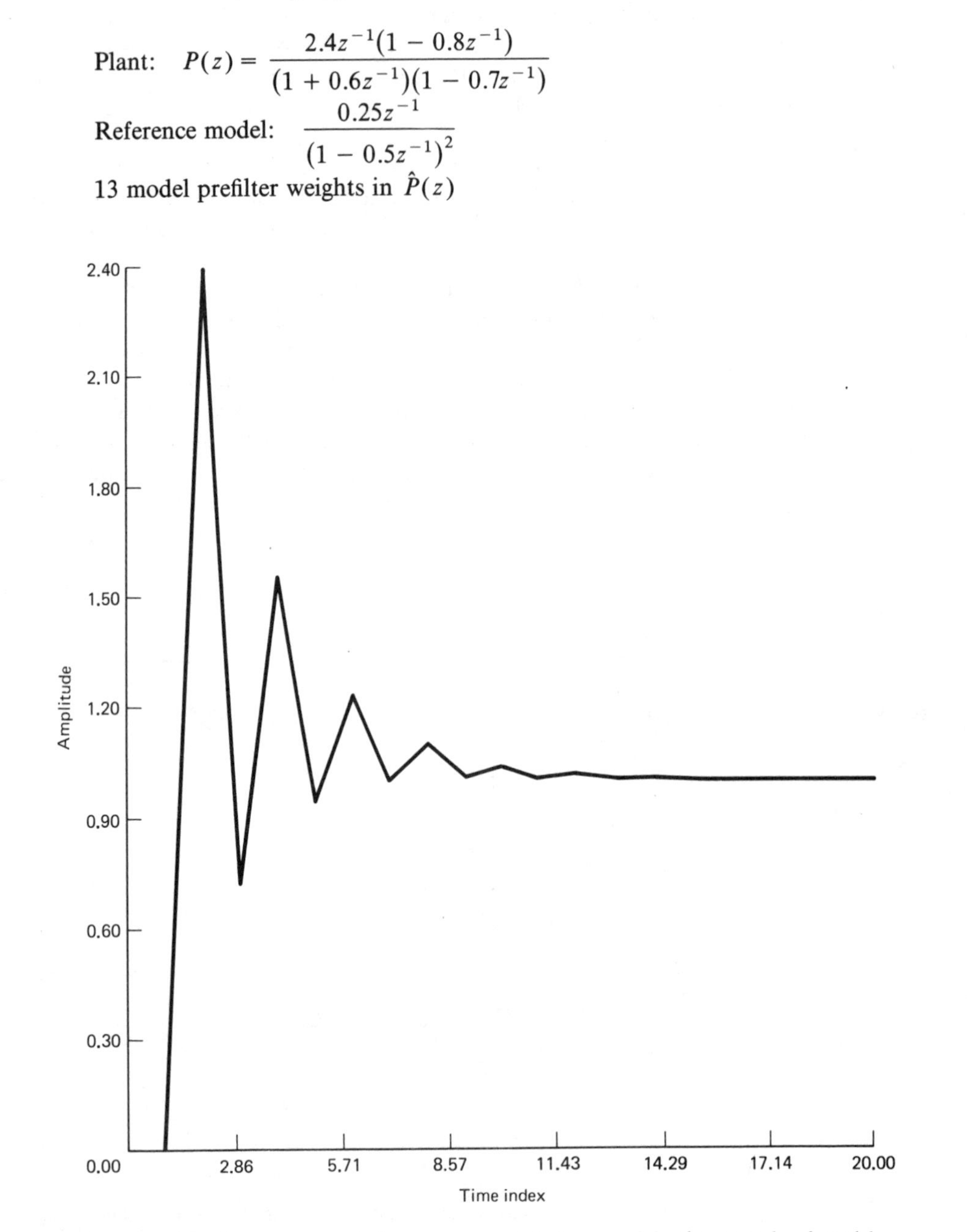

Figure 11.29 Step response of the uncompensated plant used in the example of model reference control.

45 controller weights ($L = 44$)
$\mu = 0.0005$
20,000 iterations for the adaptive process

The step response (to a unit step function) of the uncompensated plant is given in Figure 11.29, and Figure 11.30 shows the step response of the compensated plant superimposed on the step response of the reference model. It is evident that a very close match has been obtained.

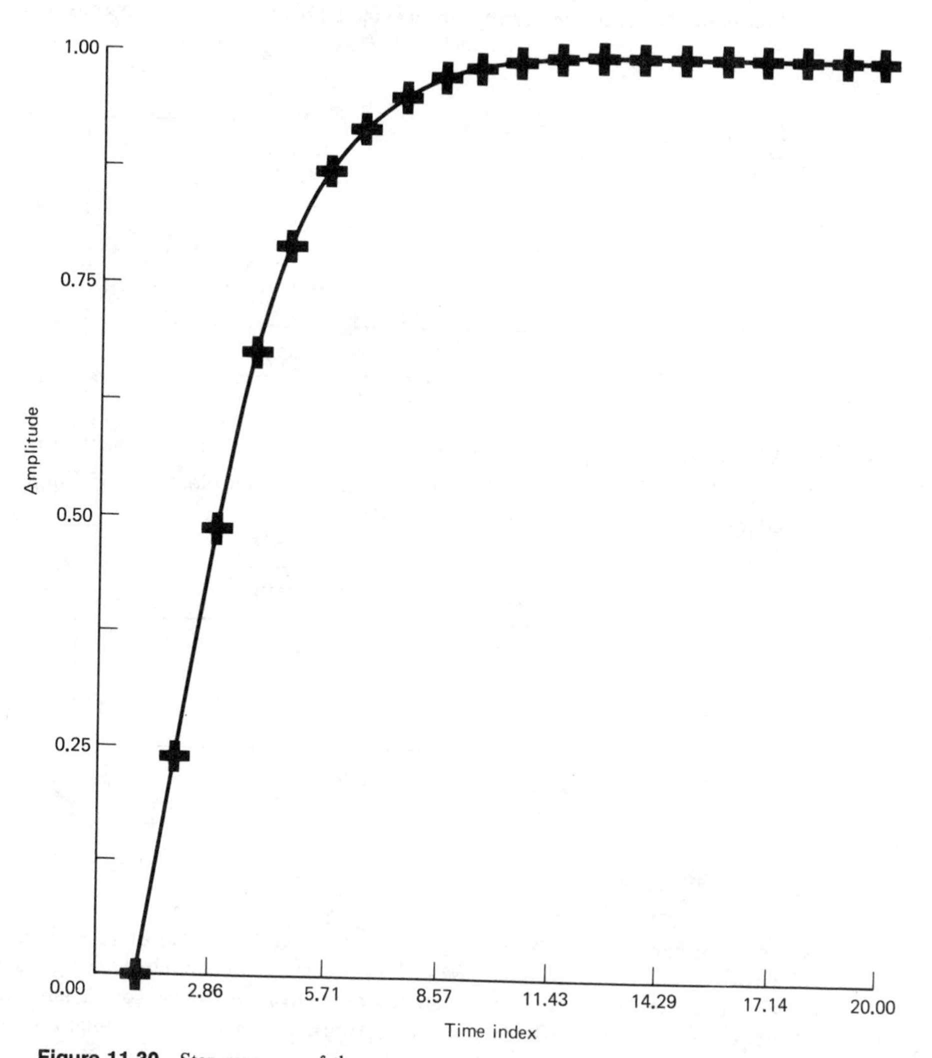

Figure 11.30 Step response of the compensated plant (+) superimposed on the desired step response of the reference model, illustrating successful adaptation.

EXERCISES

1. Given the transfer function $H(z) = \sum_{l=0}^{L} w_l z^{-l}$ with input x_k and output y_k, derive an inverse algorithm, that is, a formula for the input in terms of the output. Check your answer with Equation (11.2).
2. Explain in 100 words or less the difference between adaptive model control and adaptive inverse control as described in this chapter.
3. Given the system below with u_k a white random signal, program the LMS algorithm and show convergence of the adaptive inverse model by averaging 100 learning curves.

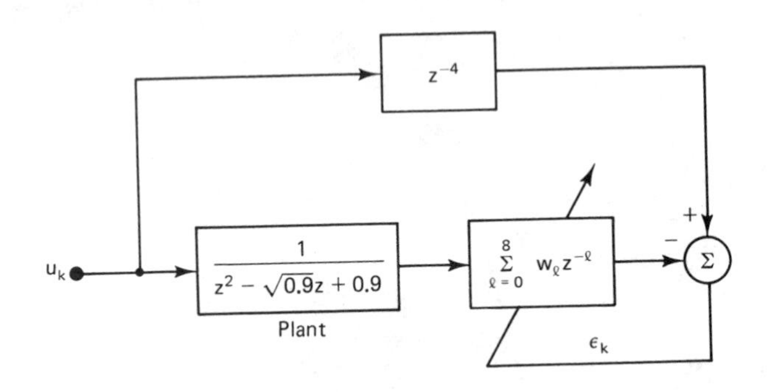

4. After convergence, plot the step response from u_k to y_k of the system in Exercise 3.
5. Use the LMS algorithm to adapt the system below and plot the model impulse response after convergence. Use a white random signal for u_k.

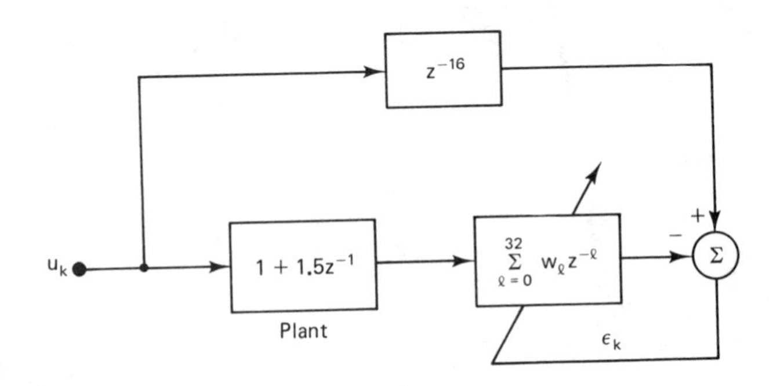

6. In the situation shown below, let r_k again be a white random signal with power σ_r^2, and let n_k be independent white noise with power σ_n^2. Let the model converge using the LMS algorithm and plot the overall step response, from r_k to c_k, for signal-to-noise ratios, σ_r^2/σ_n^2, equal to 0.0, 0.01, 0.1, and 1.0. Adjust μ to give near-optimal results. Compare and explain the four plots.

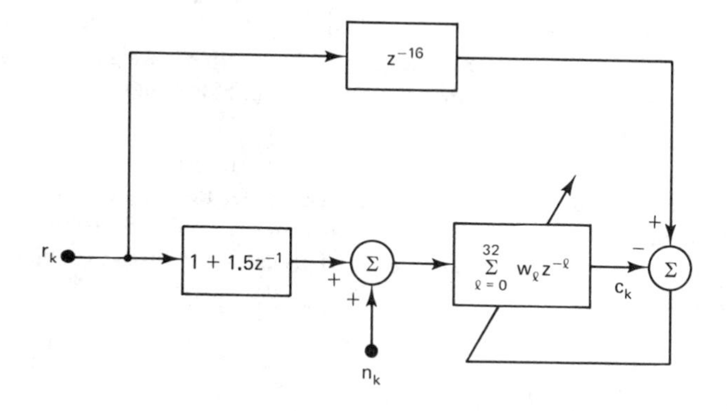

7. Prove that $E[\varepsilon_k^2]$ in Figure 11.23 is a quadratic function of the inverse model weights, given that the model is an FIR filter.

8. In the system with plant noise shown in Figure 11.25(c), show that the optimum weight vector is the same as when there is no noise.

9. One can use the filtered-x concept in Figure 11.25(b) to obtain the adaptive recursive filter shown below, with transfer function $H_k(z) = A_k(z)/[1 - B_k(z)]$. Try to relate the diagram below to Figure 11.25(b) and explain the inputs to each of the LMS algorithms.

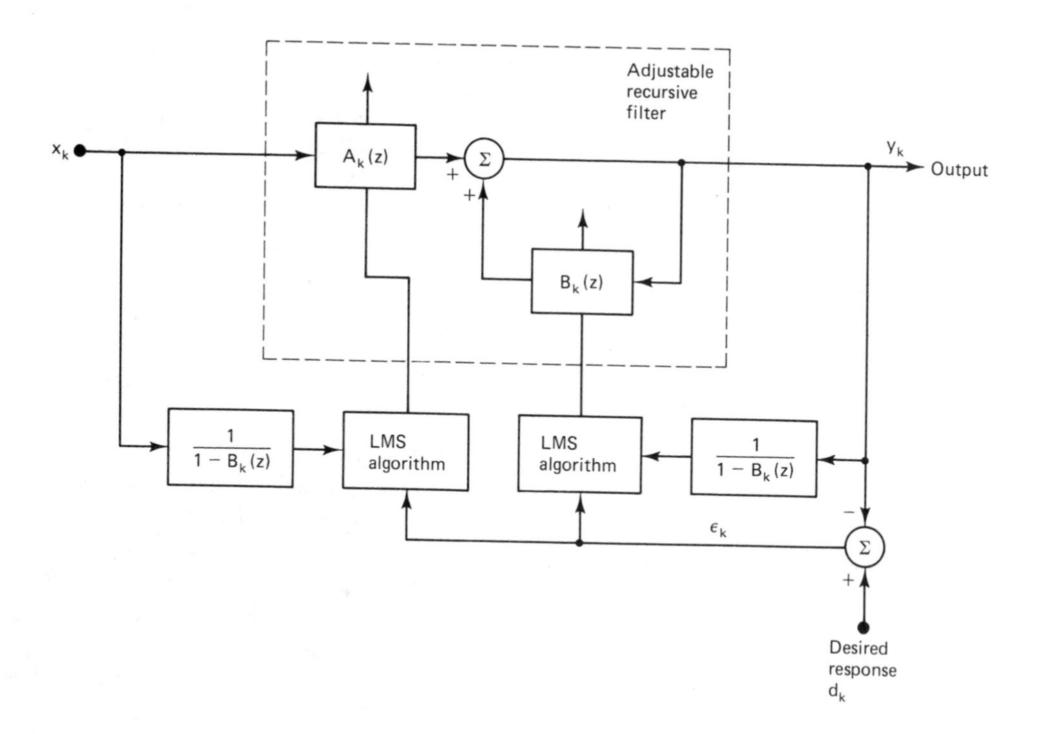

10. Write an algorithm similar to Equation (8.55) for the adaptive recursive filter above. Assume that both $A_k(z)$ and $B_k(z)$ have $L + 1$ weights as in (8.55) and that the convergence parameters are the same as in (8.55). Comment on any differences between the two algorithms.

11. Suppose that we construct the system below to establish a "zone of silence" in a room containing noisy machinery. The idea is to get the loudspeaker to cancel the machinery noise in the vicinity of the second microphone. Show that this is a filtered-x system by relating it in detail to Figure 11.26. Discuss some of the problems that might occur in making it work.

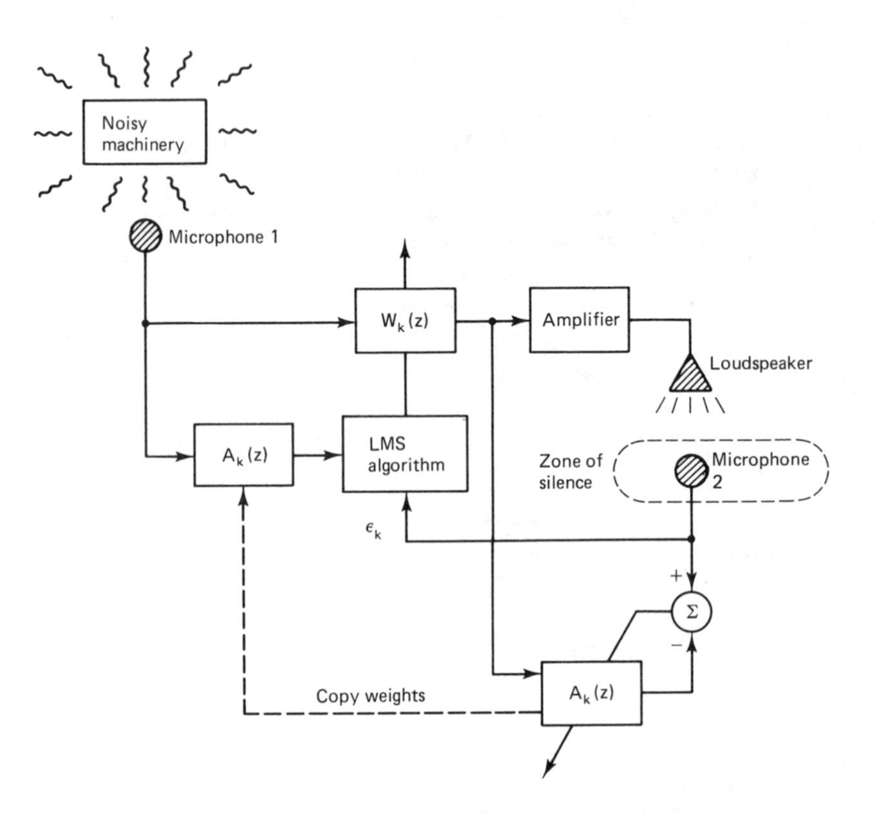

12. Derive an expression for the time constant of the bias weight in Figure 11.27. Let the zero-frequency plant gain be P_0. Assume that w_0' is adjusted using the LMS algorithm with $\mu = \mu_0$, and assume that the plant drift is constant and equal to d.

13.. Assuming that μ is chosen properly for the bias weight in Figure 11.27 and that the weight converges, prove that the mean output, $E[c_k]$, equals the mean input, $E[r_k]$, regardless of plant drift.

14. The inclusion of a bias weight accomplishes the same result as the feedback process below, in terms of plant output. What are α and β in terms of P_0, μ_0, and d in Exercise 12?

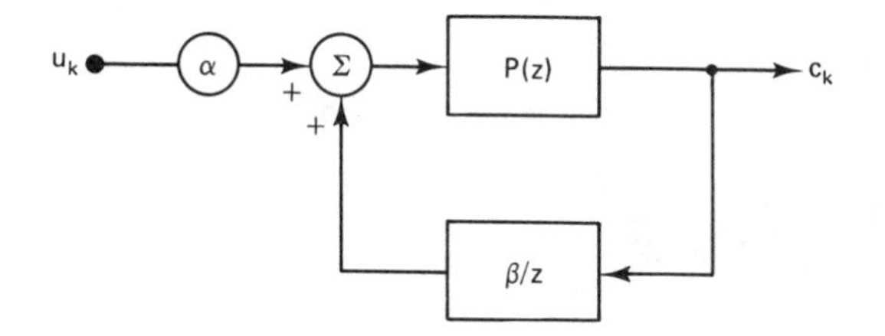

15. What is the range of μ for stable operation of the bias weight, given that the zero-frequency plant gain is P_0?
16. Using the data from the example of the model-reference adaptive inverse control system at the end of this chapter, run the system shown in Figure 11.26. Use zero dither and plant noise, let r_k be a white random signal, and use $\Delta = 22$. Plot a learning curve of ε_k^2 versus k to get a rough idea of the convergence of the filtered-x process, and plot a similar learning curve for the prefilter, $\hat{P}(z)$. Using the final weights, make a plot like Figure 11.30.

ANSWERS TO SELECTED EXERCISES

12. $\tau = 1/(2\mu_0 P_0)$
14. $\alpha = -2\mu d/P_0$; $\beta = 1/P_0 - 2\mu$
15. $0 < \mu < 1/(2P_0)$

REFERENCES AND ADDITIONAL READINGS

1. I. D. Landau, *Control and Systems Theory*, Vol. 8: *Adaptive Control*. New York: Marcel Dekker, 1969.
2. G. F. Franklin and J. D. Powell, *Digital Control of Dynamic Systems*. Reading, Mass.: Addison-Wesley, 1980.
3. B. C. Kuo, *Automatic Control Systems*. Englewood Cliffs, N.J.: Prentice-Hall, 1962.
4. E. Mishkin and L. Braun, Jr. (Eds.), *Adaptive Control Systems*. New York: McGraw-Hill, 1961.
5. D. Sworder, *Optimal Adaptive Control Systems*. New York: Academic Press, 1966.
6. B. C. Kuo, *Digital Control Systems*. Champaign, Ill.: SRL Publishing Co., 1977.
7. S. M. Shinners, *Modern Control System Theory and Application*. Reading, Mass.: Addison-Wesley, 1973.
8. General Reference, *IEEE Transactions on Automatic Control*.
9. General Reference, *Proceedings of the IEEE Conference on Decision and Control*.

chapter 12

Adaptive Interference Canceling

The usual method of estimating a signal corrupted by additive noise† is to pass the composite signal through a filter that tends to suppress the noise while leaving the signal relatively unchanged. The design of such filters is the domain of optimal filtering, which originated with the pioneering work of Wiener and was extended and enhanced by the work of Kalman, Bucy, and others [1–5].

Filters used for the foregoing purpose can be fixed or adaptive. The design of fixed filters must be based on prior knowledge of both the signal and the noise, but adaptive filters have the ability to adjust their own parameters automatically, and their design requires little or no prior knowledge of signal or noise characteristics.

Noise canceling is a variation of optimal filtering that is highly advantageous in many applications. It uses an auxiliary or reference input derived from one or more sensors located at points in the noise field where the signal is weak or undetectable. This input is filtered and subtracted from a primary input containing both signal and noise. As a result, the primary noise is attenuated or eliminated by cancellation.

At first glance, subtracting the noise from a received signal would seem dangerous. If done improperly it could result in an increase in output noise power. If, however, filtering and subtraction are controlled by an appropriate adaptive process, noise reduction can in many cases be accomplished with little risk of distorting the signal or increasing the output noise level. In circumstances where adaptive noise canceling is applicable, we can often achieve a degree of noise rejection that would be difficult or impossible to achieve by direct filtering.

Our purpose in this chapter is to describe the concept of adaptive noise canceling, to provide a theoretical treatment of its advantages and limitations, and to describe some of the applications where it is most useful. The discussion here is based on the December 1975 *Proceedings of the IEEE* paper, "Adaptive Noise Cancelling: Principles and Applications," by B. Widrow, J. Glover, Jr., J. M.

† For simplicity the term "noise" is used here to signify all forms of interference, deterministic as well as stochastic.

McCool, J. Kaunitz, C. S. Williams, R. H. Hearn, J. R. Zeidler, E. Dong, M.D., and R. C. Goodlin, M.D.

EARLY WORK IN ADAPTIVE INTERFERENCE CANCELING

Some of the earliest work in adaptive interference canceling was performed by Howells and Applebaum and their colleagues at the General Electric Company between 1957 and 1960. They designed and built a system for antenna sidelobe canceling using a reference input derived from an auxiliary antenna and a simple two-weight adaptive filter [6].

At the time of this early work, only a handful of people were interested in adaptive systems, and the development of the multiweight adaptive filter was just beginning. In 1959, Widrow and Hoff at Stanford University were devising the least-mean-square (LMS) adaptive algorithm and the pattern recognition scheme known as Adaline (for "adaptive linear threshold logic element") [7, 8]. Rosenblatt had recently built his Perceptron at the Cornell Aeronautical Laboratory [9–11]. Aizermann and his colleagues at the Institute of Automatics and Telemechanics in Moscow, USSR, were constructing an automatic gradient searching machine. In Great Britain, D. Gabor and his associates were developing adaptive filters [12]. Each of these efforts was proceeding independently.

In the early and middle 1960s, work on adaptive systems intensified. Hundreds of papers on adaptation, adaptive controls, adaptive filtering, and adaptive signal processing appeared in the literature. An important commercial application of adaptive filtering in digital communications grew from the work during this period of Lucky at the Bell Laboratories [13, 14] (see Figure 10.13).

In 1965 an adaptive noise-canceling system was built at Stanford University. Its purpose was to cancel the 60-hertz interference at the output of an electrocardiographic amplifier and recorder. A description of the system, which made use of a two-weight analog adaptive filter, is presented below together with analysis and results recently obtained by computer implementation.

Since 1965 adaptive noise canceling has been successfully applied to a number of additional problems, including other aspects of electrocardiography, the elimination of periodic interference in general [15], and the elimination of echoes on long-distance telephone transmission lines [16, 17]. A paper on adaptive antennas by Riegler and Compton [18] generalizes the work originally performed by Howells and Applebaum. More of the latter topic will be presented in Chapter 13.

THE CONCEPT OF ADAPTIVE NOISE CANCELING

The basic noise-canceling situation is illustrated in Figure 12.1. A signal is transmitted over a channel to a sensor that receives the signal plus an uncorrelated noise, n_0. The combined signal and noise, $s + n_0$, form the "primary input" to the canceler. A second sensor receives a noise n_1 which is uncorrelated with the signal but correlated in some unknown way with the noise n_0. This sensor provides the "reference input" to the canceler. The noise n_1 is filtered to produce an output y that is a close replica of n_0. This output is subtracted from the primary input $s + n_0$ to produce the system output, $s + n_0 - y$.

If one knew the characteristics of the channels over which the noise was transmitted to the primary and reference sensors, one could, in general, design a fixed filter capable of changing n_1 into $y = n_0$. The filter output could then be subtracted from the primary input, and the system output would be the signal alone. Since, however, the characteristics of the transmission paths are assumed to be unknown or known only approximately and not of a fixed nature, the use of a fixed filter is not feasible. Moreover, even if a fixed filter were feasible, its characteristics would have to be adjusted with a precision difficult to attain, and the slightest error could result in increased output noise power.

In the system shown in Figure 12.1, the reference input is processed by an adaptive filter that automatically adjusts its own impulse response through a least-squares algorithm such as LMS that responds to an error signal dependent, among other things, on the filter's output. Thus with the proper algorithm, the filter can operate under changing conditions and can readjust itself continuously to minimize the error signal.

We have seen in previous chapters that the error signal used in an adaptive process depends on the nature of the application. In noise-canceling systems the practical objective is to produce a system output, $s + n_0 - y$, that is a best fit in the least-squares sense to the signal s. This objective is accomplished by feeding the system output back to the adaptive filter and adjusting the filter through an adaptive algorithm to minimize the total system output power. In an adaptive noise-canceling

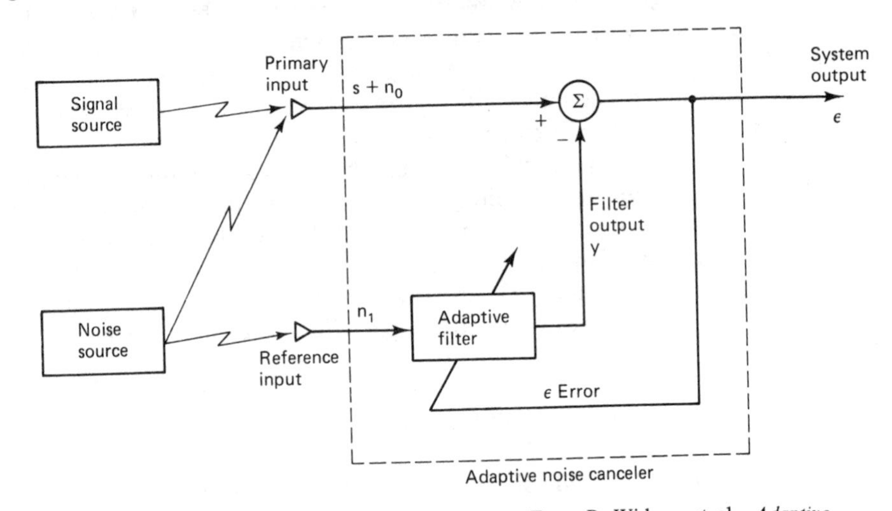

Figure 12.1 Adaptive noise-canceling concept. From B. Widrow et al., *Adaptive Noise Canceling: Principles and Applications*, © December 1975, IEEE.

system, in other words, the system output serves as the error signal for the adaptive process.

One might think that some prior knowledge of the signal s or of the noises n_0 and n_1 would be necessary before the filter could be designed, or before it could adapt to produce the noise-canceling signal y. A simple argument will show, however, that little or no prior knowledge of s, n_0, or n_1, or of their interrelationships, either statistical or deterministic, is required.

Assume that s, n_0, n_1, and y are statistically stationary and have zero means. Assume that s is uncorrelated with n_0 and n_1, and suppose that n_1 is correlated with n_0. The output is

$$\varepsilon = s + n_0 - y \tag{12.1}$$

Squaring, one obtains

$$\varepsilon^2 = s^2 + (n_0 - y)^2 + 2s(n_0 - y) \tag{12.2}$$

Taking expectations of both sides of (12.2), and realizing that s is uncorrelated with n_0 and with y, yields

$$\begin{aligned} E[\varepsilon^2] &= E[s^2] + E\left[(n_0 - y)^2\right] + 2E[s(n_0 - y)] \\ &= E[s^2] + E\left[(n_0 - y)^2\right] \end{aligned} \tag{12.3}$$

The signal power $E[s^2]$ will be unaffected as the filter is adjusted to minimize $E[\varepsilon^2]$. Accordingly, the minimum output power is

$$E_{\min}[\varepsilon^2] = E[s^2] + E_{\min}\left[(n_0 - y)^2\right] \tag{12.4}$$

When the filter is adjusted so that $E[\varepsilon^2]$ is minimized, $E[(n_0 - y)^2]$ is therefore also minimized. The filter output y is then a best least-squares estimate of the primary noise n_0. Moreover, when $E[(n_0 - y)^2]$ is minimized, $E[(\varepsilon - s)^2]$ is also minimized, since, from (12.1),

$$(\varepsilon - s) = (n_0 - y) \tag{12.5}$$

Adjusting or adapting the filter to minimize the total output power is thus tantamount to causing the output ε to be a best least-squares estimate of the signal s for the given structure and adjustability of the adaptive filter and for the given reference input.

The output ε will generally contain the signal s plus some noise. From (12.1), the output noise is given by $(n_0 - y)$. Since minimizing $E[\varepsilon^2]$ minimizes $E[(n_0 - y)^2]$, minimizing the total output power minimizes the output noise power and, since the signal in the output remains constant, minimizing the total output power maximizes the output signal-to-noise ratio.

We see from (12.3) that the smallest possible output power is $E_{\min}[\varepsilon^2] = E[s^2]$. When this is achievable, $E[(n_0 - y)^2] = 0$. Therefore, $y = n_0$ and $\varepsilon = s$. In this case, minimizing output power causes the output signal to be perfectly free of noise.

On the other hand, when the reference input is completely uncorrelated with the primary input, the filter will "turn itself off" and will not increase output noise.

In this case the filter output y will be uncorrelated with the primary input. The output power will be

$$E[\varepsilon^2] = E\left[(s + n_0)^2\right] + 2E[-y(s + n_0)] + E\left[y^2\right]$$
$$= E\left[(s + n_0)^2\right] + E\left[y^2\right] \tag{12.6}$$

Minimizing output power requires that $E[y^2]$ be minimized, which is accomplished by making all weights zero, bringing $E[y^2]$ to zero.

These arguments can readily be extended to the case where the primary and reference inputs contain, in addition to n_0 and n_1, additive random noises uncorrelated with each other and with s, n_0, and n_1. They can also readily be extended to the case where n_0 and n_1 are deterministic rather than stochastic. Note that in making the arguments above, we have not assumed that the adaptive filter necessarily converges to a linear filter; that is, we have not used the Wiener filter theory described in Chapter 2.

STATIONARY NOISE-CANCELING SOLUTIONS

In this section optimal unconstrained Wiener solutions to certain stationary noise-canceling problems are derived. Our purpose is to demonstrate the increase in signal-to-noise ratio and other advantages of the noise-canceling technique as opposed to the noise filtering technique, when using adaptive filters.

As previously noted, fixed filters are for the most part inapplicable in noise canceling because the correlation and cross-correlation functions of the primary and reference inputs are generally unknown and often variable with time. Adaptive filters are required to "learn" the statistics initially and to track them if they vary slowly. For stationary stochastic inputs, however, the steady-state performance of slowly adapting adaptive filters closely approximates that of fixed Wiener filters, and Wiener filter theory thus provides a convenient mathematical analysis of statistical noise-canceling problems.

Figure 12.2 shows a classic single-input, single-output Wiener filter. The input signal is x_k, the output signal y_k, and the desired response d_k. The input and

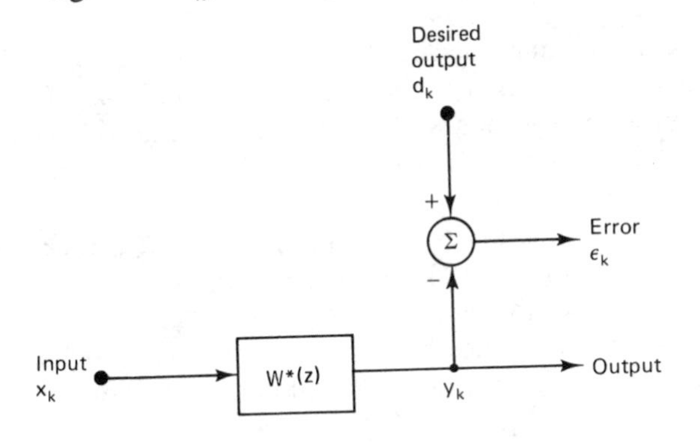

Figure 12.2 Single-channel Wiener filter. From B. Widrow et al., *Adaptive Noise Canceling: Principles and Applications*, © December 1975, IEEE.

output signals are assumed to be discrete in time, and the input signal and desired response are assumed to be statistically stationary. The error signal is $\varepsilon_k = d_k - y_k$. The filter is linear and discrete, and is designed to be optimal in the minimum mean-square-error sense. For our analysis here we will assume that it is an infinitely long, two-sided adaptive transversal filter.

The error surface for such a filter is of course quadratic. It is described in (7.67) and (10.1).

$$\xi = \phi_{dd}(0) + \sum_{l=-\infty}^{\infty} \sum_{m=-\infty}^{\infty} w_l w_m \phi_{xx}(l-m) - 2 \sum_{l=-\infty}^{\infty} w_l \phi_{xd}(l) \tag{12.7}$$

The minimum point on this performance surface corresponds with the optimum weight vector, $\mathbf{W}^*$, that is, the weights of the optimum Wiener filter with transfer function $W^*(z)$. In (10.3) we derived this optimal transfer function as a ratio of power spectra,

$$W^*(z) = \frac{\Phi_{xd}(z)}{\Phi_{xx}(z)} \tag{12.8}$$

This result represents the unconstrained, noncausal solution to the Wiener filtering problem. The Shannon–Bode realization [2], by contrast, is constrained to be a causal filter. The causality constraint generally leads to a loss of performance and, as shown below, can usually be avoided in adaptive noise-canceling applications.

Let us now consider how to apply the result in (12.8) in adaptive noise canceling. Figure 12.3 shows a somewhat more detailed picture of the single-channel

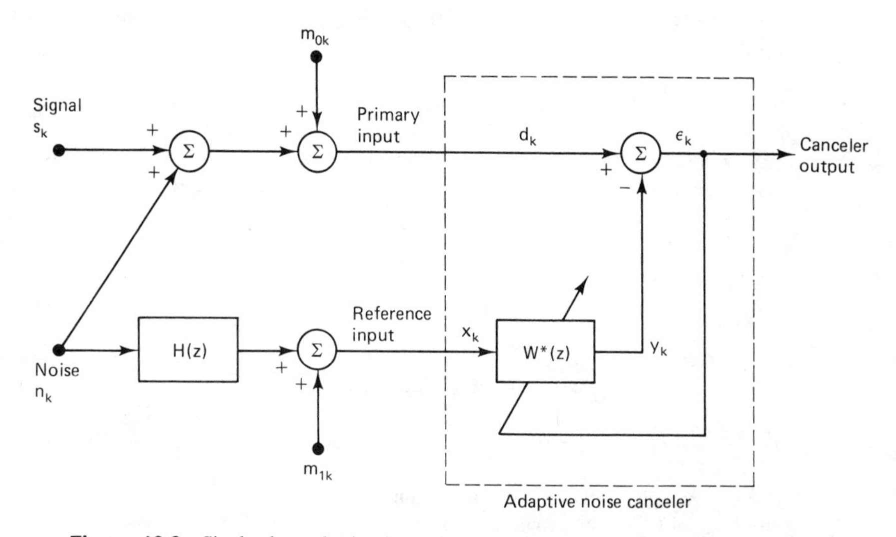

Figure 12.3 Single-channel adaptive noise canceler with correlated and uncorrelated noises in the primary and reference inputs. From B. Widrow et al., *Adaptive Noise Canceling: Principles and Applications*, © December 1975, IEEE.

adaptive noise canceler in Figure 12.1. It includes a model of how the signal and noise inputs might have arisen. The primary input consists of a signal s_k plus the sum of two noises, n_k and m_{0k}. The reference input consists of the sum of two other noises, m_{1k} and n_k convolved with h_k, the impulse response of the channel whose transfer function is $H(z)$.† The two noise inputs, n_k and n_k convolved with h_k, have a common origin, are correlated with each other, and are uncorrelated with the signal, s_k. They are further assumed to have finite power spectra at all frequencies. The noises m_{0k} and m_{1k} are uncorrelated with each other, with s_k, and with n_k and n_k convolved with h_k. For the purposes of analysis all noise propagation paths are assumed to be equivalent to linear, time-invariant filters.

The noise canceler of Figure 12.3 includes an adaptive filter whose input x_k, the reference input to the canceler, is $m_{1k} + n_k$ convolved with h_k and whose desired response d_k, the primary input to the canceler, is $s_k + m_{0k} + n_k$. The error signal ε_k is the noise canceler's output. If one assumes that the adaptive process has converged and the minimum-mean-square-error solution has been found, the adaptive filter is equivalent to the Wiener filter in (12.8). A least-squares solution for the interference canceler of Figure 12.3 is attractive for the following reason. The output is the error of the Wiener filter and, as shown in (2.39), the error ε_k is uncorrelated with the filter input x_k. Therefore, all primary noise components which are correlated with reference noise components will be perfectly canceled. Other noise components will not be canceled and could appear at the output, however.

The optimal unconstrained transfer function, $W^*(z)$, of the adaptive filter is the Wiener solution (12.8), which we may now expand as follows. The spectrum of the filter's input $\Phi_{xx}(z)$ can be expressed in terms of the spectra of its two mutually uncorrelated additive components. The spectrum of the noise m_1 is $\Phi_{m_1m_1}(z)$, and that of the noise n arriving via $H(z)$ is $\Phi_{nn}(z)|H(z)|^2$. The filter's input spectrum is thus

$$\Phi_{xx}(z) = \Phi_{m_1m_1}(z) + \Phi_{nn}(z)|H(z)|^2 \tag{12.9}$$

The cross power spectrum between the filter's input and the desired response depends only on the mutually correlated primary and reference components, and is given by

$$\Phi_{xd}(z) = \Phi_{nn}(z)H(z^{-1}) \tag{12.10}$$

Using the results in (12.8), the Wiener transfer function is thus

$$W^*(z) = \frac{\Phi_{nn}(z)H(z^{-1})}{\Phi_{m_1m_1}(z) + \Phi_{nn}(z)|H(z)|^2} \tag{12.11}$$

† To simplify notation, the transfer function of the noise path from n_k to the primary input has been set at unity. This procedure does not restrict the analysis, since by a suitable choice of $H(z)$ and of statistics for n_k any combination of mutually correlated noises can be made to appear at the primary and reference inputs. Although $H(z)$ may consequently be required to have poles inside and outside the unit circle in the z-plane, a stable two-sided impulse response h_k will always exist to do this.

Note that $W^*(z)$ is independent of the primary signal spectrum $\Phi_{ss}(z)$ and of the primary uncorrelated noise spectrum $\Phi_{m_0m_0}(z)$.

An interesting special case occurs when the additive noise m_1 in the reference input is zero. Then $\Phi_{m_1m_1}(z)$ is zero and the optimal transfer function in (12.11) becomes

$$W^*(z) = 1/H(z) \tag{12.12}$$

This result is intuitively correct, because the adaptive filter, as in the balancing of a bridge, causes the noise n_k to be perfectly nulled at the noise-canceler output. The primary uncorrelated noise m_{0k} remains uncanceled and appears with the signal at the system output.

The performance of the single-channel noise canceler can be evaluated more generally in terms of the ratio of the signal-to-noise power density ratio at the output, $\rho_{\text{out}}(z)$, to the signal-to-noise power density ratio at the primary input, $\rho_{\text{pri}}(z)$.† Assuming that the signal spectrum is greater than zero at all frequencies and canceling out the signal power spectrum, we have

$$\frac{\rho_{\text{out}}(z)}{\rho_{\text{pri}}(z)} = \frac{\text{primary noise power density}}{\text{output noise power density}}$$

$$= \frac{\Phi_{nn}(z) + \Phi_{m_0m_0}(z)}{\Phi_{\text{output noise}}(z)} \tag{12.13}$$

The canceler's output noise power spectrum, as in Figure 12.3, is the sum of three components, one due to the propagation of m_{0k} directly to the output, another due to the propagation of m_{1k} to the output via the transfer function $-W^*(z)$, and another due to the propagation of n_k to the output via the transfer function $1 - H(z)W^*(z)$. The output noise power spectrum is thus

$$\Phi_{\text{output noise}}(z) = \Phi_{m_0m_0}(z) + \Phi_{m_1m_1}(z)|W^*(z)|^2$$

$$+\Phi_{nn}(z)\left|[1 - H(z)W^*(z)]\right|^2 \tag{12.14}$$

For convenience, we now define the ratios of the spectra of the uncorrelated to the spectra of the correlated noises ("noise-to-noise density ratios") at the primary and reference inputs as

$$A(z) \triangleq \frac{\Phi_{m_0m_0}(z)}{\Phi_{nn}(z)} \tag{12.15}$$

and

$$B(z) \triangleq \frac{\Phi_{m_1m_1}(z)}{\Phi_{nn}(z)|H(z)|^2} \tag{12.16}$$

† The signal-to-noise density ratio is here defined as the ratio of signal power density to noise power density and is thus a function of frequency. These definitions apply on the unit circle in the z-plane, where we always have $z = e^{j\Omega T}$, and T is the sampling period, as described in Chapter 7.

With these definitions the transfer function (12.11) can be written as

$$W^*(z) = \frac{1}{H(z)[B(z)+1]} \tag{12.17}$$

The output noise power spectrum (12.14) can accordingly be rewritten as

$$\begin{aligned}\Phi_{\text{output noise}}(z) &= \Phi_{m_0m_0}(z) + \frac{\Phi_{m_1m_1}(z)}{|H(z)|^2|B(z)+1|^2} + \Phi_{nn}(z)\left|1 - \frac{1}{B(z)+1}\right|^2 \\ &= \Phi_{nn}(z)A(z) + \Phi_{nn}(z)\frac{B(z)}{B(z)+1}\end{aligned} \tag{12.18}$$

The ratio of the primary to the output signal-to-noise power density ratios in (12.13) is now

$$\begin{aligned}\frac{\rho_{\text{out}}(z)}{\rho_{\text{pri}}(z)} &= \frac{\Phi_{nn}(z)[1+A(z)]}{\Phi_{\text{output noise}}(z)} \\ &= \frac{1+A(z)}{A(z)+\dfrac{B(z)}{B(z)+1}} \\ &= \frac{[A(z)+1][B(z)+1]}{A(z)+A(z)B(z)+B(z)}\end{aligned} \tag{12.19}$$

This expression is a general representation of the ideal noise-canceler performance with single primary and reference inputs and stationary signals and noises. It allows one to estimate the level of noise reduction to be expected with an ideal noise-canceling system containing a two-sided unconstrained Wiener filter. In such a system the signal propagates to the output in an undistorted fashion, with a transfer function of unity.† Classical configurations of Wiener, Kalman, and adaptive filters, in contrast, generally introduce some signal distortion in the process of noise reduction.

It is apparent from (12.19) that the ability of a noise canceling system to reduce noise is limited by the uncorrelated-to-correlated noise density ratios at the primary and reference inputs. When $A(z)$ and $B(z)$ are relatively small, $\rho_{\text{out}}(z)/\rho_{\text{pri}}(z)$ is relatively large and the action of the canceler is relatively more effective. The desirability of low levels of uncorrelated noise in both inputs is made still more evident by considering the following special cases:

1. Small $A(z)$:

$$\frac{\rho_{\text{out}}(z)}{\rho_{\text{pri}}(z)} \simeq \frac{1+B(z)}{B(z)} \tag{12.20}$$

†Some signal cancellation is possible when adaptation is rapid (i.e., when the value of the convergence parameter μ is large) because of the dynamic response of the weight vector, which approaches but does not equal the Wiener solution. In most cases this effect is negligible; a particular case where it is not negligible is described later in this chapter.

2. Small $B(z)$:

$$\frac{\rho_{\text{out}}(z)}{\rho_{\text{pri}}(z)} \simeq \frac{1 + A(z)}{A(z)} \tag{12.21}$$

3. Small $A(z)$ and $B(z)$:

$$\frac{\rho_{\text{out}}(z)}{\rho_{\text{pri}}(z)} \simeq \frac{1}{A(z) + B(z)} \tag{12.22}$$

Infinite improvement is implied by these relations when both $A(z)$ and $B(z)$ are zero. In this case there is complete removal of noise at the system output, resulting in perfect signal reproduction. When $A(z)$ and $B(z)$ are small, however, other factors become important in limiting system performance. These factors include the finite length of the adaptive filter in practical systems and "misadjustment" caused by gradient estimation noise in the adaptive process, discussed in Chapter 5 and in References 19 and 20. A third factor, involving signal components sometimes leaking into the reference input, is discussed in the following section.

EFFECTS OF SIGNAL COMPONENTS IN THE REFERENCE INPUT

In certain instances the available reference input to an adaptive noise canceler may contain low-level signal components in addition to the usual correlated and uncorrelated noise components, and of course these signal components will cause some cancellation of the primary input signal. We must then ask whether they will cause sufficient cancellation to render the application of noise canceling useless. To provide an answer we will use the unconstrained Wiener solution as in the preceding section, and we will derive expressions for the signal-to-noise density ratio, signal distortion, and noise spectrum at the canceler output.

Figure 12.4 shows an adaptive noise canceler whose reference input contains signal components and whose primary and reference inputs contain additive correlated noises. Additive uncorrelated noises have been omitted to simplify the analysis. The signal components leaking into the reference input are assumed to be propagated through a channel with the transfer function $J(z)$. The other terminology is the same as that of Figure 12.3.

The input signal and noise power spectra in Figure 12.4 are $\Phi_{ss}(z)$ and $\Phi_{nn}(z)$, respectively. The spectrum of the reference input, which is identical to the spectrum of the input x_k to the adaptive filter, is thus

$$\Phi_{xx}(z) = \Phi_{ss}(z)|J(z)|^2 + \Phi_{nn}(z)|H(z)|^2 \tag{12.23}$$

The cross-spectrum between the reference and primary inputs, identical to the

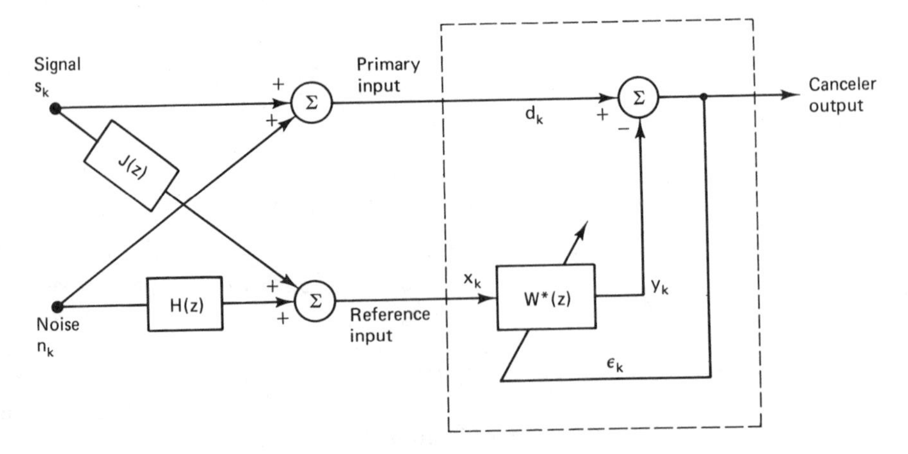

Figure 12.4 Adaptive noise canceler with signal components leaking into the reference input. From B. Widrow et al., *Adaptive Noise Canceling: Principles and Applications*, © December 1975, IEEE.

cross-spectrum between the filter's input x_k and desired response d_k, is similarly

$$\Phi_{xd}(z) = \Phi_{ss}(z)J(z^{-1}) + \Phi_{nn}(z)H(z^{-1}) \tag{12.24}$$

When the adaptive process has converged, the unconstrained Wiener transfer function of the adaptive filter, given by (12.8), is thus

$$W^*(z) = \frac{\Phi_{ss}(z)J(z^{-1}) + \Phi_{nn}(z)H(z^{-1})}{\Phi_{ss}(z)|J(z)|^2 + \Phi_{nn}(z)|H(z)|^2} \tag{12.25}$$

The first objective of our analysis is to find the signal-to-noise density ratio $\rho_{\text{out}}(z)$, at the noise-canceler output. The transfer function of the propagation path from the signal input to the noise-canceler output is $1 - J(z)W^*(z)$ and that of the path from the noise input to the canceler output is $1 - H(z)W^*(z)$. The spectrum of the signal component in the output is thus

$$\begin{aligned}\Phi_{ss_{\text{out}}}(z) &= \Phi_{ss}(z)|1 - J(z)W^*(z)|^2 \\ &= \Phi_{ss}(z)\left|\frac{[H(z) - J(z)]\Phi_{nn}(z)H(z^{-1})}{\Phi_{ss}(z)|J(z)|^2 + \Phi_{nn}(z)|H(z)|^2}\right|^2\end{aligned} \tag{12.26}$$

and that of the noise component is similarly

$$\begin{aligned}\Phi_{nn_{\text{out}}}(z) &= \Phi_{nn}(z)|1 - H(z)W^*(z)|^2 \\ &= \Phi_{nn}(z)\left|\frac{[J(z) - H(z)]\Phi_{ss}(z)J(z^{-1})}{\Phi_{ss}(z)|J(z)|^2 + \Phi_{nn}(z)|H(z)|^2}\right|^2\end{aligned} \tag{12.27}$$

The output signal-to-noise density ratio is thus

$$\rho_{\text{out}}(z) = \frac{\Phi_{ss}(z)}{\phi_{nn}(z)} \left| \frac{\Phi_{nn}(z)H(z^{-1})}{\Phi_{ss}(z)J(z^{-1})} \right|^2$$

$$= \frac{\Phi_{nn}(z)|H(z)|^2}{\Phi_{ss}(z)|J(z)|^2} \tag{12.28}$$

The output signal-to-noise density ratio can be conveniently expressed in terms of the signal-to-noise density ratio at the reference input, $\rho_{\text{ref}}(z)$, as follows. The spectrum of the signal component in the reference input is

$$\Phi_{ss_{\text{ref}}}(z) = \Phi_{ss}(z)|J(z)|^2 \tag{12.29}$$

and that of the noise component is similarly

$$\Phi_{nn_{\text{ref}}}(z) = \Phi_{nn}(z)|H(z)|^2 \tag{12.30}$$

The signal-to-noise density ratio at the reference input is thus

$$\rho_{\text{ref}}(z) = \frac{\Phi_{ss}(z)|J(z)|^2}{\Phi_{nn}(z)|H(z)|^2} \tag{12.31}$$

The output signal-to-noise density ratio (12.28) is, therefore,

$$\boxed{\rho_{\text{out}}(z) = \frac{1}{\rho_{\text{ref}}(z)}} \tag{12.32}$$

This result is exact and somewhat surprising. It shows that, assuming the adaptive solution to be unconstrained and the noises in the primary and reference inputs to be mutually correlated, the signal-to-noise density ratio at the noise-canceler output is simply the reciprocal at all frequencies of the signal-to-noise density ratio at the reference input. The process in (12.32) is called *power inversion*.

With this result, our next objective is to derive an expression for signal distortion at the noise-canceler output. The most useful reference input is one composed almost entirely of noise correlated with the noise in the primary input. When signal components are present some signal distortion will generally occur. The amount will depend on the amount of signal propagated through the adaptive filter, which may be determined as follows. The transfer function of the propagation path through the filter is, from (12.25),

$$-J(z)W^*(z) = -J(z)\frac{\Phi_{ss}(z)J(z^{-1}) + \Phi_{nn}(z)H(z^{-1})}{\Phi_{ss}(z)|J(z)|^2 + \Phi_{nn}(z)|H(z)|^2} \tag{12.33}$$

When $|J(z)|$ is small compared with $|H(z)|$, this function can be approximated as

$$-J(z)W^*(z) \simeq \frac{J(z)}{H(z)} \tag{12.34}$$

The spectrum of the signal component propagated to the noise-canceler output through the adaptive filter is thus approximately

$$\Phi_{ss}(z)\left|\frac{J(z)}{H(z)}\right|^2 \tag{12.35}$$

The summing of the component represented in (12.35) with the signal component in Figure 12.4 in the primary input results in signal distortion. The worst case, bounding the distortion to be expected in practice, occurs when the two signal components are of opposite phase.

Let "signal distortion" $D(z)$ be defined† as a dimensionless ratio of the spectrum of the output signal component propagated through the adaptive filter in (12.35) to the spectrum of the signal component at the primary input:

$$\begin{aligned} D(z) &\approx \frac{\Phi_{ss}(z)|J(z)/H(z)|^2}{\Phi_{ss}(z)} \\ &\approx \left|\frac{J(z)}{H(z)}\right|^2 \end{aligned} \tag{12.36}$$

This expression may be rewritten in a more useful form by combining the expression for the signal-to-noise density ratio at the primary input which, in Figure 12.4, would be

$$\rho_{\text{pri}}(z) \triangleq \frac{\Phi_{ss}(z)}{\Phi_{nn}(z)} \tag{12.37}$$

with the signal-to-noise density ratio at the reference input in (12.31). The result is

$$\boxed{D(z) \cong \rho_{\text{ref}}(z)/\rho_{\text{pri}}(z)} \tag{12.38}$$

Equation (12.38) shows that, with an unconstrained adaptive solution and mutually correlated noises at the primary and reference inputs, low signal distortion results from a high signal-to-noise density ratio at the primary input and a low signal-to-noise density ratio at the reference input. This conclusion is intuitively reasonable.

Our final objective in this section is to derive an expression for the spectrum of the output noise. In Figure 12.4, the noise n_k propagates to the output with a transfer function

$$\begin{aligned} 1 - H(z)W^*(z) &= 1 - H(z)\left[\frac{\Phi_{ss}(z)J(z^{-1}) + \Phi_{nn}(z)H(z^{-1})}{\Phi_{ss}(z)|J(z)|^2 + \Phi_{nn}(z)|H(z)|^2}\right] \\ &= \frac{\Phi_{ss}(z)J(z^{-1})[J(z) - H(z)]}{\Phi_{ss}(z)|J(z)|^2 + \Phi_{nn}(z)|H(z)|^2} \end{aligned} \tag{12.39}$$

†Note that signal distortion as defined here is a linear phenomenon related to alteration of the signal waveform as it appears at the noise-canceler output and is not to be confused with nonlinear harmonic distortion.

When $|J(z)|$ is small compared with $|H(z)|$, (12.39) reduces to

$$1 - H(z)W^*(z) \simeq \frac{-\Phi_{ss}(z)J(z^{-1})}{\Phi_{nn}(z)H(z^{-1})} \tag{12.40}$$

The output noise spectrum, from Figure 12.4, is

$$\Phi_{\text{output noise}} = \Phi_{nn}(z)|1 - H(z)W^*(z)|^2 \tag{12.41}$$

Again, when $|J(z)|$ is small compared with $|H(z)|$, (12.41) reduces to

$$\Phi_{\text{output noise}}(z) \simeq \Phi_{nn}(z)\left|\frac{\Phi_{ss}(z)J(z^{-1})}{\Phi_{nn}(z)H(z^{-1})}\right|^2 \tag{12.42}$$

This equation can be expressed more conveniently in terms of the signal-to-noise density ratios at the reference input (12.31) and primary input (12.37):

$$\boxed{\Phi_{\text{output noise}}(z) \simeq \Phi_{nn}(z)|\rho_{\text{ref}}(z)||\rho_{\text{pri}}(z)|} \tag{12.43}$$

This final result, which may appear strange at first glance, can be understood intuitively as follows. The first factor implies that the output noise spectrum is proportional to the input noise spectrum and is readily accepted. The second factor implies that, if the signal-to-noise density ratio at the reference input is low, the output noise will be low; that is, the smaller the signal component in the reference input, the more perfectly the noise will be canceled. The third factor implies that if the signal-to-noise density ratio in the primary input (the desired response of the adaptive filter) is low, the filter will be trained most effectively to cancel the noise rather than the signal and consequently output noise will be low.

Thus we have shown that small signal components in the reference input, though undesirable, do not render the application of adaptive noise canceling useless.† For an illustration of the level of performance attainable in practical circumstances, consider the following example. Figure 12.5 shows a multiple input (Chapter 2) adaptive noise-canceling system designed to pass a plane-wave signal received in the main beam of an antenna array and to discriminate against strong interference in the near field or in a minor lobe of the array. If one assumes that the signal and interference have overlapping and similar power spectra and that the interference power density is 20 times greater than the signal power density at the individual array elements, the signal-to-noise ratio at the reference input, ρ_{ref}, is $1/20$. If one further assumes that, because of array gain, the signal power equals the interference power at the array output, the signal-to-noise ratio at the primary input, ρ_{pri}, is 1. After convergence of the adaptive filter the signal-to-noise ratio at the

† Note that if the reference input contained signal components but no noise components, correlated or uncorrelated, the signal would be completely canceled. When the reference input is properly derived, however, this condition should not occur.

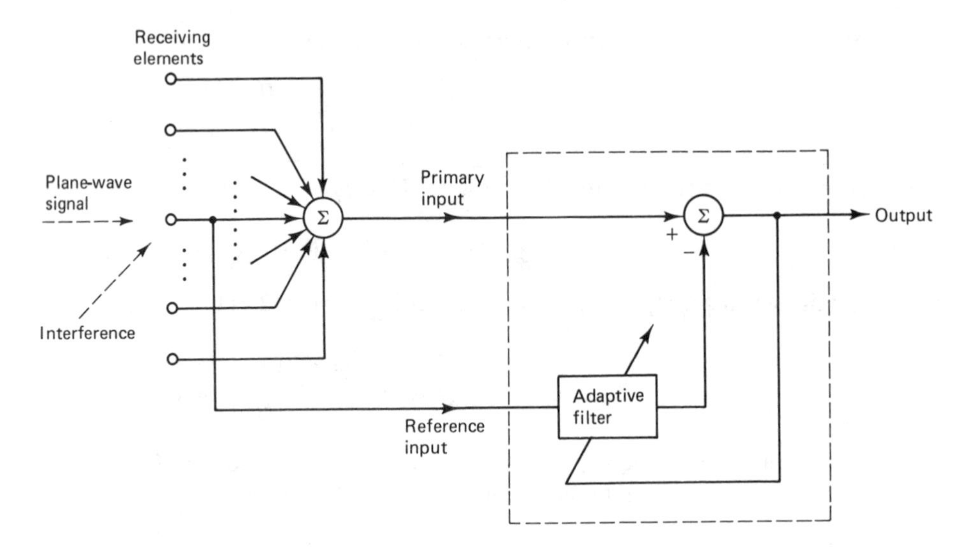

Figure 12.5 Adaptive noise canceling applied to a receiving array. From B. Widrow et al., *Adaptive Noise Canceling: Principles and Applications*, © December 1975, IEEE.

system output given in (12.32) will thus be

$$\rho_{\text{out}} = \frac{1}{\rho_{\text{ref}}} = 20$$

The maximum signal distortion from (12.38) will similarly be

$$D = \frac{\rho_{\text{ref}}}{\rho_{\text{pri}}} = \frac{1/20}{1} = 5 \text{ percent}$$

In this case, therefore, adaptive noise canceling improves the signal-to-noise ratio 20-fold and introduces only a small amount of signal distortion. Adaptive arrays are discussed further in Chapters 13 and 14.

THE ADAPTIVE INTERFERENCE CANCELER AS A NOTCH FILTER

In certain situations a primary input is available consisting of a signal component with an additive undesired sinusoidal interference. The conventional method of eliminating such interference is through the use of a notch filter. In this section an unusual notch filter, realized by an adaptive noise canceler, is described. The advantages of this form of notch filter are that it offers easy control of bandwidth, an infinite null, and the capability of adaptively tracking the exact frequency and phase of the interference. In this section we analyze the adaptive formation of a notch at a single frequency. The results can easily be extended to show that if more than one

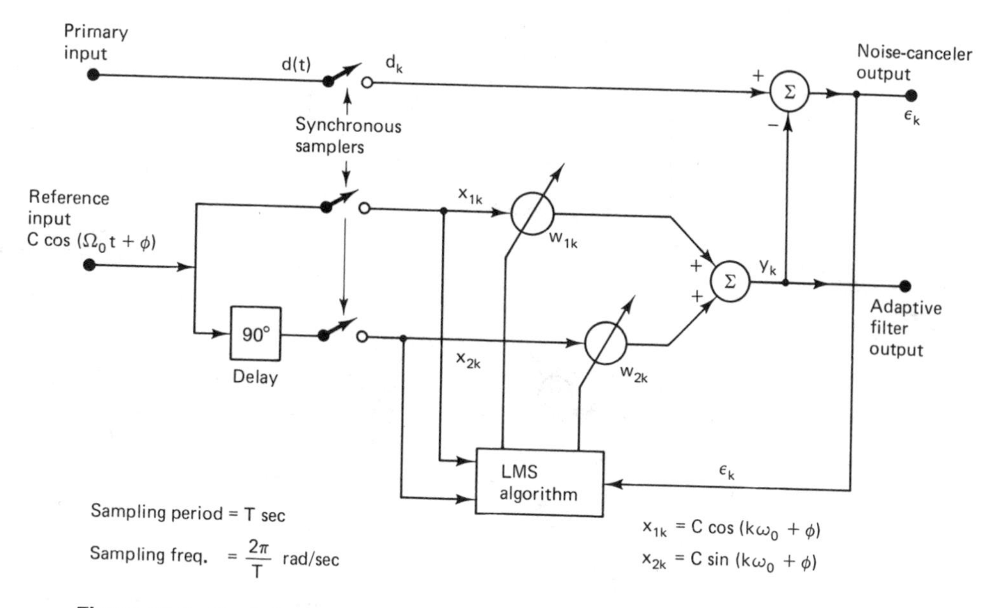

Figure 12.6 Single-frequency adaptive notch filter. From B. Widrow et al., *Adaptive Noise Canceling: Principles and Applications*, © December 1975, IEEE.

frequency is present in the reference input, a notch for each component will be formed [21, 48], and also extended to cases where there is noise in the reference input [49].

Figure 12.6 shows a single-frequency noise canceler with two adaptive weights. The primary input is assumed to be any kind of signal—stochastic, deterministic, periodic, transient, and so on—or any combination of signals. The reference input is a pure cosine wave, $C\cos(\Omega_0 t + \phi)$. The primary and reference inputs are sampled at intervals of T seconds as described in Chapter 7. The reference input is sampled directly, giving x_{1k}, and after undergoing a 90° phase shift, giving x_{2k}.

A linear transfer function for the noise canceler of Figure 12.6 may be obtained by analyzing signal propagation from the primary input to the system output.† For this purpose the flow diagram of Figure 12.7, showing the operation of the LMS algorithm in detail, is constructed. Note that the procedure for updating the weights, as indicated in the diagram, is given by (6.3):

$$\left.\begin{aligned} w_{1,k+1} &= w_{1k} + 2\mu\varepsilon_k x_{1k} \\ w_{2,k+1} &= w_{2k} + 2\mu\varepsilon_k x_{2k} \end{aligned}\right\} \tag{12.44}$$

If we let $\omega_0 = 2\pi f_0 T$ as in (7.16), the sampled reference inputs are

$$\left.\begin{aligned} x_{1k} &= C\cos(k\omega_0 + \phi) \\ x_{2k} &= C\sin(k\omega_0 + \phi) \end{aligned}\right\} \tag{12.45}$$

†Although the 2-weight adaptive filter in Figure 12.6 is inherently nonlinear and time-variable, when used with a sinusoidal reference the signal path from d_k to ϵ_k is shown by the subsequent analysis to be linear and time-invariant [21, 48].

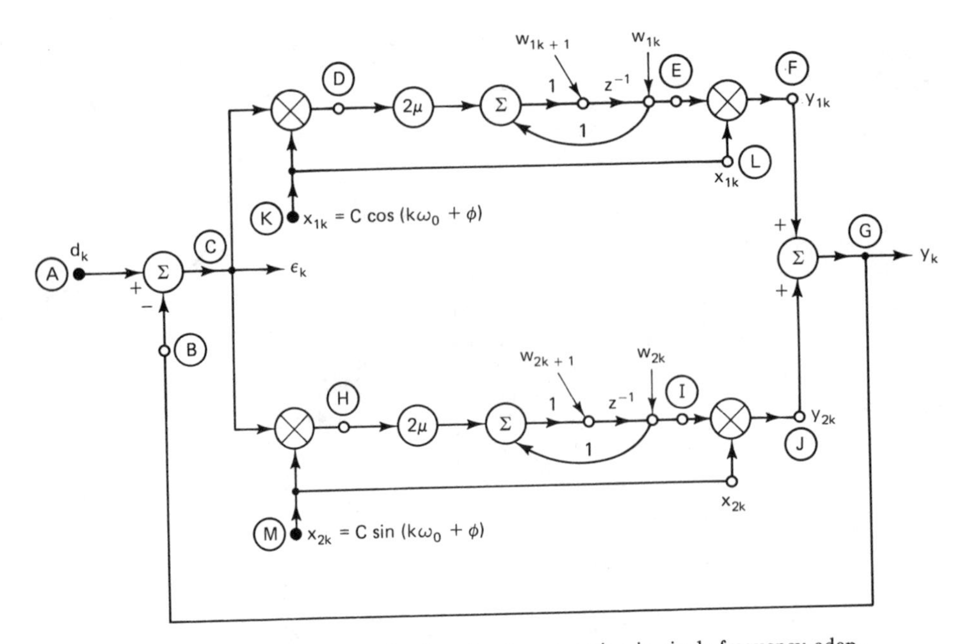

Figure 12.7 Flow diagram showing signal propagation in single-frequency adaptive notch filter. From B. Widrow et al., *Adaptive Noise Canceling: Principles and Applications*, © December 1975, IEEE.

The first step in the analysis is to obtain the isolated impulse response from the error ε_k, point C, to the filter output, point G, with the feedback loop from point G to point B broken. Let an impulse of amplitude α be applied at point C at discrete time $k = m$; that is,

$$\varepsilon_k = \alpha\delta(k - m) \tag{12.46}$$

where $\delta(k)$ is a unit impulse at $k = 0$, described in (7.21). The response at point D is then

$$\varepsilon_k x_{1k} = \begin{cases} \alpha C\cos(m\omega_0 + \phi) & \text{for } k = m \\ 0 & \text{for } k \neq m \end{cases} \tag{12.47}$$

which is the input impulse scaled in amplitude by the instantaneous value of x_{1k} at $k = m$. The signal flow path from point D to point E is that of a digital integrator with transfer function $2\mu/(z - 1)$ and impulse response $2\mu u(k - 1)$, where $u(k)$ is the discrete unit step function

$$u(k) = \begin{cases} 0 & \text{for } k < 0 \\ 1 & \text{for } k \geq 0 \end{cases} \tag{12.48}$$

Convolving $2\mu u(k-1)$ with $\varepsilon_k x_{1k}$ yields the response at point E:

$$w_{1k} = 2\mu\alpha C \cos(m\omega_0 + \phi) \tag{12.49}$$

where $k \geq m+1$. When the scaled and delayed step function is multiplied by x_{1k}, the response at point F is obtained:

$$y_{1k} = 2\mu\alpha C^2 \cos(k\omega_0 + \phi)\cos(m\omega_0 + \phi) \tag{12.50}$$

where $k \geq m+1$. The corresponding response at point J, obtained in a similar manner, is

$$y_{2k} = 2\mu\alpha C^2 \sin(k\omega_0 + \phi)\sin(m\omega_0 + \phi) \tag{12.51}$$

where $k \geq m+1$. Combining (12.50) and (12.51) yields the response at the filter output, point G:

$$\begin{aligned} y_k &= 2\mu\alpha C^2 \cos\left[(k-m)\omega_0\right] \quad \text{for } k \geq m+1 \\ &= 2\mu\alpha C^2 u(k-m-1)\cos\left[(k-m)\omega_0\right] \end{aligned} \tag{12.52}$$

Note that (12.52) is a function only of $(k-m)$ and is thus a true impulse response, proportional to the input impulse.

Having a formula for the output y_k, a linear transfer function for the noise canceler may now be derived in the following manner. If the impulse time m is set equal to zero, the unit impulse response of the linear time-invariant signal flow path from point C to point G is

$$y_k = 2\mu C^2 u(k-1)\cos(k\omega_0) \tag{12.53}$$

and the transfer function of this path is then the z-transform of y_k, or [Exercise 1(e) in Chapter 7]

$$\begin{aligned} G(z) &= 2\mu C^2 \left[\frac{z(z - \cos\omega_0)}{z^2 - 2z\cos\omega_0 + 1} - 1 \right] \\ &= \frac{2\mu C^2 (z\cos\omega_0 - 1)}{z^2 - 2z\cos\omega_0 + 1}. \end{aligned} \tag{12.54}$$

If the feedback loop from point G to point B is now closed, the transfer function $H(z)$ from the primary input, point A, to the noise-canceler output, point C, can be obtained from the feedback formula:

$$H(z) = \frac{1}{1 + G(z)} = \frac{z^2 - 2z\cos\omega_0 + 1}{z^2 - 2(1 - \mu C^2)z\cos\omega_0 + 1 - 2\mu C^2} \tag{12.55}$$

Equation (12.55) shows that the single-frequency noise canceler has the properties of a notch filter at the reference frequency ω_0. The zeros of the transfer function are located in the z-plane at

$$z = e^{\pm j\omega_0} \tag{12.56}$$

and are precisely on the unit circle at angles of $\pm\omega_0$ rad. The poles are located at

$$z = (1 - \mu C^2)\cos\omega_0 \pm j\left[(1 - 2\mu C^2) - (1 - \mu C^2)^2\cos^2\omega_0\right]^{1/2} \tag{12.57}$$

The poles are inside the unit circle at a radial distance $(1 - 2\mu C^2)^{1/2}$, approximately equal to $1 - \mu C^2$, from the origin and at angles of

$$\pm\cos^{-1}\left[(1 - \mu C^2)(1 - 2\mu C^2)^{-1/2}\cos\omega_0\right] \tag{12.58}$$

For slow adaptation (i.e., small values of μC^2) these angles depend on the factor

$$\begin{aligned}\frac{1 - \mu C^2}{(1 - 2\mu C^2)^{1/2}} &= \left(\frac{1 - 2\mu C^2 + \mu^2 C^4}{1 - 2\mu C^2}\right)^{1/2} \\ &= (1 + \mu^2 C^4 + \cdots)^{1/2} \\ &= 1 + \tfrac{1}{2}\mu^2 C^4 + \cdots\end{aligned} \tag{12.59}$$

which differs only slightly from a value of 1. The result is that, in practical instances, *the angles of the poles are almost identical to those of the zeros.*

The locations of the poles and zeros and the half-power points of the transfer function are shown in Figure 12.8. Since the zeros lie on the unit circle, the depth in dB of the notch in the transfer function is infinite at the frequency $\omega = \omega_0$. The sharpness of the notch is determined by the closeness of the poles to the zeros. Corresponding poles and zeros are separated by a distance approximately equal to μC^2. The arc length along the unit circle (or the angle) spanning the distance between half-power points is the "bandwidth" of the notch filter, and is seen to be

$$\text{BW} = 2\mu C^2 \text{ rad} = \frac{\mu C^2}{\pi T} \text{ Hz} \tag{12.60}$$

A common measure of the sharpness of a notch is the "quality factor," Q, which is determined by the ratio of the center frequency to the bandwidth:

$$Q \triangleq \frac{\text{center frequency}}{\text{bandwidth}} = \frac{\omega_0}{2\mu C^2} \tag{12.61}$$

The single-frequency noise canceler is thus equivalent to a stable notch filter when the reference input is a pure cosine wave. The depth of the null is generally superior to that of a fixed filter because the adaptive process maintains the correct phase relationships for canceling, even if the reference frequency changes slowly.

Figure 12.9 shows the results of two experiments performed to demonstrate the characteristics of the adaptive notch filter. In the first the primary input was a cosine wave of unit power stepped through 512 discrete frequencies. The reference input was a cosine wave with a frequency ω_0 of $\pi/2$ rad. The value of C was 1, and the value of μ was 0.0125. The frequency resolution of the discrete Fourier transform [Equation (7.43)] was $\Delta\omega = \pi/512$. The output power at each frequency is shown in Figure 12.9(a). As the primary frequency approaches the reference frequency, significant cancellation occurs. When ω is near but not at ω_0 the weights

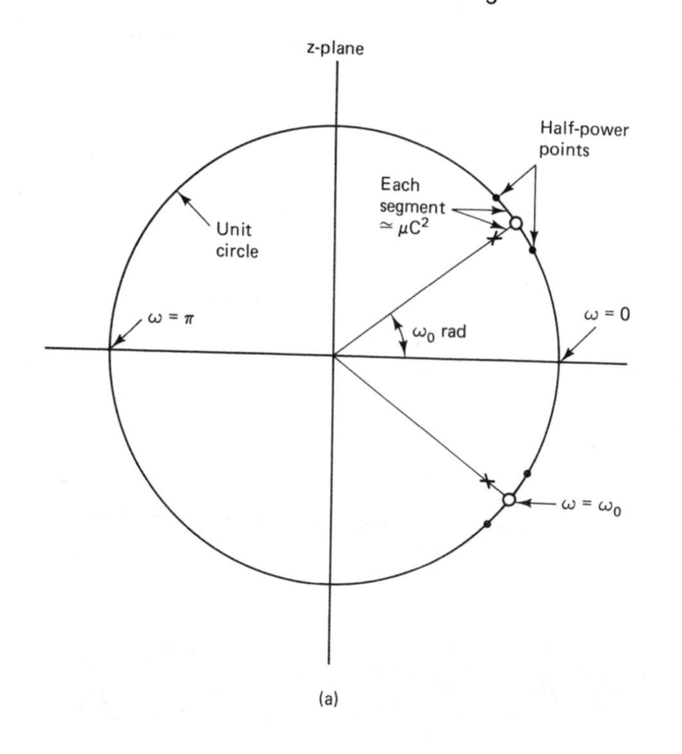

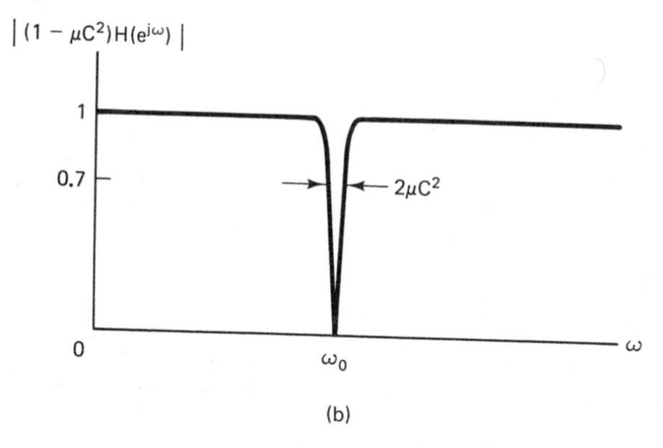

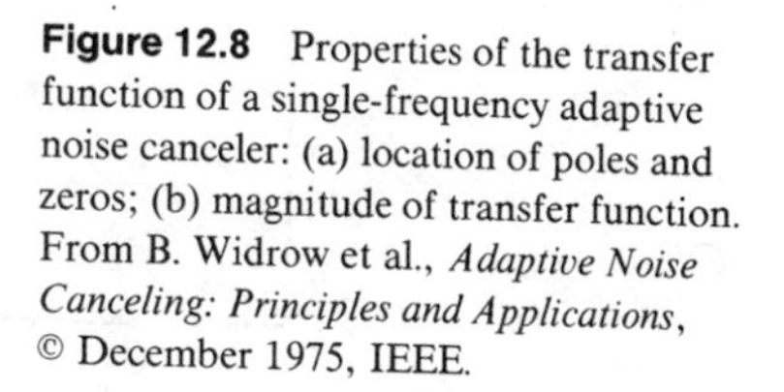

Figure 12.8 Properties of the transfer function of a single-frequency adaptive noise canceler: (a) location of poles and zeros; (b) magnitude of transfer function. From B. Widrow et al., *Adaptive Noise Canceling: Principles and Applications*, © December 1975, IEEE.

do not converge to stable values but "tumble" at the difference frequency,† and the adaptive filter behaves like a modulator, converting the reference frequency into the

† When the primary and reference frequencies are held at a constant difference, the weights develop a sinusoidal steady state at the difference frequency. In other words, they converge on a dynamic rather than a static solution. This is an unusual form of adaptive behavior that comes from the LMS algorithm itself. Although the weight vector converges on a dynamic rather than static solution, it is in a sense forming a nonstationary least-squares solution. This is called "non-Wiener" behavior. The Wiener solution under these circumstances would require the weights to be zero.

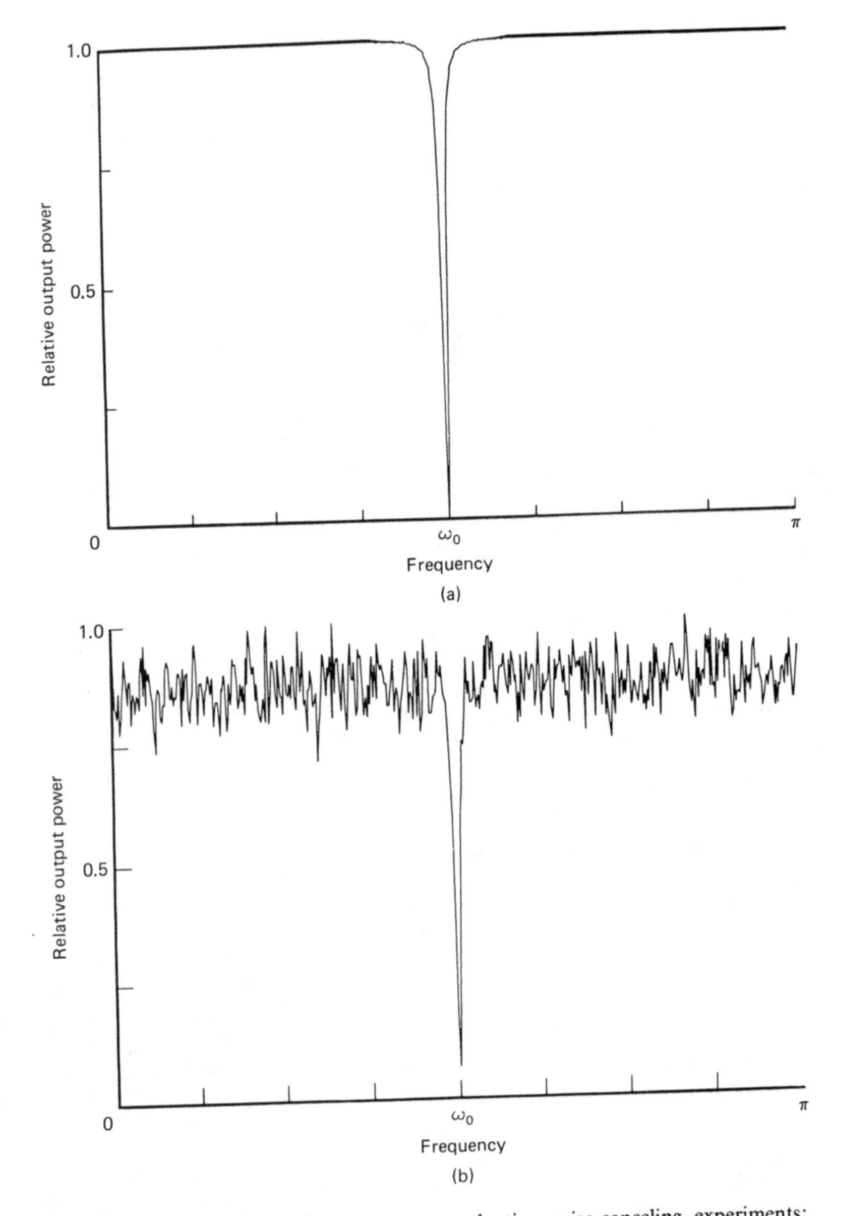

Figure 12.9 Results of single-frequency adaptive noise-canceling experiments: (a) primary input was a cosine wave stepped through 512 discrete frequencies; (b) primary input composed of samples of white noise. From B. Widrow et al., *Adaptive Noise Canceling: Principles and Applications*, © December 1975, IEEE.

primary frequency. The theoretical notch width between half-power points, 0.0250 rad, compares closely with the measured notch width of 0.0255 rad.

In the second experiment, the primary input was composed of uncorrelated samples of white noise at unit power. The reference input and the processing parameters were the same as in the first experiment. An ensemble average of 4096 power spectra at the noise-canceler output is shown in Figure 12.9(b). An infinite null was not obtained in this experiment because of the finite frequency resolution of the spectral analysis algorithm.

In these experiments, filtering with a reference cosine wave at a given frequency caused cancellation of primary input components at adjacent frequencies. This result indicates that, under some circumstances, primary input components may be partially canceled and distorted even though the reference input is uncorrelated with them. In practice this kind of cancellation is of concern only when the adaptive process is rapid, that is, when it is effected with large values of μ. When the adaptive process is slow, the weights converge to values that are nearly fixed, and although signal cancellation as described in this section occurs it is generally not significant. The weights approximate the Wiener solution.

Additional experiments have recently been conducted with reference inputs containing more than one sinusoid. The formation of multiple notches has been achieved by using an adaptive filter with multiple weights (typically, an adaptive transversal filter). Two weights are required for each sinusoid to achieve the necessary filter gain and phase. Uncorrelated broadband noise imposed on the reference input creates a need for additional weights. A full analysis of the multiple-notch problem is contained in Reference 21.

THE ADAPTIVE INTERFERENCE CANCELER AS A HIGH-PASS FILTER

The use of a bias weight in an adaptive filter to cancel low-frequency drift in the primary input is a special case of notch filtering with the notch at zero frequency. The method of incorporating only a bias weight is shown in Figure 12.10. Because there is no need to match the phase of the signal, only one weight is needed. The reference input may be set to a constant value of 1.

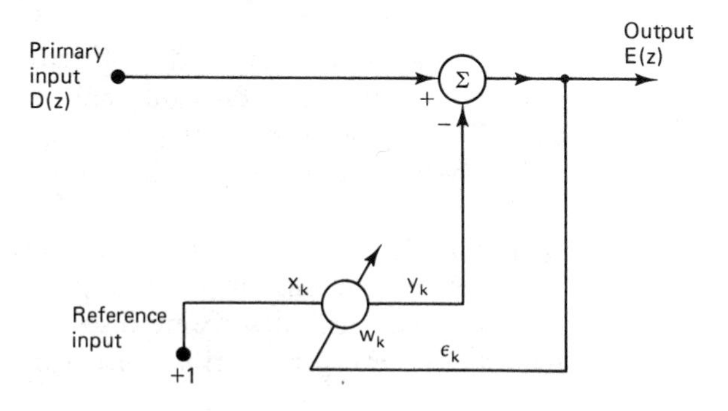

Figure 12.10 Setting the reference input to one produces an adaptive high-pass filter.

The transfer function from the primary input to the noise-canceler output is derived as follows. In Figure 12.10 we see that y_k is equal to w_k, so the LMS algorithm (6.3) in this case gives

$$\begin{aligned} y_{k+1} &= y_k + 2\mu\varepsilon_k \\ &= y_k + 2\mu(d_k - y_k) \end{aligned} \tag{12.62}$$

Taking the z-transform of (12.62) yields the steady-state solution,

$$Y(z) = \frac{2\mu D(z)}{z - (1 - 2\mu)} \tag{12.63}$$

If we now substitute $Y(z) = D(z) - E(z)$ into (12.63), we obtain the transfer function from D to E in Figure 12.10,

$$H(z) = \frac{E(z)}{D(z)} = \frac{z - 1}{z - (1 - 2\mu)} \tag{12.64}$$

Equation (12.64) shows that the bias-weight filter is a high-pass filter with a zero on the unit circle at zero frequency and a pole on the real axis at a distance 2μ to the left of the zero. Note that this corresponds to a single-frequency notch filter, described by (12.55), for the case where $\omega_0 = 0$ and $C = 1$. The half-power frequency of the notch is at $\omega = 2\mu$ rad.

The single-weight noise canceler acting as a high-pass filter is capable of removing not only a constant bias but also a slowly varying drift in the primary input. Moreover, though it is not demonstrated here, bias or drift removal can be accomplished simultaneously with cancellation of low-frequency stochastic interference, using the type of zero-frequency reference shown in Figure 12.10.

EFFECTS OF FINITE LENGTH AND CAUSALITY

In the analyses above, questions of the physical realizability of the adaptive filters were not considered. The expressions derived were ideal, based on the assumption of an infinitely long, two-sided (noncausal) adaptive transversal filter. Although such a filter cannot in reality be implemented, we will show that its performance can be closely approximated.

Typical impulse responses of ideal filters approach amplitudes of zero exponentially over time. Approximate realizations are thus possible with finite-length FIR filters. The more weights used in the FIR filter, the closer its impulse response will be to that of the ideal filter. Increasing the number of weights, however, also increases the cost of implementation.

Noncausal filters, of course, are not physically realizable in real-time systems. In many cases, however, they can be realized approximately in delayed form, providing an acceptable delayed real-time response. Such cases were discussed in Chapters 10 and 11. In practical circumstances excellent performance can be

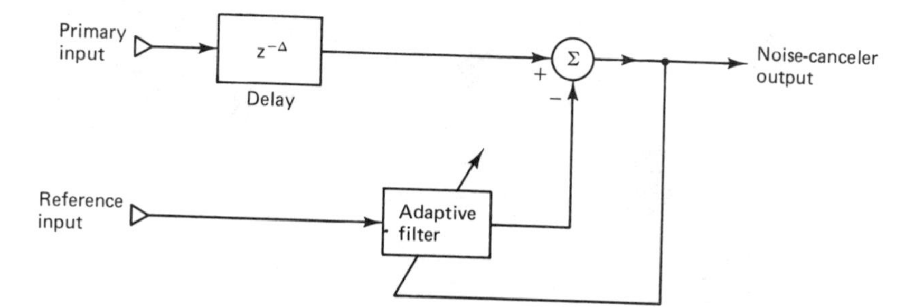

Figure 12.11 Adaptive noise canceller with delay in primary input path. From B. Widrow et al., *Adaptive Noise Canceling: Principles and Applications*, © December 1975, IEEE.

obtained with two-sided filter impulse responses even when they are truncated in time to the left and right. Using a delay, the truncated response can be made causal and physically realizable.

Figure 12.11 shows an adaptive noise-canceling system with a delay Δ inserted in the primary input. This delay causes an equal delay to develop in the unconstrained optimal filter impulse response, which remains otherwise unchanged. In practical, finite-length adaptive transversal filters, on the other hand, the optimal impulse response generally changes shape with changes in the value of Δ, which is chosen to cause the peak of the impulse response to center along the delay line in the adaptive filter.

As in Chapter 10, the value of Δ is not critical within a certain optimal range; that is, the curve showing minimum mean-square error as a function of Δ generally has a very broad minimum. A value typically equal to about half the time delay of the adaptive filter produces the least minimum output noise power. (Compare Figure 10.10, for example.)

Figure 12.12 shows the results of a computer-simulated noise-canceling experiment with an unconstrained optimal filter response that was noncausal. The primary input consisted of a periodic triangular wave and additive colored noise. The reference input consisted of colored noise correlated with the primary noise.† The unconstrained optimal impulse response and the causal, finite-length adaptive impulse response obtained without a delay in the primary input are plotted in Figure 12.12(a). The large difference in these impulse responses indicates that the noise-canceler output will be a poor approximation to the signal. The corresponding

† Except for the delay in the primary input, the simulated noise-canceling system was identical with the system shown in Figure 12.3. The transfer function $H(z)$ was a non-minimum phase, low-pass FIR filter with two zeros,

$$H(z) = 2z^{-1}\left(1 - \tfrac{1}{2}z^{-1}\right)\left(1 - \tfrac{1}{2}z\right)$$

The optimal unconstrained adaptive filter solution, in this case given by (12.13), is the reciprocal of $H(z)$. It has one pole inside and one pole outside the unit circle in the z-plane. A stable realization of $W^*(z)$ must, therefore, be two-sided.

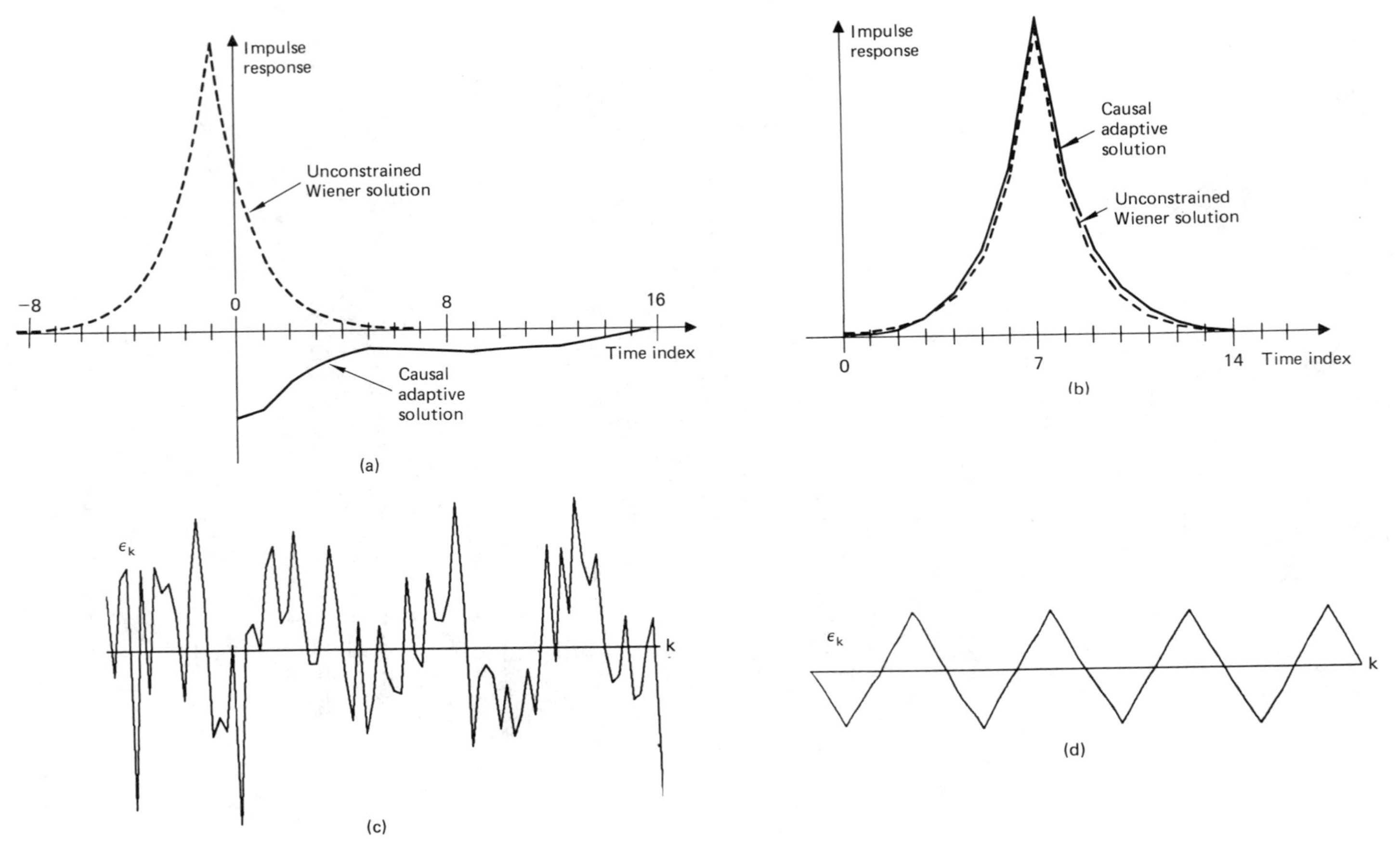

Figure 12.12 Results of noise-canceling experiment with delay in primary input path: (a) optimal solution and adaptive solution without time delay; (b) optimal solution and adaptive solution with delay of eight time units; (c) noise-canceler output without delay; (d) noise-canceler output with delay. From B. Widrow et al.,

optimal and adaptive impulse responses obtained with a delay of eight time units (half the length of the adaptive filter) are shown in Figure 12.12(b). These solutions are similar, indicating that performance of the adaptive filter will be close to optimal. Typical noise-canceler outputs with and without the delay are shown in Figure 12.12(c) and (d). The waveform obtained with the delay is very close to that of the original triangular signal, whereas that obtained with no delay still contains a great amount of noise.

MULTIPLE-REFERENCE NOISE CANCELING

When there is more than one noise or interference to be canceled and a number of linearly independent reference inputs containing mixtures of each can be obtained, we can construct a multiple-reference noise-canceling system by expanding the single-reference noise canceler analyzed above. In the model shown in Figure 12.13 the M inputs, ψ_1 through ψ_M, represent mutually uncorrelated sources of either signal or noise. The transfer functions $G_i(z)$ represent the propagation paths from these sources to the primary input. The $F_{ij}(z)$ similarly represent the propagation paths to the reference inputs and allow for cross-coupling. This model permits treatment not only of multiple noise sources but also of signal components in the reference inputs and uncorrelated noises in the reference and primary inputs. In other words, it is a general representation of the adaptive noise canceler in Figure 12.4.

The unconstrained optimal transfer function of the multiple-reference canceler is the matrix equivalent of (12.8) and is derived in the following manner. The source power spectral matrix is defined as

$$[\Phi_{\psi\psi}(z)] = \begin{bmatrix} \Phi_{\psi_1\psi_1}(z) & & & 0 \\ & \Phi_{\psi_2\psi_2}(z) & & \\ & & \ddots & \\ 0 & & & \Phi_{\psi_M\psi_M}(z) \end{bmatrix} \tag{12.65}$$

With this definition, the spectral matrix of the N reference inputs to the adaptive filters becomes

$$[\Phi_{xx}(z)] = [F(z^{-1})]^T [\Phi_{\psi\psi}(z)][F(z)] \tag{12.66}$$

where

$$[F(z)] = \begin{bmatrix} F_{11}(z) & \cdots & F_{1N}(z) \\ \vdots & & \vdots \\ F_{M1}(z) & \cdots & F_{MN}(z) \end{bmatrix} \tag{12.67}$$

and again $F_{ij}(z)$ is the transfer function from source i to input j.

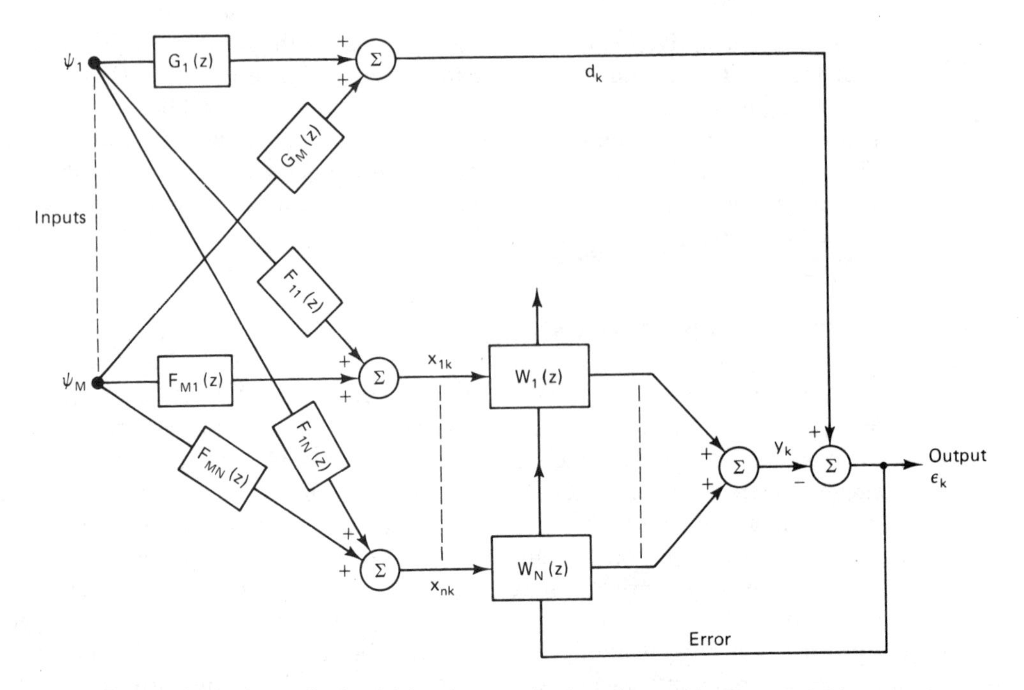

Figure 12.13 Generalized multiple-reference adaptive noise canceler. From B. Widrow et al., *Adaptive Noise Canceling: Principles and Applications*, © December 1975, IEEE.

The cross-spectral vector from the reference inputs to the primary input in Figure 12.13 is given similarly by

$$[\Phi_{xd}(z)] = [F(z^{-1})]^T[\Phi_{\psi\psi}(z)][G(z)] \tag{12.68}$$

where

$$[G(z)] = [G_1(z) \quad \cdots \quad G_M(z)]^T \tag{12.69}$$

and again $G_i(z)$ is the transfer function from source i to the primary input.

With these relationships for Figure 12.13, the optimal weight vector solution is similar to (12.8). If we define

$$[W^*(z)] = [W_1^*(z) \quad \cdots \quad W_N^*(z)] \tag{12.70}$$

as the set of optimal transfer functions (each containing an optimal weight vector), we then have, similar to (12.8),

$$\begin{aligned}[W^*(z)] &= [\Phi_{xx}(z)]^{-1}\Phi_{xd}(z)\\ &= \left[[F(z^{-1})]^T[\Phi_{\psi\psi}(z)][F(z)]\right]^{-1}[F(z^{-1})]^T[\Phi_{\psi\psi}(z)][G(z)]\end{aligned} \tag{12.71}$$

This solution represents the entire set of optimal weight vectors, $\mathbf{W}_1^*$ through $\mathbf{W}_N^*$, which define the minimum of the overall error surface given by $E[\varepsilon_k^2]$ as a function of all of the weights. We do not need mathematics to show that this surface is quadratic in all of the weights, because we can see in Figure 12.13 that y_k is a linear combination of adaptive linear combiner outputs, and thus y_k^2 (and ε_k^2) must contain only linear and quadratic weight terms.

If $[F(z)]$ in (12.67) is square (i.e., $N = M$) and has an inverse, then (12.71) simplifies to†

$$[W^*(z)] = [F(z)]^{-1}[G(z)] \tag{12.72}$$

which is the matrix equivalent of (12.12). The inverse of $[F(z)]$ may not exist at all frequencies.

The expressions above can be used to derive steady-state optimal solutions to multiple-source, multiple-reference noise canceling problems more general than those discussed previously. An application of the multiple-reference problem in the field of fetal electrocardiography is given below.

The foregoing generalization completes our discussion of the principles of adaptive noise canceling, which included a description of the concept and theoretical analyses of performance with various kinds of signal and noise. The following sections describe a variety of practical applications of the technique. These applications include the canceling of several kinds of interference in electrocardiography, of noise in speech signals, of antenna sidelobe interference, and of periodic or narrowband interference for which there is no external reference source. Experimental results are presented that demonstrate the performance of adaptive noise canceling in these applications.

CANCELING 60-Hz INTERFERENCE IN ELECTROCARDIOGRAPHY

Huhta and Webster [22] have pointed out that a major problem in the recording of electrocardiograms (ECGs) is "the appearance of unwanted 60-Hz interference in the output." They analyze the various causes of such power-line interference, including magnetic induction, displacement currents in leads or in the body of the patient, and equipment interconnections and imperfections. They also describe a number of techniques for minimizing it that can be effected in the recording process itself, such as proper grounding and the use of twisted pairs. Another method capable of reducing 60-Hz ECG interference is adaptive noise canceling, which can be used separately or in conjunction with more conventional approaches.

Figure 12.14 shows the application of adaptive noise canceling in electrocardiography. The primary input is taken from the ECG preamplifier; the 60-Hz reference input is taken from a wall outlet with proper attenuation. The adaptive filter contains two variable weights, applied in a manner similar to that of Figure 12.6. The two weighted versions of the reference are summed to form the filter's

†See Exercise 1(c) in Chapter 2.

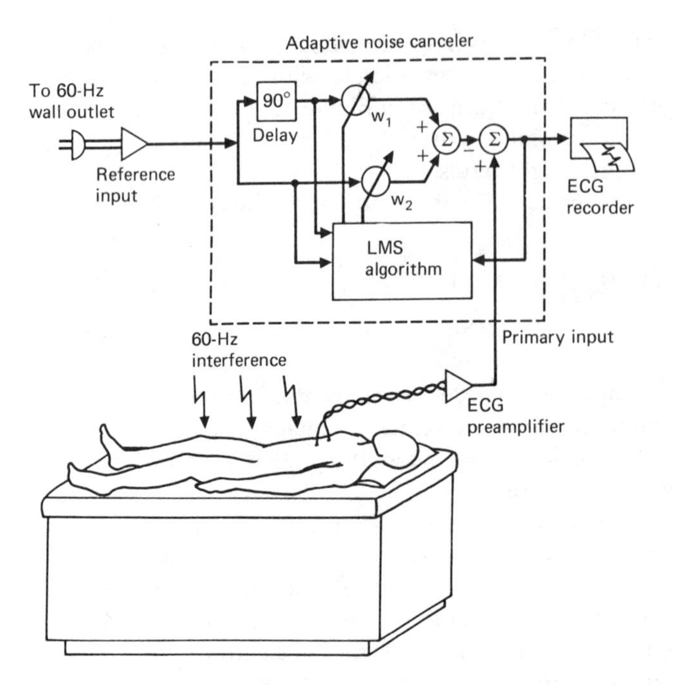

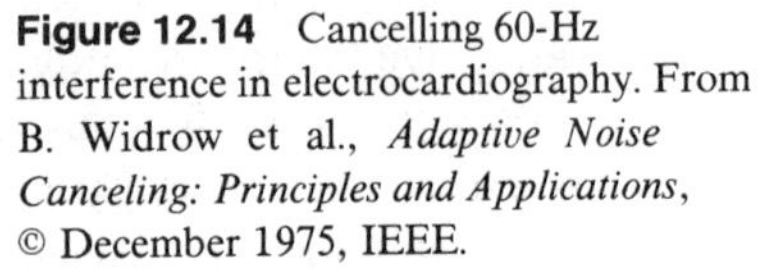
Figure 12.14 Cancelling 60-Hz interference in electrocardiography. From B. Widrow et al., *Adaptive Noise Canceling: Principles and Applications*, © December 1975, IEEE.

output, which is subtracted from the primary input. Selected combinations of the values of the weights allow the reference waveform to be changed in magnitude and phase in any way required for cancellation. The two variable weights, or two "degrees of freedom," are required to cancel the single pure sinusoid.

A typical result of a group of experiments performed with real-time processing is shown in Figure 12.15. The samples had 10 bits and the sampling rate was 1000 Hz. Figure 12.15(a) shows the primary input, an ECG waveform with an excessive amount of 60-Hz interference, and Figure 12.15(b) shows the reference input from the wall outlet. Figure 12.15(c) shows the noise-canceler output. Note the absence of interference and the clarity of detail after the adaptive process has converged.

CANCELING DONOR-HEART INTERFERENCE IN HEART-TRANSPLANT ELECTROCARDIOGRAPHY

The electrical depolarization of the ventricles of the human heart is triggered by a group of specialized muscle cells known as the atrioventricular (AV) node. Although capable of independent, asynchronous operation, this node is normally controlled by a similar cell group, the sinoatrial (SA) node, whose depolarization initiates an electrical impulse transmitted by conduction through the atrial heart muscle to the AV node. The SA node is connected through the vagus and sympathetic nerves to the central nervous system, which by controlling the rate of depolarization controls the frequency of the heartbeat [23, 24].

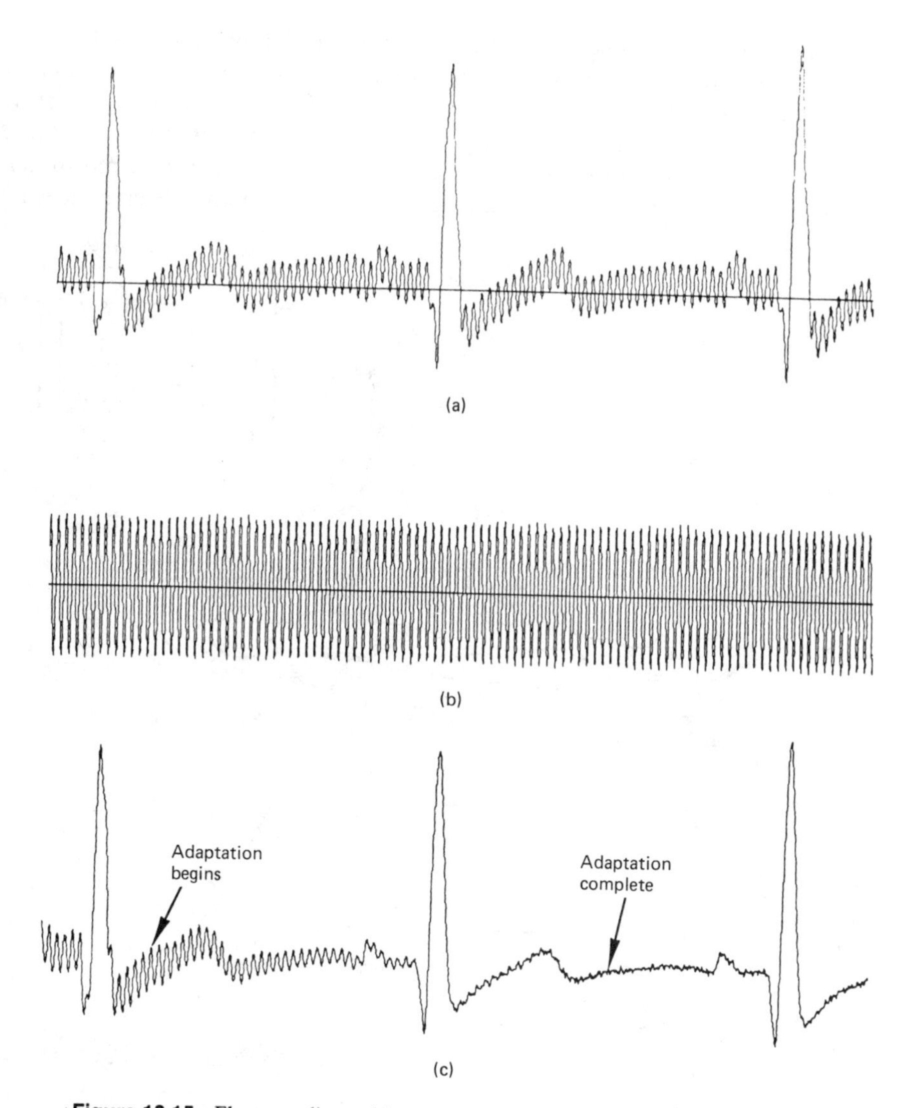

Figure 12.15 Electrocardiographic noise canceling: (a) primary unit; (b) reference input; (c) noise canceler output. From B. Widrow et al., *Adaptive Noise Canceling: Principles and Applications*, © December 1975, IEEE.

The cardiac transplantation technique developed by Norman Shumway of the Stanford University Medical Center involves the suturing of the "new" or donor heart to a portion of the atrium of the patient's "old" heart [25]. Scar tissue forms at the suture line and electrically isolates the small remnant of the old heart, containing only the SA node, from the new heart, containing both SA and AV nodes. The SA node of the old heart remains connected to the vagus and sympathetic nerves, and the old heart continues to beat at a rate controlled by the central nervous system. The SA node of the new heart, which is not connected to the central nervous system because the severed vagus nerve cannot be surgically reattached, generates a spontaneous pulse that causes the new heart to beat at a separate self-pacing rate.

In cardiac transplant research, as well as in cardiac research in general, one would like to determine the firing rate of the old heart and observe the waveforms of

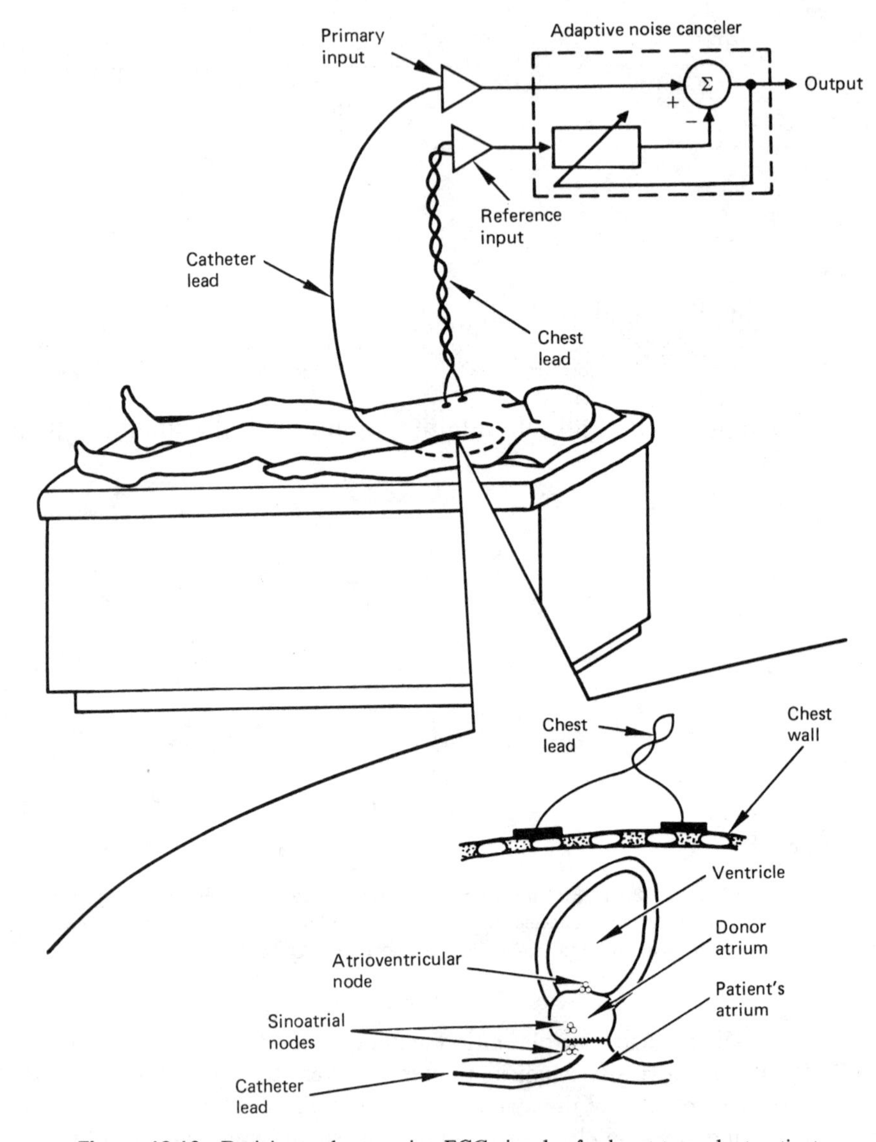

Figure 12.16 Deriving and processing ECG signals of a heart-transplant patient. From B. Widrow et al., *Adaptive Noise Canceling: Principles and Applications*, © December 1975, IEEE.

its electrical output. These waveforms, which cannot be obtained by ordinary electrocardiographic means because of interference from the beating of the new heart, are readily obtained with adaptive noise canceling.

Figure 12.16 shows the method of applying adaptive noise canceling in heart-transplant electrocardiography. The reference input is provided by a pair of ordinary chest leads. These leads receive a signal that comes mainly from the new heart, the source of interference. The primary input is provided by a catheter consisting of a small coaxial cable threaded through the left brachial vein and the vena cava to a position in the atrium of the old heart. The tip of the catheter, a few millimeters long, is an exposed portion of the center conductor that acts as an antenna and is capable of receiving cardiac electrical signals. When it is in the most favorable position, the desired signal from the old heart and the interference from the new heart are received in about equal proportion.

Figure 12.17 shows typical reference and primary inputs and the corresponding noise-canceler output. The reference input contains the strong QRS waves that, in a

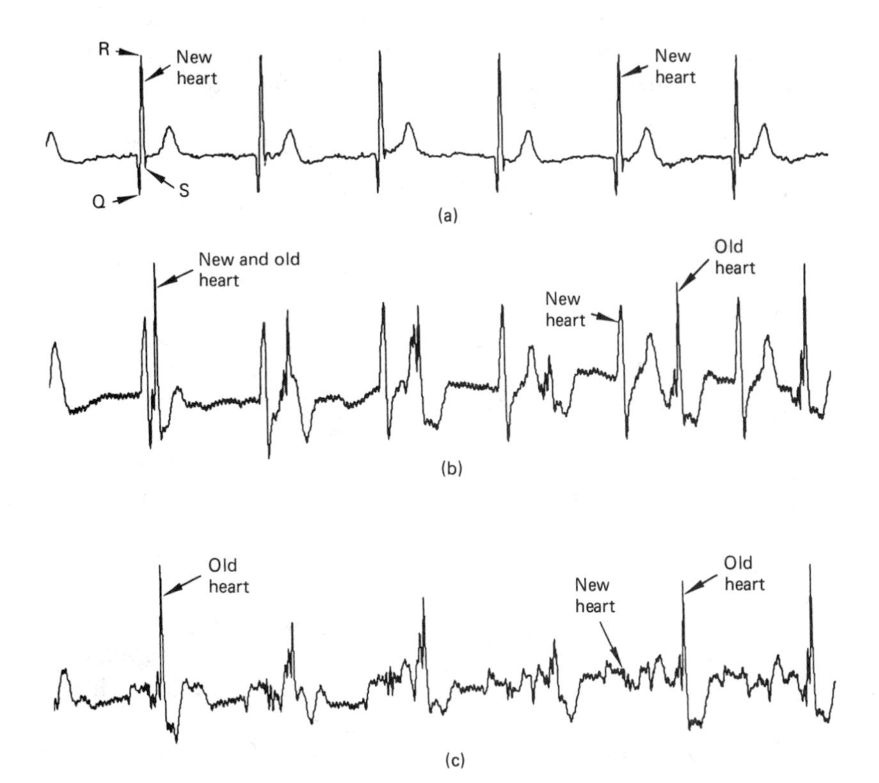

Figure 12.17 ECG waveforms of a heart-transplant patient. (a) reference input (new heart); (b) primary input (new and old heart); (c) noise-canceler output (old heart). From B. Widrow et al., *Adaptive Noise Canceling: Principles and Applications*, © December 1975, IEEE.

normal electrocardiogram, indicate the firing of the ventricles. The primary input contains pulses that are synchronous with the QRS waves of the reference input and indicate the beating of the new heart. The other waves seen in this input are due to the old heart, which is beating at a separate rate. When the reference input is adaptively filtered and subtracted from the primary input, one obtains the waveform shown in Figure 12.17(c), which is that of the old heart together with very weak residual pulses originating in the new heart. Note that the pulses of the two hearts are easily separated, even when they occur at the same instant. Note also that the electrical waveform of the new heart is steady and precise, whereas that of the old heart varies significantly from beat to beat.

For this experiment the noise canceler was implemented in software with an adaptive transversal filter containing 48 weights. The sampling rate was 500 Hz.

CANCELING THE MATERNAL ECG IN FETAL ELECTROCARDIOGRAPHY

Abdominal electrocardiograms make it possible to determine the fetal heart rate and to detect multiple fetuses, and are often used during labor and delivery [26–28]. The background noise due to muscle activity and fetal motion, however, often has an amplitude equal to or greater than that of the fetal heartbeat [29–31]. A still more serious problem is the mother's heartbeat, which has an amplitude 2 to 10 times greater than that of the fetal heartbeat and often interferes with its recording [32].

At Stanford University, some experiments† were performed to demonstrate the usefulness of adaptive noise canceling in fetal electrocardiography. The objective was to derive as clear a fetal ECG as possible, so that one could observe not only the heart rate but also the actual waveform of the electrical output.

Four ordinary chest leads were used to record the mother's heartbeat and provide multiple reference inputs to the canceler.‡ A single abdominal lead was used to record the combined maternal and fetal heartbeats that served as the primary input. Figure 12.18 shows the cardiac electric field vectors of mother and fetus and the positions in which the leads were placed. Each lead terminated in a pair of electrodes. The chest and abdominal inputs were prefiltered, digitized, and recorded on tape. The multichannel adaptive noise canceler, shown in Figure 12.19, is a special case of Figure 12.13. Each reference-channel filter had 32 taps with nonuniform (log periodic) spacing and a total delay of 129 ms.

Figure 12.20 shows typical reference and primary inputs together with the corresponding noise-canceler output. The prefiltering bandwidth was 3 to 35 Hz and the sampling rate was 256 Hz. The maternal heartbeat, which dominates the

†A similar attempt to cancel the maternal heartbeat had previously been made by Walden and Birnbaum [33] without the use of an adaptive processor. Some reduction of the maternal interference was achieved by the careful placement of leads and adjustment of amplifier gain, but apparently better results can be obtained with adaptive processing.

‡More than one reference input was used to make the interference filtering task easier. The number of reference inputs required to eliminate the maternal ECG is still under investigation.

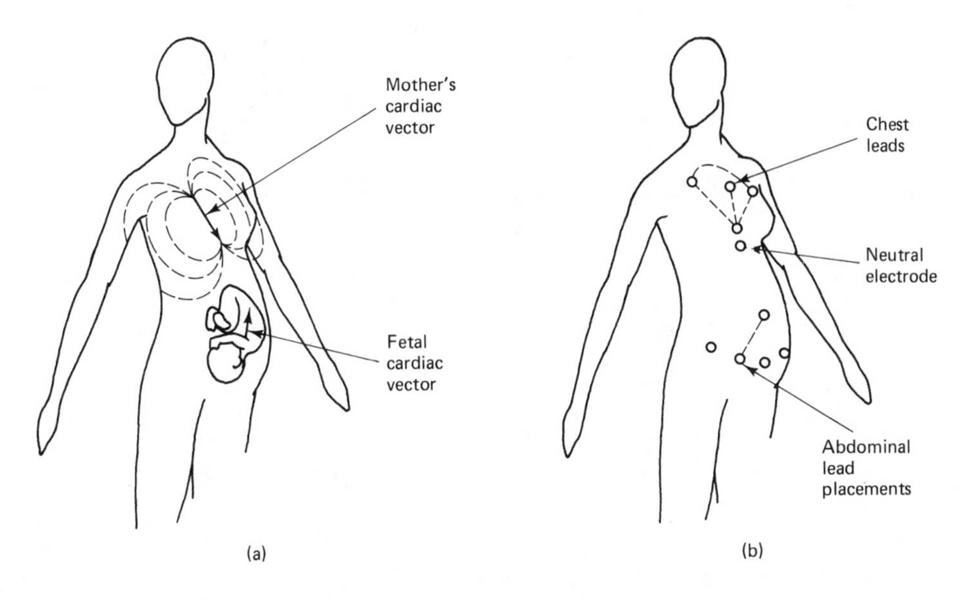

Figure 12.18 Canceling maternal heartbeat in fetal electrocardiography: (a) cardiac electric field vectors of mother and fetus; (b) placement of leads. From B. Widrow et al., *Adaptive Noise Canceling: Principles and Applications*, © December 1975, IEEE.

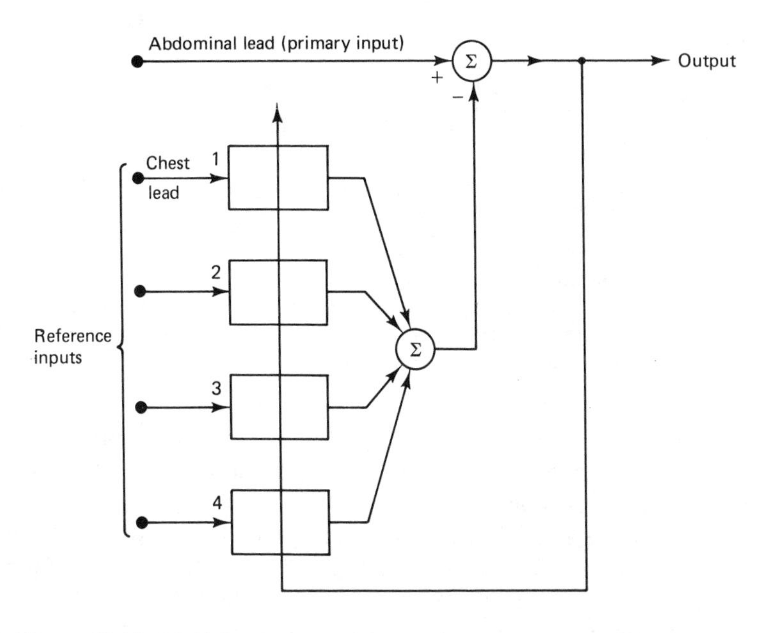

Figure 12.19 Multiple-reference noise canceler used in fetal ECG experiment. From B. Widrow et al., *Adaptive Noise Canceling: Principles and Applications*, © December 1975, IEEE.

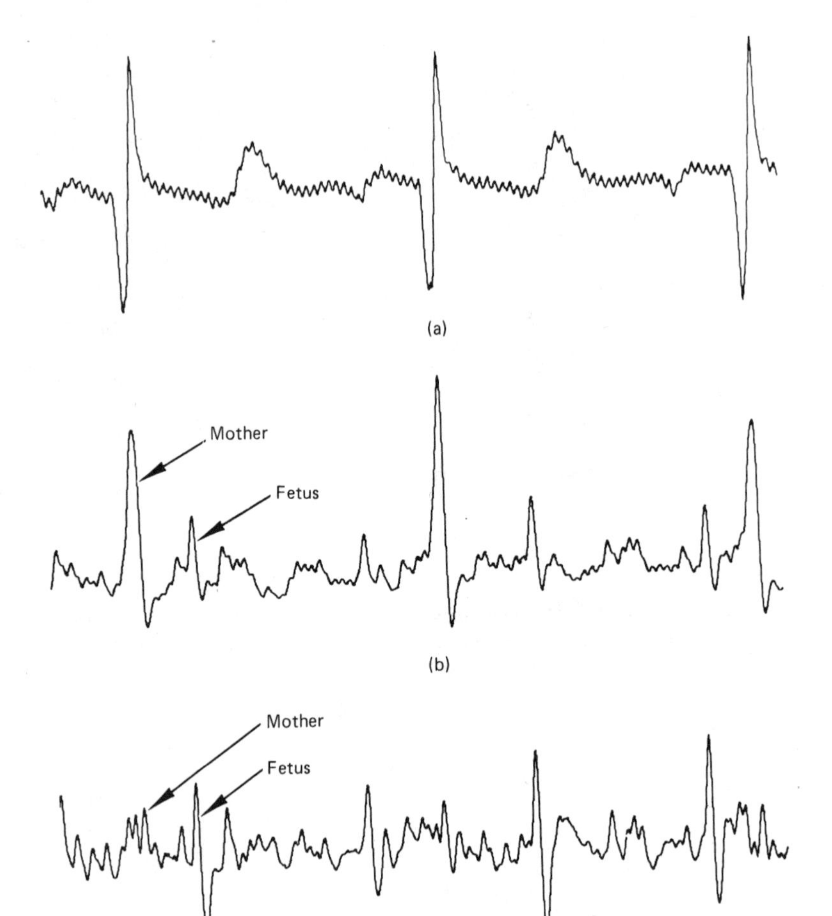

Figure 12.20 Result of fetal ECG experiment (bandwidth, 3–35 Hz; sampling rate, 256 Hz): (a) reference input (chest lead); (b) primary input (abdominal lead); (c) noise-canceler output. From B. Widrow et al., *Adaptive Noise Canceling: Principles and Applications*, © December 1975, IEEE.

primary input, is suppressed in the noise-canceler output. (The voltage scale of the noise-canceler output [Figure 12.20(c)] is approximately two times greater than that of the primary input [Figure 12.20(b)].)

Figure 12.21 shows corresponding results for a prefiltering bandwidth of 0.3 to 75 Hz and a sampling rate of 512 Hz. Baseline drift and 60-Hz interference are clearly present in the primary input, obtained from the abdominal lead. The interference is so strong that it is almost impossible to detect the fetal heartbeat. The inputs obtained from the chest leads contained the maternal heartbeat and a 60-Hz component sufficient to serve as a reference for both of these interferences. In the

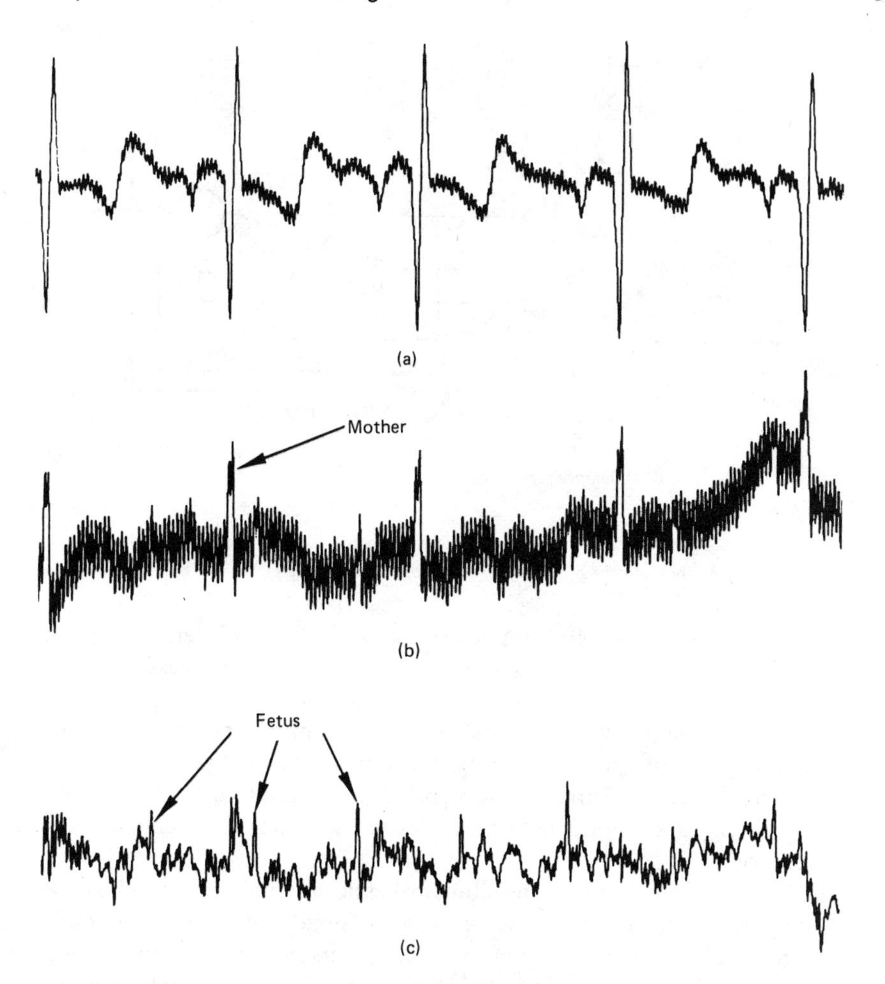

Figure 12.21 Result of wide-band fetal ECG experiment (bandwidth, 0.3–75 Hz; sampling rate, 512 Hz): (a) reference input (chest lead); (b) primary input (abdominal lead); (c) noise canceller output. From B. Widrow et al., *Adaptive Noise Canceling: Principles and Applications*, © December 1975, IEEE.

noise-canceler output both interferences have been significantly reduced, and the fetal heartbeat is clearly discernible.

CANCELING NOISE IN SPEECH SIGNALS

Consider the situation of a pilot communicating by radio from the cockpit of an aircraft where a high level of engine noise is present. The noise contains, among other things, strong periodic components, rich in harmonics, that occupy the same frequency band as the pilot's speech. These components are picked up by the

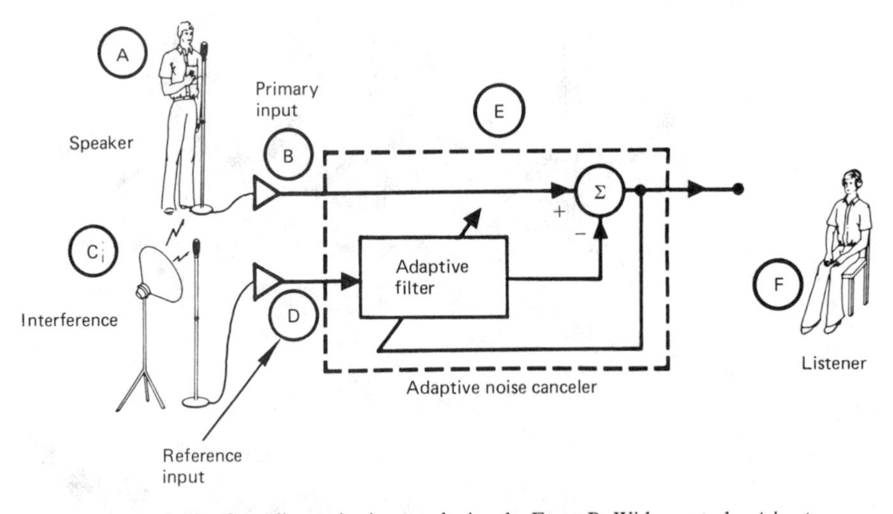

Figure 12.22 Canceling noise in speech signals. From B. Widrow et al., *Adaptive Noise Canceling: Principles and Applications*, © December 1975, IEEE.

microphone into which the pilot speaks and severely interfere with the intelligibility of the radio transmission. It would be impractical to process the received signal with a conventional filter because the frequency and intensity of the noise components vary with engine speed, the position of the pilot's head, and so on. By placing a second microphone at a suitable location in the cockpit, however, a sample of the ambient noise field free of the pilot's speech could be obtained. This sample could be filtered and subtracted from the transmission, significantly reducing the interference.

To demonstrate the feasibility of canceling noise in speech signals, experiments simulating the cockpit noise problem in simplified form were conducted. In these experiments, as shown in Figure 12.22, a person (A) spoke into a microphone (B) in a room where strong acoustic interference (C) was present. A second microphone (D) was placed in the room away from the speaker. The output of microphones (B) and (D) formed the primary and reference inputs, respectively, of a noise canceler (E), whose output was monitored by a remote listener (F). The canceler included an adaptive filter with 16 weights, and the rate of adaptation was approximately 5 kHz. A typical learning curve, showing output power as a function of number of adaptation cycles, is shown in Figure 12.23. Convergence was complete after about 5000 adaptations or 1 second of real time.

In a typical experiment the interference was a triangular wave containing many harmonics that, because of multipath effects, varied in amplitude and phase, from point to point in the room. The periodic nature of the interference made it possible to ignore the difference in time delay caused by the different transmission paths to the two sensors. The noise canceler was able to reduce the output power of this interference, which otherwise made the speech unintelligible, by 20 to 25 dB, rendering the interference barely perceptible to the remote listener. No noticeable

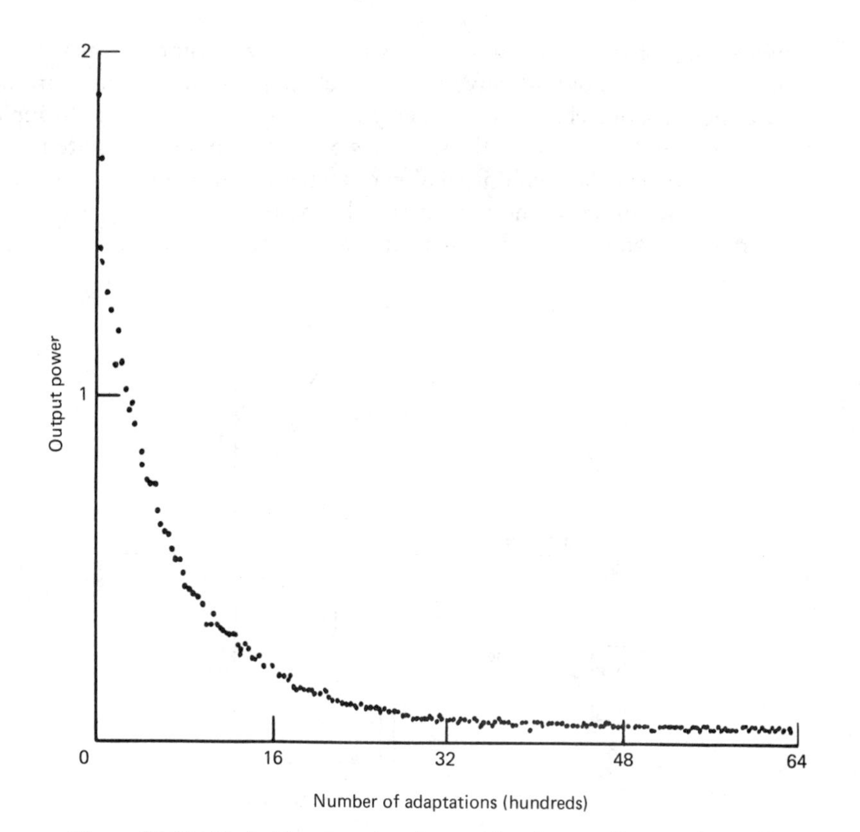

Figure 12.23 Typical learning curve for speech noise-canceling experiment. From B. Widrow et al., *Adaptive Noise Canceling: Principles and Applications*, © December 1975, IEEE.

distortion was introduced into the speech signal. Convergence times were on the order of seconds, and the processor was readily able to readapt when the position of the microphones was changed or when the frequency of the interference was varied over a range from 100 to 2000 Hz.

CANCELING ECHOES IN LONG-DISTANCE TELEPHONE CIRCUITS

Long-distance telephone circuits have generally been impaired by echo effects [16, 17]. Echo suppressors developed at the Bell Telephone Laboratories have been highly perfected during the past decade. Transcontinental calls via satellite links have not been as successfully handled by conventional echo suppression techniques and for this reason, adaptive echo cancelers have been developed. The time delay involved in radio transmission to a synchronous satellite and back to earth is around 250 ms and the two-way round-trip time is near half a second. Switching, which is involved in conventional echo suppression, is very disruptive when the path delays are this long. Without adaptive echo cancellation, satellite circuits are best utilized like radiotelephone channels, one party talking at a time (i.e., "half-duplex").

We shall briefly review the workings of the telephone as required to understand the echo problem. We omit discussion of many of its functions, such as switching, in order to concentrate on the flow of signals. A simple two-party telephone system is pictured in Figure 12.24. In each direction, a carbon transmitter is connected in

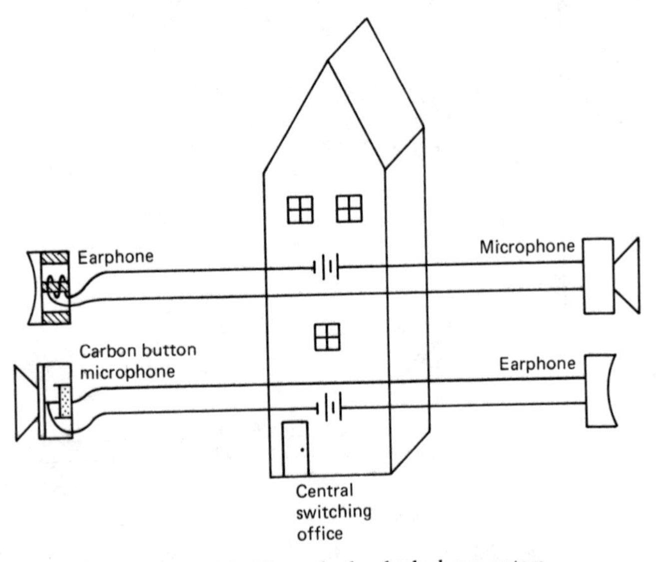

Figure 12.24 Four-wire local telephone system.

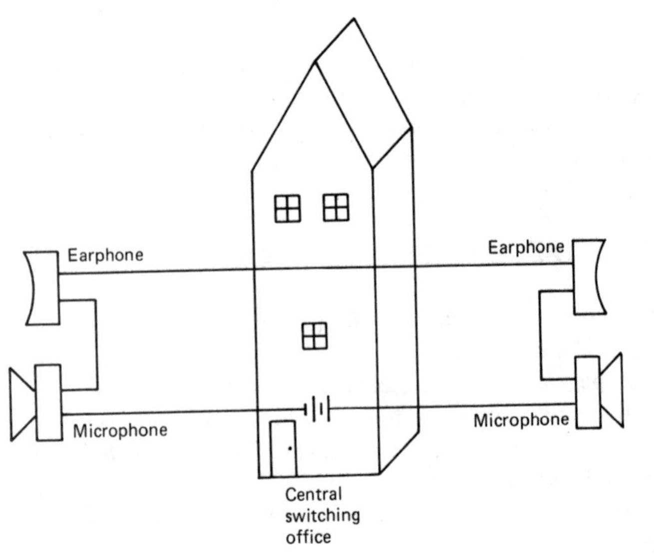

Figure 12.25 Two-wire local telephone system.

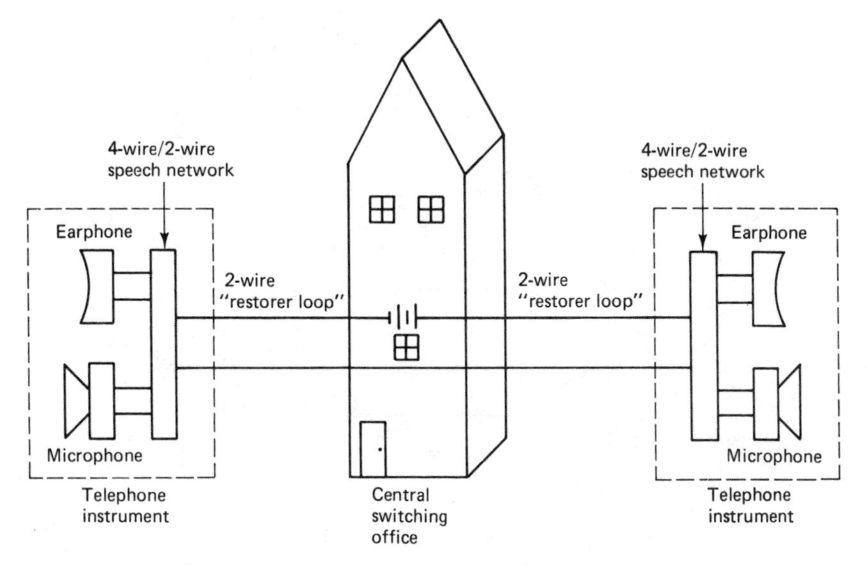

Figure 12.26 Improved two-wire local telephone system.

series with an earphone and a battery. An independent circuit is used in each direction. This approach is very simple, but has some drawbacks. The system of Figure 12.25 has the advantage of using two wires from each customer to the central office (where batteries are housed and local switching is done). The system of Figure 12.25 also introduces what is called "sidetone," making each user's speech available to his or her own earphone. This makes the telephone sound "alive." The feedback allows the user to adjust the speaking level. A modern telephone is somewhat more complicated than that of Figure 12.25. A highly simplified diagram is shown in Figure 12.26. A "two-wire/four-wire speech network," incorporated in the base of the telephone, consists of a transformer-coupled network that permits direct current (dc) to flow from the central office battery through the carbon microphone, but blocks the flow of dc in the earphone. The earphone contains a permanent magnet and is designed to work with an ac signal. Although the speech network blocks dc flow through the earphone, it permits the injection of a controlled amount of sidetone from the microphone to the earphone. The four wires of the microphone and earphone are thus coupled to a two-wire "loop" connecting the customer to the central office. This is the basic telephone circuit used today for local calling.

Long-distance calling is based on the use of repeater amplifiers which require separate circuits for the two directions, as shown in Figure 12.27. An important element of the system is the "hybrid" device, a circuit element that has the capability of applying the incoming long-distance trunk signal to the customer loop so that this signal is heard at the customer's earphone, while the signal from the customer's microphone is taken from the same loop and applied to the outgoing trunk.

The hybrid converts the two-wire loop circuits (where the incoming and outgoing signals are together) to a four-wire circuit where the incoming and outgoing

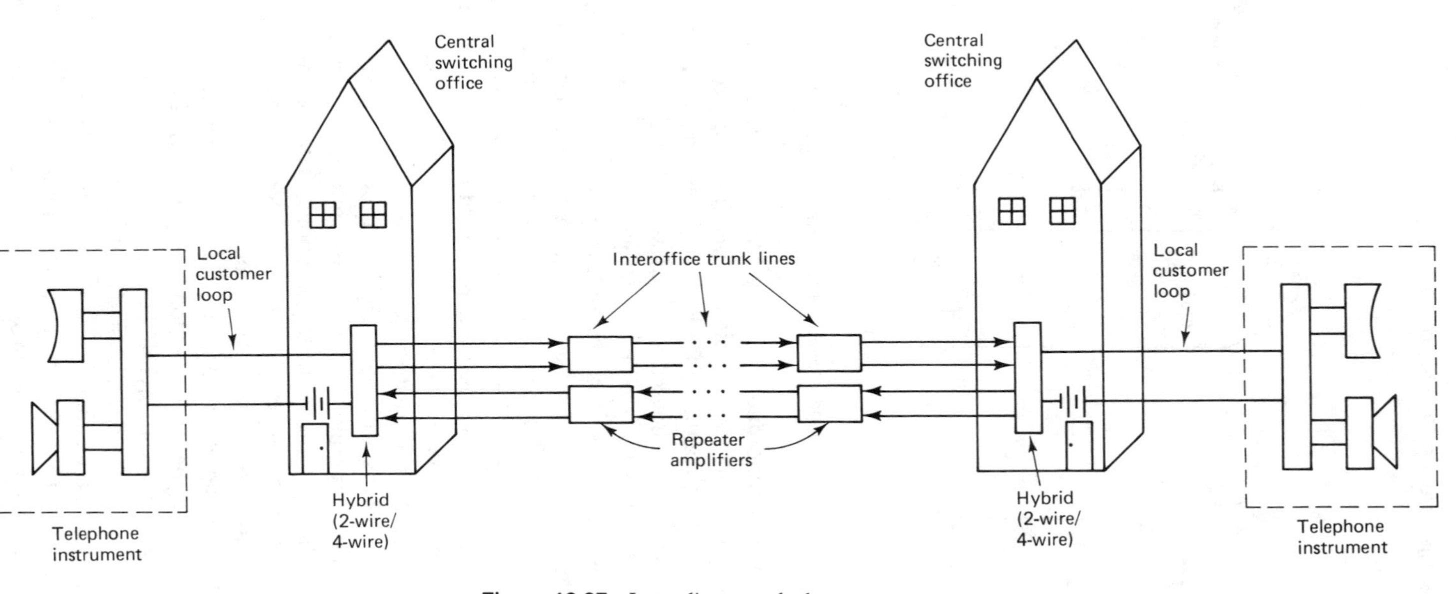

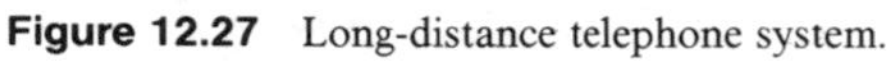

Figure 12.27 Long-distance telephone system.

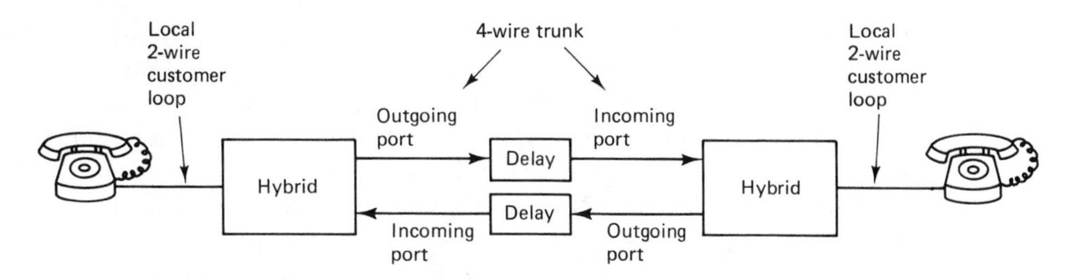

Figure 12.28 Simplified version of the long-distance system.

signals are separated. The ideal hybrid passes an incoming signal from its "incoming" port to its "two-wire" port, attenuating it by 3 dB, and does not pass anything to its "outgoing" port. Furthermore, it passes signals from its "two-wire" port, attenuated by 3 dB, to its "outgoing" port, without reflecting any energy back into the two-wire line. A simplified long-distance diagram is shown in Figure 12.28, where the repeaters and trunk lines are represented as delays.

For many reasons, the hybrid devices do not provide ideal performance. Variations in instrument characteristics and in loop length and impedance cause imperfect separation of incoming and outgoing signals at the hybrid. Thus some incoming signal components leak into the outgoing port of the hybrid. Typically, the hybrid reduces the amplitude of the "leakage" signal by about 15 dB relative to the incoming signal. The "leakage path" is essentially linear and has an unknown amplitude and phase characteristic. The leakage on each end of the long-distance circuit causes echoes which may be very apparent to some users even when the echo is as much as 40 dB below the incoming signal.

During the past 50 years, echo suppressors have been in use on long-distance circuits in the United States, and they have successfully alleviated the problem when the round-trip delay is 100 ms or less. Figure 12.29 shows a two-way long-distance telephone system equipped with echo suppressors. On a given end of the circuit, a relay in the echo suppressor cuts the outgoing hybrid port when a signal-level detector determines that there is an incoming signal. On the other hand, the relay action is overruled and the outgoing hybrid path is reestablished when another signal detector determines that a microphone signal is present originating from the same end of the line. Thus the echo path is cut when receiving only. The outgoing

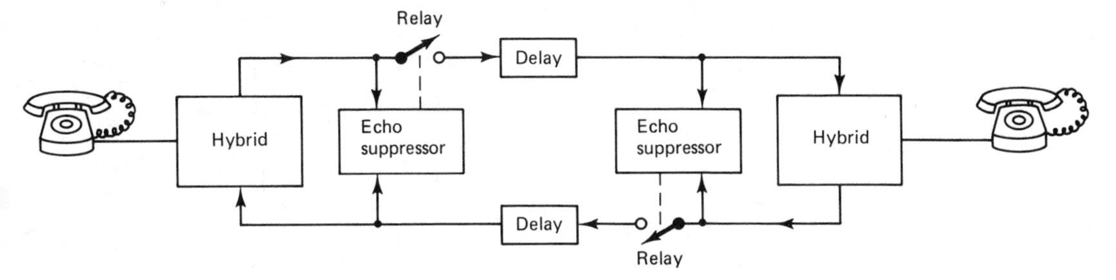

Figure 12.29 Long-distance system with echo suppressors.

path is established when transmitting, even during double talking, that is, simultaneous talking at both ends of the line.

When the round-trip delay exceeds 100 ms, as with satellite circuits, the choppiness and missing pieces of words due to switching in the echo suppressors is quite annoying and this has stimulated the development of adaptive echo cancelers [17]. A long-distance system with adaptive echo cancellation is depicted in Figure 12.30. The principle of operation is simple. At a given end of the long-distance line, the incoming signal is applied both to the hybrid and to the adaptive filter. The output of the adaptive filter is subtracted from the hybrid output. The difference, the error signal for the adaptive filter, is used to adjust the weights. Hybrid leakage causes a hybrid outgoing component which is correlated with the incoming signal. The adaptive filter will remove the leakage component as best possible in the least-squares sense. Its ability to cancel the hybrid leakage is limited only by its ability to match the unknown hybrid leakage path. Economic considerations place limits on the fineness of sampling in time and on the fineness of quantization in the adaptive filter realization, but technological improvements are relaxing these limits. Fundamental limits come from misadjustment in the adaptive filter and from nonlinearities in the hybrid leakage path. A successful adaptive echo canceler for speech might, for example, use an 8-kHz sampling rate and 128 weights.

Some of the problems in adaptive echo cancellation are due to the effects of noise in the weight vector. The diagram of Figure 12.31 shows the adaptive filters and the essential signals in the telephone system. We assume for this discussion that the adaptive filters contain a sufficient number of degrees of freedom to match the hybrid leakage transfer functions, that the μ-settings in the adaptive filters ensure stability, and that the local loop noises, $n_{k\mathrm{AL}}$ and $n_{k\mathrm{BL}}$, and the microphone signals, $s_{k\mathrm{A}}$ and $s_{k\mathrm{B}}$, are of zero mean. We also assume that the convergence mechanisms of the two adaptive filters act independently. This assumption is not strictly correct,

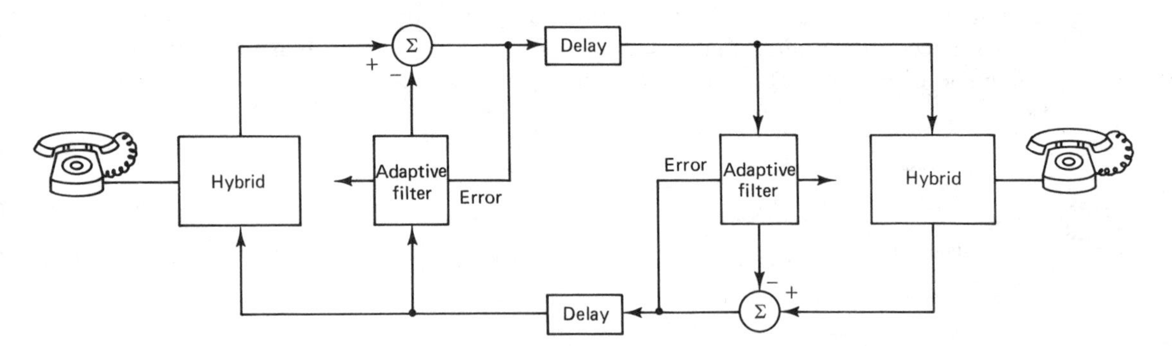

Figure 12.30 Long-distance system with adaptive echo cancellation.

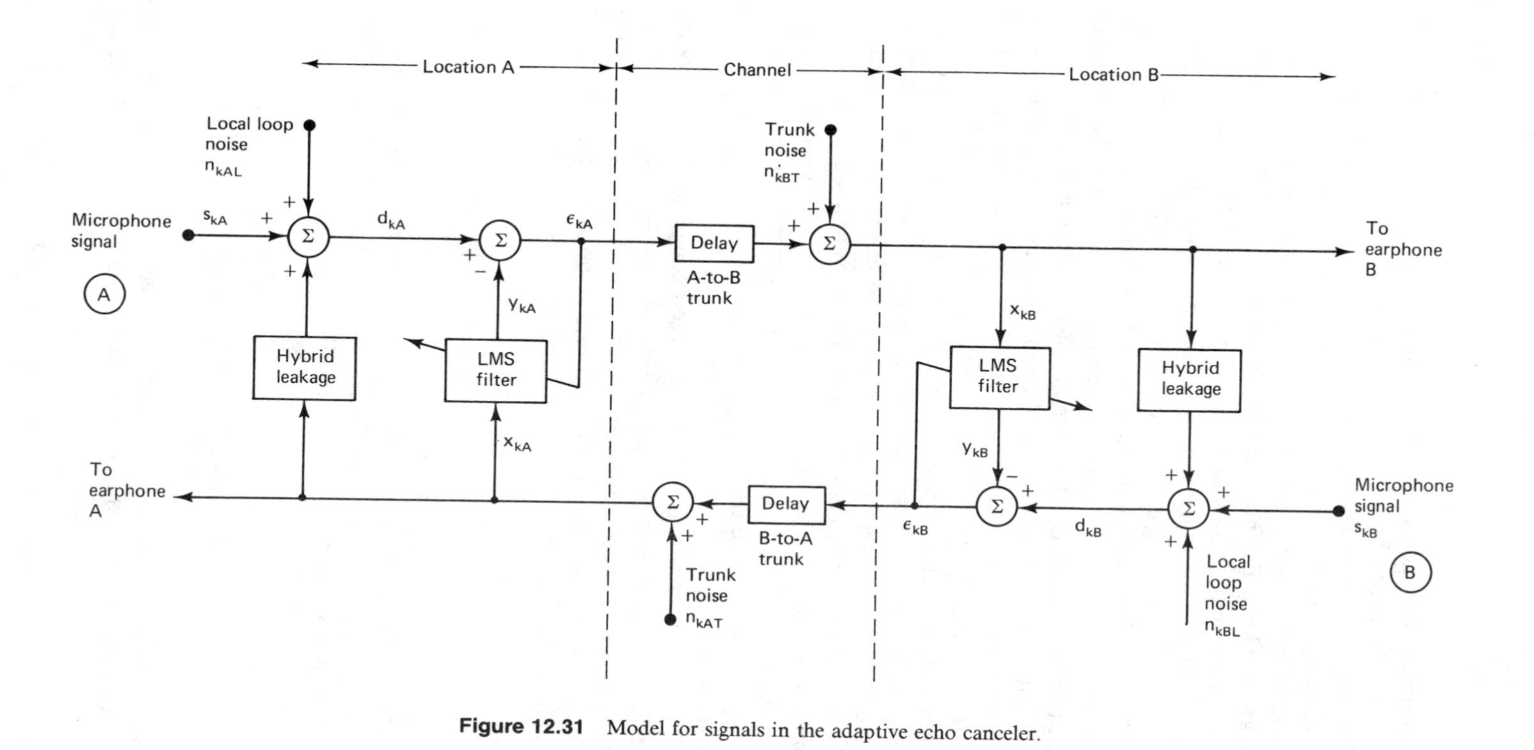

Figure 12.31 Model for signals in the adaptive echo canceler.

but experimental evidence has indicated that good approximate analyses of the adaptive processes can be obtained by treating them separately. Actually, the two adaptive filters and the telephone network comprise a feedback network. The input to the adaptive filter at terminus A comes partially from the output of the adaptive filter at terminus B, and vice versa. However, the microphone signals, the local loop noises, and the long-distance trunk noises are independent of the adaptive processes. Let us assume that signals and noises in Figure 12.31 are all independent and stationary. This assumption was made for purposes of analysis [16, 17], and the usefulness of the resulting analysis has been confirmed by experience.

With all of the assumptions above, one concludes that upon convergence the mean weight vectors of the adaptive filters perfectly cancel the corresponding hybrid leakage transfer functions. However, another result of adaptation is that steady-state noise in the weights, randomly modulating the adaptive filter input signals, causes additional random interference components to be added to the microphone signals, which then propagate via the long-distance trunks.

For purposes of analysis, suppose that the value of μ in the adaptive filter at terminus A is very small, so that the noise in the weight vector is negligible. The adaptive filter then perfectly cancels the hybrid leakage at end A and introduces no other "side effects." The net signal from side A as applied to the A-to-B trunk is the microphone signal $s_{k\mathrm{A}}$ plus the local loop noise $n_{k\mathrm{AL}}$. Assuming a gain of unity on the trunk circuit, the incoming waveform available at terminus B is given by $s_{k\mathrm{A}} + n_{k\mathrm{AL}} + n_{k\mathrm{BT}}$. The term $n_{k\mathrm{BT}}$ is the noise at terminus B originating along the trunk from A to B.

In practice, μ would need to be large enough to allow the adaptive filter to converge in a reasonable amount of time and to be able to track changes in the hybrid leakage path. Accordingly, there will be noise in the weights. Let the misadjustment of the adaptive filter at end A be designated M_{A}. In Figure 12.31, the minimum mean square of $\varepsilon_{k\mathrm{A}}$ is seen to be

$$\xi_{\mathrm{A}_{\min}} = E\left[s_{k\mathrm{A}}^2\right] + E\left[n_{k\mathrm{AL}}^2\right] \tag{12.73}$$

We have tacitly assumed a double-talking situation in making this determination, with $x_{k\mathrm{A}} = n_{k\mathrm{AT}} + (s_{k\mathrm{B}} + n_{k\mathrm{BL}})_{\mathrm{delayed}}$ and $d_{k\mathrm{A}} = s_{k\mathrm{A}} + n_{k\mathrm{AL}}$ being the adaptive filter input and desired response, respectively. The average excess mean-square error is $M_{\mathrm{A}}\xi_{\mathrm{A}_{\min}}$, or

$$\text{excess MSE} = M_{\mathrm{A}}\left(E\left[s_{k\mathrm{A}}^2\right] + E\left[n_{k\mathrm{AL}}^2\right]\right) \tag{12.74}$$

This is the power of the additive random independent interference caused by noise in the weights. Accordingly, the incoming noise power at terminus B is

$$E\left[n_{k\mathrm{AL}}^2\right] + E\left[n_{k\mathrm{BT}}^2\right] + M_{\mathrm{A}}\left(E\left[s_{k\mathrm{A}}^2\right] + E\left[N_{k\mathrm{AL}}^2\right]\right) \tag{12.75}$$

The signal-to-noise ratio at terminus B is therefore

$$\text{SNR}_\text{B} = \frac{E\left[s_{k\text{A}}^2\right]}{M_\text{A} E\left[s_{k\text{A}}^2\right] + (1 + M_\text{A}) E\left[n_{k\text{AL}}^2\right] + E\left[n_{k\text{BT}}^2\right]} \tag{12.76}$$

In cases of interest, M is substantially less than 1 and the power level of the local loop noise is very low compared with the noise level on the long-distance trunk. Neglecting the local loop noise power, we have

$$\text{SNR}_\text{B} = \frac{E\left[s_{k\text{A}}^2\right]}{M_\text{A} E\left[s_{k\text{A}}^2\right] + E\left[n_{k\text{BT}}^2\right]} \tag{12.77}$$

By making μ small and adapting slowly, the misadjustment M_A can be made small. When the product of M_A times the signal power is negligible compared to the noise power, the maximum signal-to-noise ratio is obtained:

$$\text{SNR}_{\text{B}_{\text{max}}} = \frac{E\left[s_{k\text{A}}^2\right]}{E\left[n_{k\text{BT}}^2\right]} \tag{12.78}$$

This is the signal-to-noise ratio that one would obtain with an echo suppressing system in the absence of double talk or with an adaptive echo canceler in the absence of double talk. Because of misadjustment, the adaptive echo canceler with double talk will provide the SNR given by (12.77), which may be written as

$$\text{SNR}_\text{B} = \frac{\text{SNR}_{\text{B}_{\text{max}}}}{1 + M_\text{A} \cdot \text{SNR}_{\text{B}_{\text{max}}}} \tag{12.79}$$

Thus one should keep the misadjustment low in order to obtain an SNR close to the maximum during periods of double talk.

CANCELING ANTENNA SIDELOBE INTERFERENCE

Strong unwanted signals incident on the sidelobes of an antenna array can severely interfere with the reception of weaker signals in the main beam. When the number of spatially discrete interference sources is small, adaptive noise canceling can often provide a simple method of dealing with this problem.

To demonstrate the level of sidelobe reduction achievable with adaptive noise canceling, an interference canceling problem was simulated on the computer. The method is very similar to that of Howells and Applebaum [6, 35] described in the next chapter. As shown in Figure 12.32, an array consisting of a circular pattern of 16 equally spaced omnidirectional elements was chosen. The outputs of the elements were properly delayed and summed to form a main beam steered at a relative angle

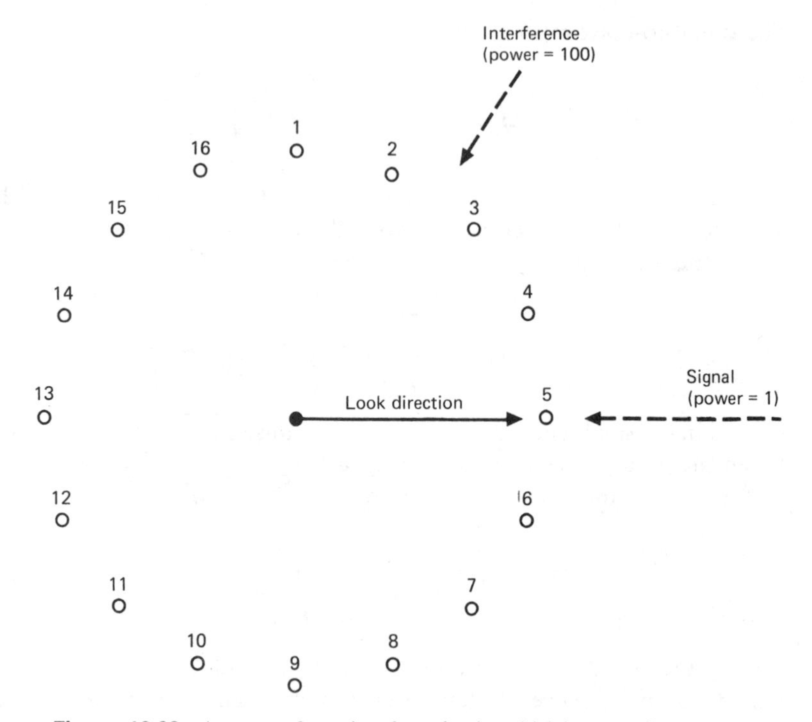

Figure 12.32 Array configuration for adaptive sidelobe canceling experiment. From B. Widrow et al., *Adaptive Noise Canceling: Principles and Applications*, © December 1975, IEEE.

of 0°. Simulated interference with the same bandwidth and with a power of 100 was incident at a relative angle of 58°. The array was connected to an adaptive noise canceler in the manner shown in Figure 12.5. The output of the beamformer served as the canceler's primary input, and the output of a single element (4) was chosen as the reference input. The canceler included an adaptive filter with 14 weights; the adaptation constant in the LMS algorithm was set at $\mu = 7 \times 10^{-6}$.

Figure 12.33 shows two series of computed directivity patterns, one representing a single frequency of $\frac{1}{4}$ of the sampling frequency and the other an average of eight frequencies of from $\frac{1}{8}$ to $\frac{3}{8}$ of the sampling frequency. These patterns indicate the evolution of the main beam and sidelobes as observed by stopping the adaptive process after the specified number of iterations. Note the deep nulls that develop in the direction of the interference, indicated by arrow. At the start of adaptation all weights were set at zero, providing a conventional 16-element beam pattern.

The signal-to-noise ratio at the system output, averaged over the eight frequencies, was found after convergence to be +20 dB. The signal-to-noise ratio at the single array element was −20 dB. This result bears out the expectation arising from (12.32) that the signal-to-noise ratio at the system output would be the reciprocal of the ratio at the reference input, which is derived from a single element.

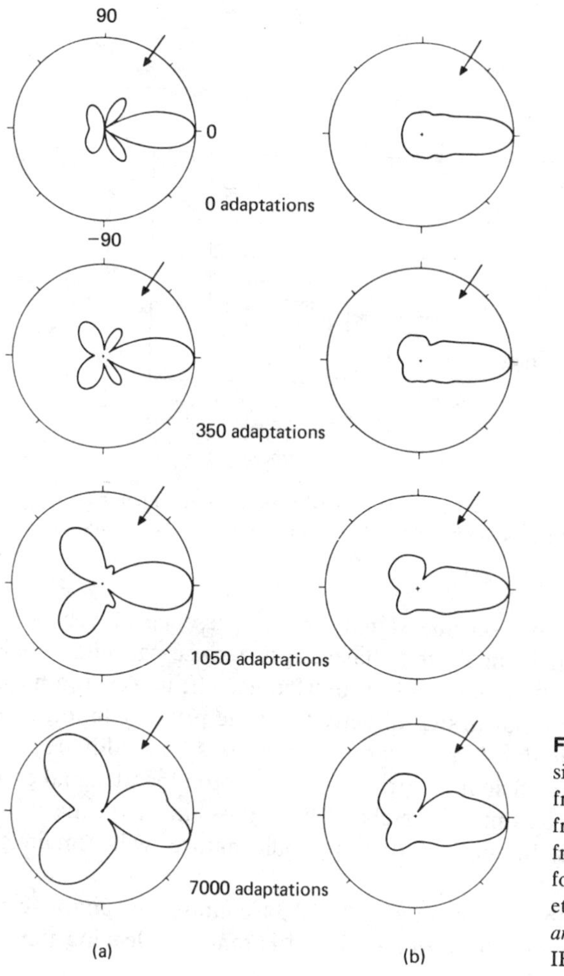

Figure 12.33 Results of adaptive sidelobe canceling experiment: (a) single frequency (0.5 relative to folding frequency); (b) average of eight frequencies (0.25 to 0.75 relative to folding frequency). From B. Widrow et al., *Adaptive Noise Canceling: Principles and Applications*, © December 1975, IEEE.

A small amount of signal cancellation occurred, as evidenced by the changes in sensitivity of the main beam in the direction of the signal (i.e., 0°). These changes were not unexpected, since the main-lobe pattern was not constrained by the adaptive process. A method of LMS adaptation with constraints that could have been used to prevent this loss of sensitivity has been developed by Frost [37] and is described in Chapter 13.

CANCELING PERIODIC INTERFERENCE WITH AN ADAPTIVE PREDICTOR

There are a number of circumstances where a broadband signal is corrupted by periodic interference and no external reference input free of the signal is available. Examples include the playback of speech or music in the presence of tape hum or

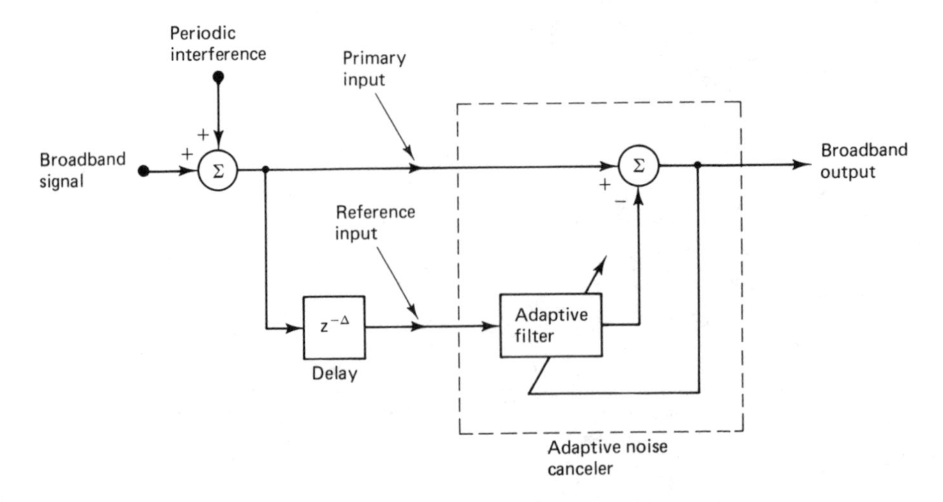

Figure 12.34 Canceling periodic interference without an external reference source. From B. Widrow et al., *Adaptive Noise Canceling: Principles and Applications*, © December 1975, IEEE.

turntable rumble, or sensing seismic signals in the presence of vehicle engine or powerline noise. It might seem at first that adaptive noise canceling could not be applied to reduce or eliminate this kind of interference. If, however, a fixed delay Δ is inserted in a reference input drawn directly from the primary input, as shown in Figure 12.34, the periodic interference can in many cases be readily canceled.† The delay chosen must be of sufficient length to cause the broadband signal components in the reference input to become decorrelated from those in the primary input. The interference components, because of their periodic nature, will remain correlated with each other.

One can see that the system in Figure 12.34 contains an adaptive predictor. The predictable component of the primary input is removed, leaving the unpredictable component at the output.

Figure 12.35 presents the results of a computer simulation performed to demonstrate the canceling of periodic interference without an external reference. Figure 12.35(a) shows the primary input to the canceler. This input is composed of colored Gaussian noise representing the signal and a sine wave representing the interference. Figure 12.35(b) shows the output at the noise canceler. Since the problem was simulated, the exact nature of the broadband input was known and is plotted together with the output. Note the close correspondence in form and registration. The correspondence is not perfect only because the filter was of finite length and had a finite rate of adaptation.

† The delay Δ may be inserted in the primary instead of the reference input if its total length is greater than the total delay of the adaptive filter. Otherwise, the filter will converge to match it and cancel both signal and interference.

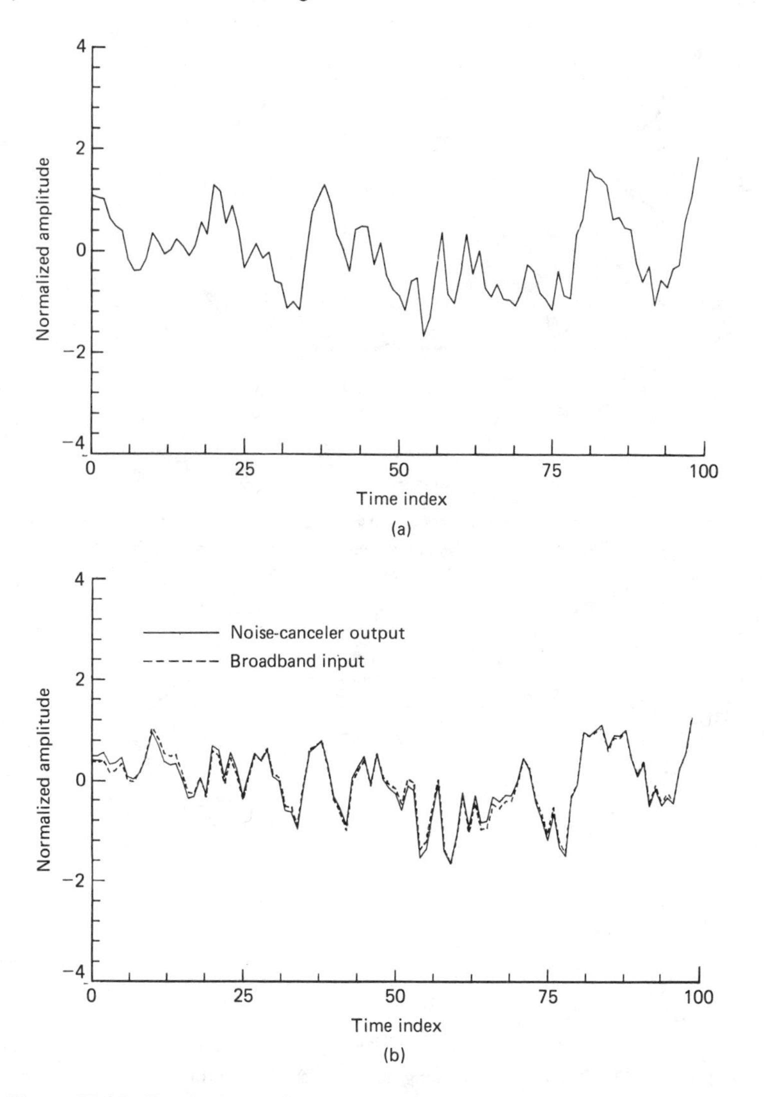

Figure 12.35 Result of periodic interference canceling experiment: (a) input signal (correlated Gaussian noise and sine wave); (b) noise-canceler output (correlated Gaussian noise). From B. Widrow et al., *Adaptive Noise Canceling: Principles and Applications*, © December 1975, IEEE.

AN ADAPTIVE SELF-TUNING FILTER

The previous experiment can also be used to demonstrate another important application of the adaptive noise canceler. In many instances where an input signal consisting of mixed periodic and broadband components is available, the periodic rather than the broadband components are of interest. If the system output of the noise canceler of Figure 12.34 is taken from the adaptive filter, the result is an

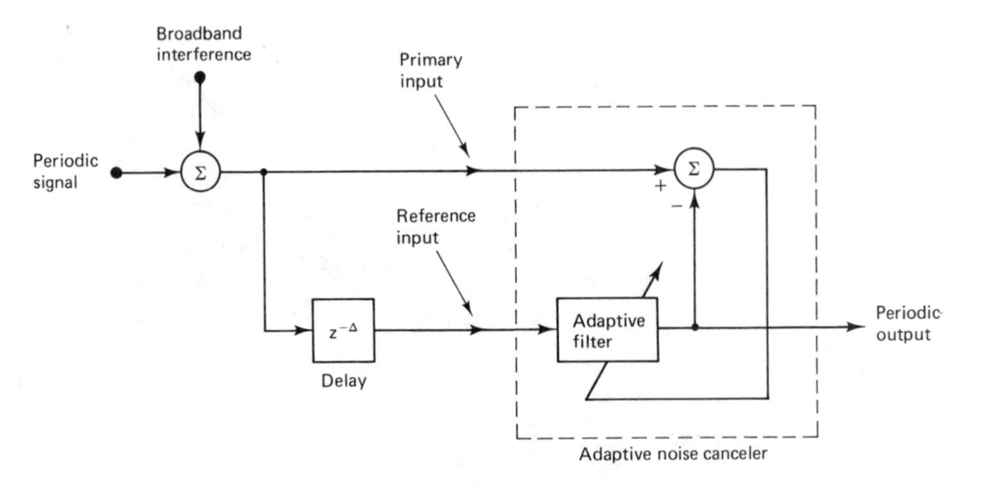

Figure 12.36 Adaptive noise canceler as a self-tuning filter. From B. Widrow et al., *Adaptive Noise Canceling: Principles and Applications*, © December 1975, IEEE.

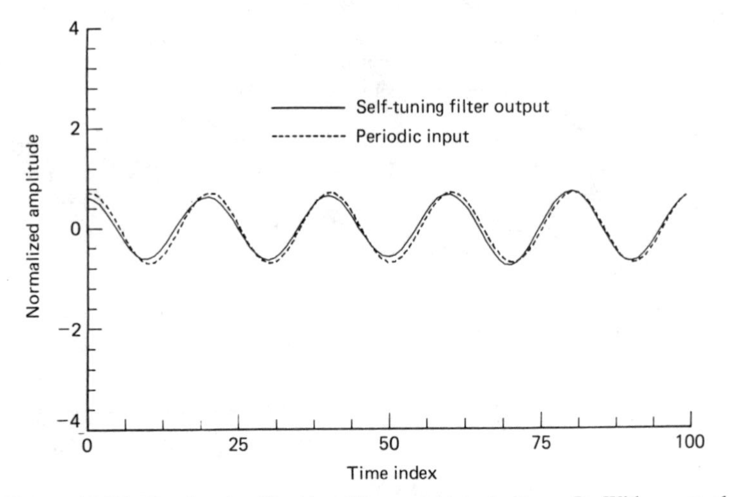

Figure 12.37 Result of self-tuning filter experiment. From B. Widrow et al., *Adaptive Noise Canceling: Principles and Applications*, © December 1975, IEEE.

adaptive self-tuning filter capable of extracting a periodic signal from broadband noise.

Figure 12.36 shows the adaptive noise canceler as a self-tuning filter. The output of this system was simulated on the computer with the input of sine wave and correlated Gaussian noise used in the previous experiment and shown in Figure 12.35(a). The resulting approximation of the input sine wave is shown in Figure 12.37 together with the actual input sine wave. Note once again the close agreement in form and registration. The error is a small-amplitude stochastic process.

Figure 12.38 shows the impulse response and transfer function of the adaptive filter after convergence. The impulse response, shown in Figure 12.38(a), closely

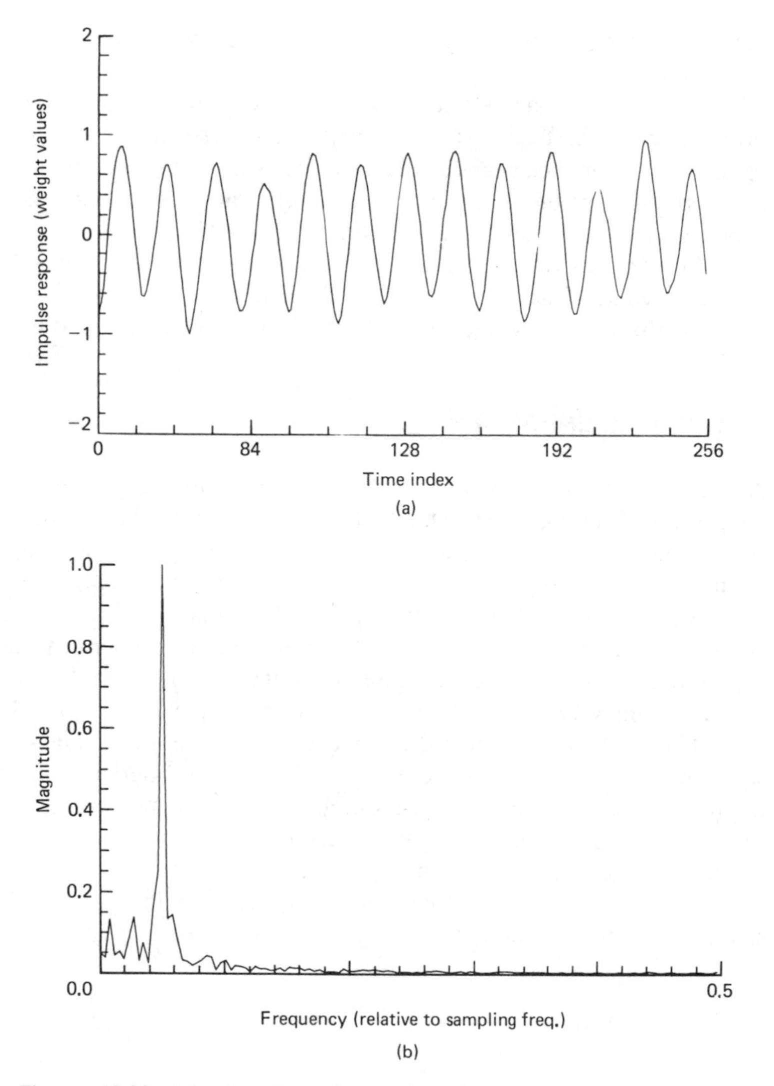

Figure 12.38 Adaptive filter characteristic in self-tuning filter experiment: (a) impulse response of adaptive filter after convergence; (b) magnitude of transfer function of adaptive filter after convergence. From B. Widrow et al., *Adaptive Noise Canceling: Principles and Applications*, © December 1975, IEEE.

resembles a sine wave. If the broadband input component had been white noise, the optimal estimator would have been a matched filter, and the impulse response would have been sinusoidal.

The transfer function, shown in Figure 12.38(b) is the digital Fourier transform of the impulse response. Its magnitude at the frequency of the interference is nearly 1, the value required for perfect cancellation. The phase shift at this frequency is not zero but when added to the phase shift caused by the delay Δ forms an integral multiple of 360°.

Similar experiments have been conducted with sums of sinusoidal signals in broadband stochastic interference. In these experiments the adaptive filter developed sharp resonance peaks at the frequencies of all the spectral line components of the periodic portion of the primary input. The system thus shows considerable promise as an automatic signal seeker.

Further experiments have shown the ability of the adaptive self-tuning filter to be employed as a line enhancer for the detection of extremely low-level sine waves in noise. An introductory treatment of this application is provided in the next section.

THE ADAPTIVE LINE ENHANCER

A classical detection problem is that of finding a low-level sine wave in noise. The adaptive self-tuning filter, whose ability to separate the periodic and stochastic components of a signal was illustrated above (where these components were of comparable level), is able to serve as an "adaptive line enhancer" for the detection of extremely low-level sine waves in noise. The adaptive line enhancer becomes a competitor of the fast Fourier transform algorithm as a sensitive detector and has capabilities that may exceed those of conventional spectral analyzers when the unknown sine wave has finite bandwidth or is frequency modulated.

The method is illustrated in Figure 12.39. The input consists of a sinusoidal signal plus some noise. The "output" is the discrete Fourier tranform of the filter's impulse response. Detection is accomplished when a spectral peak is evident above the background noise. A modified version of this method has been proposed by Griffiths [42] and is discussed in Exercise 17.

Note that the filter output (y) is also available in Figure 12.39, and that this signal could be used directly or as an input to a spectral analyzer or phase-locked

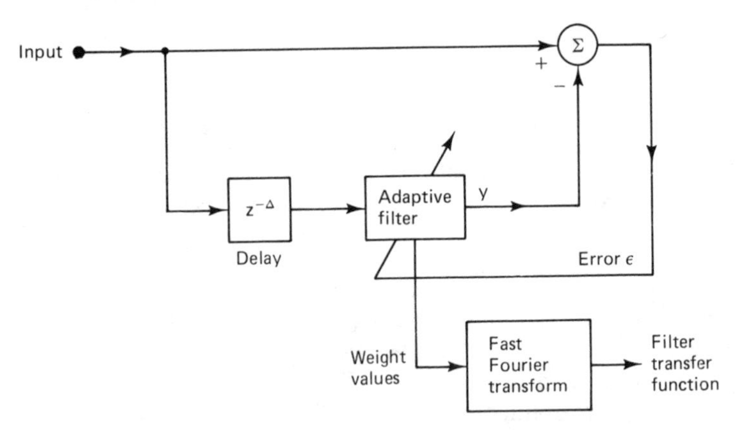

Figure 12.39 Adaptive line enhancer. From B. Widrow et al., *Adaptive Noise Canceling: Principles and Applications*, © December 1975, IEEE.

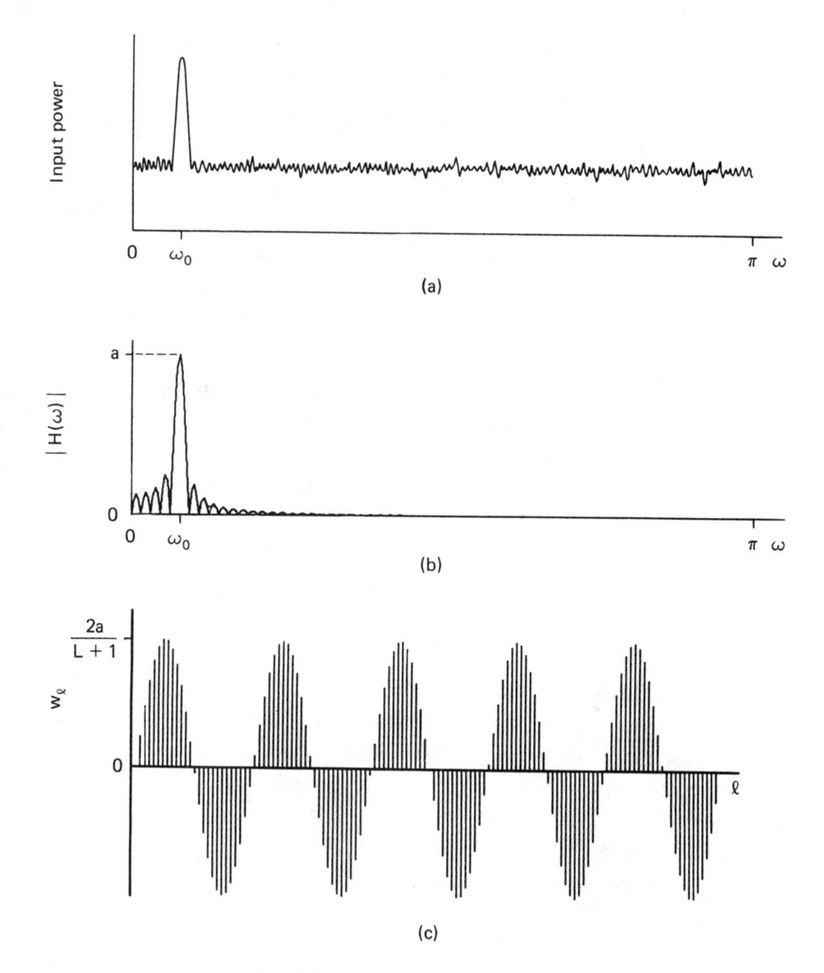

Figure 12.40 Ideal adaptive filter impulse response and transfer function of adaptive line enhancer for a given input spectrum: (a) input spectrum; (b) transfer function magnitude; (c) impulse response. From B. Widrow et al., *Adaptive Noise Canceling: Principles and Applications*, © December 1975, IEEE.

loop. Furthermore, the method of Figure 12.39 could be used for the simultaneous detection of multiple sine waves. We do not consider these possibilities here, but discuss only the detection of single low-level sine waves in noise.

Figure 12.40 shows the ideal impulse response and transfer function of the adaptive line enhancer for a given input spectrum. We assume that the input noise is white, with a total power of ν^2, and that the input signal has a power of $C^2/2$ at normalized frequency ω_0 [see (7.16)]. The ideal impulse response, equivalent to the matched filter response, is a sampled sinusoid whose frequency is ω_0.† The phase

† This assertion was proved analytically for arbitrary input signal-to-noise ratio in J. R. Zeidler and D. M. Chabries, "An Analysis of the LMS Adaptive Filter Used as a Spectral Line Enhancer," Naval Undersea Center, Tech. Note 1476, Feb. 1975.

shift of this response at frequency ω_0 when added to that of the delay is an integral multiple of 360°. If the peak value of the transfer function is a, the peak value of the weights is approximately $2a/(L+1)$, where $L+1$ is the number of weights.

The adaptive process minimizes the mean-square error, $E[\varepsilon_k^2]$. In the present case this is the sum of three components: the primary input noise power, the noise power at the output of the adaptive filter, and the sinusoidal signal power. Accordingly, the MSE may be expressed as

$$\text{MSE} = \nu^2 + \nu^2 \frac{2a^2}{L+1} + \frac{C^2}{2}(1-a)^2 \tag{12.80}$$

In this result ν^2 is the input noise power, the second term is ν^2 multiplied by the sum of the squared filter weights, and the third term is the power due to the input and filtered sine waves, which are assumed to subtract coherently.

The optimal value of a that minimizes error power, a^*, is obtained by setting the derivative of (12.80) with respect to a equal to zero:

$$a^* = \frac{\dfrac{C^2/2}{\nu^2}\dfrac{L+1}{2}}{1 + \dfrac{C^2/2}{\nu^2}\dfrac{L+1}{2}} = \frac{(\text{SNR})(L+1)/2}{1+(\text{SNR})(L+1)/2} \tag{12.81}$$

At high signal-to-noise ratios, a^* is close to 1. At low signal-to-noise ratios, a^* is less than 1. To keep a^* near 1 with low signal-to-noise ratios, one may use a large number of adaptive weights, although other problems could result because of weight-vector noise. The ability to detect peaks in the transfer function due to the presence of sinusoidal signals is limited by the presence of spurious peaks caused by noise in the weight vector. The methods of spectral analysis based on the use of the discrete Fourier transform (DFT) have analogous difficulties. Spurious spectral peaks develop and fade due to fluctuations in the input data. While present, such peaks may be confused with true signals.

In Figure 12.41 we compare some experimental results, obtained by computer simulation, of a discrete Fourier transform power spectral density measurement process and an adaptive line enhancer. In each of three cases, the same data were analyzed by DFT and then fed to the adaptive line enhancer. In Figure 12.41(a), the input noise was white and the signal was sinusoidal at $\frac{1}{8}$ times the sampling frequency. The signal-to-noise ratio was 0.01562. The DFT was the average of 256 transforms, each with 128 points. The signal peak is easily distinguished above the background noise baseline. The adaptive line enhancer with 128 weights produced comparable results. The time constant and number of weights were selected so that both the DFT and the adaptive line enhancer would have the same frequency resolution and would use the same amount of input data, allowing a critical but fair comparison between the two approaches. The signal peak that developed in the transfer function magnitude is easily seen above the baseline, which hovers slightly above zero. The ability to discern the real signal peak is about equal for both approaches. In Figure 12.41(b), colored noise is present in the data to be analyzed.

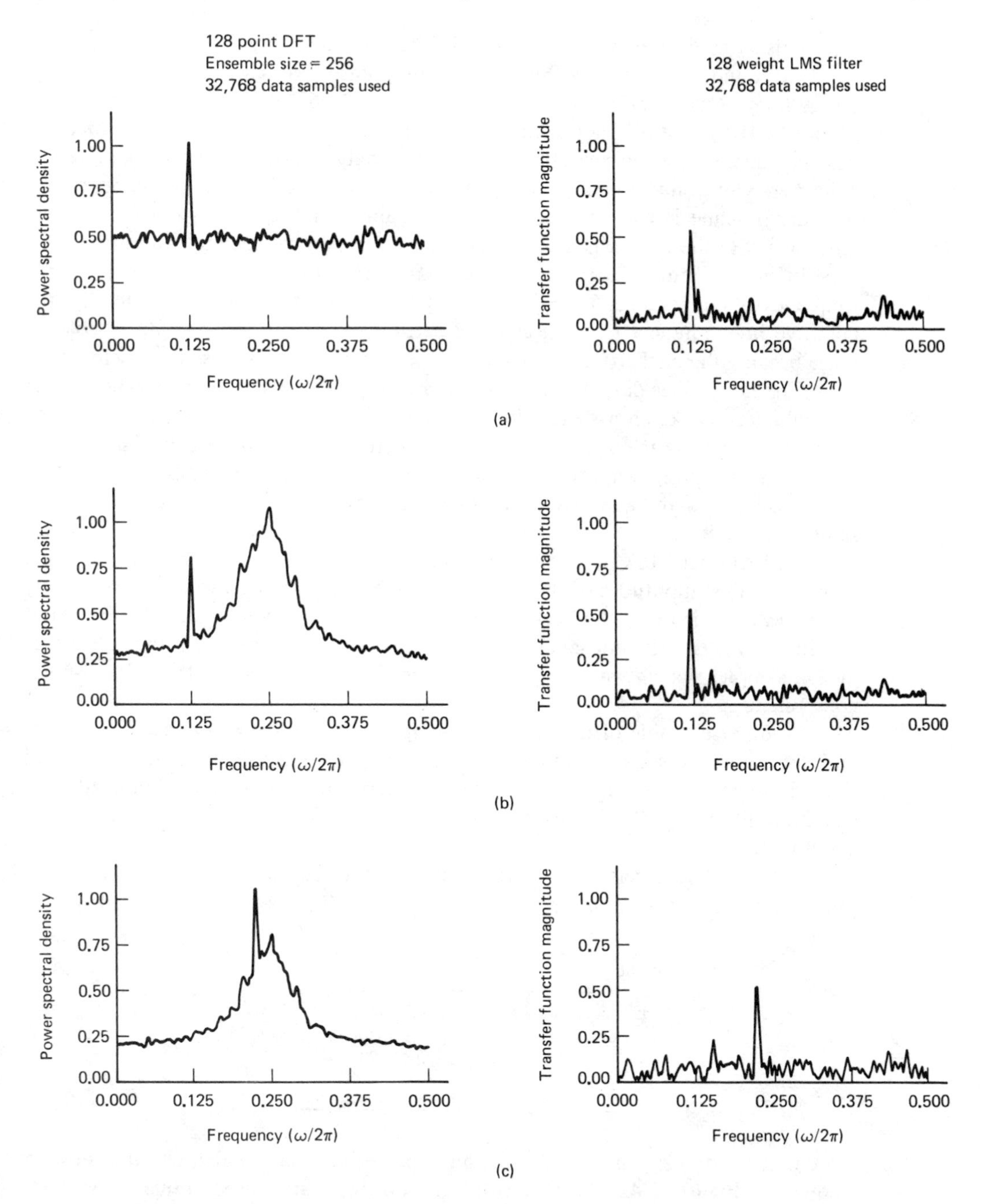

Figure 12.41 Classical spectral analysis (DFT)—left, compared with adaptive line enhancing—right: (a) single line in white noise; (b) single line in 50 percent white, 50 percent colored noise; (c) single line in 50 percent white, 50 percent colored noise; different signal frequency. From Proceedings of IEEE letter, January 1977.

This noise was 50% white, 50% colored. The colored component had 25 percent bandwidth and was generated by passing white noise through a four-pole bandpass digital filter. Again, the signal-to-noise ratio was 0.01562. The signal peak is seen in Figure 12.41(b) on a rolling baseline when analysis is done by the DFT, and on a flat baseline, again hovering just above zero, when analysis is done by the adaptive line enhancer. Our ability to detect the signal above the noise is about the same, except that the baseline is close to zero with the adaptive line enhancer and completely unknown with the DFT when the input noise has unknown spectral coloration. The experiment of Figure 12.41(c) is similar to the previous case except that the signal frequency is much closer to that of the spectral peak of the noise. Generally, the adaptive line enhancer is the preferable detector when the input contains colored noise because the background baseline is close to zero and thus easy to anticipate.

The DFT takes data in blocks and uniformly weights the samples, whereas the line enhancer works on a steady-flow basis and exponentially weights the data over time. In comparing the "data consumption" of the two techniques, the data "used" by the line enhancer were defined as the number of samples processed during four time constants of the adaptive process. The total amount of data used was 32,768 samples in each case.

Inspection of the plots of Figure 12.41, all drawn to the same linear scale, shows that the amplitudes of the signal components (the spikes) were approximately the same, and also that the background noise levels were similar, for the DFT and the line enhancer. The line enhancer in each case was implemented with its delay set at 256 samples, which was adequate to decorrelate and eliminate the colored-noise components.

Another example of the use of the adaptive line enhancer in spectral analysis and detection of weak signals in the presence of both strong signals and background noise is illustrated in Figure 12.42, where a large-amplitude signal summed with small-amplitude signals in noise causes difficulties in detecting and/or resolving the small signals.

Figure 12.42(a) shows the formation of "input A" as the sum of white noise plus three sinusoidal signals as follows:

Input	Frequency (ω)	Power
1. Sinusoid	0.17969π	125.0
2. Sinusoid	0.15625π	0.125
3. Sinusoid	0.42187π	0.5
4. Random noise	White	1.0

Note that the first signal is 1000 times more powerful than signal 2, which is close in frequency. Figure 12.42(b) is a block diagram of the system used in this experiment. Its primary input is indicated by A and its outputs are B and C, which represent, respectively, the "error" of the adaptive process and the adaptive-filter output. An additional output is the weight vector of the adaptive filter, comprising its impulse

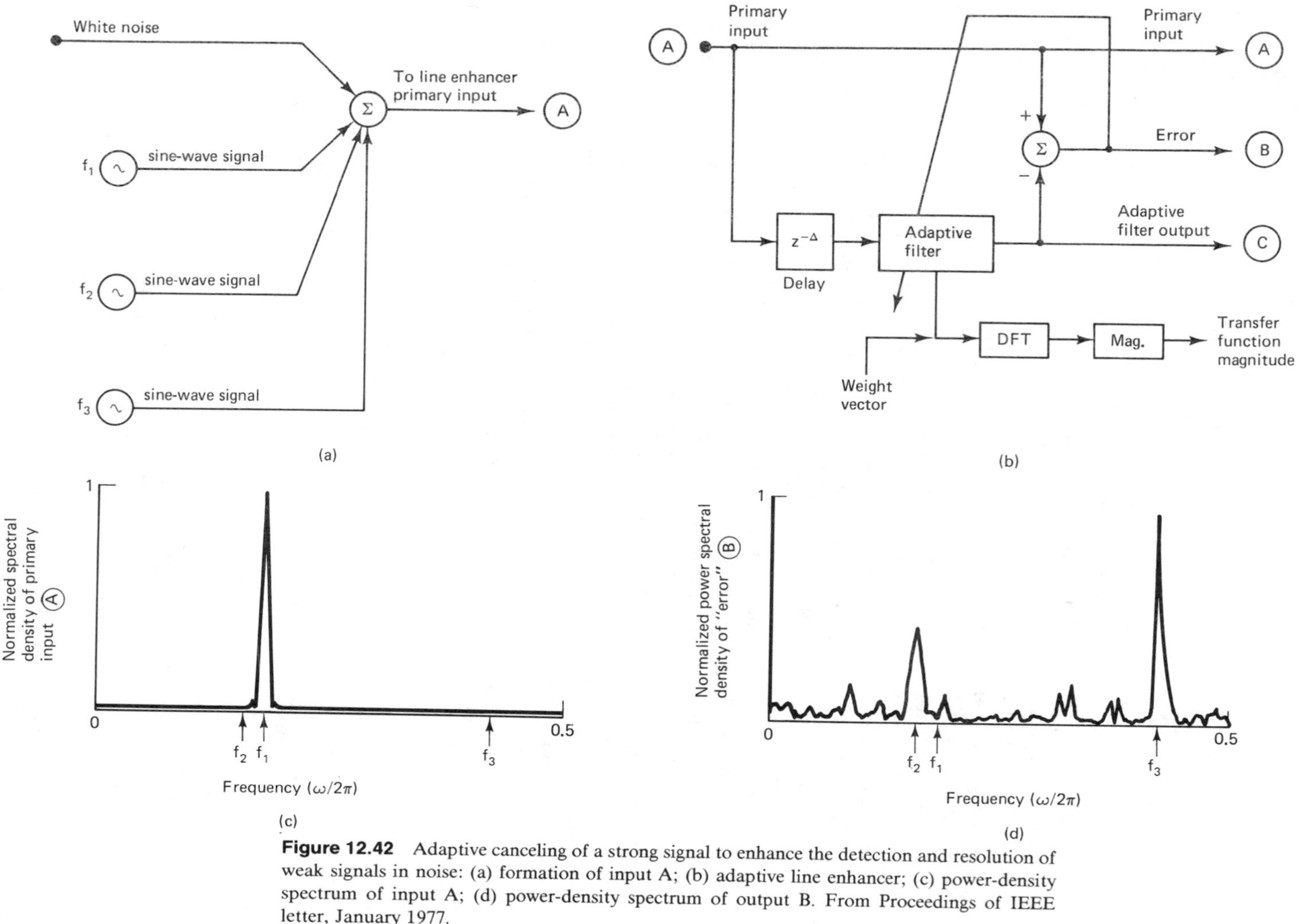

Figure 12.42 Adaptive canceling of a strong signal to enhance the detection and resolution of weak signals in noise: (a) formation of input A; (b) adaptive line enhancer; (c) power-density spectrum of input A; (d) power-density spectrum of output B. From Proceedings of IEEE letter, January 1977.

response, which can be transformed to provide a transfer function as discussed above.

Figure 12.42(c) is a linear plot of the DFT power spectrum of input A. The frequencies of the three signals are indicated by the arrows. Notice that signal 2 is not resolvable; it is buried in the second sidelobe of signal 1. Even if this spectrum were plotted on a log scale, signal 3 would be smaller in amplitude than many of the sidelobes of signal 1, and signal 2 would be undetectable.

When the DFT of the error, B in Figure 12.42(b), is taken instead, a power spectrum is obtained that clearly shows the weak signal at frequency f_2. This spectrum, plotted on a linear scale in Figure 12.42(d), also shows the weak signal at frequency f_3 and the broadband background noise. The strong signal 1 was totally canceled as a result of the adaptive process.

The plots of Figure 12.42(c) and (d) were normalized so that full scale corresponds to the largest-amplitude point of each plot. The spectra of outputs A and B were taken from 128 data points, with no ensemble averaging.

The 64-weight adaptive filter in this example canceled the strong signal within about five cycles of frequency f_1 (i.e., within about 30 sample periods). The DFT of error B was then taken so that the amount of data used in forming it was only slightly greater than the quantity of data used in forming the DFT of input A. In this case the line enhancer was used in a way that caused the strong signal to cancel itself, thus improving the capability of the DFT to resolve weak signals when they are close in frequency to a strong interference. An equivalent result can be obtained using the DFT alone, but generally more data are required. The principle illustrated in Figure 12.42 could also be used in a radio receiver to enhance reception of weak signals that would otherwise be overwhelmed by strong signals adjacent in frequency. One approach is diagrammed in Figure 12.43.

We have seen that the adaptive line enhancer can be used as an alternative to the DFT to detect and estimate weak signals in noise, and that it also provides useful

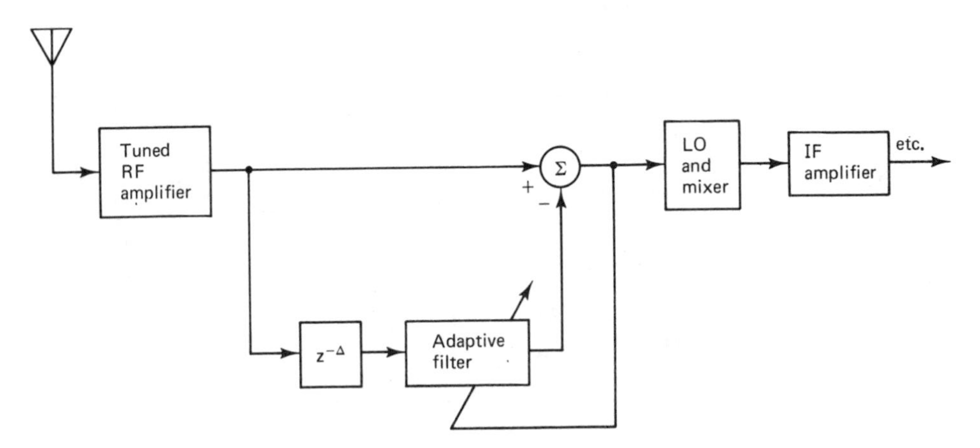

Figure 12.43 Radio receiver with an adaptive strong-signal eliminator.

output signals and can function as a self-tuning filter to tune in or tune out other signals automatically. It embodies a methodology for spectral analysis and is related to the maximum entropy technique [42, 43]. The line enhancer is entirely different in structure from the DFT, and is more easily implemented in some cases. Research is currently proceeding toward a more extensive understanding of the behavior of the line enhancer with inputs of the types described above.

CONCLUSION

Adaptive noise canceling is a method of optimal filtering that can be applied whenever a suitable reference input is available. The principal advantages of the method are its adaptive capability, its low output noise, and its low signal distortion. The adaptive capability allows the processing of inputs whose properties are unknown and in some cases nonstationary. It leads to a stable system that automatically turns itself off when no improvement in the signal-to-noise ratio can be achieved. The output noise and signal distortion are generally lower than that achievable with conventional optimal filter configurations.

The examples presented in this chapter demonstrate the ability of adaptive noise canceling to reduce additive periodic or stationary random interference in both periodic and random signals. In each instance canceling was accomplished with little signal distortion even though the frequencies of the signal and the interference overlapped. The examples indicate the wide range of applications in which adaptive noise canceling has potential use.

EXERCISES

1. In a given interference canceler, suppose that the reference signal is derived as shown below. If the independent noise m_{1k} is zero, what is the optimal $W^*(z)$?

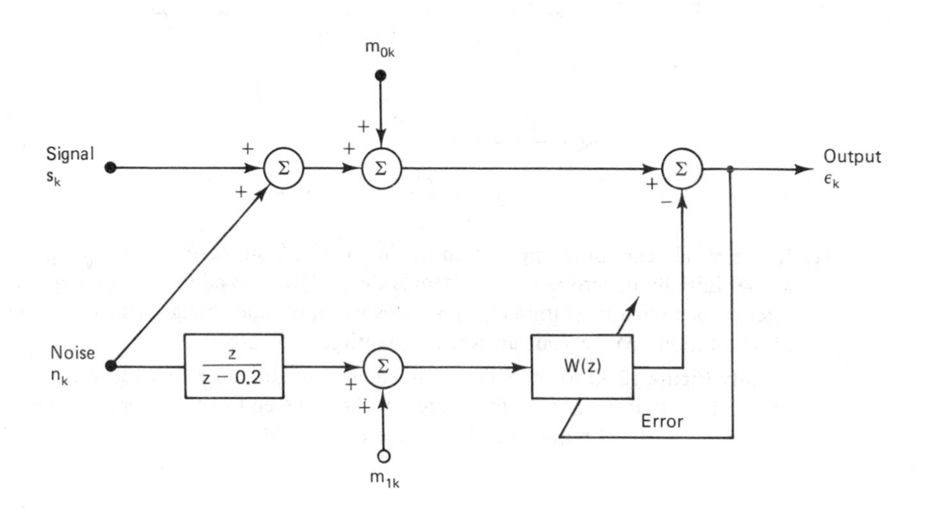

2. In the diagram for Exercise 1, suppose that n, m_0, and m_1 are independent white noise processes with total power equal to N, M_0, and M_1, respectively. Find a rational expression for the optimal transfer function, $W^*(z)$.
3. Using the data in Exercise 2, express the ratio of the output SNR to the primary SNR.
4. How is the answer to Exercise 3 affected when m_0 is insignificant, when m_1 is insignificant, and when both m_0 and m_1 are insignificant?
5. In the figure below we have the same situation as above, but with $m_0 = m_1 = 0$ and a fraction of the signal leaking into the reference channel. Given that n_k is independent white noise with total power N, express the optimal $W^*(z)$ in terms of the input power spectrum $\Phi_{ss}(z)$.

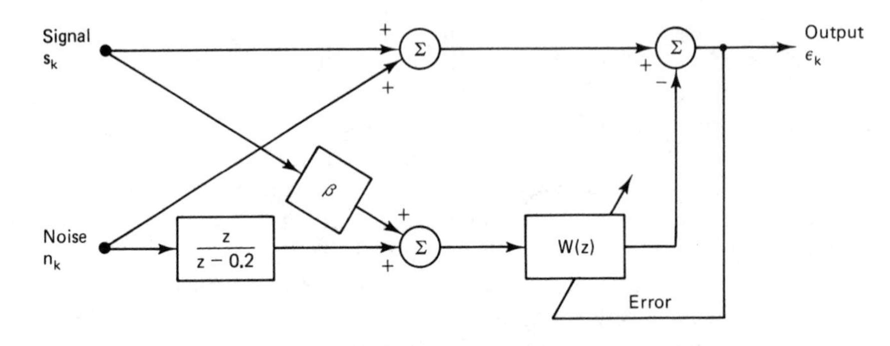

6. In Exercise 5, what is the output SNR in terms of the input power spectrum, $\Phi_{ss}(z)$?
7. In Exercise 5, what is the "signal distortion," $D(z)$?
8. In Exercise 5, let s_k be a sine wave with 16 samples per cycle, let $\beta = 0$, and let the input SNR = 1. Using an adaptive linear combiner with three weights and the LMS algorithm with $\mu = 0.03$, plot an experimental learning curve as the average of 100 runs. Estimate the output SNR from the learning curve.
9. Do Exercise 8 with 10 percent leakage (i.e., $\beta = 0.1$).
10. What is the equivalent transfer function of the diagram below?

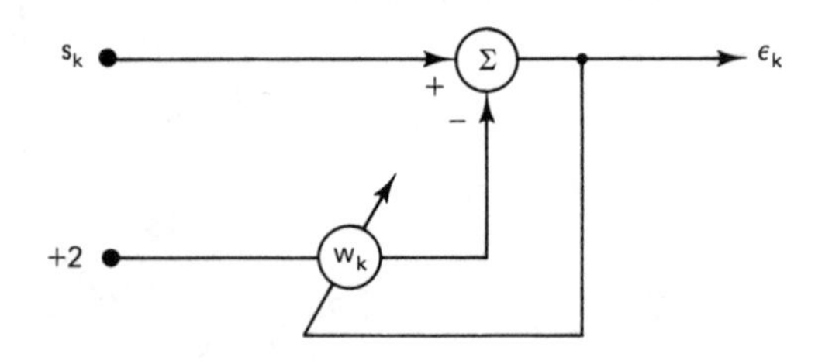

11. In the 60-Hz canceling application in Figure 12.14, suppose that the weights w_1 and w_2 are set initially to zero and the system is then allowed to adapt. Assuming that the 60-Hz interference remains stationary, how does its amplitude change with time from the onset of adaptation? Make your answer as quantitative as possible.
12. Modify Figure 12.30 to show how you would configure an echo canceling system to allow both cancelers to be physically located at the same end of the telephone circuit. Assume that one-way travel requires 270 milliseconds.

13. Assume that noise and signal power levels are symmetric in both directions in Figure 12.31. Let the sampling rate be 8 kHz and let the adaptive filters have 128 weights each. Assume that local loop noises are negligible, that trunk noises have power levels of 1, and that microphone signal levels have power of 5. Choose a value of μ so that the adaptive process will converge as rapidly as possible but that the resulting SNR will be at least half as great as the maximum possible SNR. For this value of μ, find τ_{mse}. Assume that all eigenvalues are equal.

14. The figure below shows a block diagram of a canceling system that allows a transmission line or a coaxial cable to carry signals in two directions simultaneously without using modulated carriers. The characteristic impedance of the cable is R_c. Each operational amplifier is assumed to have a zero output impedance and an infinite input impedance. Both ends of the transmission line are properly terminated to prevent reflections.
 (a) Explain how the single-weight adaptive canceler works.
 (b) Calculate the optimal value of the weight w for the nonadaptive canceler. In practice, what would be troubling about the nonadaptive approach in spite of the fact that it would be simpler to implement than the adaptive approach? (*Note:* Normally, one would use an adaptive canceler on both ends of the cable.)

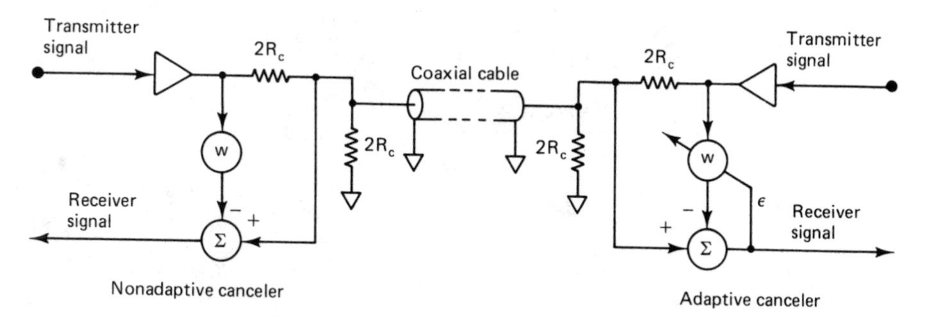

15. Using the signal transmission methodology suggested by the system of Exercise 14, design a "repeater" amplifying system that would allow two-way signal transmission. Start with the diagram below and fill in the missing components. Use the adaptive canceling technology suggested in Exercise 14.

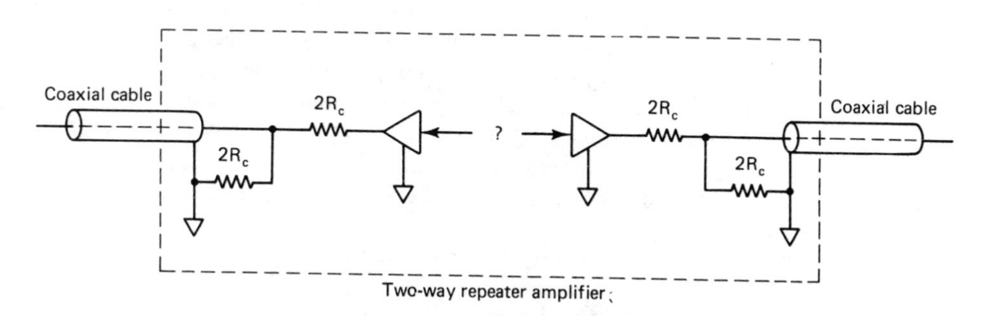

16. The figure below describes a system used in earthquake prediction. The primary sensor, geophone 1 (a microphone for receiving noise and vibration from the earth), is placed close to the San Andreas fault in California to pick up the signal, consisting of rock-

crushing noise which may be indicative of an impending earthquake. But geophone 1 also picks up independent seismic noise plus a strong directional noise due to auto traffic, and so on, from the city of San Jose. A second identical sensor, geophone 2, is placed on a straight-line path between geophone 1 and the city of San Jose. Geophone 2 is far enough away from the fault so that it does not pick up rock crushing, but it does pick up San Jose noise plus independent seismic noise. The Wiener filter attempts to cancel the San Jose noise from the output of geophone 1, hopefully without adding too much earth noise to it. The outputs of geophones 1 and 2 contain mutually uncorrelated white noises, both of variance $0.25\ \sigma^2$. The San Jose noise picked up by both geophones has, after sampling, an autocorrelation of $\sigma^2(0.9)^{|k|}$. The delay in propagation of San Jose noise from geophone 2 to geophone 1 is two sampling periods.

(a) Design the digital Wiener filter to minimize the noise in the output. A noncausal filter is permissible.

(b) In the absence of signal, what is the total output noise if the Wiener filter is disconnected?

(c) In the absence of signal, what is the total output noise when the Wiener filter is installed as shown?

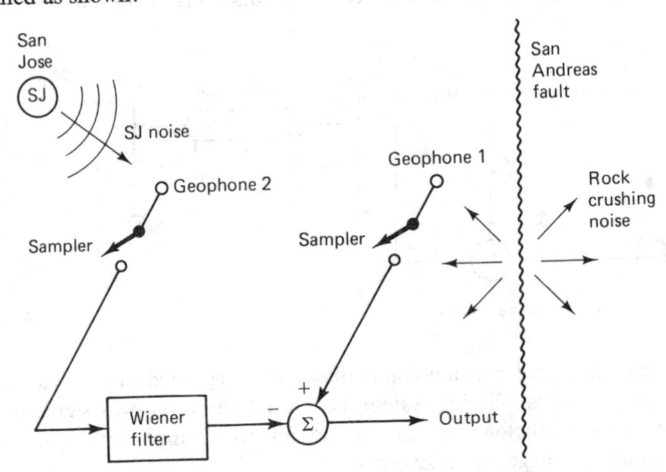

17. In the adaptive line enhancer (Figure 12.39), suppose that the input consists of two sine waves, one with 16 and one with 17 samples per cycle, plus additive white noise with SNR = 0.25. Using $\Delta = 1$, $L = 63$, and $\mu = 0.04$, run the line enhancer for 1000 iterations and then plot the "instantaneous spectrum," that is, the magnitude of the filter transfer function. Use the LMS algorithm and generate the noise as in Appendix A.

18. A modified instantaneous spectral measure, also known as "maximum entropy spectral estimation" [42, 43, 53–56], used with the adaptive line enhancer is

$$\text{instantaneous spectrum} = \frac{\xi_{\min}}{|H(\omega)|^2}$$

where $H(\omega)$ is the overall transfer function, from input to error output, of the one-step ($\Delta = 1$) adaptive predictor in Figure 12.39. The scale factor $\xi_{\min}$ is generally unknown, but this function often gives a sharper indication of the spectral line than the function in Figure 12.39. Do Exercise 17 using this function with $\xi_{\min} = 1$ and compare results.

REFERENCES AND ADDITIONAL READINGS

1. N. Wiener, *Extrapolation, Interpolation and Smoothing of Stationary Time Series, with Engineering Applications*. New York: Wiley, 1949.
2. H. Bode and C. Shannon, "A simplified derivation of linear least squares smoothing and prediction theory," *Proc. IRE*, vol. 38, pp. 417–425, Apr. 1950.
3. R. Kalman, "On the general theory of control," in *Proc. First IFAC Congress*. London: Butterworth, 1960.
4. R. Kalman and R. Bucy, "New results in linear filtering and prediction theory," *Trans. ASME, Ser. D. J. Basic Eng.*, vol. 83, pp. 95–107, Dec. 1961.
5. T. Kailath, "A view of three decades of linear filtering theory,"*IEEE Trans. Inf. Theory*, vol. IT-20, pp. 145–181, Mar. 1974.
6. P. Howells, "Intermediate frequency side-lobe canceller," U.S. Patent 3,202,990, Aug. 24, 1965.
7. B. Widrow and M. Hoff, Jr., "Adaptive switching circuits," *IRE WESCON Conv. Rec.*, pt. 4, pp. 96–104, 1960.
8. J. Koford and G. Groner, "The use of an adaptive threshold element to design a linear optimal pattern classifier," *IEEE Trans. Inf. Theory*, vol. IT-12, pp. 42–50, Jan. 1966.
9. F. Rosenblatt, "The Perceptron: A perceiving and recognizing automaton, Project PARA," Cornell Aeronaut. Lab., Rep. 85-460-1, Jan. 1957.
10. F. Rosenblatt, *Principles of Neurodynamics: Perceptrons and the Theory of Brain Mechanisms*. Washington, D.C.: Spartan Books, 1961.
11. N. Nilsson, *Learning Machines*. New York: McGraw-Hill, 1965.
12. D. Gabor, W. P. L. Wilby, and R. Woodcock, "A universal nonlinear filter predictor and simulator which optimizes itself by a learning process," *Proc. Inst. Electr. Eng.*, vol. 108B, July 1960.
13. R. Lucky, "Automatic equalization for digital communication," *Bell Syst. Tech. J.*, vol. 44, pp. 547–588, Apr. 1965.
14. R. Lucky, J. Salz, and E. J. Weldon, Jr., *Principles of Data Communication*. New York: McGraw-Hill, 1968.
15. J. Kaunitz, "Adaptive filtering of broadband signals as applied to noise cancelling," Stanford Electronics Lab., Stanford Univ., Stanford, Calif., Rep. SU-SEL-72-038, Aug. 1972 (Ph.D. dissertation).
16. M. Sondhi, "An adaptive echo canceller," *Bell Syst. Tech. J.*, vol. 46, pp. 497–511, Mar. 1967.
17. J. Rosenberger and E. Thomas, "Performance of an adaptive echo canceller operating in a noisy, linear, time-invariant environment," *Bell Syst. Tech. J.*, vol. 50, pp. 785–813, Mar 1971.
18. R. Riegler and R. Compton, Jr., "An adaptive array for interference rejection," *Proc. IEEE*, vol. 61, pp. 748–758, June 1973.
19. B. Widrow, P. Mantey, L. Griffiths, and B. Goode, "Adaptive antenna systems," *Proc. IEEE*, vol. 55, pp. 2143–2159, Dec. 1967.

20. B. Widrow, "Adaptive filters," in *Aspects of Network and System Theory*, R. Kalman and N. DeClaris (Eds.). New York: Holt, Rinehart and Winston, pp. 563–587, 1971.

21. J. Glover, "Adaptive noise cancelling of sinusoidal interferences," Stanford Univ., Stanford, Calif., May 1975. (Ph.D. dissertation).

22. J. C. Huhta and J. G. Webster, "60-Hz interference in electrocardiography," *IEEE Trans. Biomed. Eng.*, vol. BME-20, pp. 91–101, Mar. 1973.

23. W. Adams and P. Moulder, "Anatomy of heart," in *Encycl. Britannica*, vol. 11, pp. 219–229, 1971.

24. G. von Anrep and L. Arey, "Circulation of blood," in *Encycl. Britannica*, vol. 5, pp. 783–797, 1971.

25. R. R. Lower, R. C. Stofer, and N. E. Shumway, "Homovital transplantation of the heart," *J. Thoracic Cardiovasc. Surg.*, vol. 41, p. 196, 1961.

26. T. Buxton, I. Hsu, and R. Barter, "Fetal electrocardiography," *JAMA*, vol. 185, pp. 441–444, Aug. 10, 1963.

27. J. Roche and E. Hon, "The fetal electrocardiogram," *Am. J. Obstet. Gynecol.*, vol. 92, pp. 1149–1159, Aug. 15, 1965.

28. S. Yeh, L. Betyar, and E. Hon, "Computer diagnosis of fetal heart rate patterns," *Am. J. Obstet. Gynecol.*, vol. 114, pp. 890–897, Dec. 1, 1972.

29. E. Hon and S. Lee, "Noise reduction in fetal electrocardiography," *Am. J. Obstet. Gynecol.*, vol. 87, pp. 1087–1096. Dec. 15, 1963.

30. J. Van Bemmel, "Detection of weak foetal electrocardiograms by autocorrelation and crosscorrelation of envelopes," *IEEE Trans. Biomed. Eng.*, vol. BME-15, pp. 17–23, Jan. 1968.

31. J. R. Cox, Jr., and L. N. Medgyesi-Mitschang, "An algorithmic approach to signal estimation useful in fetal electrocardiography," *IEEE Trans. Biomed. Eng.*, vol. BME-16, pp. 215–219, July 1969.

32. J. Van Bemmel, L. Peeters, and S. Hengeveld, "Influence of the maternal ECG on the abdominal fetal ECG complex," *Am. J. Obstet. Gynecol.*, vol. 102, pp. 556–562, Oct. 15, 1968.

33. W. Walden and S. Birnbaum, "Fetal electrocardiography with cancellation of maternal complexes." *Am. J. Obstet. Gynecol.*, vol. 94, pp. 596–598, Feb. 15, 1966.

34. J. Capon, R. J. Greenfield, and R. J. Kolker, "Multidimensional maximum likelihood processing of a large aperture seismic array," *Proc. IEEE*, vol. 55, pp. 192–211, Feb. 1967.

35. S. P. Applebaum, "Adaptive arrays," Special Projects Lab., Syracuse Univ. Res. Corp., Rep. SPL 769.

36. L. J. Griffiths, "A simple adaptive algorithm for real-time processing in antenna arrays," *Proc. IEEE*, vol. 57, pp. 1696–1704, Oct. 1969.

37. O. L. Frost III, "An algorithm for linearly constrained adaptive array processing," *Proc. IEEE*, vol. 60, pp. 926–935, Aug. 1972.

38. K. Senne, "Adaptive linear discrete-time estimation," Stanford Electronics Lab., Stanford Univ., Stanford, Calif., Rep. SEL-68-090, June 1968 (Ph.D. dissertation).

39. T. Daniell, "Adaptive estimation with mutually correlated training samples," Stanford Electronics Lab., Stanford Univ., Stanford, Calif., Rep. SEL-68-083, Aug. 1968 (Ph.D. dissertation).
40. J. K. Kim and L. D. Davisson, "Adaptive linear estimation for stationary M-dependent processes," *IEEE Trans. Inf. Theory*, vol. IT-21, pp. 23–31, Jan. 1975.
41. B. Widrow, "Adaptive filters 1: Fundamentals," Stanford Electronics Lab., Stanford Univ., Stanford, Calif., Rep. SU-SEL-66-126, Dec. 1966.
42. L. J. Griffiths, "Rapid measurement of instantaneous frequency," *IEEE Trans. Acoust. Speech Signal Process.*, vol. ASSP-23, pp. 209–222, Apr. 1975.
43. J. P. Burg, "Maximum entropy spectral analysis," presented at the 37th Annual Meeting, Soc. Exploration Geophysicists, Oklahoma City, Okla., 1967.
44. P. M. Woodward, *Probability and Information Theory with Applications to Radar*, 2nd ed. London: Pergamon Press, 1964.
45. M. I. Skolnik, *Introduction to Radar Systems*. New York: McGraw-Hill, 1962.
46. P. Swerling, "Probability of detection for fluctuating targets," *IRE Trans. Inf. Theory*, vol. IT-6, pp. 269–308, Apr. 1960.
47. J. I. Marcum, "A statistical theory of target detection by pulsed radar: Mathematical appendix," *IRE Trans. Inf. Theory*, vol. IT-6, pp. 145–267, Apr. 1960.
48. J. R. Glover, "Adaptive noise cancelling applied to sinusoidal interferences," *IEEE Trans. Acoust. Speech Signal Process.*, vol. ASSP-25, p. 484, Dec. 1977.
49. M. J. Shensha, "Non-Wiener solutions of the adaptive noise canceller with a noisy reference," *IEEE Trans. Acoust. Speech Signal Process.*, vol. ASSP-28, p. 468, Aug. 1980.
50. T. R. Rosenburger and E. J. Thomas, "Performance of an adaptive echo canceller operating in a noisy, linear time-invariant environment," *Bell Syst. Tech. J.*, vol. 50, p. 785, Mar. 1971.
51. J. R. Zeidler et al., "Adaptive enhancement of multiple sinusoids in uncorrelated noise," *IEEE Trans. Acoust. Speech Signal Process.*, vol. ASSP-26, p. 240, June 1978.
52. D. R. Morgan, "Response of a delay-constrained adaptive linear predictor filter to a sinusoid in white noise," *Proc.* 1981 *ICASSP*, p. 271.
53. J. A. Edward and M. M. Fitelson, "Notes on maximum-entropy processing," *IEEE Trans. Inf. Theory*, vol. IT-19, p. 232, Mar. 1973.
54. R. T. Lacoss, "Data adaptive spectral analysis methods," *Geophysics*, p. 661, Aug. 1971.
55. A. Van Den Bos, "Alternative interpretation of maximum entropy spectral analysis," *IEEE Trans. Inf. Theory*, vol. IT-17, p. 493, July 1971.
56. A. Popoulis, "Maximum entropy and spectral estimation: a review," *IEEE Trans. Acoust. Speech Signal Process.*, vol. ASSP-29, p. 1176, Dec. 1981.
57. M. R. Sambur, "Adaptive noise cancelling for speech signals," *IEEE Trans. Acoust. Speech Signal Process.*, vol. ASSP-26, p. 419, Oct. 1978.

chapter 13

Introduction to Adaptive Arrays and Adaptive Beamforming

So far, we have applied adaptive techniques to the problems of signal processing or signal conditioning in the time and frequency domains. In these final chapters, we develop further the idea of adaptive signal conditioning in the spatial domain. In order to achieve spatial selectivity, it is necessary to receive signals with an antenna array comprising two or more independent sensor elements that are spatially separated. Depending on the application, the antenna elements could be dipoles intended for the reception of electromagnetic signals, they could be hydrophones placed in the ocean to receive acoustic signals, they could be seismometers or geophones buried in the earth listening for seismic signals, or they could be other types of sensors. Many kinds of sensor elements exist for the reception of wavelike signals, and these sensors can be mounted in a spatial array to achieve signal reception with variable directional sensitivity. We will be concerned primarily with adaptive receiving arrays, but adaptive transmitting arrays are also of interest.

Adaptive arrays are essentially multichannel adaptive signal processors, which were discussed in Chapter 12. In one practical example (Figure 12.19), an array of maternal chest-lead reference inputs was applied to a bank of adaptive filters for conditioning, combining, and subtraction from the abdominal-lead primary input. The result was a subtraction of the maternal ECG interference using an adaptive array.

An adaptive antenna or adaptive beamforming system consists of a set of spatially disposed sensor elements connected to a single-channel or to a multichannel adaptive signal processor. It is natural to discuss adaptive arrays for all types of signals (electromagnetic, acoustic, seismic) at the same time. Adding the spatial dimension to the signal-processing picture introduces new capabilities leading to a wide range of unusual applications and unusual algorithms. It is impossible to describe them all here, because the technology is rapidly evolving. Instead, we will focus on several basic concepts and present a detailed bibliography to the current literature.

Adaptive receiving arrays can be used to reduce or eliminate directional interference by adaptive canceling or adaptive nulling, thus improving the signal-to-noise ratio. Using different kinds of adaptive algorithms, adaptive receiving arrays can "self-cohere," that is, steer themselves automatically to pick out the signal without knowing beforehand its direction of arrival, and separate this signal from directional interferences as long as their directions of arrival are different from that of the signal. Using other adaptive algorithms, it is possible to determine the direction of arrival of the signal in the presence of interference. Using still other algorithms, one can separate weak signals from strong signals as long as the signals have different arrival angles. Adaptive receiving arrays can be made sensitive to signals originating nearby and highly insensitive to distantly arriving signals, or vice versa. Or they can be made very sensitive to infrequent transient signals and insensitive to frequent or stationary signals, or vice versa. The number of possibilities is very large. We will begin by discussing receiving arrays that are capable of nulling directional interference.

SIDELOBE CANCELLATION

The simplest form of adaptive antenna is the sidelobe canceler originally devised by Howells in the late 1950s [27] and subsequently developed by Howells and Applebaum. Figure 13.1 is a block diagram of the Howells–Applebaum scheme. Two omnidirectional (equally sensitive in all directions) antenna elements are used, one the "primary," the other the "reference," in the fashion of adaptive noise canceling (see Figure 12.1, for example). Suppose that one signal is present, and at the same time one interferer (or jammer) is present. Both the primary and reference omnis receive the signal and the interference. Being at separated locations, their outputs are not identical but are related time functions.

An important case occurs where the interference is much stronger than the signal. As such, the adaptive filter weights are almost completely controlled by the

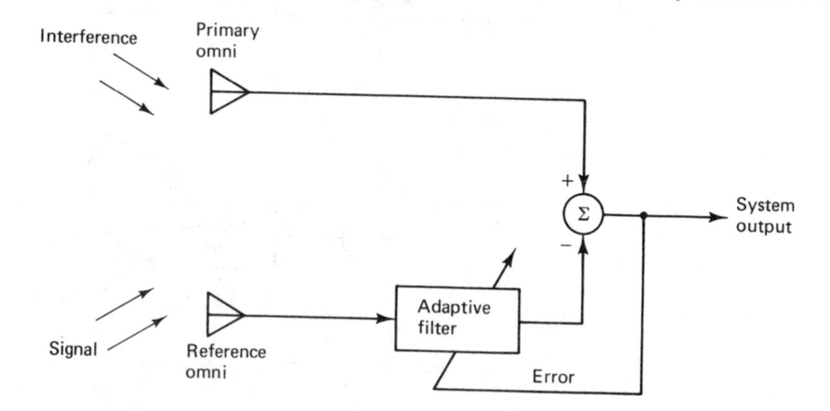

Figure 13.1 Adaptive sidelobe cancellation with two array elements.

interference. Upon convergence, the adaptive filter output contains an interference component which is close to an exact match to the interference component of the primary omni. The system output therefore contains almost no interference, but it does contain the signal, as described in Chapter 12. The output signal is equal to the primary signal component minus the adaptively filtered reference signal component. The amount of output signal to be expected is generally uncertain, since the converged adaptive filter transfer function is unpredictable without prior detailed knowledge of the character and directions of arrival of the interference and the signal. However, the output signal-to-interference ratio is generally improved by the scheme in Figure 13.1 when there is strong interference.

In a practical system, each omni signal is typically fed to an individual receiver for amplification, tuning (bandpass filtering), and signal detection. The receivers and the antennas themselves add independent noises to the incoming signal components. We will designate these noises as "receiver noises." Their effects on adaptive antenna performance are significant. For directional interference to be able to be nulled by the system in Figure 13.1, it must be strong compared to receiver noise. For the useful signal not to be nulled, it must be weak compared to receiver noise. These conditions are frequently met in practice.

Figure 13.2 shows the adaptive sidelobe canceler with receiver noises present at the outputs of the two omnis. The omnis are spaced apart by a distance l. A single signal is assumed to arrive from an angle θ_0. The signal component at the primary omni is

$$\text{primary signal} = C\cos k\omega_0 \tag{13.1}$$

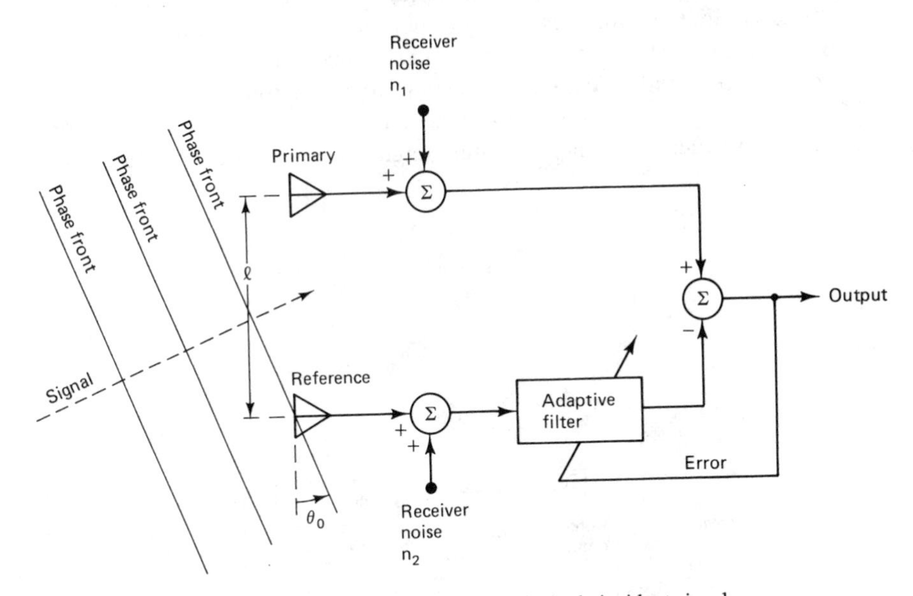

Figure 13.2 Sidelobe canceler with single incident signal.

where C is a constant amplitude, k designates the sample number, and ω is in rad as in (7.16). Phase fronts perpendicular to the direction of signal arrival pass over the two-element array as indicated in Figure 13.2. If we use c to represent the propagation speed, the arrival of a given phase front at the reference omni is earlier than at the primary omni by a number of time steps equal to

$$\delta_0 = \frac{l \sin \theta_0}{cT} = \frac{2\pi l \sin \theta_0}{\lambda_0 \omega_0} \tag{13.2}$$

where T is the time step size in seconds and λ_0 is the wavelength. Therefore, the signal component at the reference omni is

$$\text{reference signal} = C \cos[(k + \delta_0)\omega_0] \tag{13.3}$$

In the absence of receiver noise, the weights of the adaptive filter would adjust themselves to cancel the signal exactly. With receiver noise present, however, we wish to show that the signal is not canceled unless it is strong compared to the receiver noise, and that the sidelobe canceler performs satisfactorily when the useful incoming signal is small compared with the interfering or jamming signal.

To do this we will analyze the behavior of the adaptive sidelobe canceler with receiver noise present and with a single incoming signal. The primary and reference signal components are given by (13.1) and (13.3), respectively. For this analysis, we will use the narrowband tuned sidelobe canceler shown in Figure 13.3(a). The inclusion of tuned bandpass filters in the signal flow paths indicates that we are dealing with a tuned, narrowband receiving system. Each of these filters is assumed to have unit gain at the center of the receiver passband. As in Figure 12.6, the adaptive filter needs only a direct and a quadrature weight to control the magnitude and phase at the signal frequency, ω_0, at the center of the receiver passband.

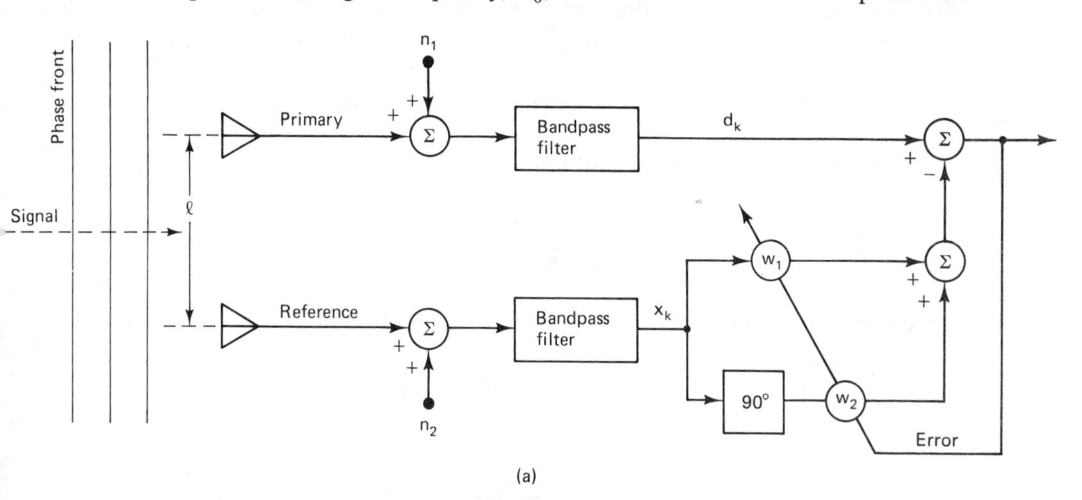

Figure 13.3(a) Tuned sidelobe canceler with signal incident at $\theta_0 = 0$.

We assume that the receiver noises, n_1 and n_2 in Figure 13.3(a), are independent of each other and of the signal components, and we use σ_r^2 to represent the in-band power of either n_1 or n_2. Then the autocorrelation vector of the reference signal is, from (13.3),

$$[\phi_{xx}(n)] = \left[\left(\sigma_r^2 + \frac{C^2}{2}\right) \quad 0\right]^T \tag{13.4}$$

We note that the second component, $\phi_{xx}(1)$, is zero because it represents the correlation of (13.3) with its quadrature (sine) component, and of course the average product of these is zero. Similarly, assuming an arbitrary arrival angle θ_0 as before, the cross-correlation vector is

$$\mathbf{P} = \begin{bmatrix} \phi_{dx}(0) \\ \phi_{dx}(1) \end{bmatrix} = \begin{bmatrix} E[C\cos k\omega_0 \cdot C\cos(k+\delta_0)\omega_0] \\ E[C\cos k\omega_0 \cdot C\sin(k+\delta_0)\omega_0] \end{bmatrix}$$

$$= \frac{C^2}{2}\begin{bmatrix} \cos\delta_0\omega_0 \\ \sin\delta_0\omega_0 \end{bmatrix} \tag{13.5}$$

With these results, (2.17) gives us the optimum weight vector in Figure 13.3a as

$$\begin{bmatrix} w_1^* \\ w_2^* \end{bmatrix} = \frac{C^2}{2\sigma_r^2 + C^2}\begin{bmatrix} \cos\delta_0\omega_0 \\ \sin\delta_0\omega_0 \end{bmatrix} \tag{13.6}$$

This is the general solution. If we let $\theta_0 = 0$ as in Figure 13.3a, then $\delta_0 = 0$, and the optimum weight vector is

$$\begin{bmatrix} w_1^* \\ w_2^* \end{bmatrix}_{\theta_0 = 0} = \begin{bmatrix} \dfrac{C^2}{2\sigma_r^2 + C^2} \\ 0 \end{bmatrix} \tag{13.7}$$

We can define an input signal-to-noise ratio in this case as the ratio of either primary or reference signal power to the passband receiver noise power:

$$\text{SNR} \triangleq \frac{C^2/2}{\sigma_r^2} \tag{13.8}$$

Accordingly, the least-squares solution expressed in terms of the SNR is

$$w_1^* = \frac{\text{SNR}}{1 + \text{SNR}}$$

$$w_2^* = 0 \tag{13.9}$$

This narrowband result is similar to the broadband result of (12.81) for the adaptive line enhancer.

The output signal can be expressed using (13.1) and (13.7):

$$\text{output signal} = C\cos k\omega_0 - w_1^* C\cos k\omega_0$$

$$= \frac{2\sigma_r^2 C}{2\sigma_r^2 + C^2}\cos k\omega_0 \tag{13.10}$$

When the input signal is large, that is, when

$$\frac{C^2}{2\sigma_r^2} \gg 1 \tag{13.11}$$

the output signal can be written as

$$\text{output signal} = \frac{2\sigma_r^2}{C} \cos k\omega_0 \tag{13.12}$$

From the analysis above, one can see that if the input signal is weak compared to the receiver noise in the reference channel, then the SNR is low, the weight w_1^* is near zero, the output signal approximates the primary signal, and there is little or no signal cancellation. On the other hand, if the input signal is strong compared to the receiver noise, then the SNR is high, the weight w_1^* has a value close to 1, and the output signal is small, approximately equal to the primary signal divided by the SNR. In this case, the output signal amplitude is inversely proportional to the input signal amplitude as seen in (13.12), and substantial signal cancellation will occur.

If the angle of arrival of the signal is other than zero, the least-squares solution is given by (13.5). Weight w_2 does not go to zero and weight w_1 goes to a different value, but the amplitude of the output signal would be similar to that predicted by (13.10).

It is also of interest to determine the sensitivity of the receiving system in Figure 13.3(a) at all angles of incidence, assuming that $\theta_0 = 0$. Let the weights be frozen in accordance with (13.9) after adapting to a signal arriving at angle zero. Figure 13.3(b) is a simplified diagram of the sidelobe canceler useful in the determination of sensitivity versus arrival angle. Such a sensitivity pattern is often called a "directivity pattern" for the converged array.

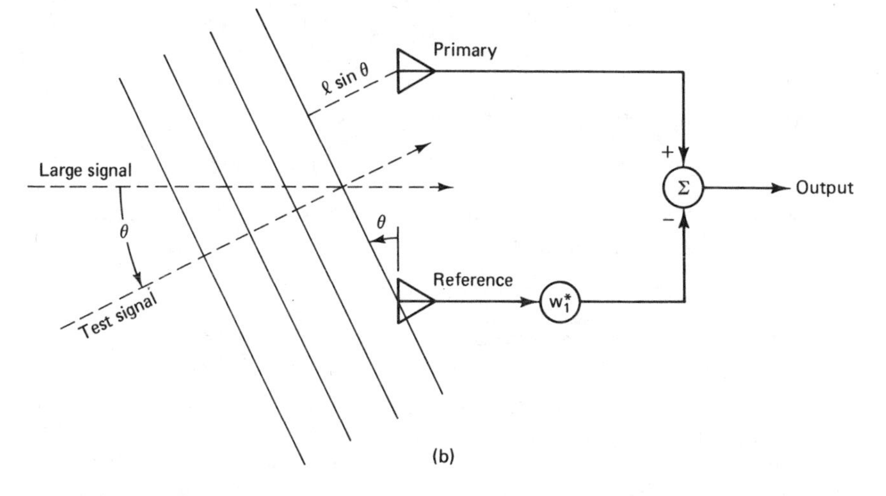

Figure 13.3(b) Simplified version of part (a) for determining the directivity pattern of the converged sidelobe canceler.

A test signal at frequency ω_0 is assumed to arrive at angle θ. If we let $\delta = (l \sin\theta)/cT$ as in (13.2) and let the primary component of the test signal be $\cos k\omega_0$ as before with $C = 1$, then, in accordance with Figure 13.3(b), the reference component is

$$\cos[(k + \delta)\omega_0] \tag{13.13}$$

and the output signal is

$$\begin{aligned} \text{output signal} &= \cos k\omega_0 - w_1^* \cos(k + \delta)\omega_0 \\ &= A\cos(k\omega_0 - \alpha) \end{aligned} \tag{13.14}$$

where

$$A = \left[1 + w_1^{*2} - 2w_1^* \cos(\delta\omega_0)\right]^{1/2} \tag{13.15}$$

and

$$\alpha = \tan^{-1} \frac{w_1^* \sin(\delta\omega_0)}{1 - w_1^* \cos(\delta\omega_0)} \tag{13.16}$$

We define the "array amplitude gain" as the ratio of the output amplitude to the input amplitude. Since the amplitude of the test input, $\cos k\omega_0$, is 1, the array amplitude gain is

$$\begin{aligned} \text{array amplitude gain} &= \frac{\text{output amplitude}}{\text{input amplitude}} \\ &= A \text{ in (13.14)} \end{aligned} \tag{13.17}$$

We also define the "array power gain" as the square of the above:

$$\begin{aligned} \text{array power gain} &\triangleq \left(\frac{\text{output amplitude}}{\text{input amplitude}}\right)^2 = A^2 \\ &= 1 + w_1^{*2} - 2w_1^* \cos(\delta\omega_0) \end{aligned} \tag{13.18}$$

As in (7.20), the array power gain in dB is

$$\text{dB} = 10\log_{10}(A^2) = 20\log_{10} A \tag{13.19}$$

We also define the "array output signal phase" as the output phase relative to the input phase which, from (13.14), is

$$\text{array output signal phase} \triangleq \alpha \text{ in (13.16)} \tag{13.20}$$

When interpreting the response formulas for the converged beamformer, that is, the array gain in (13.18) and phase in (13.20), we recall that w_1^* is a function of SNR in accord with (13.9), and that the original input signal that the array had adapted to was represented by (13.1) and had a direction of arrival of $\theta = 0$. The array gain has minimum value at $\theta = 0$, and this amounts to a spatial dip or spatial notch at the arrival angle of the original signal. The adaptive process in effect tries to eliminate the incoming signal. The stronger this signal, the deeper the notch. As the SNR of

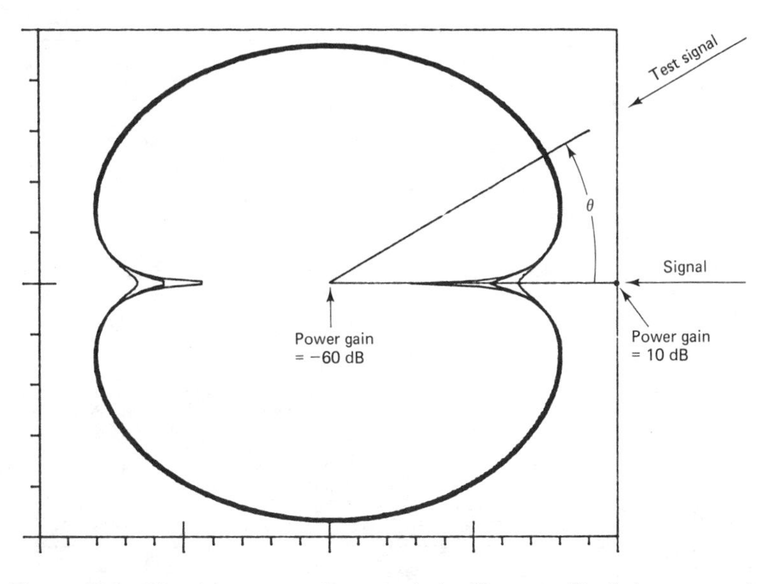

Figure 13.4 Directivity pattern (power gain in dB versus θ) of the converged two-element sidelobe canceler with $l\omega_0/T = c\pi$. Input SNR is 6 dB for outer curve, 10 dB for middle curve, and 100 dB for inner curve. Main signal arrival at $\theta = 0$.

the original signal increases past unity, the adaptive notch deepens rapidly and, as indicated by (13.12), the gain becomes inversely proportional to the signal strength.

It is often useful to make a polar plot of the array power gain versus angle θ, as in Figure 13.4. The arrival angle of the original signal is assumed to be $\theta_0 = 0$, and plots are shown for various values of the SNR in accordance with (13.9) and (13.18). For these plots, the array aperture is adjusted to $l = c\pi T/\omega_0 = \lambda_0/2$. Note that this value of l is in a sense optimum, giving two distinct notches and a maximum variation in (13.18).

The array power gain at the bottom of the notch is obtained from (13.9) and (13.18) by setting $\theta = 0$. Thus

$$\begin{aligned}(\text{array power gain})_{\theta=0} &= \left(1 - w_1^*\right)^2 \\ &= \frac{1}{(1 + \text{SNR})^2}\end{aligned} \tag{13.21}$$

This minimum array power gain is plotted versus the SNR in Figure 13.5. The adaptive sidelobe canceler is in fact a power separator, as can be seen in this figure. Strong signals eliminate themselves, while weak signals pass through at other angles essentially with unit gain. We again note that "weak" and "strong" are relative to the receiver noise level.

The Howells–Applebaum sidelobe canceler will work well as long as useful input signals are weak and the interfering or jamming signals are strong. Sometimes, however, the useful signal can be strong enough compared to receiver noise to delete

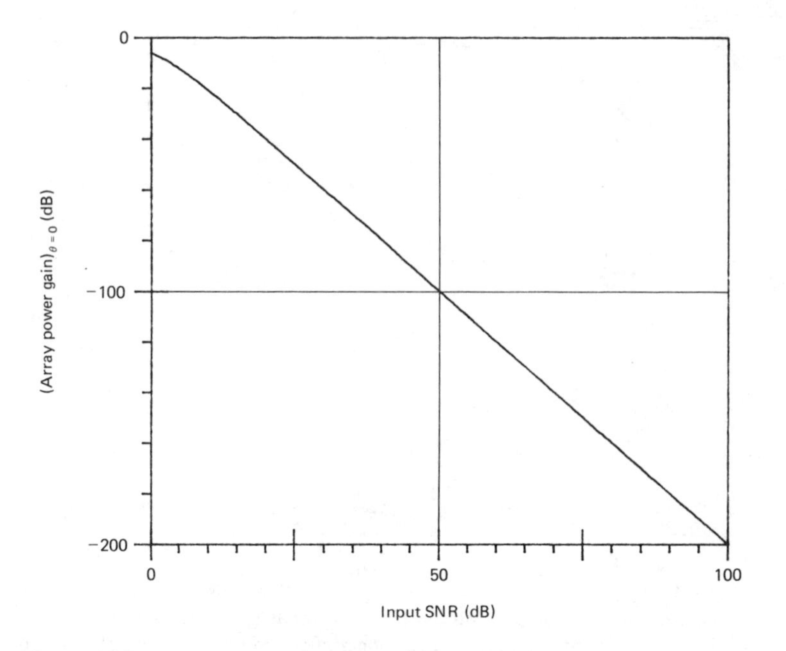

Figure 13.5 Array power gain of two-element sidelobe canceler at bottom of notch.

itself. To prevent this, one could deliberately add random independent noise to the reference omni signal to augment the receiver noise, but this procedure has the disadvantage of adding noise to the system that will ultimately cause the output to be more noisy. A better method for obtaining a least-squares solution corresponding to a noisier input without actually adding input noise is the "leaky LMS algorithm."

We can develop the leaky LMS algorithm by starting with the LMS algorithm itself. As in (6.3), the LMS algorithm can be written

$$\begin{aligned} \mathbf{W}_{k+1} &= \mathbf{W}_k + 2\mu\varepsilon_k\mathbf{X}_k \\ &= \mathbf{W}_k + 2\mu\mathbf{X}_k\left(d_k - \mathbf{X}_k^{\mathrm{T}}\mathbf{W}_k\right) \\ &= \left(\mathbf{I} - 2\mu\mathbf{X}_k\mathbf{X}_k^{\mathrm{T}}\right)\mathbf{W}_k + 2\mu\mathbf{X}_k d_k \end{aligned} \tag{13.22}$$

Assuming that $\mathbf{X}_k$ is uncorrelated with $\mathbf{W}_k$, we may take expected values of both sides of (13.22) as follows:

$$E[\mathbf{W}_{k+1}] = (\mathbf{I} - 2\mu\mathbf{R})E[\mathbf{W}_k] + 2\mu\mathbf{P} \tag{13.23}$$

The expected converged weight vector with μ in the stable range is, of course,

$$\lim_{k\to\infty} E[\mathbf{W}_k] = \mathbf{W}^* = \mathbf{R}^{-1}\mathbf{P} \tag{13.24}$$

An adaptive filter driven by the LMS algorithm is shown in Figure 13.6(a). White noise with power σ^2 has been added to the input x_k. For this situation,

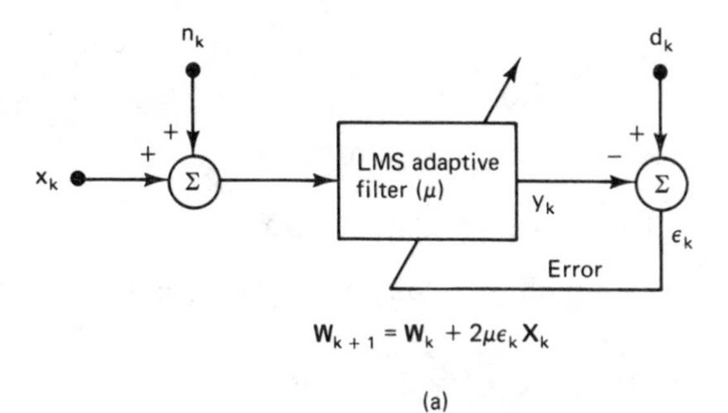

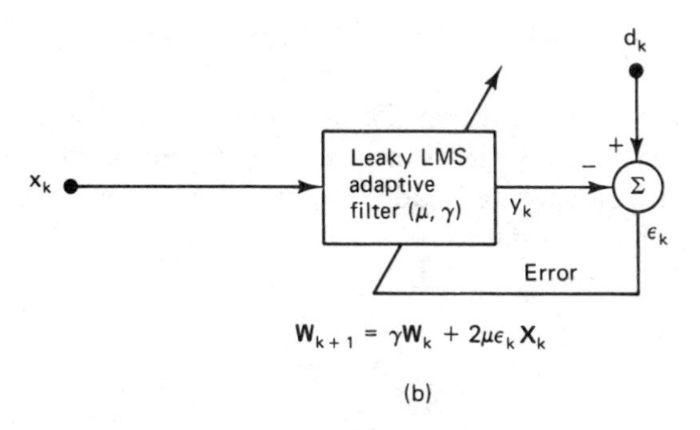

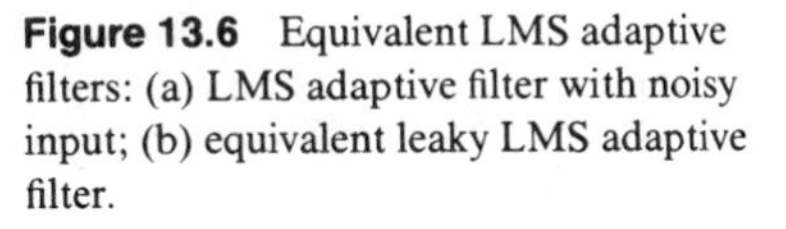

Figure 13.6 Equivalent LMS adaptive filters: (a) LMS adaptive filter with noisy input; (b) equivalent leaky LMS adaptive filter.

(13.23) becomes

$$E[\mathbf{W}_{k+1}] = \left(\mathbf{I} - 2\mu(\mathbf{R} + \sigma^2\mathbf{I})\right)E[\mathbf{W}_k] + 2\mu\mathbf{P} \tag{13.25}$$

and the expected converged solution becomes

$$\lim_{k\to\infty} E[\mathbf{W}_k] = (\mathbf{R} + \sigma^2\mathbf{I})^{-1}\mathbf{P} \tag{13.26}$$

The leaky LMS algorithm is specified as follows:

$$\mathbf{W}_{k+1} = \gamma\mathbf{W}_k + 2\mu\varepsilon_k\mathbf{X}_k \tag{13.27}$$

The parameter γ is a positive constant chosen in the range

$$|\gamma| < 1 \tag{13.28}$$

With the conventional LMS algorithm in (13.22), if μ were suddenly set to zero, the weight vector would remain constant indefinitely. With the leaky LMS algorithm, on the other hand, with γ just under 1, if μ were suddenly set to zero, the weight vector would gradually leak away, undergoing a geometric decay toward zero. The leaky algorithm has to adapt to "stay alive" so to speak, like a shark needing to swim in order to breathe.

The effect of leakage on the converged solution can be determined in the following way. A set of equations analogous to (13.22) for the leaky LMS algorithm is

$$\begin{aligned}\mathbf{W}_{k+1} &= \gamma\mathbf{W}_k + 2\mu\varepsilon_k\mathbf{X}_k \\ &= \gamma\mathbf{W}_k + 2\mu\mathbf{X}_k\left(d_k - \mathbf{X}_k^{\mathrm{T}}\mathbf{W}_k\right) \\ &= \left(\gamma\mathbf{I} - 2\mu\mathbf{X}_k\mathbf{X}_k^{\mathrm{T}}\right)\mathbf{W}_k + 2\mu\mathbf{X}_k d_k \end{aligned} \tag{13.29}$$

Once again, with $\mathbf{X}_k$ uncorrelated with $\mathbf{W}_k$, taking expected values of both sides yields

$$\begin{aligned}E[\mathbf{W}_{k+1}] &= (\gamma\mathbf{I} - 2\mu\mathbf{R})E[\mathbf{W}_k] + 2\mu\mathbf{P} \\ &= \left(\mathbf{I} - 2\mu\left(\mathbf{R} + \frac{1-\gamma}{2\mu}\mathbf{I}\right)\right)E[\mathbf{W}_k] + 2\mu\mathbf{P}\end{aligned} \tag{13.30}$$

Comparing (13.30) with (13.25), it is clear that the same expected converged solution can be obtained with the leaky LMS algorithm as with the noisy case in Figure 13.6(a). To obtain the effect of a specified added input noise power σ, the leakage parameter γ should be chosen according to

$$\frac{1-\gamma}{2\mu} = \sigma^2$$

or

$$\gamma = 1 - 2\mu\sigma^2 \tag{13.31}$$

From (13.26) and (13.31), the converged solution for the leaky LMS algorithm is

$$\lim_{k\to\infty} E[\mathbf{W}_k] = \left(\mathbf{R} + \frac{1-\gamma}{2\mu}\mathbf{I}\right)^{-1}\mathbf{P} \tag{13.32}$$

In the sidelobe canceler, although it will increase the output noise, the leaky LMS algorithm can be used to establish an equivalent input noise power level. The power of the incoming signal can be strong or weak compared to this equivalent noise power and the ratio determines the extent of acceptance or cancellation in accordance with (13.21) and with Figures 13.4 and 13.5. The equivalent input noise power is

$$\sigma_r^2 + \sigma^2 = \sigma_r^2 + \frac{1-\gamma}{2\mu} \tag{13.33}$$

where σ_r^2 is the actual receiver noise power.

The equivalent input noise power will be positive with γ selected in range (13.28). In cases where the receiver noise level is high, causing a least-squares solution that is heavily biased by this noise, one could allow γ to be greater than unity and cause the equivalent input noise power to be less than σ_r^2. However, with $\gamma > 1$, the direction of the leakage is reversed, and making μ equal to zero in this case causes the adaptive algorithm to be unstable. It is necessary to adapt to remain stable, if possible. The leaky LMS algorithm is usable in the reverse mode ($\gamma > 1$), but in this case we should call it the "dangerous LMS algorithm." When stable, the leaky LMS algorithm used in the presence of receiver noise develops an expected converged solution similar to (13.32):

$$\lim_{k\to\infty} E[\mathbf{W}_k] = \left[\mathbf{R} + \left(\frac{1-\gamma}{2\mu} + \sigma_r^2\right)\mathbf{I}\right]^{-1}\mathbf{P} \tag{13.34}$$

The leaky LMS algorithm gives a weight vector solution like that obtained with a noisy input, but without requiring the injection of input noise. The algorithm itself accomplishes this, without causing additional noise in the output signal or the weights.

Some directivity patterns for a sidelobe canceler driven by the leaky LMS algorithm are shown in Figure 13.7. They have been computed just like the patterns of Figure 13.4, except that the SNR in (13.8) is now taken as the ratio of signal power to the equivalent input noise power, $\sigma_r^2 + \sigma^2$. To obtain the patterns of Figure 13.7, the receiver noise power has been assumed to be unity and the input signal power has been assumed to be 10; thus the SNR without leakage is 10 dB. The value of γ is varied to control the threshold of acceptance or rejection of the signal. A near-omnidirectional directivity pattern, indicating nonrejection of the input signal, is the desired result here.

Next we consider the practical case for the sidelobe canceling array, where signal and interferers are present at the same time. Assuming that a single narrowband interferer is present and uncorrelated with the narrowband signal, we can add the R-matrices and the P-vectors given for either signal by (13.4) and (13.5).

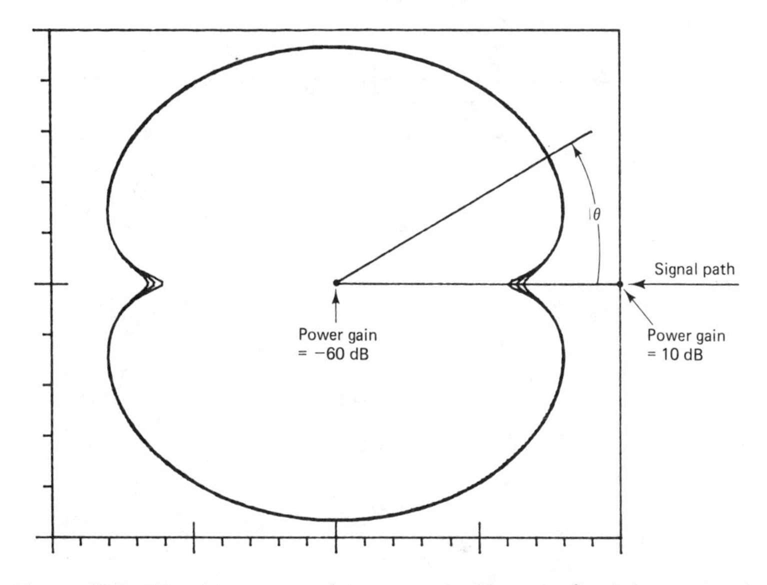

Figure 13.7 Directivity patterns (power gain in dB versus θ) of the converged two-element sidelobe canceler with $l = \lambda_0/2$, driven by the leaky LMS algorithm. Main signal arrival at $\theta = 0$, and input SNR = 10. Values of σ^2/σ_r^2 are 0.5, 1.0, and 1.5 for the inner, middle, and outer curves.

Suppose that a large jamming signal is present at $\theta = 0$ with power σ_j^2, and a smaller useful or "target" signal arrives at angle θ_0 with power σ_s^2. Then, as in (13.4) and (13.5), we have

$$\left(\sigma_r^2 + \sigma_s^2 + \sigma_j^2\right)\begin{bmatrix} 1 & 0 \\ 0 & 1 \end{bmatrix}\begin{bmatrix} w_1^* \\ w_2^* \end{bmatrix} = \begin{bmatrix} \sigma_j^2 + \sigma_s^2\cos(\delta_0\omega_0) \\ \sigma_s^2\sin(\delta_0\omega_0) \end{bmatrix} \tag{13.35}$$

where $\delta_0 = (l\sin\theta_0)/cT$ as in (13.2). The solution is

$$w_1^* = \frac{\sigma_j^2 + \sigma_s^2\cos\delta_0\omega_0}{\sigma_r^2 + \sigma_s^2 + \sigma_j^2}$$

$$w_2^* = \frac{\sigma_s^2\sin\delta_0\omega_0}{\sigma_r^2 + \sigma_s^2 + \sigma_j^2} \tag{13.36}$$

If we now assume that the weights are converged and fixed, and that a unit input test signal $\cos k\omega_0$ is arriving at some other angle θ, then, similar to (13.14), the output is

$$\begin{aligned} \text{output test signal} &= \cos k\omega_0 - w_1^*\cos[(k+\delta)\omega_0] - w_2^*\sin[(k+\delta)\omega_0] \\ &= A\cos(k\omega_0 - \alpha) \end{aligned} \tag{13.37}$$

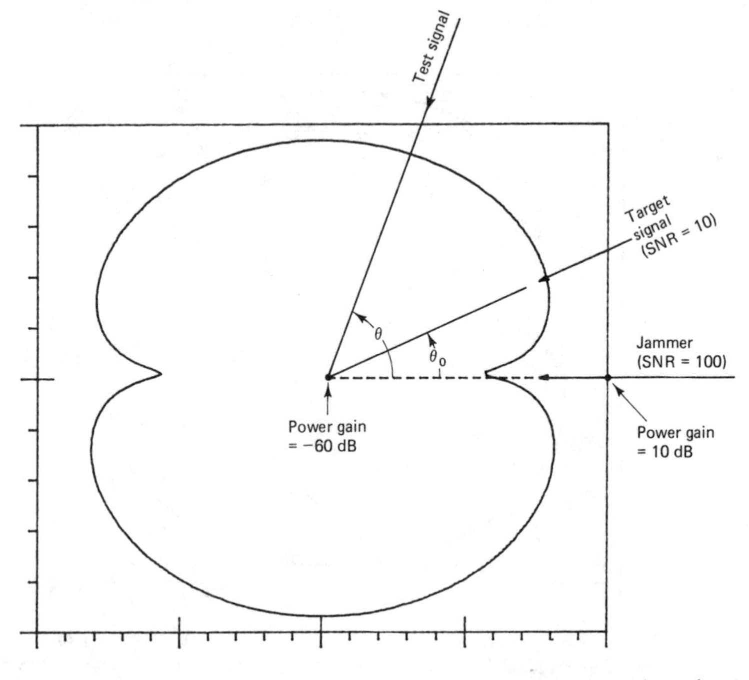

Figure 13.8 Directivity pattern for the "signal plus jammer" case with $l = \lambda_0/2$. The small signal has the effect of shifting slightly the interference notch.

where again $\delta = (l \sin\theta)/cT$ as in (13.2). Thus the array power gain, similar to (13.18), is

$$\begin{aligned}\text{array power gain} &\triangleq A^2 \\ &= 1 + w_1^{*2} + w_2^{*2} - 2\left(w_1^{*}\cos\delta\omega_0 + w_2^{*}\sin\delta\omega_0\right) \quad (13.38)\end{aligned}$$

A directivity pattern for this "signal-plus-jammer" case is shown in Figure 13.8. Here the jammer has 10 times the power of the signal arriving at $\theta_0 = 30°$, and the signal has the effect of shifting both the notches slightly around toward its direction of arrival.

Another practical case occurs where there are multiple jammers. To cope with more than one jammer at a time, it is necessary to have more than one reference omni. A system with two spatially separated reference omnis is shown in Figure 13.9. Note that the error signal is formed by subtracting both reference outputs from the primary signal.

Because the reference omnis are spatially separate, the system in Figure 13.9 can form two separate notches in its directivity pattern. As in (13.2), let $\delta_i = (l_i \sin\theta)/cT$ be the delay caused by the ith omni for a signal arriving at angle θ, where l_i is the primary-to-reference omni separation as before, and let w_{i1} and w_{i2} be the filter weights, with $i = 1$ or 2. Then the version of (13.37) applying to this case is

$$\begin{aligned}\text{output signal} = \cos k\omega_0 &- w_{11}^{*}\cos[(k+\delta_1)\omega_0] - w_{12}^{*}\sin[(k+\delta_1)\omega_0] \\ &- w_{21}^{*}\cos[(k+\delta_2)\omega_0] - w_{22}^{*}\sin[(k+\delta_2)\omega_0] \quad (13.39)\end{aligned}$$

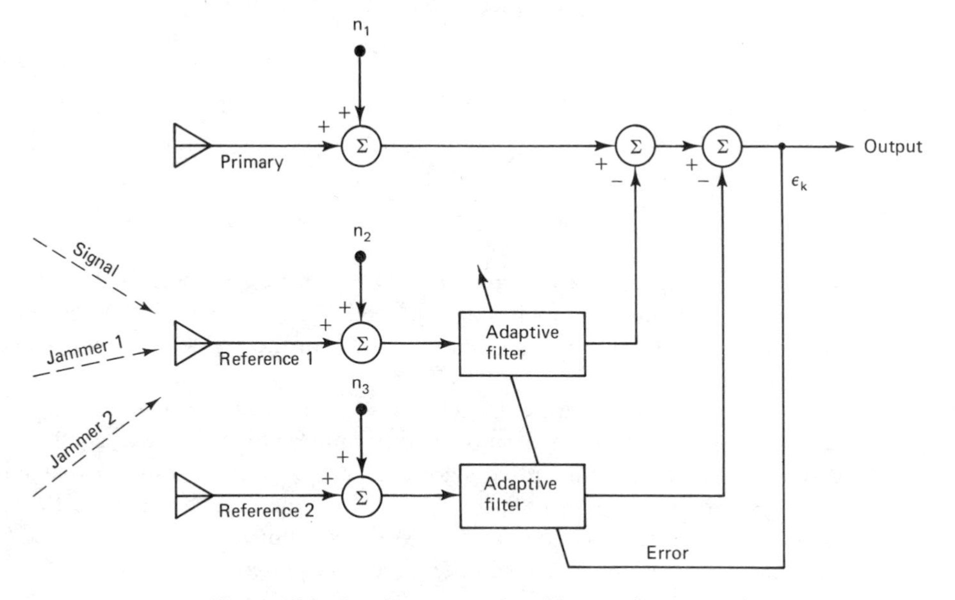

Figure 13.9 Interference canceler with two reference omnis.

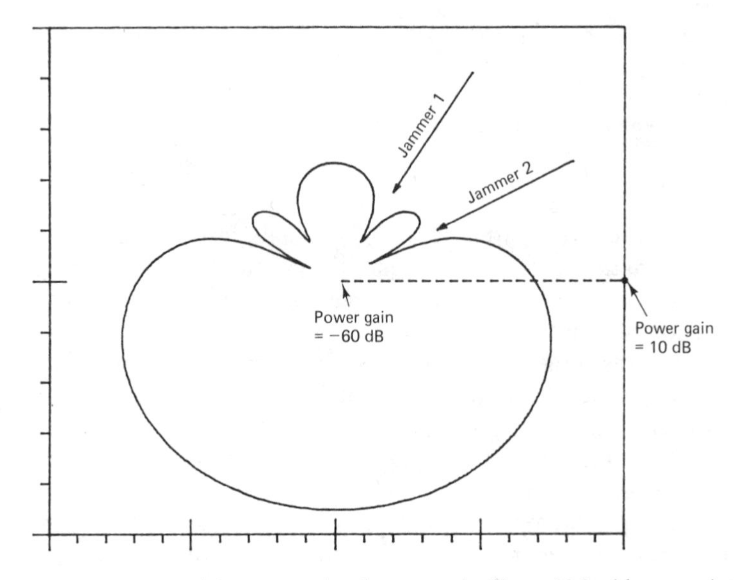

Figure 13.10 Directivity pattern for the system in Figure 13.9 with strong jamming signals arriving at $\theta = 30°$ and $60°$.

An example of the directivity pattern formed by plotting the power gain using (13.39) with converged weights is shown in Figure 13.10. The data used to obtain the pattern in Figure 13.10 are as follows:

$$\text{Jammer 1 at } \theta = 60°: \quad l_1 = \frac{\lambda_0}{4}$$

$$\text{Jammer 2 at } \theta = 30°: \quad l_2 = \frac{\lambda_0}{8}$$

$$w_{11}^* = -0.48 \qquad w_{12}^* = -0.87$$

$$w_{21}^* = 1.70 \qquad w_{22}^* = \ \ 1.00$$

We can see in Figure 13.10 the two distinct notches at jamming angles 30° and 60°, together with symmetric notches at 150° and 120°, respectively. We also note that the circular gain pattern is distorted more than in the previous examples due to the two jammers.

If more interferers were expected, one would need to include more reference omnis, at least one for each jammer. If the number of jammers exceeded the number of reference omnis, the adaptive algorithm would find the solution that minimizes output power, as in Figure 13.8, for example. On the other hand, when the reference omnis outnumber the jammers, the LMS algorithm has no problem locating the appropriate number of notches in the directivity pattern.

BEAMFORMING WITH A PILOT SIGNAL

We have seen how the adaptive interference canceler is used to form nulls in the directivity pattern. Another type of adaptive beamformer, developed by B. Widrow, L. J. Griffiths; P. E. Mantey, and B. B. Goode [1], is based on a "pilot-signal" algorithm. Unlike the Howells–Applebaum beamformer which starts out being omnidirectional and loses sensitivity to strong (presumed jamming) signals as a result of the adaptive process, the pilot-signal adaptive beamformer forms a beam toward a specified "look direction" and uses adaptivity to support this beam while simultaneously forming notches to null interferences arriving outside the look direction. The nulling process is determined by direction of arrival and power level.

While the system is adapting, an injected pilot signal simulates a received signal from a look direction chosen by the system operator. The same pilot signal is used as the desired response for the adaptive processor attached to the antenna array elements. Using the pilot signal, the adaptive beamformer is trained so that its directivity pattern has a main lobe in the specified look direction as well as notches corresponding with incident interference signals whose directions of arrival differ from the look direction. The array thus adapts to form the main lobe with its direction and bandwidth determined by the pilot signal; meanwhile it rejects interference signals or noise uncorrelated with the pilot signal which arrive outside of the main lobe. These characteristics are achieved as well as possible in the minimum-mean-square-error sense.

Adaptive antenna algorithms which produce results similar to the pilot signal algorithm but which in fact are simpler to implement and in some cases perform better have been devised by Griffiths [28] and Frost [29]. In many applications, the pilot signal algorithm has been supplanted by the Griffiths and Frost algorithms, but the pilot signal algorithm was an important springboard for the development of the other algorithms. However, Compton et al. [30] have found specific uses for it and additional uses will be described here that could not be accomplished with other algorithms. Practical real-time adaptive arrays first came into being in the forms of the Howells–Applebaum sidelobe canceler and the pilot-signal adaptive beamformer. The latter is described in Reference 1; here we will summarize and expand on that description.

Many arrays of sensors are "linear" in that the antenna elements (omnis) are arranged along a line, or the arrays are "planar" with the elements arranged in a plane. Antenna arrays are often configured this way. An example of a conventional linear-array receiving antenna is shown in Figure 13.11. The antenna of Figure 13.11(a) consists of seven isotropic elements spaced $\lambda_0/2$ apart along a straight line, where λ_0 is the wavelength of the center frequency ω_0 of the array. The received signals are summed to produce an array output signal. The directivity pattern, that is, the relative sensitivity of response to signals from various directions, is plotted in this figure in a plane over an angular range of $-\pi/2 < \theta < \pi/2$ for frequency ω_0. This pattern is symmetric about $\theta = 0$ as well as $\theta = 90^0$, and the main lobe is

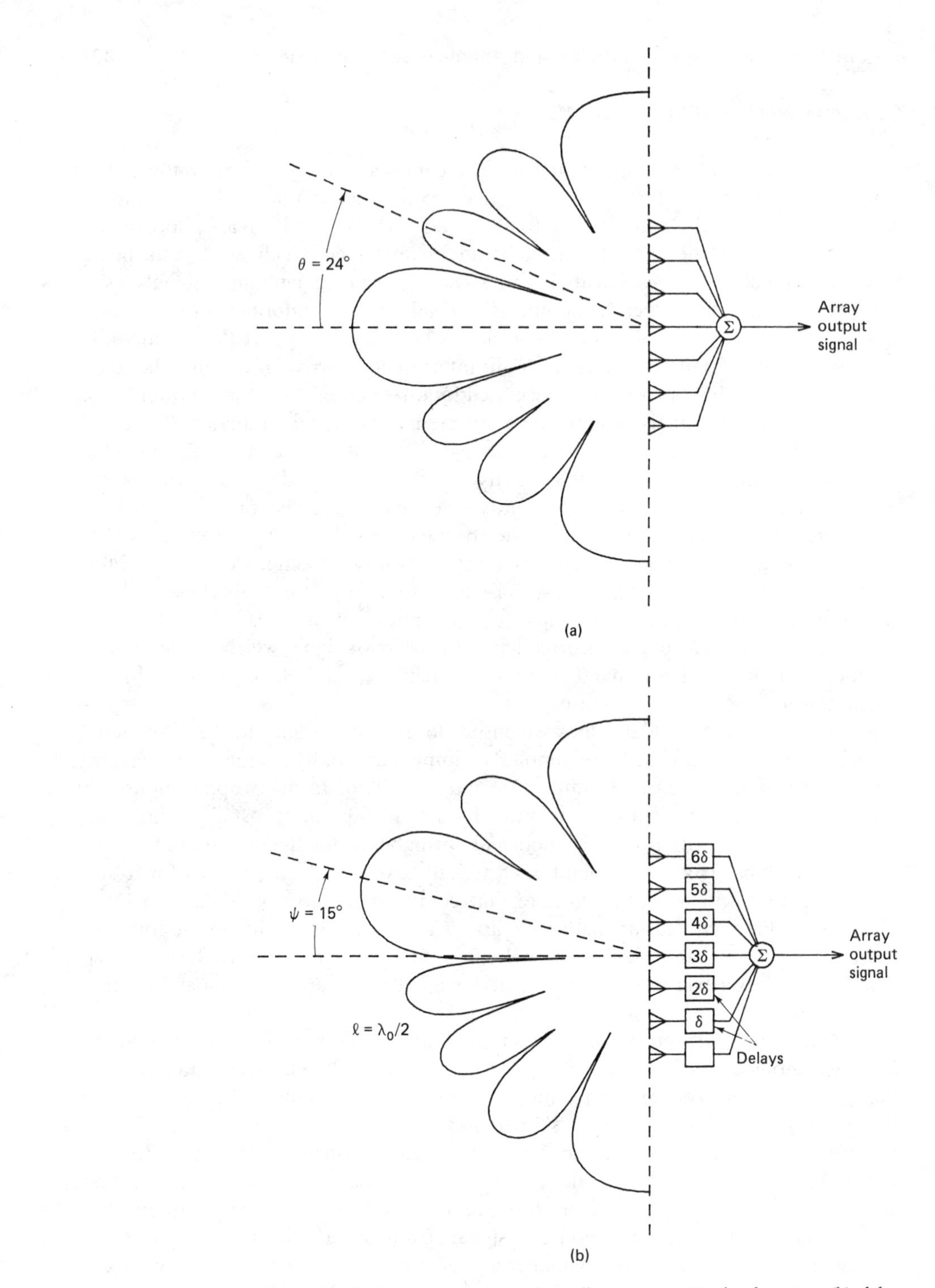

Figure 13.11 Directivity pattern for a linear array: (a) simple array; (b) delays added. From B. Widrow et al., *Adaptive Antenna Systems*, © December 1967, IEEE.

centered at $\theta = 0$. The largest-amplitude sidelobe, at $\theta = 24°$, has a maximum sensitivity which is 12.5 dB below the maximum main-lobe sensitivity. This pattern would be different if it were plotted for frequencies other than ω_0.

The same array configuration is shown in Figure 13.11(b); however, in this case the output of each element is delayed in time before being summed. The resulting directivity pattern now has its main lobe at an angle of ψ radians, where, similar to (13.2),

$$\psi = \sin^{-1}\frac{\lambda_0\delta\omega_0}{2\pi dT} = \sin^{-1}\frac{c\delta}{d} \tag{13.40}$$

where ω_0 = normalized frequency of received signal (rad)

λ_0 = wavelength at frequency ω_0 (m)

δ = time-delay difference between neighboring element outputs (samples)

d = spacing between antenna elements (m)

c = signal propagation velocity $= \dfrac{\lambda_0\omega_0}{2\pi T}$ (m/s)

T = time step (s)

The sensitivity is maximum at angle ψ because signals received from a plane-wave source incident at this angle, and delayed as in Figure 13.11(b), are in phase with one another and produce the maximum output signal. For the example illustrated in the figure, $d = \lambda_0/2$, $\delta = (0.8131/\omega_0)$, and therefore $\psi = \sin^{-1}(\delta\omega_0/\pi) = 15°$. Derivation of the plots in Figure 13.11 is straightforward and is left to Exercises 13 through 16.

There are many possible configurations for phased arrays. Figure 13.12(a) shows one such configuration where each of the antenna-element outputs is weighted by two weights in parallel, one being preceded by a time delay of a quarter of a cycle at frequency ω_0 (i.e., a 90° phase shift, or $\pi T/2\omega_0$ seconds). The output signal is the sum of all weighted signals, and since all weights are set to unit values, the directivity pattern at frequency ω_0 is by symmetry the same as that of Figure 13.11(a). For purposes of illustration, an interfering directional sinusoidal "noise" of frequency ω_0 incident on the array is shown in Figure 13.12(a), indicated by the dashed arrow. The angle of incidence (45°) of this noise is such that it would be received on one of the sidelobes of the directivity pattern with a sensitivity only 17 dB less than that of the main lobe at $\theta = 0°$.

If the weights are now set as indicated in Figure 13.12(b), the directivity pattern at frequency ω_0 becomes as shown in that figure. In this case, the main lobe is almost unchanged from that shown in Figures 13.11(a) and 13.12(a), while the particular sidelobe that previously intercepted the sinusoidal noise in Figure 13.12(a)

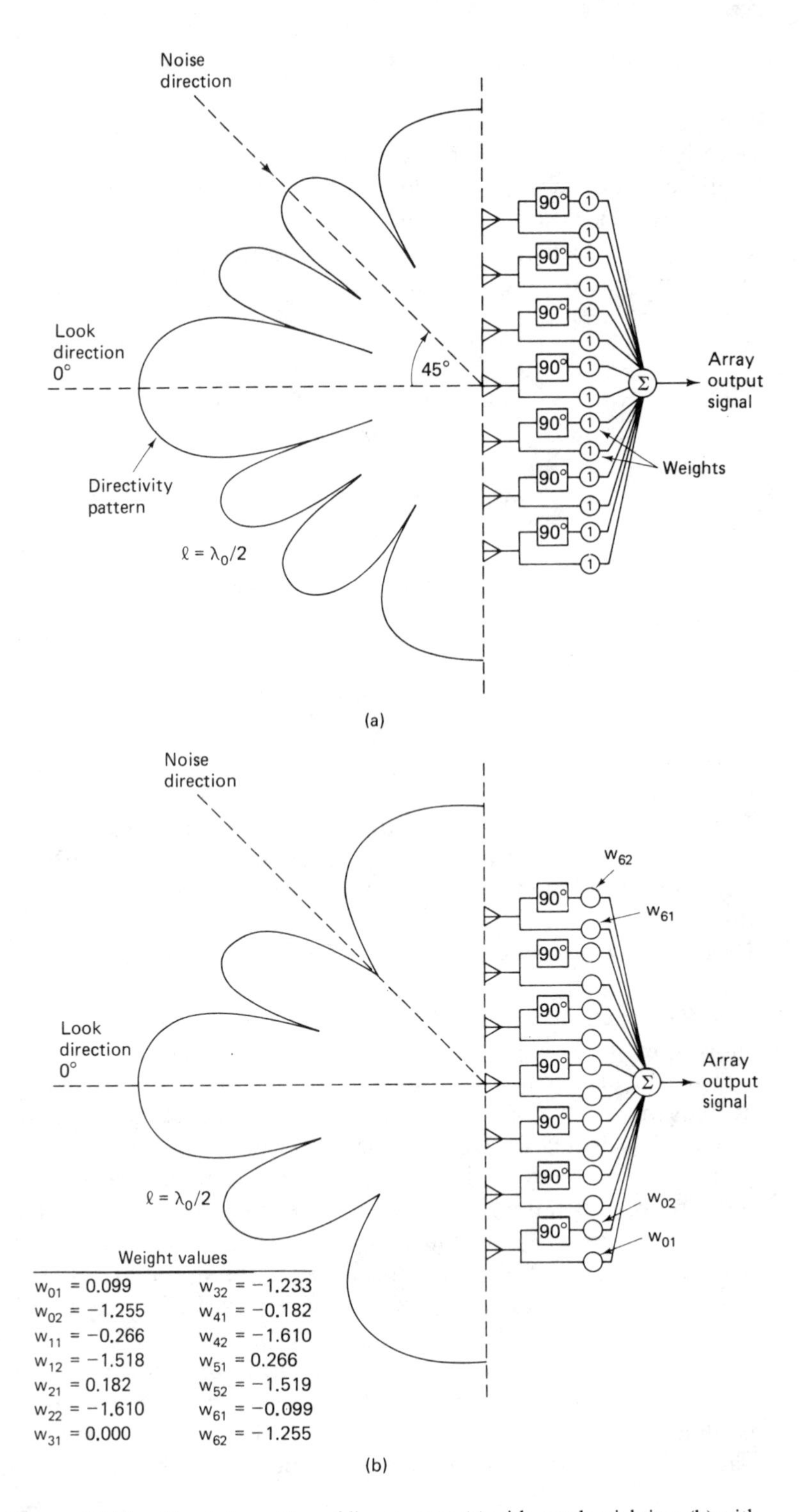

Figure 13.12 Directivity pattern of linear array: (a) with equal weighting; (b) with weighting for noise elimination. From B. Widrow et al., *Adaptive Antenna Systems*, © December 1967, IEEE.

has been shifted so that a null is now placed in the direction of that noise. The sensitivity in the noise direction is 77 dB below the main-lobe sensitivity, improving the noise rejection by 60 dB.

Let us now consider a simple example to illustrate the existence and calculation of a set of weights which will cause a signal from a desired direction to be accepted while a "noise" from a different direction is rejected. Such an example is illustrated in Figure 13.13. Let the signal arriving from the desired direction $\theta = 0°$ be called the "pilot" signal $p_k = P \sin k\omega_0$, and let the noise, $n_k = N \sin k\omega_0$, be incident to the receiving array at $\theta = \pi/6$ radians. Both the pilot signal and the noise signal are assumed for this example to be at exactly the same frequency ω_0. At a point in space midway between the antenna array elements, the signal and the noise are assumed to be in phase. In the example shown, there are two identical omnis spaced $\lambda_0/2$ apart. The signals received by each element are fed to two variable weights, one weight being preceded by a quarter-wave time delay of $\pi T/2\omega_0$. The four weighted signals are then summed to form the array output.

The problem in Figure 13.13 is to obtain a set of weights to accept p_k and reject n_k. Note that with any set of nonzero weights, the output is of the form

$$A \sin(k\omega_0 + \phi),$$

and a number of solutions exist which will make the output be p_k. However, the output of the array must be independent of the amplitude and phase of the noise if the array is to be regarded as rejecting the noise. Satisfaction of this constraint leads to a unique set of weights determined as follows. The array output due to the pilot signal is

$$P\left[(w_1 + w_3)\sin k\omega_0 + (w_2 + w_4)\sin\left(k\omega_0 - \frac{\pi}{2}\right)\right] \tag{13.41}$$

For this output to equal the desired output $p_k = P \sin k\omega_0$ (which is the pilot signal

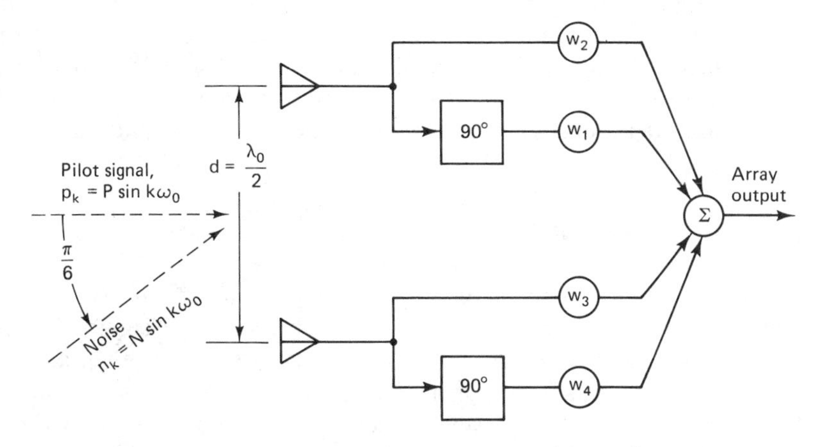

Figure 13.13 Pilot signal arriving at $\theta = 0$ and noise arriving at $\theta = \pi/6$.

itself), it is necessary that

$$\begin{aligned} w_1 + w_3 &= 1 \\ w_2 + w_4 &= 0 \end{aligned} \tag{13.42}$$

With respect to the midpoint between the antenna elements, the relative time delays of the noise at the two elements are $\pm(d/2c)\sin(\pi/6) = \pm d/4c = \pm \pi T/4\omega_0$ seconds, which correspond to phase shifts of $\pm\pi/4$ at frequency ω_0. The array output due to the incident noise at $\theta = \pi/6$ is then

$$\begin{aligned} N\Big[& w_1 \sin\left(k\omega_0 - \frac{\pi}{4}\right) + w_2 \sin\left(k\omega_0 - \frac{3\pi}{4}\right) \\ & + w_3 \sin\left(k\omega_0 + \frac{\pi}{4}\right) + w_4 \sin\left(k\omega_0 - \frac{\pi}{4}\right)\Big] \end{aligned} \tag{13.43}$$

For this response to equal zero, it is necessary that

$$\begin{aligned} w_1 + w_4 &= 0 \\ w_2 - w_3 &= 0 \end{aligned} \tag{13.44}$$

Thus the set of weights that satisfies the signal and noise response requirements can be found by solving (13.42) and (13.44) simultaneously. The solution is

$$w_1 = \tfrac{1}{2},\ w_2 = \tfrac{1}{2},\ w_3 = \tfrac{1}{2},\ w_4 = -\tfrac{1}{2} \tag{13.45}$$

With these weights, the array will have the desired properties in that it will accept a signal from the desired direction, while rejecting noise which is at the same frequency ω_0 as the signal, but arriving from a different direction than that of the signal.

The foregoing method of calculating the weights is more illustrative than practical. This method is usable when there is only a small number of directional noise sources, when the noises are monochromatic, and when the directions of the noises are known a priori. A practical processor should not require detailed information about the number and the nature of the noises, and the adaptive processor described in the following meets this requirement. It solves recursively a sequence of simultaneous equations, which are generally overspecified, and it finds solutions that minimize the mean-square error between the pilot signal and the total array output.

SPATIAL CONFIGURATIONS

Before discussing methods of adaptive filtering and signal processing to be used in the adaptive array, we need to extend the preceding discussion and consider some types of spatial array configurations. An adaptive array configuration for processing narrowband signals is shown in Figure 13.14. Each individual antenna element is shown connected to a variable weight and to a quarter-period time delay whose output is in turn connected to another variable weight. The weighted signals are

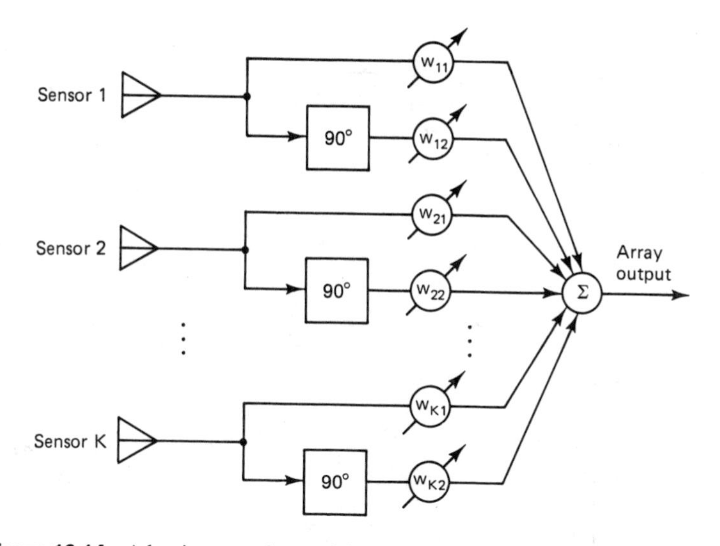

Figure 13.14 Adaptive array for receiving narrowband signals. From B. Widrow et al., *Adaptive Antenna Systems*, © December 1967, IEEE.

summed, as shown in the figure. The signal at each sensor, assumed to be either monochromatic or narrowband, is thus weighted by a complex gain factor $Ae^{j\phi}$. Any phase angle $\phi = -\tan^{-1}(w_2/w_1)$ can be chosen by setting the two weight values, and the magnitude of this complex gain factor $A = \sqrt{w_1^2 + w_2^2}$ can take on a range of values that is limited only by the range limitations of the two individual weights. Thus the two weights and the time delay (90° phase lag) provide completely adjustable linear processing for narrowband signals received by each individual antenna element.

The full array of Figure 13.14 represents a completely general way to combine the antenna-element signals in an adjustable linear structure when the received signals and noises are narrowband. We note that the same generality (for narrowband signals) can be achieved even when the time delays do not result in a phase shift of exactly $\pi/2$ at the center frequency ω_0. Keeping the phase shifts close to $\pi/2$ is desirable for keeping required weight values small, but is not necessary in principle.

When one is interested in receiving signals over a range of frequencies, each of the phase shifters in Figure 13.14 can be replaced by an adaptive transversal filter as shown in Figure 13.15. This tapped delay line permits adjustment of gain and phase as desired at a number of frequencies over the band of interest. If the weight spacing (Δ) is sufficiently small, this network approximates the ideal filter which would allow complete control of the gain and phase at each frequency in the passband.

Once the network connected to each array element has been chosen as in Figure 13.14 or 13.15, the next step is to develop an adaptation procedure so that the weights can adjust to achieve the desired spatial and frequency filtering. The

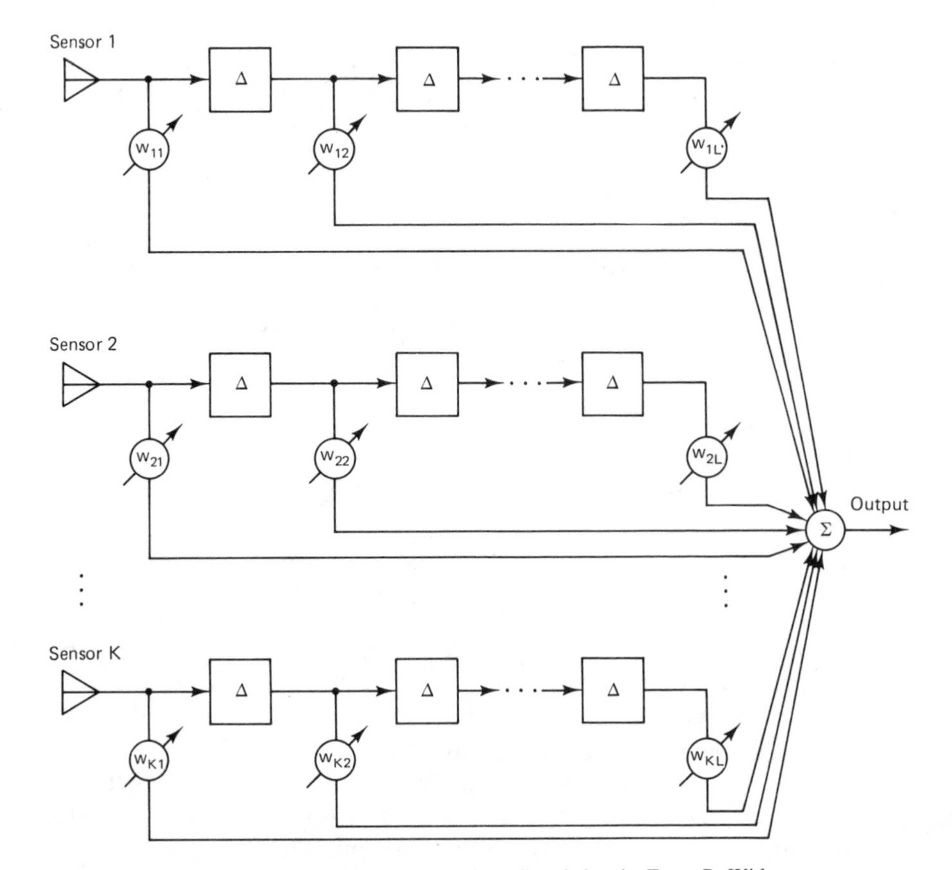

Figure 13.15 Adaptive array for receiving broadband signals. From B. Widrow et al., *Adaptive Antenna Systems*, © December 1967, IEEE.

procedure should produce a given array gain in the specified look direction while simultaneously nulling out interfering noise sources.

If the radiated signals received by the elements of an adaptive array consist of signal components plus undesired noise, the signal will be reproduced (and the noise eliminated) as best as possible in the least-squares sense if the desired response of the adaptive processor is made to be the signal itself. The signal is not generally available for adaptation purposes, however. If it were available, there would be no need for a receiver and a receiving array.

In the adaptive antenna systems to be described here, the desired response signal is provided through the use of an artificially injected signal, the "pilot signal" discussed above, which is completely known at the receiver and usually generated there. The pilot signal is constructed to have spectral and directional characteristics similar to those of the incoming signal of interest. These characteristics may, in some cases, be known a priori but, in general, represent estimates of the parameters of the signal of interest.

Adaptation with the pilot signal causes the array to form a beam in the pilot-signal direction having essentially flat spectral response and linear phase shift within the passband of the pilot signal. Moreover, directional noises impinging on the adaptive array will cause reduced array response (nulling) in their directions within their passbands. These notions are demonstrated by experiments that are described below.

Injection of the pilot signal would normally render the beamformer output useless. To circumvent this, "one-mode" and "two-mode" adaptation algorithms have been devised. The two-mode process alternately adapts on the pilot signal to form the main beam and then, with the pilot signal off, adapts on the natural inputs to eliminate noise. The array output is usable during the second mode, while the pilot signal is off. The one-mode algorithm permits listening at all times, but requires more equipment for its implementation.

ADAPTIVE ALGORITHMS

Figure 13.16 illustrates a method for providing the pilot signal wherein the latter is actually transmitted by an antenna located some distance from the array in the desired look direction. Figure 13.17 shows a self-contained method for providing a local pilot signal that was originally suggested by M. E. Hoff, Jr. The inputs to the processor in Figure 13.17 are connected either to the actual antenna element outputs (during "mode A"), or to a set of delayed signals derived from the pilot-signal generator (during "mode P"). The filters $\delta_1, \ldots, \delta_K$ (ideal time delays if the array elements are identical) are chosen to result in a set of input signals identical with

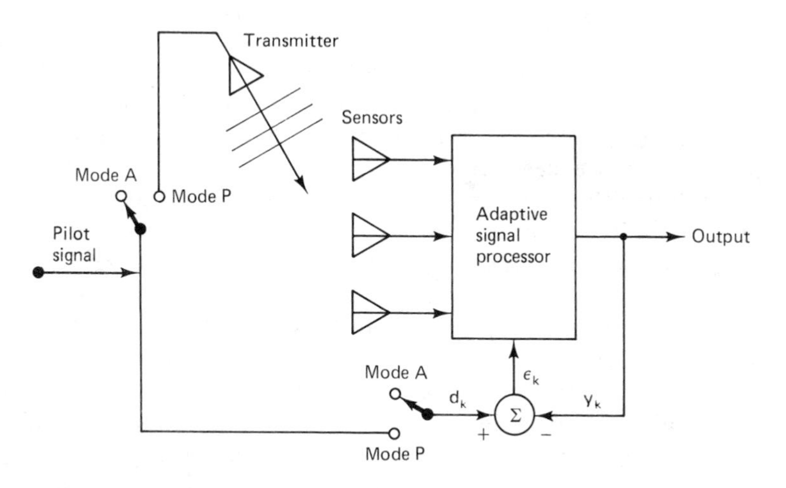

Figure 13.16 Two-mode adaptation with distant external pilot signal. Mode P: adaptation with pilot signal present; mode A: adaptation with pilot signal absent. From B. Widrow et al., *Adaptive Antenna Systems*, © December 1967, IEEE.

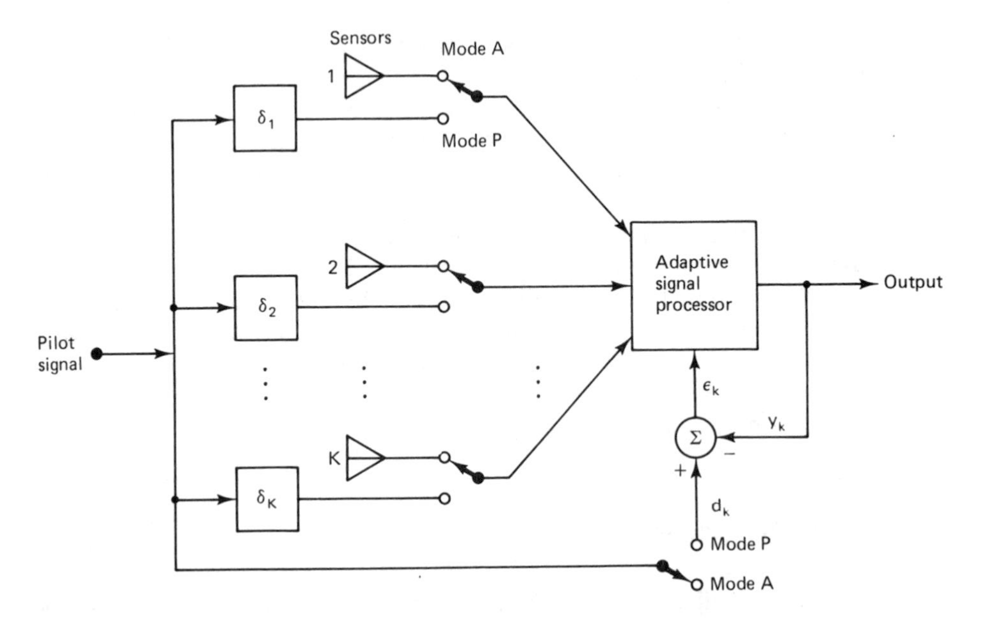

Figure 13.17 Two-mode adaptation with a local internal pilot signal. Mode P: adaptation with pilot signal present; mode A: adaptation with pilot signal absent. From B. Widrow et al., *Adaptive Antenna Systems*, © December 1967, IEEE.

those that would appear if the array were actually receiving a distant radiated plane-wave pilot signal from the desired "look" direction, the direction intended for the main lobe of the receiving directivity pattern.

During adaptation in mode P, the input signals to the adaptive processor are derived from the pilot signal, and the desired response of the adaptive processor is the pilot signal itself. If a sinusoidal pilot signal at frequency ω_0 is used, for example, adapting the weights to minimize mean-square error will force the gain of the antenna array in the look direction to have a specific amplitude and a specific phase shift at frequency ω_0.

During adaptation in mode A, all signals applied to the adaptive processor are received by the antenna elements from the actual noise field. In this mode, the adaptation process proceeds to eliminate all received signals, since the desired response is set to zero. Continuous operation in mode A would cause all the weight values to tend to zero however, and the system would shut itself off. However, by alternating frequently between mode P and mode A and causing only small changes in the weight vector during each mode of adaptation, it is possible to maintain a unit-gain beam approximately in the desired look direction and, in addition, to minimize approximately the reception of incident-noise power.

The pilot signal can be chosen as the sum of several sinusoids with differing frequencies so that adaptation in mode P will constrain the antenna gain and phase in the look direction to have specific values at each of these frequencies. Furthermore, if several pilot signals of different directions are added together, it is possible to constrain the array gain simultaneously at various angles as well as frequencies when adapting in mode P. This feature affords some control of the bandwidth and beamwidth in the various look directions. The two-mode adaptive process approximately minimizes the mean-square value (the total power) of all signals received by the antenna elements which are uncorrelated with the pilot signals, subject to the constraint that the gain and phase in the beam approximate predetermined values at the frequencies and angles dictated by the pilot-signal components.

In two-mode adaptation the beam is formed and supported during mode P and the noises are eliminated in the least-squares sense (subject to the pilot-signal constraints) in mode A. Signal reception during mode P is impossible because the processor is connected to the pilot-signal generator. Reception can therefore take place only during mode A. This difficulty is eliminated in the system of Figure 13.18, in which the actions of modes P and A can be accomplished simultaneously. The pilot signals and the received signals enter into an adaptive processor, just as

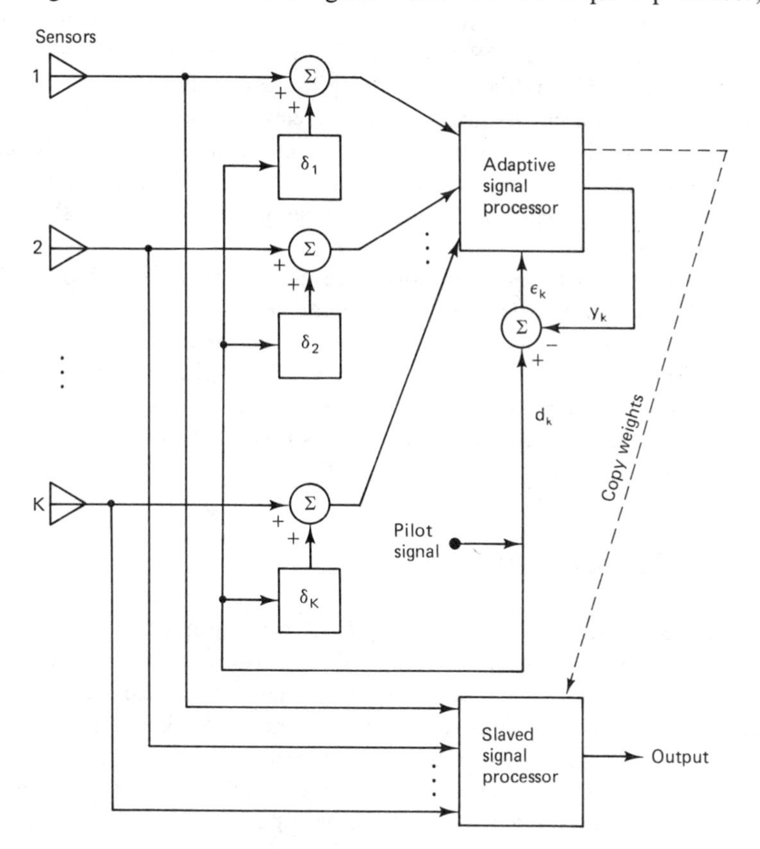

Figure 13.18 Single-mode adaptation with a pilot signal. From B. Widrow et al., *Adaptive Antenna Systems*, © December 1967, IEEE.

described previously. For this processor, the desired response is the pilot signal. A second slaved processor generates the actual array output signal, but it performs no adaptation. Its input signals do not contain the pilot signal. It is slaved to the adaptive processor in such a way that its weights are exact copies of the corresponding weights of the adapting system, so that it never needs to receive the pilot signal.

In the single-mode system of Figure 13.18, the pilot signal is on continuously. Adaptation to minimize the mean-square error will force the adaptive processor to reproduce the pilot signal as closely as possible and, at the same time, to reject as well as possible (in the mean-square sense) all signals received by the antenna elements which are uncorrelated with the pilot signal. Thus the adaptive process forces a directivity pattern having approximately the proper main lobe sensitivity in the look direction in the pilot signal passband (satisfying approximately the pilot signal constraints), and it forces nulls in the directions of the noises and in their frequency bands. Usually, the stronger the noises compared to the pilot signal, the deeper are the corresponding nulls.

The single-mode system of Figure 13.18 can be analyzed by LMS design theory in the following way. The converged weights of the adaptive processor can be obtained as $\mathbf{R}^{-1}\mathbf{P}$. The R-matrix is the sum of the R-matrix of the antenna-derived signals and the R-matrix of the set of pilot signals, assuming that the pilot signal and the incoming antenna signals are uncorrelated. The vector $\mathbf{P}$ is determined only by the pilot signal. Its elements are values of the autocorrelation function of the pilot signal, with lags determined by the corresponding pilot signal delays. Since the main lobe is steered by choice of the pilot signal delays, the vector $\mathbf{P}$ is directly related to the steering angle.

To demonstrate the performance characteristics of adaptive antenna systems, many simulation experiments, involving a wide variety of array geometries and signal- and noise-field configurations, have been carried out.

For simplicity of presentation, the examples outlined in the following are restricted to planar arrays composed of ideal isotropic sensors, or omnis. In every case, the LMS adaptation algorithm was used. All experiments were begun with all weight values initially equal.

NARROWBAND EXPERIMENTS

Figure 13.19 shows a 12-element circular array and signal processor which was used to demonstrate the performance of the narrowband system shown in Figure 13.14. The pilot signal was a unit-amplitude sine wave with frequency ω_0 and power $\sigma_p^2 = 0.5$, which was used to train the array to look in the $\theta = 0°$ direction. The noise field consisted of a sinusoidal noise signal (with the same frequency and power as the pilot signal) incident at angle $\theta = 40°$, and a small amount of random, uncorrelated, zero-mean white noise with power $\sigma_n^2 = 0.1$ at each antenna element. In our simulation, the weights were adapted using the two-mode LMS algorithm.

Figure 13.20 shows the sequence of directivity patterns which evolved during

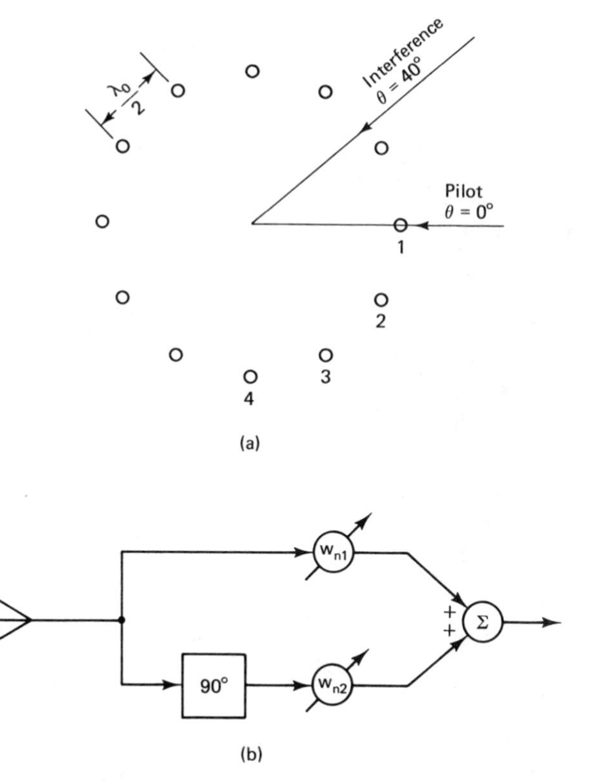

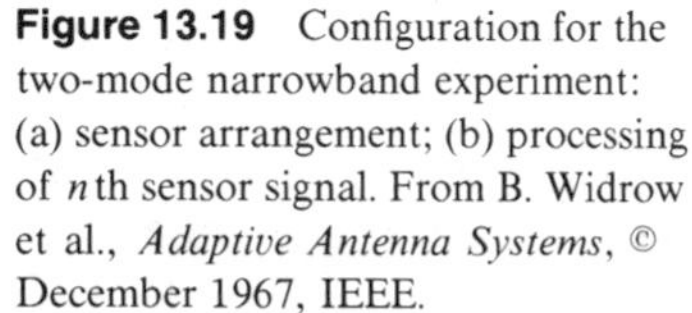
Figure 13.19 Configuration for the two-mode narrowband experiment: (a) sensor arrangement; (b) processing of nth sensor signal. From B. Widrow et al., *Adaptive Antenna Systems*, © December 1967, IEEE.

the learning process. These computer-plotted patterns represent the decibel sensitivity of the array at frequency ω_0, similar to the previous directivity patterns in this chapter. Each directivity pattern is computed from the set of weights resulting at various stages of adaptation. The solid arrow indicates the direction of arrival of the sinusoidal interference. Notice that the initial directivity pattern is essentially circular, due to the symmetry of the antenna array and the equal initial weight values. A timing indicator NC, the number of elapsed cycles at frequency ω_0, is presented with each directivity pattern. The number of adaptations equals 20NC in these experiments. Note that if $\omega_0/2T = 1$ kHz, then NC $= 1$ corresponds to 1 ms of real time; if $\omega_0/2T = 1$ MHz, then NC $= 1$ corresponds to 1 μs; and so on.

Several observations can be made from the series of directivity patterns of Figure 13.20. Notice that the sensitivity of the array in the look direction is essentially constant during the adaptation process. Also notice that the array sensitivity drops very rapidly in the direction of the sinusoidal noise source; a deep notch in the directivity pattern forms in the noise direction as the adaptation process progresses. After the adaptive transient has died out, the array sensitivity in the noise direction is 27 dB below that of the array in the desired look direction. The notch was created by the adaptive algorithm without prior knowledge of the noise frequency content or direction of arrival.

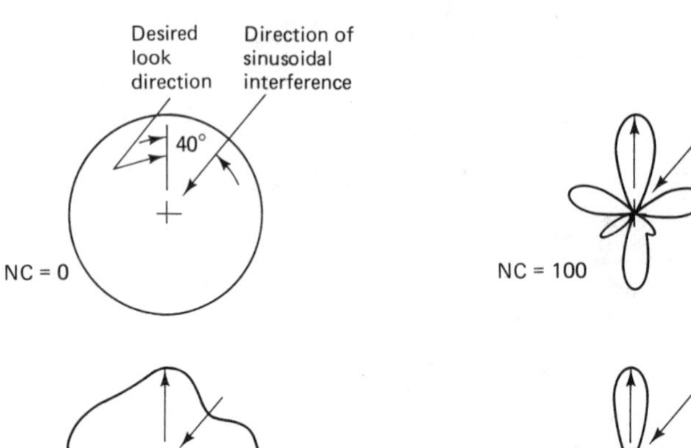

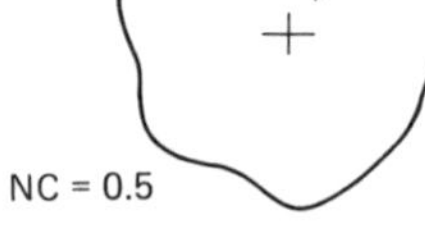

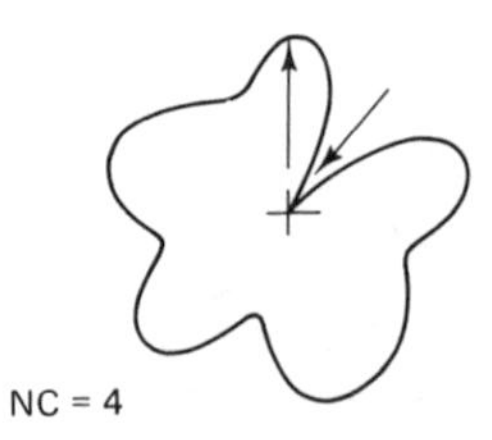

Figure 13.20 Evolution of the directivity pattern while learning to eliminate directional noise as well as uncorrelated noise (array configuration in Figure 13.19). NC = number of elapsed cycles of frequency ω_0; number of adaptations = 20 NC. From B. Widrow et al., *Adaptive Antenna Systems*, © December 1967, IEEE.

The total noise power in the array output is the sum of the sinusoidal noise power due to the directional noise source plus the power due to the white, mutually uncorrelated noise-input signals. The total noise power at the output generally drops during the adaptation process until it reaches an irreducible level.

A plot of the total output noise power as a function of NC is shown in Figure 13.21. This plot is a learning curve for the adaptive beamformer. Starting with the initial weights, the total output noise power was 0.65, as shown in the figure. After adaptation, the total output noise power was 0.01. In this noise field, the output signal-to-noise ratio after adaptation was better than that of a single isotropic receiving element by a factor of about 60.

A second experiment using the same array configuration in Figure 13.19 and the two-mode adaptive process was performed to investigate adaptive array perform-

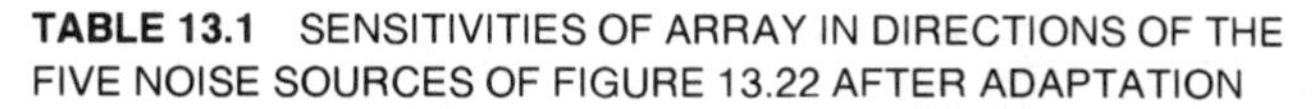

TABLE 13.1 SENSITIVITIES OF ARRAY IN DIRECTIONS OF THE FIVE NOISE SOURCES OF FIGURE 13.22 AFTER ADAPTATION

Noise direction (deg)	Noise frequency (times ω_0)	Array sensitivity in noise direction, relative to sensitivity in desired look direction (dB)
67	1.10	−26
134	0.95	−30
191	1.00	−28
236	0.90	−30
338	1.05	−38

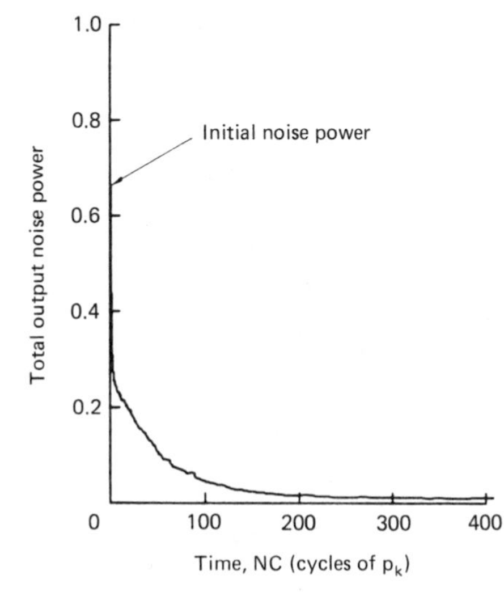

Figure 13.21 Learning curve for narrowband system of Figure 13.19 with noise from one direction only. From B. Widrow et al., *Adaptive Antenna Systems*, © December 1967, IEEE.

ance in the presence of several interfering directional noise sources. In this example, the noise field was composed of five directional sinusoidal noises, each with amplitude 0.5 and power 0.125, acting simultaneously and, in addition, superposed uncorrelated white noises of power 0.5 at each of the antenna elements. The frequencies of the five directional noises are shown in Table 13.1.

Figure 13.22(a) shows the evolution of the directivity pattern, plotted at frequency ω_0, from the initial conditions to the finally converged (adapted) state. The latter was achieved after NC = 682 cycles at frequency ω_0. The learning curve for this experiment is shown in Figure 13.22(b). The final array sensitivities in the five noise directions relative to the array sensitivity in the desired look direction are shown in Table 13.1. The signal-to-noise ratio was improved by a factor of about 15 over that of a single isotropic radiator. In Figure 13.22(b), one can roughly discern a time constant approximately equal to 70 cycles. Since there were 20 adaptations per cycle at ω_0, the learning curve time constant was approximately $\tau_{\text{mse}} = 1400$ adaptations. Within about 400 cycles at ω_0, the adaptive process virtually converges

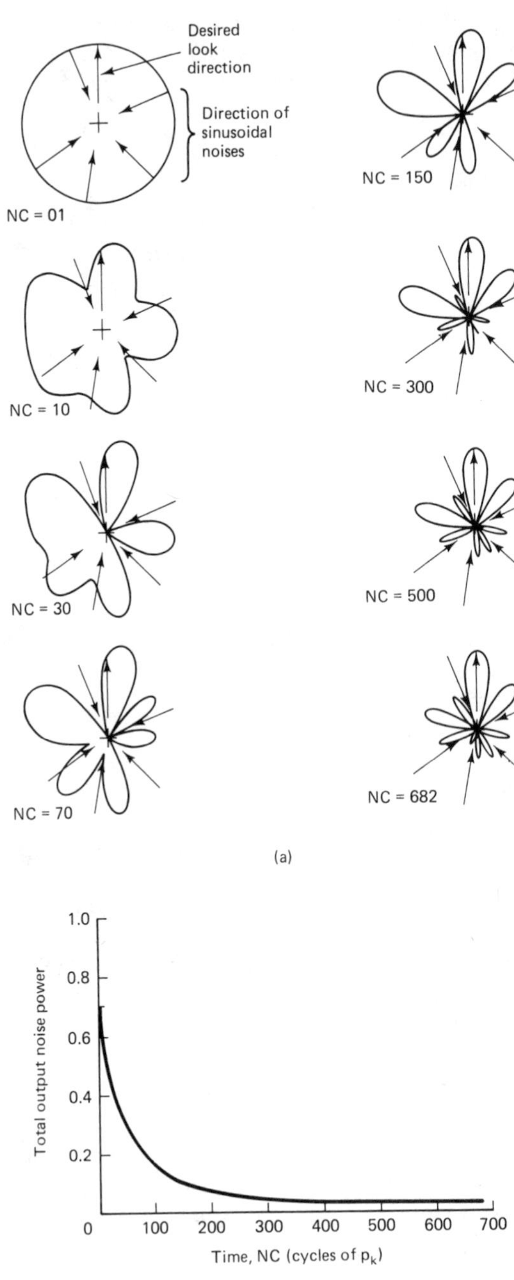

Figure 13.22 Evolution of the directivity pattern while learning to eliminate five directional noises and uncorrelated noises (array configuration of Figure 13.19): (a) sequence of directivity patterns during adaptation; (b) learning curve (total number of adaptations = 20 NC). From B. Widrow et al., *Adaptive Antenna Systems*, © December 1967, IEEE.

to steady state. If $\omega_0/2\pi T$ were 1 MHz, 400 μs would be the real settling time. The misadjustment for this process can be roughly estimated by using (6.41), although actually all eigenvalues were not equal as required by this equation:

$$M = \frac{\text{number of weights}}{4\tau_{\text{mse}}} = \frac{24}{4\tau_{\text{mse}}} = \frac{6}{1400} = 0.43 \text{ percent} \tag{13.46}$$

This is a low value of misadjustment, indicating a very slow, precise adaptive process. This is evidenced by the learning curve Figure 13.22(b) for this experiment, which is smooth and noise-free.

If all of the sinusoidal interferers were at the same frequency, they would have been just as easily attenuated by the adaptive beamformer, but generally nulls would not have formed at their angles of incidence. They would all have been accepted by the adaptive beamformer and would have been given just the right magnitude and phase characteristics so that they would add up at the output approximately to zero. If, as in the experiment above, all the frequencies are different, the adaptive beamformer must form individual nulls to eliminate the sinusoidal interferers.

BROADBAND EXPERIMENTS

Figure 13.23 shows an antenna array configuration and signal processor that was used in a series of computer-simulated broadband experiments. In these experiments, the one-mode or simultaneous adaptation process was used to adjust the weights. Each omnidirectional sensor in a five-element circular array was connected to an adaptive transversal filter having five weights, as shown in the figure. A broadband pilot signal was used, and the desired look direction was chosen (arbitrarily) to be $\theta = -13°$. The power spectrum of the pilot signal is shown in Figure 13.24(a). This spectrum is approximately one octave wide and is centered at frequency ω_0. A delay increment of $\delta = \pi/2\omega_0$ time steps was used in the adaptive filter, providing a delay between adjacent weights of a quarter cycle at frequency ω_0, and a total filter delay of one wavelength at this frequency.

The computer-simulated noise field consisted of two broadband directional noise sources† incident on the array at angles $\theta = 50°$ and $\theta = -70°$. Each source of noise had power 0.5. The noise at $\theta = 50°$ had the same frequency spectrum as the pilot signal (though with reduced power); while the noise at $\theta = -70°$ was narrower and centered at a slightly higher frequency. The noise sources were uncorrelated with the pilot signal. Figure 13.24(b) shows the power spectra. Additive white noises (mutually uncorrelated) of power 0.0625 were also present in each of the antenna-element signals.

To demonstrate the effects of adaptation rate, the experiments were performed twice, using two different values of the adaptive gain constant, μ in (6.3). Figure

† The broadband directional noises were computer-simulated by first generating a series of uncorrelated (white) pseudorandom numbers, applying these to an appropriate digital filter to achieve the proper spectral characteristics, and then applying the resulting correlated noise waveform to each of the simulated antenna elements with the appropriate delays to simulate the effect of a propagating wavefront.

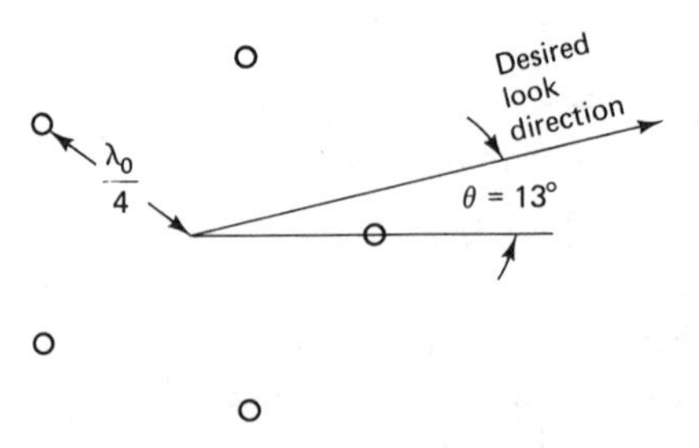

(a)

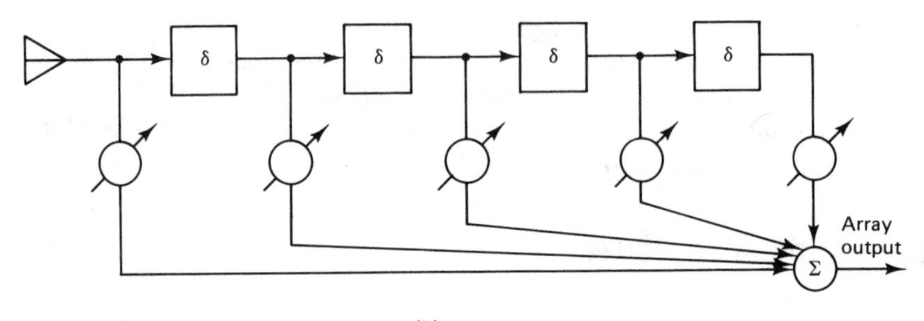

(b)

Figure 13.23 Configuration for the one-mode broadband experiment: (a) sensor arrangement; (b) processing of each sensor signal. From B. Widrow et al., *Adaptive Antenna Systems*, © December 1967, IEEE.

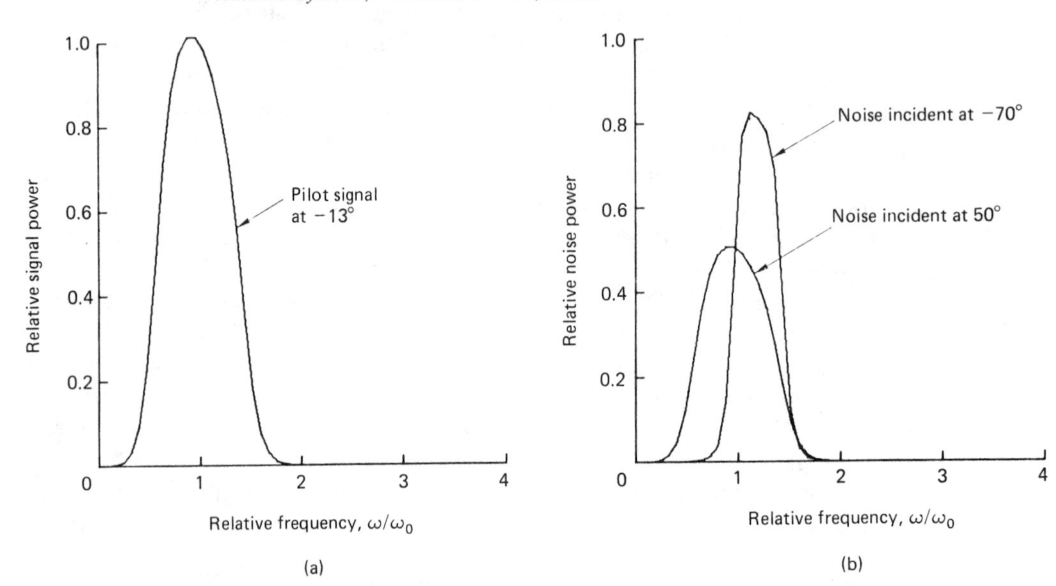

Figure 13.24 Power spectra for broadband experiments: (a) pilot signal at $\theta = -13°$; (b) incident noises at $\theta = 50°$ and $\theta = -70°$. From B. Widrow et al., *Adaptive Antenna Systems*, © December 1967, IEEE.

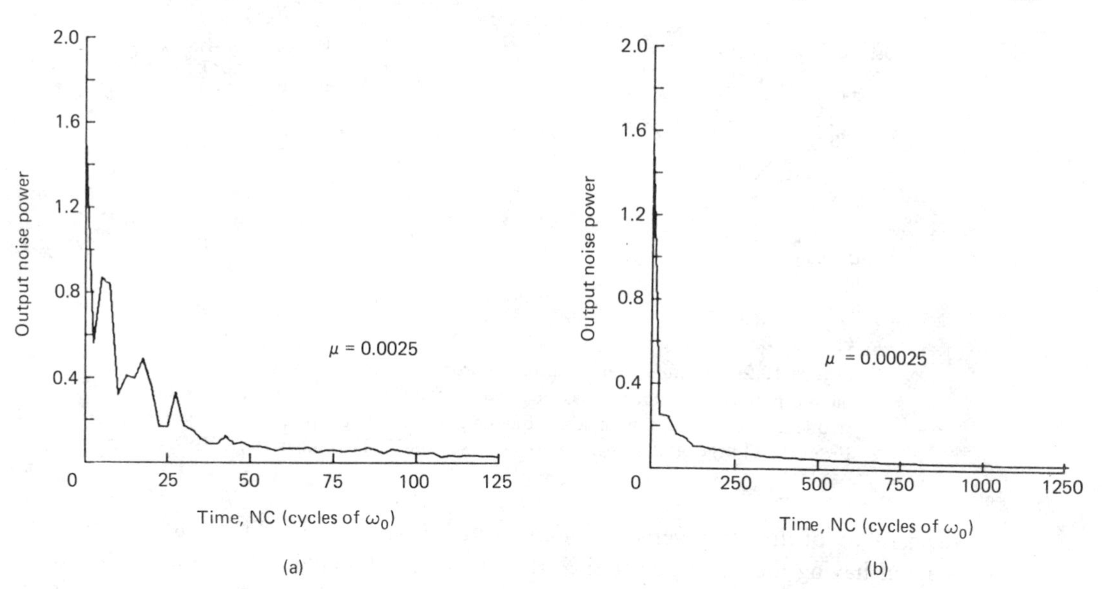

Figure 13.25 Learning curves for broadband experiments: (a) rapid learning ($M = 13\%$); (b) slow learning ($M = 1.3\%$). From B. Widrow et al., *Adaptive Antenna Systems*, © December 1967, IEEE.

13.25 shows the learning curves obtained under these conditions. The abscissa of each curve is expressed in cycles of ω_0, the array center frequency, and, as before, the array was adapted at a rate of 20 times per cycle of ω_0. Note that the faster learning curve is much noisier.

Since the statistics of the pilot signal and directional noises in this example are known (having been generated in the computer simulation) and thus the R-matrix is known, it is possible to check measured values of misadjustment against theoretical values.

Using (6.36), the misadjustment for the two values of μ is calculated to give the following comparison:

μ	Measured value of M	Theoretical value of M
0.0025	0.134	0.1288
0.00025	0.0170	0.0129

The theoretical values of misadjustment check quite well with corresponding measured values.

From the known statistics the optimum weight vector $\mathbf{W}^* = \mathbf{R}^{-1}\mathbf{P}$ can be computed. The antenna directivity pattern for this optimum weight vector is shown in Figure 13.26(a). This is a broadband directivity pattern, in which the relative

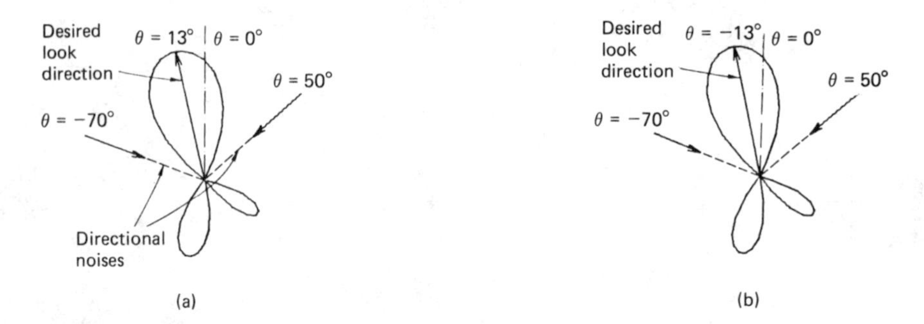

Figure 13.26 Comparison of optimum broadband directivity pattern with experimental pattern after former has been adapted during 625 cycles at ω_0; Plotted at frequency ω_0: (a) optimum least-squares pattern; (b) adapted with $\mu = 0.00025$. From B. Widrow et al., *Adaptive Antenna Systems*, © December 1967, IEEE.

sensitivity of the array versus incidence angle θ is plotted for a broadband received signal having the same power spectrum as the pilot signal. This form of directivity pattern has few sidelobes, and nulls which are generally not very deep. In Figure 13.26(b), the broadband directivity pattern resulting from adaptation (after 625 cycles of ω_0, with $\mu = 0.00025$) is plotted for comparison with the optimum broadband pattern (i.e., the least-squares solution). Note that the patterns are almost indistinguishable.

The learning curves of Figure 13.25 are composed of decaying exponentials with various time constants. When μ is set to 0.00025 in Figure 13.25(b), the misadjustment is about 1.3 percent, which is a small but practical value. With this rate of adaptation, we can see in Figure 13.25(b) that adapting transients are essentially finished after about 500 cycles at ω_0. If $\omega_0/2\pi T$ is 1 MHz, for example, adaptation could be completed (if the processor is fast enough) in about 500 μs. If $\omega_0/2\pi T$ is 1 kHz, adaptation could be completed in about one-half second. Faster adaptation is possible, but there will be more misadjustment. These figures are typical for an adaptive antenna with broadband noise inputs with 25 adaptive weights. For the same level of misadjustment, convergence times increase approximately linearly with the number of weights as shown, for example, by (6.22).

The ability of this adaptive antenna array to obtain "frequency tuning" is shown in Figure 13.27. This figure gives the sensitivities of the adapted array (after 1250 cycles at ω_0) as a function of frequency for the desired look direction [Figure 13.27(a)] and for the two noise directions [Figure 13.27(b) and (c)]. The spectra of the pilot signal and noises are also shown in the figure.

In Figure 13.27(a), the adaptive process makes the sensitivity of this simple array configuration close to unity over the band of frequencies where the pilot signal has finite power density. Improved performance could be attained by adding antenna elements and by adding more taps to each delay line; or, more simply, by limiting the output to the passband of the pilot signal. Figure 13.27(b) and (c) shows the sensitivities of the array in the directions of the noises. These plots illustrate the

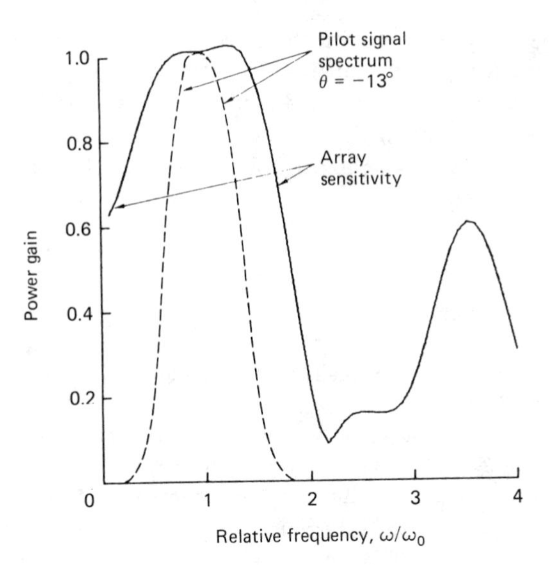

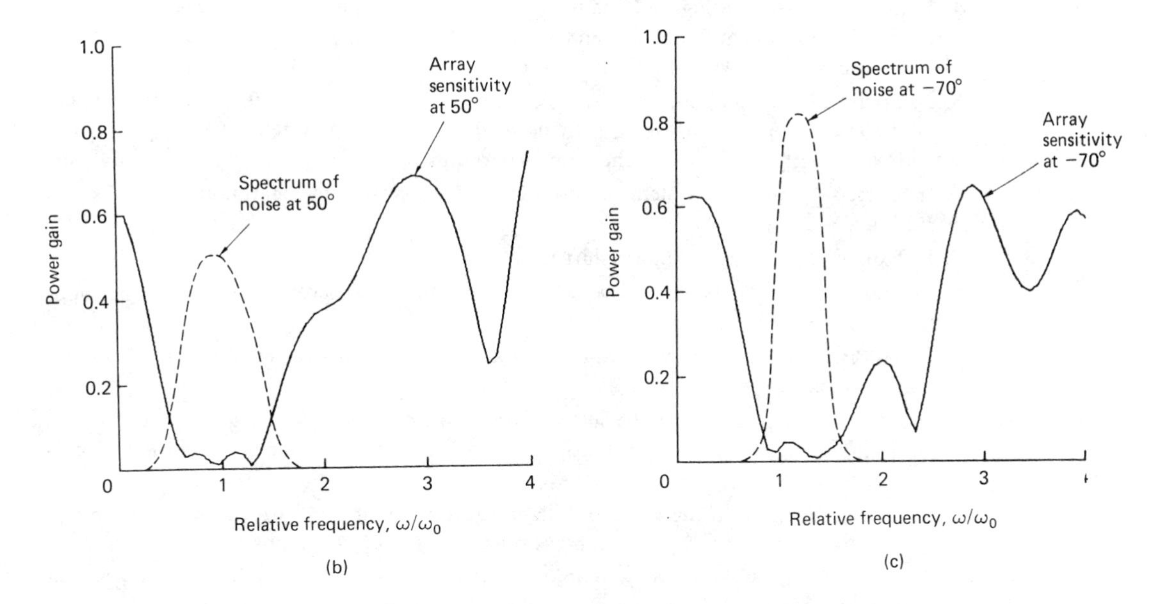

Figure 13.27 Array power gain versus frequency for the broadband experiment in Figure 13.23: (a) power gain in look direction, $\theta = -13°$; (b) power gain in one noise direction, $\theta = 50°$; (c) power gain in other noise direction, $\theta = -70°$. From B. Widrow et al., *Adaptive Antenna Systems*, © December 1967, IEEE.

reduction of the array sensitivity in the directions of the noises, within their specific passbands. The same idea is illustrated by the nulls in the broadband directivity patterns which occur in the noise directions, shown in Figure 13.26. After the adaptive transients subside in this experiment, the signal-to-noise ratio is improved by the array over that of a single isotropic sensor by a factor of 56.

EXERCISES

1. In Figure 13.2, show mathematically that cancellation is the same for signals arriving at incident angles θ and $180° - \theta$.
2. In Figure 13.2, a plane wave with a frequency of 2 Hz and a speed of 5000 m/s is incident at angle $\theta_0 = 30°$. What would the array spacing need to be to produce a phase shift of 10° from the reference input to the primary input?
3. In Figure 13.3(a), a plane wave at frequency 3 MHz and speed 3×10^8 m/s is incident at $\theta = 20°$ (rather than $\theta = 0°$ as in the figure). If the sensors are 20 m apart and if the SNR is 6 dB, what are the weight values for minimum mean-square output?
4. Using the situation in Figure 13.3a with μ equal to 0.01 times the input signal power, run the LMS algorithm and plot w_{1k} and w_{2k} versus k, showing convergence to w_1^* and w_2^*. For noise, generate white uniform noise using r_k in Appendix A.
5. In Figure 13.3(b), suppose that the weight has converged to $w_1^* = 0.9$. What is the gain of the system for a small signal arriving at $\theta = 40°$ at normalized frequency = 0.5 rad and wavelength = 0.305 m, if the omni separation is 0.229 m?
6. In Exercise 5, plot the system power gain in dB versus omni separation l, for $1 \leq l \leq 1.27$ m.
7. In Exercise 5, what is the phase shift from input to output?
8. For a two-element adaptive array, what is the minimum power gain in dB if the input SNR is 8 dB?
9. Plot a directivity pattern for a two-element array with weights $w_1 = 0.6$, $w_2 = 0.8$, and no receiver noise.
10. In an adaptive array using the leaky LMS algorithm with $\gamma = 0.99$, suppose that the entire input signal, $\mathbf{X}_k$, suddenly drops to zero. In how many time steps will the weights change by 10%?
11. In an adaptive filter using the leaky LMS algorithm with $\mu = 0.01/\sigma_x^2$, where σ_x^2 is the input power, how much equivalent noise power is added to the input by setting $\gamma = 0.94$?
12. An adaptive array with two weights spaced 0.8 wavelength apart has a strong signal arriving at $\theta = 0°$ and another signal with 25% of the strong signal power arriving at $\theta = 90°$. What are the converged weight values in this case, assuming no receiver noise?
13. In the array shown in Figure 13.11(a), an input signal, $C \cos k\omega_0$, arrives at angle θ. Give an expression for the array output signal.
14. Using the result of Exercise 13, express the array gain in dB as a function of θ.
15. Do Exercise 13 for the array in Figure 13.11(b).
16. Using the result of Exercise 15, express the array gain in dB as a function of θ.

17. In Figure 13.12, assuming that the weight values are $w_{11}, w_{12}, \ldots, w_{71}, w_{72}$ and that the omnis are separated by 1/2 wavelength, derive a formula for array gain in dB as a function of θ.

18. In Figure 13.13, what are the correct weight values if the noise arrives at $\theta = \pi/4$ instead of $\pi/6$?

19. For the leaky LMS algorithm, using the methods in Chapter 6:
(a) Determine the range of μ for stable operation.
(b) Determine the time constant of the learning curve.
(c) Give an expression for the misadjustment.

20. Express the processor output signal for the system described in Figure 13.19.

21. Assume that the array in Figure 13.12(b), with weights w_{01} through w_{62}, is to be trained using single-mode adaptation with a pilot signal of unit amplitude at a frequency of 1.0 Hz and with simultaneous interference of unit amplitude at 1.1 Hz. Starting with all weights = 1.0, run the LMS algorithm for this configuration with $\mu = 0.01$, and plot the directivity pattern after 0, 10, 100, and 1000 cycles of the pilot signal. Use time step $T = 0.02$ s.

22. A two-element pilot-signal adaptive beamformer is shown below with an incident plane wave, $B \sin k\omega_1$, at $\theta = 0°$. The pilot signal is $A \sin k\omega_0$. Beam steering is accomplished

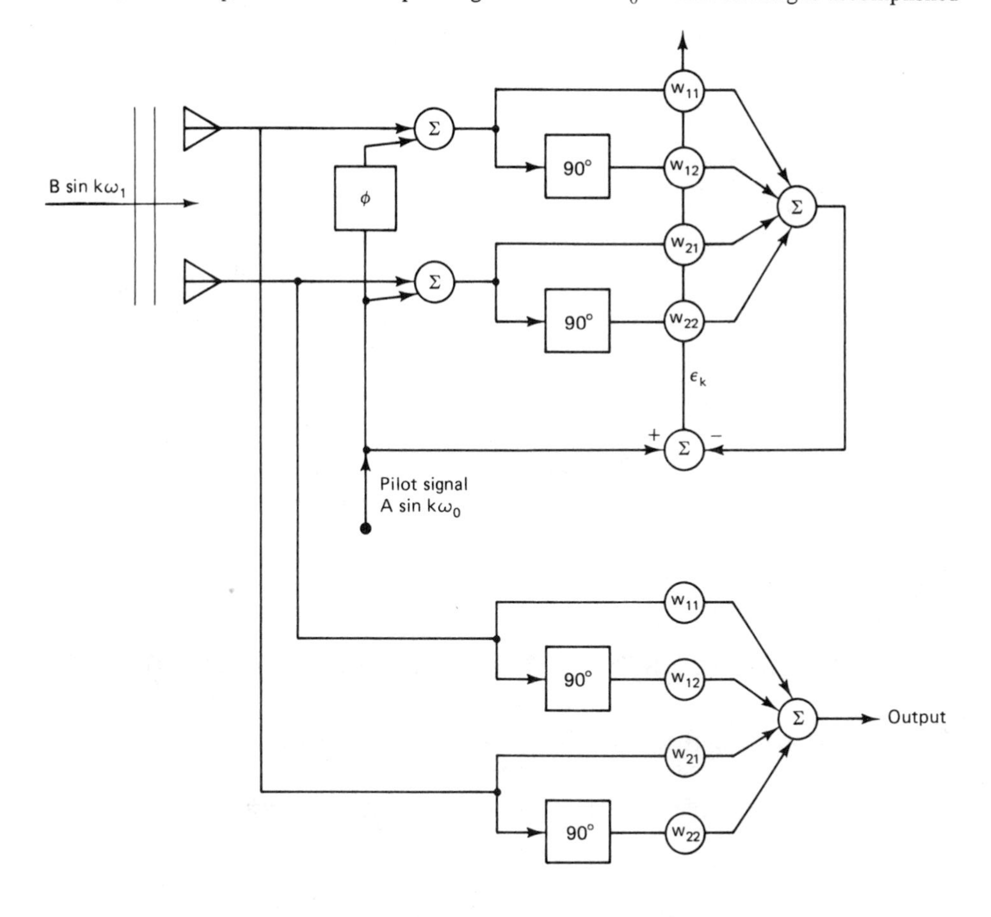

by varying the phase shift ϕ to cause the look direction to be at angle θ. Assume that ω_0 and ω_1 differ enough to decorrelate the signals but that the 90° units provide 90° phase shift for both signals.

(a) Express the R-matrix and the P-vector as functions of A, B, and ϕ, assuming noise-free reception.

(b) With $A = B$, determine the optimum weight vector for $\phi = 0$, 22.5°, 45°, 67.5°, and 90°.

(c) What is the look direction, θ, corresponding to each value of ϕ above?

(d) On a single graph, plot directivity patterns for $\phi = 45°$, 60°, and 90°. Discuss the angular resolution of this adaptive beamformer in the look direction.

23. In the adaptive beamformer of Exercise 22, suppose there is receiver noise with unit power at each omni. Repeat Exercise 22 first with $A = B = 2$, then with $A = 5$ and $B = 1$, and finally with $A = B = 5$. What can we conclude regarding the relation of angular resolution to pilot and incident signal-to-noise ratios?

ANSWERS TO SELECTED EXERCISES

1. *Hint:* See Equations (13.2) and (13.3).

2. $l = 138.9\,\text{m}$

3. $w_1^* = 0.7266,\ w_2^* = 0.3330$

13. $y_k = C \sum_{n=0}^{6} \cos(k\omega_0 + n\pi \sin\theta)$

14. $$\text{dB} = 10\log_{10}\left\{\left[\sum_{n=0}^{6} \cos(n\pi\sin\theta)\right]^2 + \left[\sum_{n=0}^{6} \sin(n\pi\sin\theta)\right]^2\right\}$$

15. $y_k = C \sum_{n=0}^{6} \cos(k\omega_0 + n\pi \sin\theta - 0.8131n)$

16. $$\text{dB} = 10\log_{10}\left\{\left[\sum_{n=0}^{6} \cos n(\pi\sin\theta - 0.8131)\right]^2 + \left[\sum_{n=0}^{6} \sin n(\pi\sin\theta - 0.8131)\right]^2\right\}$$

17. $$\text{dB} = 10\log_{10}\left\{\left[\sum_{n=0}^{6} w_{n1}\cos(n\pi\sin\theta) + w_{n2}\sin(n\pi\sin\theta)\right]^2 + \left[\sum_{n=0}^{6} w_{n2}\cos(n\pi\sin\theta) - w_{n1}\sin(n\pi\sin\theta)\right]^2\right\}$$

REFERENCES AND ADDITIONAL READINGS

1. B. Widrow, P. E. Mantey, L. J. Griffiths, and B. B. Goode, "Adaptive antenna systems," *Proc. IEEE*, vol. 55, p. 2143, Dec. 1967.
2. R. A. Monzingo and T. W. Miller, *Introduction to Adaptive Arrays*. New York: Wiley, 1980.

3. B. Widrow et al, "Adaptive noise cancelling: principles and applications," *Proc. IEEE*, vol. 63, p. 1692, Dec. 1975.
4. J. S. Koford and G. F. Groner, "The use of an adaptive threshold element to design a linear optimal pattern classifier," *IEEE Trans. Inf. Theory*, vol. IT-12, pp. 42–50, Jan. 1966.
5. K. Steinbuch and B. Widrow, "A critical comparison of two kinds of adaptive classification networks," *IEEE Trans. Electron. Comput. (Short Notes)*, vol. EC-14, pp. 737–740, Oct. 1965.
6. C. H. Mays, "The relationship of algorithms used with adjustable threshold elements to differential equations," *IEEE Trans. Electron. Comput. (Short Notes)*, vol. EC-14, pp. 62–63, Feb. 1965.
7. Special Issue on Active and Adaptive Antennas, *IEEE Trans. Antennas Propagat.*, vol. AP-12, Mar. 1964.
8. C. V. Jakowatz, R. L. Shuey, and G. M. White, "Adaptive waveform recognition," in *Fourth London Symp. on Information Theory*. London: Butterworth, Sept. 1960, pp. 317–326.
9. L. D. Davisson, "A theory of adaptive filtering," *IEEE Trans. Inf. Theory*, vol. IT-12, pp. 97–102, Apr. 1966.
10. E. M. Glaser, "Signal detection by adaptive filters," *IRE Trans. Inf. Theory*, vol. IT-7, pp. 87–98, Apr. 1961.
11. F. Bryn, "Optimum signal processing of three-dimensional arrays operating on gaussian signals and noise," *J. Acoust. Soc. Am.*, vol. 34, pp. 289–297, Mar. 1962.
12. H. Mermoz, "Adaptive filtering and optimal utilization of an antenna," U.S. Navy Bureau of Ships (translation 903 of Ph.D. thesis, Institut Polytechnique, Grenoble, France), Oct. 4, 1965.
13. S. W. W. Shor, "Adaptive technique to discriminate against coherent noise in a narrowband system," *J. Acoust. Soc. Am.*, vol. 39, pp. 74–78, Jan. 1966.
14. F. W. Smith, "Design of quasi-optimal minimum time controllers," *IEEE Trans. Autom. Control*, vol. AC-11, pp. 71–77, Jan. 1966.
15. J. P. Burg, "Three-dimensional filtering with an array of seismometers," *Geophysics*, vol. 29, pp. 693–713, Oct. 1964.
16. J. F. Claerbout, "Detection of P waves from weak sources at great distances," *Geophysics*, vol. 29, pp. 197–211, Apr. 1964.
17. H. Robbins and S. Monro, "A stochastic approximation method," *Ann. Math. Stat.*, vol. 22, pp. 400–407, Mar. 1951.
18. J. Kiefer and J. Wolfowitz, "Stochastic estimation of the maximum of a regression function," *Ann. Math. Stat.*, vol. 23, pp. 462–466, Mar. 1952.
19. A. Dvoretzky, "On stochastic approximation," *Proc. Third Berkeley Symp. on Mathematical Statistics and Probability*, J. Neyman (Ed.). Berkeley, Calif.: University of California Press, 1956, pp. 39–55.
20. B. Widrow, "Adaptive sampled data systems," *Proc. First International Congress of the International Federation of Automatic Control* (Moscow, 1960). London: Butterworth, 1960.

21. D. Gabor, W. P. L. Wilby, and R. Woodcock, "A universal non-linear filter predictor and simulator which optimizes itself by a learning process," *Proc. IEE* (London), vol. 108 B, July 1960.
22. R. V. Southwell, *Relaxation Methods in Engineering Science*. London: Oxford University Press, 1940.
23. B. B. Goode, "Synthesis of a nonlinear Bayes detector for gaussian signal and noise fields using Wiener filters," *IEEE Trans. Inf. Theory* (Correspondence), vol. IT-13, pp. 116–118, Jan. 1967.
24. J. Capon, R. J. Greenfield, and R. J. Kolker, "Multidimensional maximum-likelihood processing of a large aperture seismic array," *Proc. IEEE*, vol. 55, pp. 192–211, Feb. 1967.
25. L. J. Griffiths, "A comparison of multidimensional Wiener and maximum-likelihood filters for antenna arrays," *Proc. IEEE* (Letters), vol. 55, pp. 2045–2047, Nov. 1967.
26. B. Widrow, "Bootstrap learning in threshold logic systems," presented at the American Automatic Control Council (Theory Committee), IFAC Meeting, London, June 1966.
27. Special Issue on Adaptive Arrays, *IEEE Trans. Antennas Propagat.*, vol. Ap-24, Sept. 1976.
28. L. J. Griffiths, "A simple adaptive algorithm for real-time processing in antenna arrays," *Proc. IEEE*, vol. 57, pp. 1696–1704, Oct. 1969.
29. O. L. Frost III, "An algorithm for linearly constrained adaptive array processing," *Proc. IEEE*, vol. 60, pp. 926–935, Aug. 1972.
30. R. T. Compton, Jr., et al, "Adaptive arrays for communication systems: an overview of research at the Ohio State University," *IEEE Trans. Antennas Propagat.*, vol. PGAP-24, pp. 599–607, Sept. 1976.
31. S. P. Applebaum, "Adaptive arrays," *IEEE Trans. Antennas Propagat.*, vol. PGAP-24, pp. 585–598, Sept. 1976.

chapter 14

Analysis of Adaptive Beamformers

In Chapter 13 we discussed some of the basic properties of linear arrays, and how to apply adaptive signal processing for the arrays using the LMS algorithm. In this chapter we discuss some additional adaptive beamforming techniques and algorithms.

First, however, we discuss some additional performance characteristics of receiving arrays connected to adaptive beamformers. As we have seen, such systems are intended to receive signals from a selected look direction while nulling interferers from other directions.

PERFORMANCE CHARACTERISTICS OF RECEIVING ARRAYS

The broadband adaptive array of Figure 13.15 is the basic system configuration to be discussed here. Its signals could be either narrowband or broadband. There are K receiving elements, each connected to a tapped delay line (adaptive transversal filter) having L taps; thus the total number of weights in the system is KL.

Incident upon the receiving array is a target signal plus noise which includes not only receiver noise but all forms of interference and jamming from point sources and from spatially distributed sources. Ideally, one would like the output to be the target signal itself, with no noise. In practice this is rarely achievable and one must make compromises in the design of the array processing scheme, choosing between target signal distortion and noise elimination. Two different approaches will be discussed in this section, one that provides a system output that is a best least-squares estimate of the target signal and another that provides an output which is an undistorted version of the target signal plus added noise of minimal power. The first approach is based on a minimum-mean-square-error criterion, while the second is based on a maximum likelihood criterion. Then, in sections to follow, adaptive algorithms will be presented for real-time processing of adaptive array signals in accord with these two performance criteria. The analytical approach to be followed here is based on the work of L. J. Griffiths [3, 4] and O. L. Frost III [7, 8].

Refer once again to the adaptive array of Figure 13.15. Let an origin be defined at some point in space near the antenna elements (omnis). Imagine an omni element being placed at this origin to receive the signal plus the noise. Let the target signal component of the omni output be defined as g_k, where k is the time index as usual. The set of KL weights can be described as follows. (In this chapter, for convenience, we use a single subscript to indicate the weight number, and the second subscript, k, is again the time index.)

$$\mathbf{W}_k \triangleq \begin{bmatrix} w_{1k} \\ \vdots \\ w_{ik} \\ \vdots \\ w_{(KL)k} \end{bmatrix} \tag{14.1}$$

Each of the weights, connected to a tapped delay line, receives the target signal plus noise. The ith weight has an input of

$$x_{ik} = s_{ik} + n_{ik} \tag{14.2}$$

The target signal component is s_{ik}, which is linearly related to g_k. The linear relationships between g_k and the various input signals to the individual weights result from the passage of the target signal waveform across the array. For the ith weight,

$$s_{ik} = F_i[g_k] \tag{14.3}$$

where $F_i[\cdot]$ is a linear function. For the entire array, it is clear that

$$\mathbf{X}_k = \mathbf{S}_k + \mathbf{N}_k \tag{14.4}$$

$$\mathbf{X}_k \triangleq \begin{bmatrix} x_{1k} \\ \vdots \\ x_{ik} \\ \vdots \\ x_{KLk} \end{bmatrix} \qquad \mathbf{S}_k \triangleq \begin{bmatrix} s_{1k} \\ \vdots \\ s_{ik} \\ \vdots \\ s_{KLk} \end{bmatrix} \qquad \mathbf{N}_k \triangleq \begin{bmatrix} n_{1k} \\ \vdots \\ n_{ik} \\ \vdots \\ n_{KLk} \end{bmatrix} \tag{14.5}$$

For this system, the input correlation matrix will be the sum of that of the target signal plus that of the noise. Accordingly,

$$\mathbf{R}_{SS} \triangleq E\left[\mathbf{S}_k\mathbf{S}_k^{\mathrm{T}}\right] \qquad \mathbf{R}_{NN} \triangleq E\left[\mathbf{N}_k\mathbf{N}_k^{\mathrm{T}}\right] \tag{14.6}$$

$$\mathbf{R}_{XX} = \mathbf{R}_{SS} + \mathbf{R}_{NN} \tag{14.7}$$

The desired response at the adaptive array output is the target signal itself. The cross correlation between the desired response and the X-vector is

$$\mathbf{P}_S = E[g_k\mathbf{X}_k] = E[g_k\mathbf{S}_k] \tag{14.8}$$

To obtain an output that is a best least-squares estimate of the target signal, the optimal weight vector is

$$\mathbf{W}^* = \mathbf{R}_{XX}^{-1}\mathbf{P}_S$$

$$= [\mathbf{R}_{SS} + \mathbf{R}_{NN}]^{-1}E[g_k\mathbf{S}_k] \tag{14.9}$$

One can approximately implement this formula by estimating $\mathbf{R}$ and finding its inverse, using actual signal-plus-noise input data. Without knowing the target signal (if this were known, one would not need the receiving array), one can determine $\mathbf{P}$ with knowledge of the autocorrelation function of the target signal and its direction of arrival. The cross correlations between the target signal g_k and the various target signal components of the weight inputs s_{ik} can be computed by taking into account the array geometry, the time delays that result as the target signal passes over the array, and the time delays that accrue at the taps of the tapped delay lines. In any event, $\mathbf{R}^{-1}$ and $\mathbf{P}$ can be evaluated and the optimal least-squares processor can be constructed.

Rather than use matrix methods for computing the least-squares solution, real-time adaptive techniques are more often used. The pilot-signal algorithm described in Chapter 13 is one approach. Let us now examine the pilot-signal algorithm to determine whether its weight solution agrees with (14.9).

We will analyze the single-mode pilot-signal algorithm whose block diagram is shown in Figure 13.18. Let the pilot signal be designated p_k, and suppose that this signal is applied to the system through an adjustable gain β. The desired response for the adaptive process is then

$$d_k = \beta p_k \tag{14.10}$$

The inputs to the weights consist of target signal components and interference components, as described above, and in this case additional pilot signal components. Let these pilot signal components be designated by the vector $\beta\mathbf{T}_k$.

The target signal is of course unknown, but its direction of arrival and its statistical character are assumed to be known, together with the receiving array geometry and other characteristics. The pilot signal is designed to have the same autocorrelation function as the target signal. All of the weight input components, target signal, interference, and pilot signal, are assumed to be uncorrelated. Therefore, the input R-matrix is

$$\mathbf{R} = \mathbf{R}_{SS} + \mathbf{R}_{NN} + \beta^2\mathbf{R}_{SS} \tag{14.11}$$

The third term of (14.11) is the autocorrelation matrix of the pilot signal components at the weight inputs, or

$$E[\beta\mathbf{T}_k \cdot \beta\mathbf{T}_k^{\mathrm{T}}] = \beta^2 E[\mathbf{T}_k\mathbf{T}_k^{\mathrm{T}}] = \beta^2\mathbf{R}_{SS} \tag{14.12}$$

The cross correlation between the desired response and the weight inputs is the same as the cross correlation between the desired response and the pilot signal components of the weight inputs. Accordingly,

$$\mathbf{P} = E\begin{bmatrix} d_j x_{1k} \\ \vdots \\ d_j x_{KLk} \end{bmatrix} = E[d_k \mathbf{T}_k] = \beta^2 \mathbf{P}_S \tag{14.13}$$

The LMS algorithm used in the single-mode pilot-signal adaptive beamformer will converge to the following least-squares weight vector:

$$\begin{aligned} \mathbf{W}^* = \mathbf{R}^{-1}\mathbf{P} &= \left(\mathbf{R}_{SS} + \mathbf{R}_{NN} + \beta^2 \mathbf{R}_{SS}\right)^{-1} \beta^2 \mathbf{P}_S \\ &= \left(\mathbf{R}_{XX} + \beta^2 \mathbf{R}_{SS}\right)^{-1} \mathbf{P}_S \end{aligned} \tag{14.14}$$

This result does not exactly agree with the best least-squares solution given by (14.9). There is a bias in (14.14) caused by the introduction of the pilot signal. The bias can be kept small by making β small, however. Although it is not apparent from (14.14), making β small causes the effects of weight adaptation noise to become more severe, necessitating slow adaptation if a solution close to the true LMS solution is to be achieved.

Griffiths [3] has developed an algorithm that not only converges to the weight vector (14.9) that produces the best least-squares estimate of the target signal, but also obviates the requirement of the pilot signal. Next we describe this algorithm and analyze its properties.

THE GRIFFITHS LMS BEAMFORMER

Griffiths' algorithm is a "spin-off" of the LMS algorithm, utilizing certain a priori knowledge (when available) to create a highly effective real-time adaptation process. The algorithm is presented here in general terms, and is then applied to adaptive beamforming.

The LMS algorithm can be expressed as follows:

$$\begin{aligned} \mathbf{W}_{k+1} &= \mathbf{W}_k + 2\mu\varepsilon_k \mathbf{X}_k \\ &= \mathbf{W}_k + 2\mu(d_k - y_k)\mathbf{X}_k \\ &= \mathbf{W}_k + 2\mu d_k \mathbf{X}_k - 2\mu y_k \mathbf{X}_k \end{aligned} \tag{14.15}$$

If we now substitute the average $E[d_k \mathbf{X}_k] = \mathbf{P}$ in place of its instantaneous value in (14.15), the result is Griffiths' algorithm:

$$\mathbf{W}_{k+1} = \mathbf{W}_k + 2\mu\mathbf{P} - 2\mu y_k \mathbf{X}_k \tag{14.16}$$

This algorithm can be used when the cross-correlations between the desired response and the weight inputs are known a priori. Then, a real-time least-squares adaptive algorithm can operate without needing a real-time desired response input d_k. Such a process is illustrated in Figure 14.1.

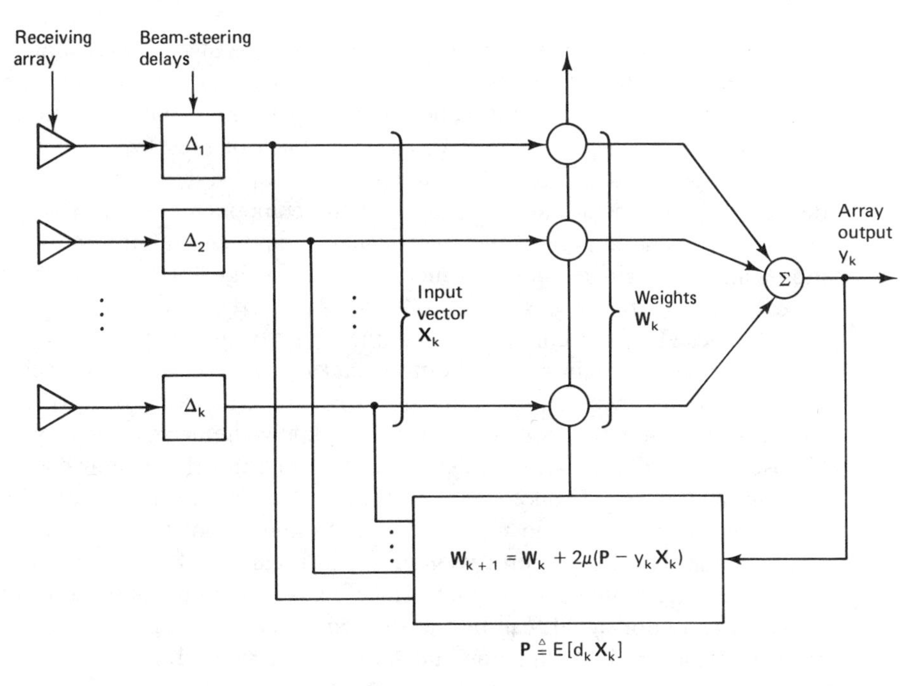

Figure 14.1 Diagram of Griffiths' algorithm. Note that there is no "desired response" input to the algorithm.

The converged solution of (14.16) can be obtained in the following manner. Equation (14.16) can be rewritten by replacing y_k with $\mathbf{X}_k^T\mathbf{W}_k$:

$$\begin{aligned}\mathbf{W}_{k+1} &= \mathbf{W}_k + 2\mu\mathbf{P} - 2\mu\left(\mathbf{X}_k^T\mathbf{W}_k\right)\mathbf{X}_k \\ &= \mathbf{W}_k + 2\mu\mathbf{P} - 2\mu\mathbf{X}_k\mathbf{X}_k^T\mathbf{W}_k \end{aligned}\tag{14.17}$$

Assuming that the stationary, zero-mean input vectors $\mathbf{X}_k, \mathbf{X}_{k-1}, \mathbf{X}_{k-2}, \ldots$ are uncorrelated, $\mathbf{W}_k$ and $\mathbf{X}_k$ are uncorrelated. Taking expectation of both sides of (14.17) yields

$$\begin{aligned}E[\mathbf{W}_{k+1}] &= E[\mathbf{W}_k] + 2\mu\mathbf{P} - 2\mu\mathbf{R}E[\mathbf{W}_k] \\ &= [I - 2\mu\mathbf{R}]E[\mathbf{W}_k] + 2\mu\mathbf{P}\end{aligned}\tag{14.18}$$

We have already solved an equation like this one in Chapter 4, starting with (4.38), and have found that

$$\lim_{k\to\infty} E[\mathbf{W}_k] = \mathbf{R}^{-1}\mathbf{P}\tag{14.19}$$

provided that, as in (4.45),

$$\frac{1}{\lambda_{\max}} > \mu > 0\tag{14.20}$$

Like the LMS algorithm, the Griffiths algorithm is unbiased and produces at convergence an expected steady-state solution that is the true least-squares solution.

An adaptive beamformer based on the Griffiths algorithm is shown in Figure 14.2. A set of bulk delays labeled $\Delta_1, \Delta_2, \Delta_3, \ldots, \Delta_K$ is used to steer the receiving beam along the desired look direction. The target signal emerges from these delays in phase, that is, in proper time registration. Each of the blocks labeled "GLMS" contains a tapped delay line whose weights are adapted in accord with (14.16). The desired response is the target signal, and the components of the P-vector are the cross-correlations between the target signal components at the individual weight inputs and the desired response, which is the target signal itself. The elements of the P-vector are therefore obtained from the autocorrelation function of the target signal, choosing lags to correspond with the delays of the delay-line taps.

In order to obtain an array output that is a best least-squares estimate of the target signal, it is necessary to know the target signal's direction of arrival as well as its autocorrelation function. One must also know the array geometry in order to choose the appropriate steering delays. Essentially, this same knowledge is required for the Widrow et al. pilot-signal algorithm, but the Griffiths algorithm has the advantages of providing an unbiased solution and of not requiring either the pilot signal or the slaved signal processor shown in Figure 13.18. On the other hand, the pilot signal algorithm is not obsolete, since it has many other applications to systems in which real pilot signals can be transmitted to the receiving array from a distance. In these cases, the look direction and array geometry need not be known.

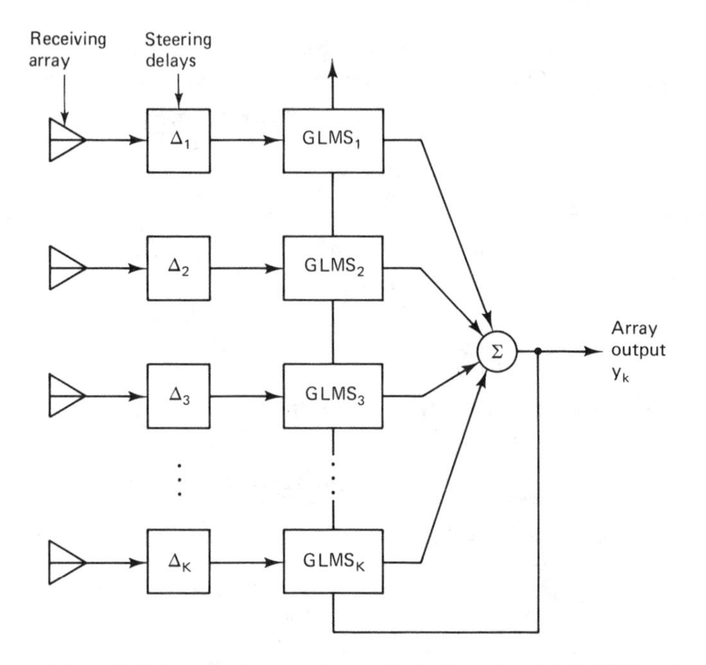

Figure 14.2 Griffiths' adaptive beamformer. Each filter labeled "GLMS" is adapted as in (14.16) and in Figure 14.1.

THE FROST ADAPTIVE BEAMFORMER

Both the pilot-signal adaptive beamformer and the Griffiths adaptive beamformer place "soft constraints" for signal reception in the look direction. A weak target signal arriving in the look direction would have almost no effect on the sensitivities of these adaptive beamformers to this signal, but a strong signal would tend to be partially rejected, even if it were to arrive exactly in the look direction. To overcome this limitation, an adaptive beamformer that maintains a "hard constraint" in the look direction was devised by O. L. Frost [7, 8]. With a hard constraint, the sensitivity of Frost's adaptive beamformer in the look direction is fixed regardless of the strength of the target signal arriving from this direction.

A block diagram of the Frost adaptive beamformer is shown in Figure 14.3. Once again, steering delays are used to synchronize the target signal components coming from the look direction at the input nodes of the tapped delay lines. The delays in all the tapped delay lines are made equal to each other to synchronize the target signal components at all corresponding nodes of the entire set of tapped delay lines.

Assume that the target signal arrives from the look direction. Then the target signal component at the system output can be considered to have originated from an "equivalent signal processor," shown in Figure 14.3. This processor is a tapped delay line having the target signal as its input. Its output will match the target signal component of the system output if, as suggested in the block diagram of Figure 14.3,

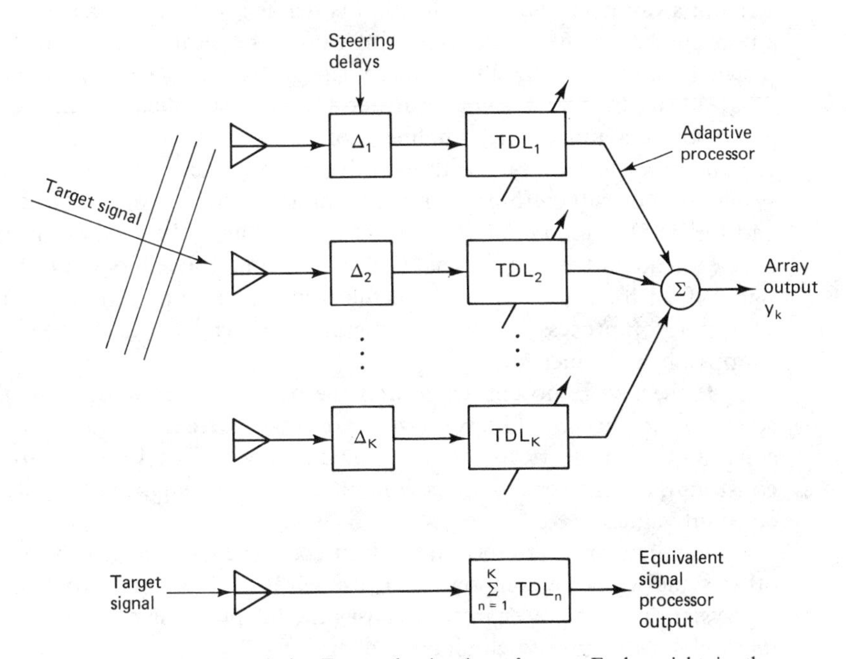

Figure 14.3 Diagram of the Frost adaptive beamformer. Each weight in the equivalent signal processor is the sum of the K corresponding weights above.

each of the weight values of the equivalent signal processor is made equal to the sum of the weight values of the corresponding weights of the adaptive processor. To visualize this more clearly, imagine a rectangular matrix of weight values arrayed to correspond with the geometrical arrangement of weights in the adaptive processor shown in the figure. Each weight of the equivalent processor should have a value equal to the sum of the elements of the corresponding matrix column.

The parameters of the equivalent processor can be chosen so that the target signal encounters linear filtration in propagating through the adaptive beamformer, with a specified frequency response. If no special filtration is required, the weights of the equivalent processor could all be set to zero except for one weight that would be set to unit value, thus producing a flat frequency response with unity gain in the look direction. The target signal components would then appear at the system output without distortion (except for additive noise).

Once the parameters or weights of the equivalent processor are chosen and fixed for a specific look-direction response, the weights of the adaptive processor can be varied as long as the sum-on-columns constraint illustrated in Figure 14.3 (the "Frost constraint") is obeyed. The weights of the adaptive processor can be adapted to minimize the system output power subject to the Frost constraint. In this way, the Frost adaptive beamformer will filter the target signal arriving from the look direction in accord with a specified transfer function (which could simply be a unit gain) and, at the same time, minimize output power. Since the signal arriving from the look direction is defined as the desirable target signal and any incoherent signal arriving away from the look direction is defined as interference, minimizing output power subject to the Frost constraint causes the system output to be a minimum-variance estimate of the filtered target signal.

When the target signal transfer function specified in the equivalent signal processor is a unit gain with linear or zero phase, the target signal component appears undistorted, but with additive interference, at the system output. The system output is therefore a minimum-variance estimate of the target signal. Kelley and Levin [5], Capon et al. [6], and Griffiths [2] have shown that this minimum-variance estimate of the target signal is also a maximum-likelihood estimate of the target signal if all interferences and also the target signal are Gaussian with zero mean. This type of processor is thus sometimes known also as a maximum-likelihood adaptive beamformer.

If there were no constraint and the output power were minimized, all of the adaptive weights would go to zero and the system output would vanish. The constraint is necessary to keep the system "alive". The Frost constraint is a linear constraint, forcing certain linear combinations of the adaptive weights to be equal to constant values.

Jammers or interferers add output components which, if uncorrelated with all other signal or noise components, cause additional system output power. Minimizing system output power in turn causes the adaptive beamformer to create nulls in its sidelobe structure to eliminate the jammers.

The Frost adaptive beamformer has a limited number of degrees of freedom and allocates these in an optimal way to minimize the total output interference power due to directional jammers, spatially distributed noises, and receiver noise. The Frost constraint causes the number of degrees of freedom to be less than the total number of adaptive weights by an amount equal to the number of constraint equations, that is, equal to the number of weights in the equivalent signal processor.

In order to adapt the weights, Frost devised a new form of the LMS algorithm that minimizes mean-square error subject to general linear constraints. A clear and comprehensive development of this algorithm is presented in References 7 and 8. The following is a less detailed explanation.

Suppose that there are K antenna elements and that each is connected to a tapped delay line having L taps. Assume that steering delays are used as in Figure 14.3 to set the look direction, so that the received target signal components are in proper time registration at corresponding points along the tapped delay lines. As described above, the adaptive weights can be arrayed in the form of a rectangular matrix:

$$\overline{\mathbf{W}}_k \triangleq \begin{bmatrix} w_{1k} & w_{2k} & \cdots & w_{Lk} \\ w_{(L+1)k} & w_{(L+2)k} & \cdots & w_{2LK} \\ w_{(2L+1)k} & \vdots & & \vdots \\ \vdots & & & \\ w_{((K-1)L+1)k} & \cdots & & w_{KLk} \end{bmatrix} \tag{14.21}$$

The input signals to the weights can also be arrayed in a corresponding rectangular matrix,

$$\overline{\mathbf{X}}_k \triangleq \begin{bmatrix} x_{1k} & x_{2k} & \cdots & x_{Lk} \\ x_{(L+1)k} & x_{(L+2)k} & \cdots & x_{2Lk} \\ x_{(2L+1)k} & \vdots & & \vdots \\ \vdots & & & \\ x_{((K-1)L+1)k} & \cdots & & x_{KLk} \end{bmatrix} \tag{14.22}$$

The fixed weights of the equivalent processor comprise a vector given by

$$\mathbf{C} \triangleq [c_1 \quad c_2 \quad \cdots \quad c_L] \tag{14.23}$$

The Frost constraint illustrated in Figure 14.3 can be expressed as

$$[1 \quad 1 \quad \cdots \quad 1]\overline{\mathbf{W}}_k = \mathbf{C} \qquad \text{for all } k \tag{14.24}$$

The Frost algorithm is an iterative process, in which each adaptation cycle can be viewed as consisting of two half-steps. The first half-step adapts in accord with the LMS algorithm to reduce output power, and the second half-step reestablishes the constraint by making a correction to the sum of each column of (14.21) so that (14.24) is satisfied. The corrections to reestablish the constraint are evenly divided among the elements of each given column of (14.21). When the constraints are all reestablished, the present adaptive cycle is complete and the system is ready for the next cycle.

Since the algorithm is minimizing output power, the output y_k serves as the "error" in this case. Half-step 1, to reduce output power in accord with the LMS algorithm, can be written

$$\overline{\mathbf{W}}_{k+1/2} = \overline{\mathbf{W}}_k + 2\mu y_k \overline{\mathbf{X}}_k \tag{14.25}$$

After half-step 1, there will be a constraint error equal to

$$\mathbf{C} - [1 \quad 1 \quad \cdots \quad 1]\overline{\mathbf{W}}_{k+1/2} \tag{14.26}$$

Half-step 2 now adjusts the weights to correct this error in the following way. Define a correction vector as

$$\frac{1}{K}\left[\mathbf{C} - [1 \quad 1 \quad \cdots \quad 1]\overline{\mathbf{W}}_{k+1/2}\right] \triangleq [e_{1k+1/2} \quad \cdots \quad e_{Lk+1/2}] \tag{14.27}$$

Next define a $K \times L$ correction matrix as

$$\overline{\mathbf{E}}_{k+1/2} \triangleq \begin{bmatrix} e_{1k+1/2} & \cdots & e_{Lk+1/2} \\ \vdots & & \vdots \\ e_{1k+1/2} & \cdots & e_{Lk+1/2} \end{bmatrix} \tag{14.28}$$

Half-step 2 now consists of adding $\overline{\mathbf{E}}_{k+1/2}$ to the matrix of weight values, that is,

$$\overline{\mathbf{W}}_{k+1} = \overline{\mathbf{W}}_{k+1/2} + \overline{\mathbf{E}}_{k+1/2} \tag{14.29}$$

With the completion of half-step 2 (i.e., with the completion of the adapt cycle), the constraint in (14.24) is reestablished. The Frost algorithm can be expressed as a combination of (14.25) and (14.29):

$$\overline{\mathbf{W}}_{k+1} = \overline{\mathbf{W}}_k + 2\mu y_k \overline{\mathbf{X}}_k + \overline{\mathbf{E}}_{k+1/2} \tag{14.30}$$

where $\overline{\mathbf{E}}_{k+1/2}$ is obtained from (14.28), (14.27), and (14.25).

The flexibility of the LMS algorithm can be noted here. Although it may seem to be an involved process, the ease with which the linear constraints were incorporated into the recursive least-squares process is remarkable. Algorithms with nonlinear constraints could also be devised, but proofs of convergence and determination of rate of convergence for such algorithms would be difficult to work out.

When the constraint is chosen to give unit array gain with zero or linear phase shift in the look direction, the array output contains an undistorted version of the target signal, plus additive noise. As such, the array output will be a maximum-likelihood estimate of the target signal.

Other versions of the Frost adaptive beamformer have been devised. One such version, similar to the adaptive interference canceler in Chapter 12, has been analyzed by Griffiths and Jim [10, 11]. Although it achieves the Frost constraint, the Griffiths–Jim beamformer does so with an unconstrained least-squares algorithm. As discussed previously, the Howells–Applebaum scheme is also essentially an adaptive interference canceler and is in this sense related to Griffiths–Jim. However,

Howells–Applebaum produces a soft constraint in every direction, whereas Griffiths–Jim produces a hard constraint only in the look direction.

A block diagram of the Griffiths–Jim beamformer is shown in Figure 14.4. As before, beam steering delays create the proper time alignment for the target signal components arriving from the look direction. In the terminology of adaptive noise canceling, the primary signal is a filtered version of the sum of these delayed antenna signals. The filter has an impulse response corresponding to the vector **C** of (14.23). If no filtering of the target signal is desired, the filter is set to have unit gain at all frequencies. In this case the primary signal will contain the target signal arriving from the look direction, plus interference.

Again referring to Figure 14.4, since the in-phase target signals from the individual omnis are subtracted in pairs, the reference signals contain no target signal components from the look direction. They contain only interference, and are applied to a bank of adaptive filters (tapped delay lines), then summed, and finally subtracted from the primary signal. The result is a system output containing an undistorted target signal (or a suitably filtered version of the target signal) plus interference. If there are K antenna elements, then there are $K-1$ adaptive filters. Since each of the L weights in each adaptive filter is not constrained, the number of

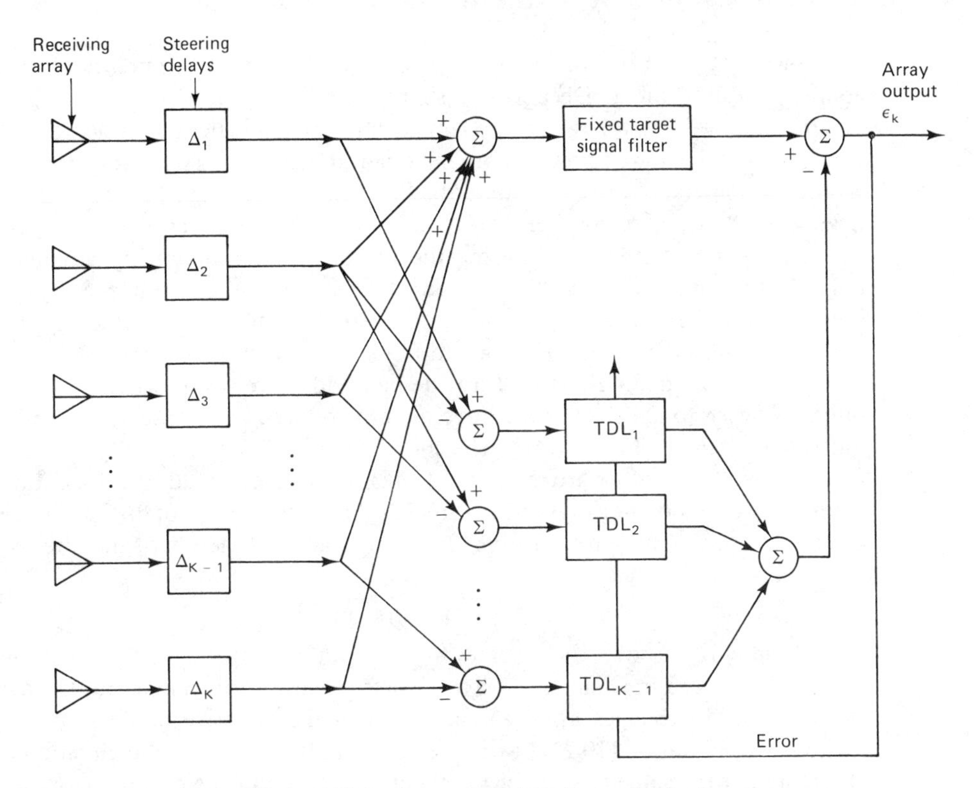

Figure 14.4 Griffiths–Jim version of the Frost adaptive beamformer.

degrees of freedom is $(K - 1)L$, the same as with the original Frost processor having KL weights.

Griffiths and Jim [11] have shown that the system of Figure 14.4 is capable of producing the same converged beamformer solution as the original Frost beamformer in Figure 14.3. The adaptive time constants may not be the same for both systems, however.

The Griffiths–Jim scheme is easy to understand and to apply. Almost any adaptive algorithm can be used in its implementation. Constraints are not required in the adaptive process, yet the system transfer function in the look direction is fixed.

The Frost beamformer, the Griffiths beamformer, and the pilot-signal beamformer presented in Chapter 13 all have different mathematical descriptions, yet in most practical cases, their behavioral characteristics are similar. They all tend to null jammers and to have small sidelobes when omnidirectional noise is present. These three beamformers are generally described as being "full" adaptive beamformers to distinguish them from sidelobe cancelers (such as Howells-Applebaum), which generally have a smaller number of weights, with of course a small number of degrees of freedom.

AN ADAPTIVE BEAMFORMER WITH POLES AND ZEROS

As shown by Gooch [12, 14], adaptive poles as well as adaptive zeros can be useful in adaptive beamforming. Deeper and sharper nulling is possible with a pole–zero system than with a zeros-only system having an equivalent number of adaptive weights. Gooch showed how to obtain a quadratic mean-square-error function using the "equation-error" approach described in Chapter 10 (e.g., Figure 10.17). Minimizing the "wrong" error function by adapting poles and zeros generally leads to better performance than minimizing the "right" error function by adapting zeros only. Minimizing the "right" error function by adapting both poles and zeros involves optimizing a nonquadratic, nonunimodal error function, and in Chapter 8 we discussed how this is an unsure and potentially unstable process.

The basic building block of a wide-band adaptive beamformer is an adaptive filter. We wish to use the IIR adaptive filter introduced in Chapters 7, 8, and 10 and illustrated again in Figure 14.5. The mean square of the error, ε_k, is a quadratic function of the feedforward weights of $A(z)$, and a nonquadratic function of the feedback weights of $B(z)$. As in Figure 7.2, the coefficient b_0 of $B(z)$ is assumed to be zero in order to have a realizable filter. The transfer function of the filter is

$$\frac{A(z)}{1 - B(z)} \tag{14.31}$$

Figure 14.6 shows, in a somewhat different arrangement, the equation error system introduced in Chapter 10. The errors ε_k and ε'_k are related, but the nature of the relationship changes as $A(z)$ changes with the adaptive process, as shown in Chapter 10, Equation (10.21). Minimizing the mean square of the equation error is therefore not the same as minimizing the mean square of the output error, as shown

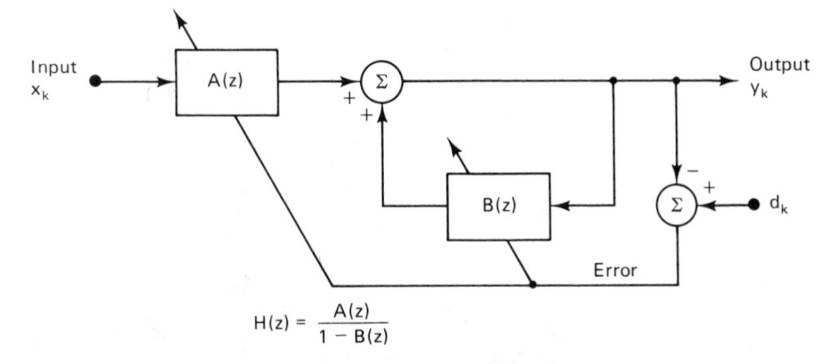

Figure 14.5 General-purpose adaptive filter with feedback. The notation is similar to that in Figures 7.2, 8.5, and 10.16.

in (10.21); however, an exceptional case occurs when values of $A(z)$ and $B(z)$ can be found that would bring the equation error to zero. Then the output error is zero, as long as the polynomial $1 - B(z)$ has no zeros on the unit circle in the z-plane. Generally, making the equation error small makes the output error small. This is the intuitive basis for equation error minimization.

The adaptation of $A(z)$ and $B(z)$ to minimize the mean square of the equation error is illustrated in Figure 14.7.[†] The figure shows a practical form of the adaptive IIR filter. It is clear from the structure in the figure that the mean square of the equation error is a quadratic function of the weights of $A(z)$ and of $B(z)$. The output signal emerges from a filter whose transfer function is

$$\frac{1}{1 - B(z)} \tag{14.32}$$

The parameters of this filter are obtained by reciprocating the known transfer function of the "$1 - B(z)$" filter, which in turn is determined by the adaptive process.

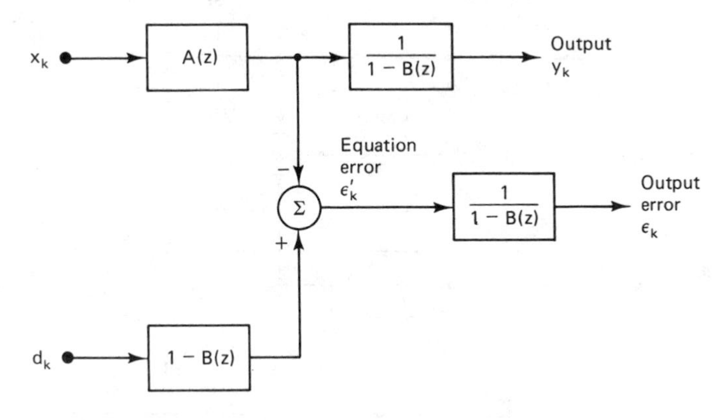

Figure 14.6 Filter diagram for the equation error method.

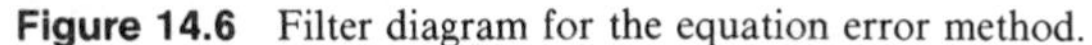

[†] The adaptive filters and the adaptive process are illustrated more explicitly in Figure 10.17.

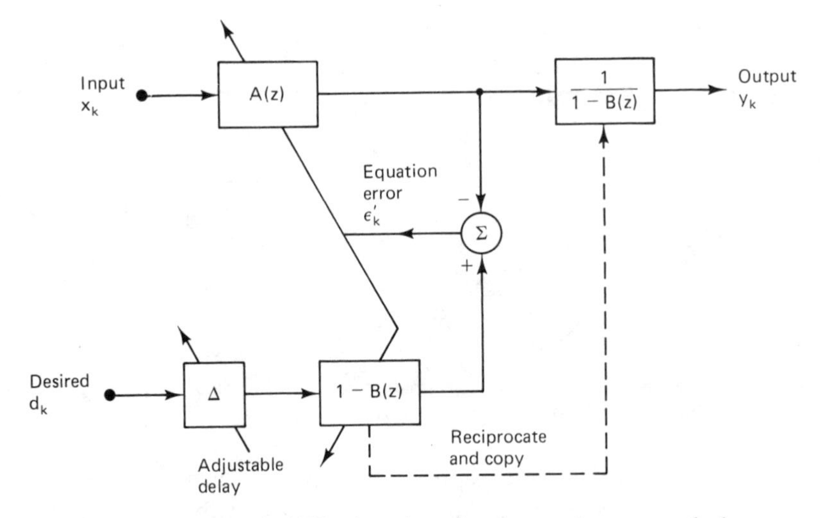

Figure 14.7 Practical IIR adaptation using the equation error method.

An implementation of the recursive transfer function (14.32) is easy and straightforward, except when one or more of its poles occur outside the unit circle. As in Chapter 10, several remedies can then be tried in order to realize a stable causal version of (14.32) as an output filter. The simplest approach uses the adjustable delay in the desired response path illustrated in Figure 14.7. A value of the delay Δ can almost always be chosen to cause $1 - B(z)$ to have minimum phase, that is, to have all zeros inside the unit circle. A potential problem in this approach is that the introduction of large values of Δ can cause the minimum mean-square

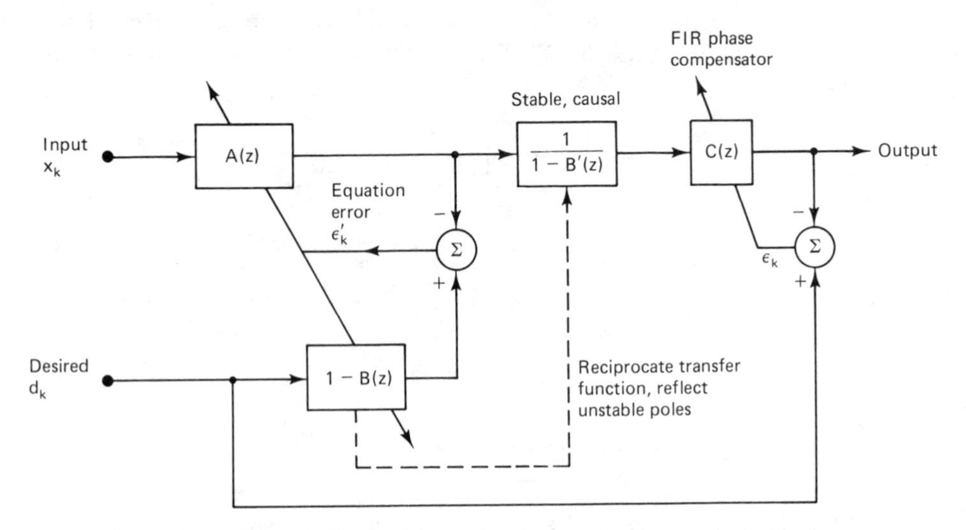

Figure 14.8 Practical IIR adaptation using the equation error method with phase compensation instead of input delay.

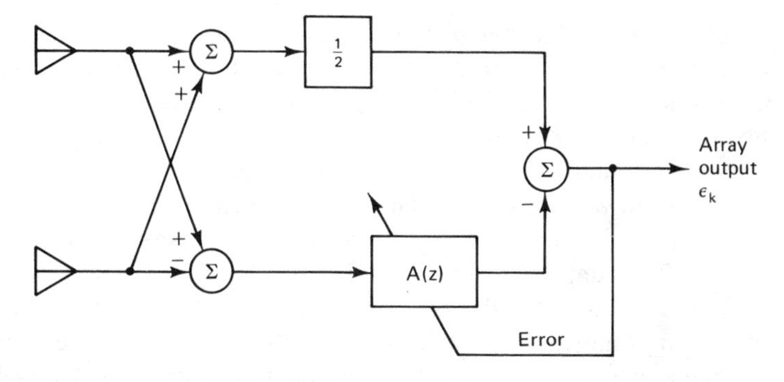

Figure 14.9 Two-element version of the Griffiths–Jim adaptive beamformer in Figure 14.4.

error to become large. Also, the output y_k is delayed by Δ, which could be a problem in some applications. In general, Δ should be kept as small as possible for best performance.

Other approaches can be pursued when the denominator of (14.32) is non-minimum-phase. As described in Chapter 10, the polynomial $1 - B(z)$ can be factored and then modified to a minimum phase form $1 - B'(z)$ by reciprocating the positions of the roots outside the unit circle, thus preserving the magnitude response but distorting the phase response. Then, as illustrated in Figure 14.8, the phase response as well as discrepancies between ε_k and ε'_k could be "patched up" by following the transfer function (14.32) with an adaptive FIR filter, $C(z)$, whose desired response is the original desired response of the entire filter. This approach

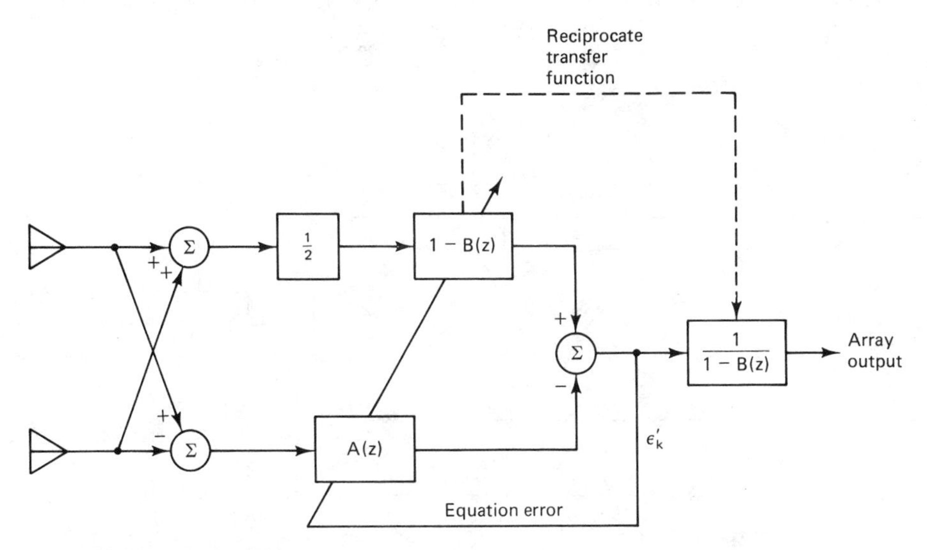

Figure 14.10 Pole–zero version of the two-element Griffiths–Jim adaptive beamformer. (Devised by Gooch [14]).

has been tried [14] and works quite well. However, as indicated in Chapter 8, more research is needed to find the best ways of dealing with a non-minimum-phase denominator of (14.32), and, in general, to gain a better understanding of the entire subject of adapting poles and zeros.

Let us now examine an application of this IIR adaptation method to the Griffiths–Jim version of the adaptive beamformer. Figure 14.9 shows a system with two omnis and unity signal gain in the look direction which, in this case, is zero degrees. The adaptive filter will, as far as possible, cancel any input arriving outside the look direction. A pole–zero version of the same system is shown in Figure 14.10. Comparing Figures 14.4, 14.6, 14.9, and 14.10, note how the equation error concept has been incorporated into this application. Next, Figure 14.11 shows how the non-minimum-phase remedies discussed above may be included in a complete system. Comparing Figures 14.4 and 14.11, one can see how the same ideas can be extended to apply to array sizes larger than 2.

Adaptive beamforming with pole–zero filtering is quite effective, as seen in the following two examples. Figure 14.12 shows the converged power gain of a conventional Frost adaptive beamformer with a look direction of 0° and a single broadband jammer incident at 225°. For this example the Frost constraint (14.24) is such that the frequency response in the signal direction is flat, and the low gain at the jammer frequencies is evident from the frequency response in the jammer direction. Figure 14.13 shows the response of a pole–zero adaptive beamformer with the same look direction, the same broadband jammer input, and the same number of weights

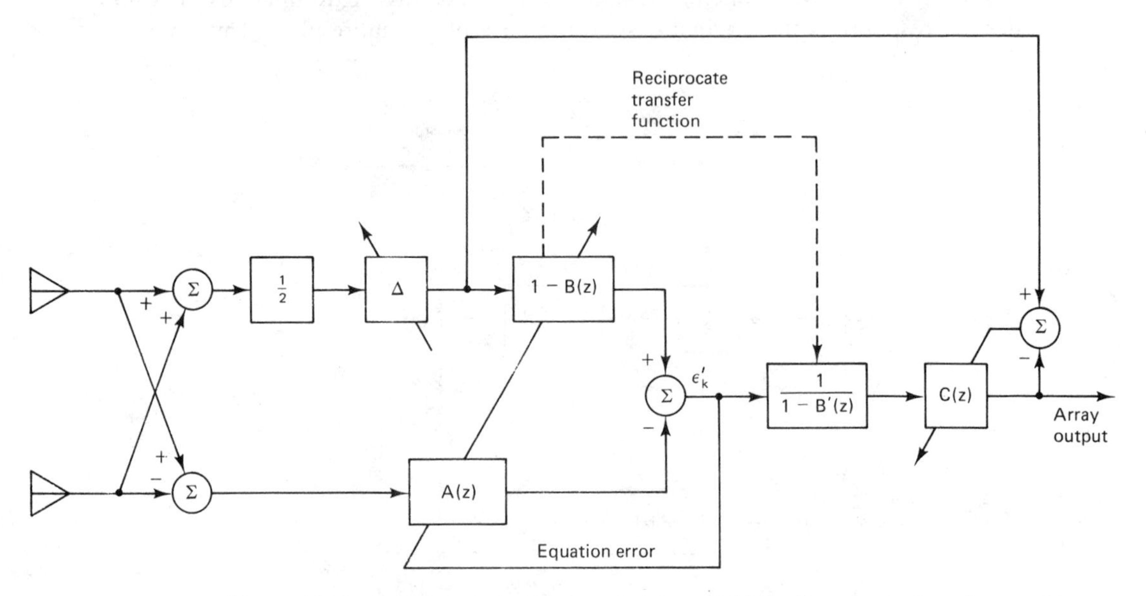

Figure 14.11 Pole–zero version of the two-element Griffiths–Jim adaptive beamformer with non-minimum-phase compensation.

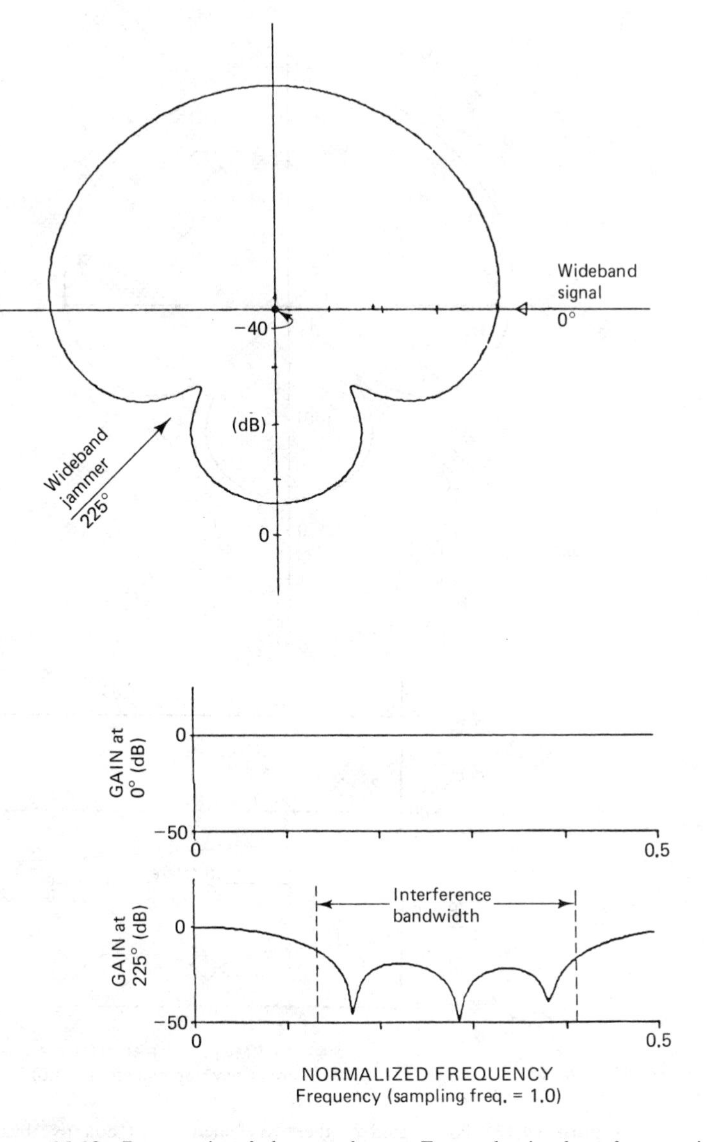

Figure 14.12 Power gain of the two-element Frost adaptive beamformer with $L = 5$ weights per element.

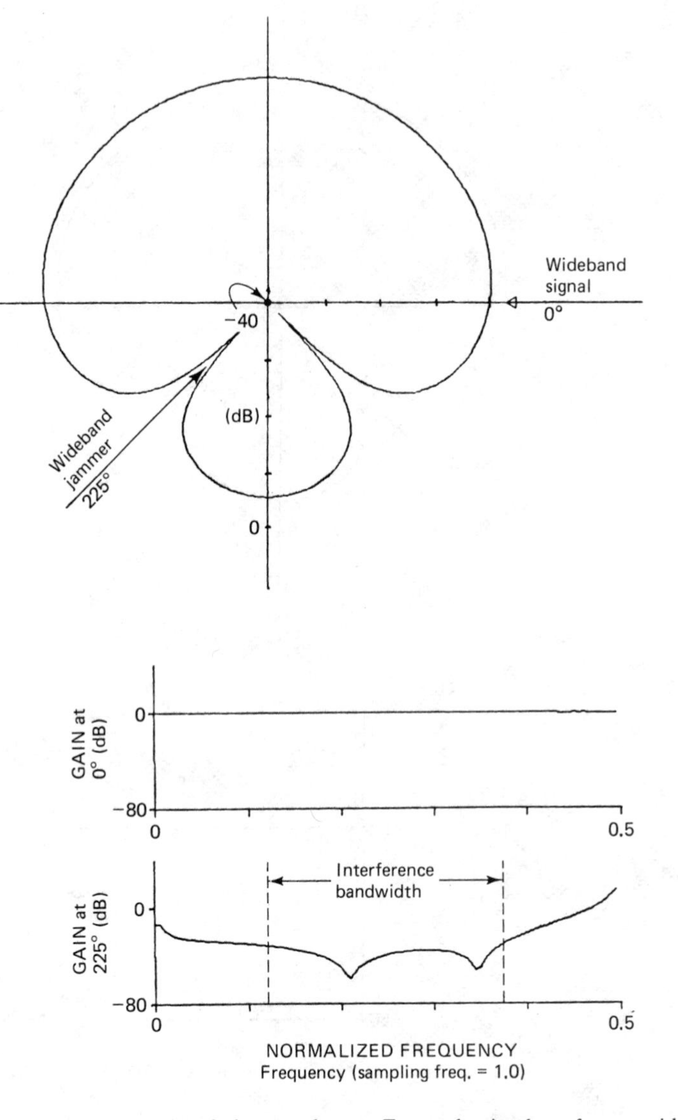

Figure 14.13 Power gain of the two-element Frost adaptive beamformer with pole–zero filtering using three zeros and two poles (five weights) per element.

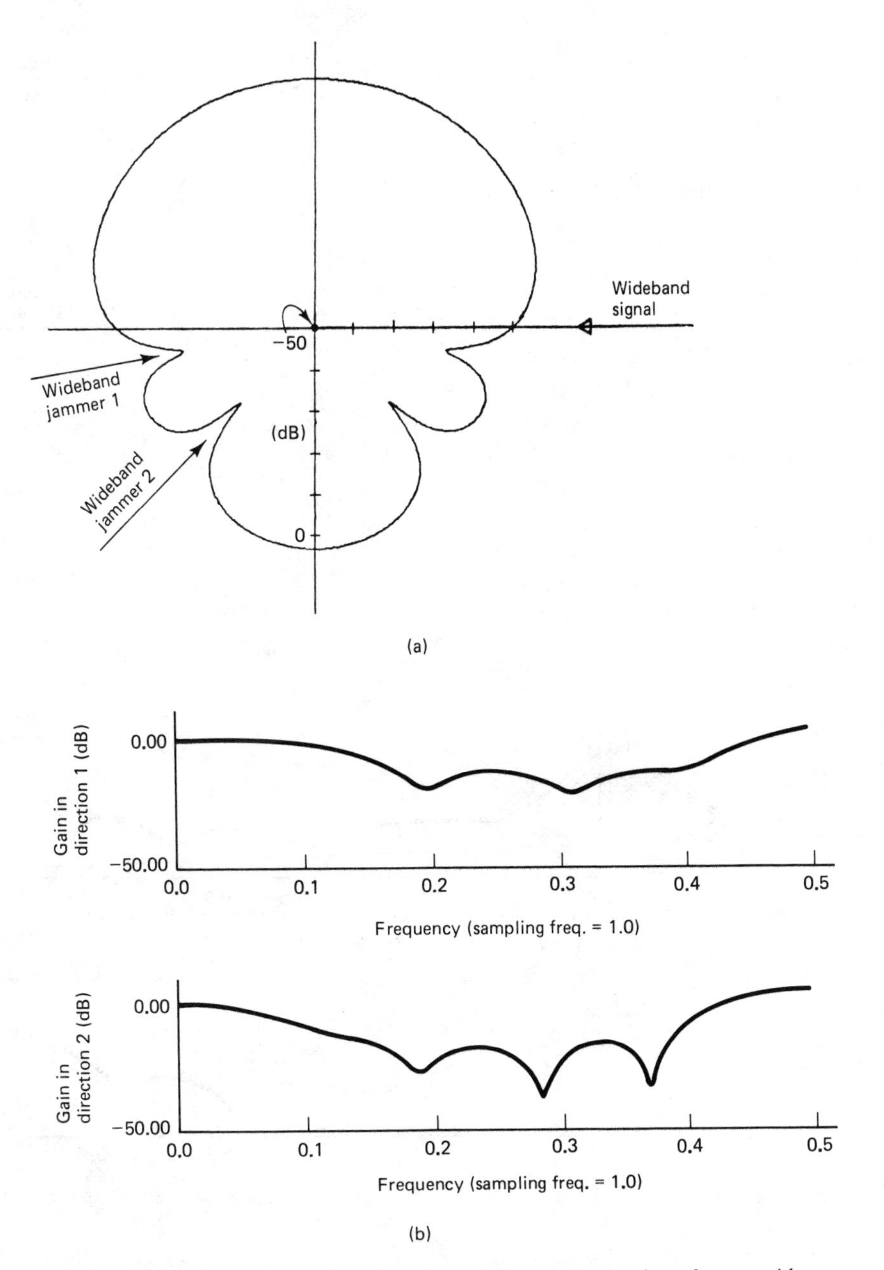

Figure 14.14 Power gain of a three-element Frost adaptive beamformer with $L = 7$.

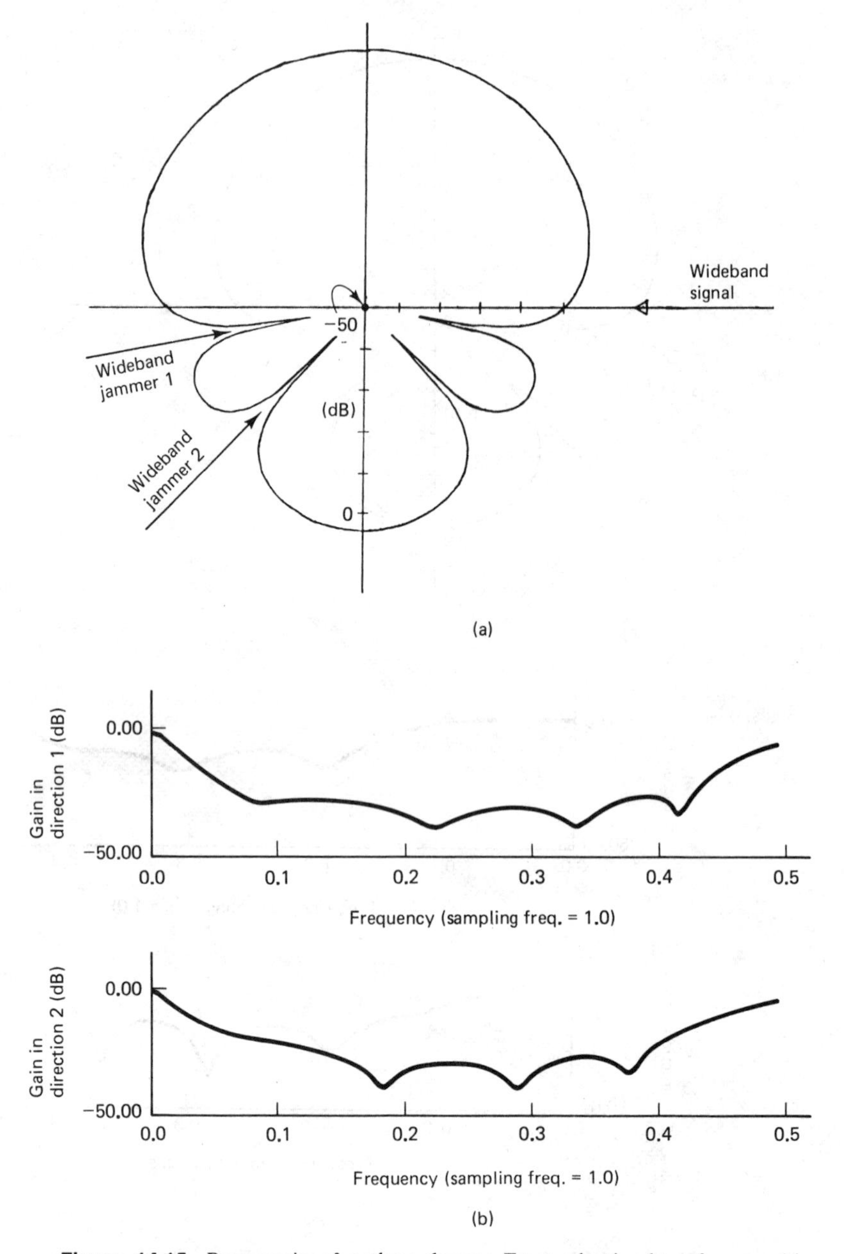

Figure 14.15 Power gain of a three-element Frost adaptive beamformer with pole–zero filtering using three zeros and four poles (7 weights) per element.

per omni. With poles as well as zeros, the nulls are much sharper spatially and much deeper, although the same number of adaptive weights is used. In the case of Figure 14.13, the denominator polynomial in (14.32) was minimum-phase, so no special measures were required for its realization. In the second example, shown in Figures 14.14 and 14.15, two simultaneous broadband jammers produced similar results with 3 omnis and $L = 7$. For a given number of adaptive weights the pole–zero method forms deeper notches in the direction of jammers.

If it were necessary to reciprocate the zeros of $1 - B(z)$ in either Figure 14.12 or 14.14, the jammer nulling process would not have been greatly affected, because nulling is determined by $A(z)$ and $B(z)$, and compensating the phase with $C(z)$ concerns primarily the phase of the target signal passing through $C(z)$ to the output in Figure 14.8.

SIGNAL CANCELLATION AND DISTORTION

In all of the adaptive beamformers described so far and in similar systems not specifically described here but based on a common set of principles obtained from least-squares optimization, an effect of the adaptive process called "signal cancellation" can take place when signals and jammers are present simultaneously. This effect is aggravated by rapid adaptation.

The behavior of adaptive beamformers has been described above in terms of least-squares theory. Least-squares solutions are achieved by real-time adaptive algorithms only in the limit as μ approaches zero and the time constants of adaptation approach infinity. On the other hand, rapid adaptation giving best least-squares solutions with relatively small amounts of input data leads to fluctuation in the weights and non-Wiener behavior. In Chapter 5 we treated noise in the weights statistically as a random process which, in many circumstances, is a simple and proper approach. There are cases, however, where the noise in the weights is more structured and can have profound effects on the signals flowing through the weights. Such cases occur in adaptive beamformers when signals and interference are present simultaneously at the antenna elements. The result is partial signal cancellation as the adaptive process and the associated weight fluctuations modulate the interference or jamming signals to make them more closely resemble the target signal. Through this mechanism, partial signal cancellation results from rapid minimization of the output power.

To understand signal cancellation in more detail, consider first a simple example. Imagine the Frost beamformer in Figure 14.3 with a sinusoidal target signal arriving from the look direction. Suppose the constraint in (14.24) is such that the signal should appear with unit gain at the beamformer output. Now suppose that a strong jammer appears suddenly at the same frequency as the signal, somewhat away from the look direction. This interference would normally be rejected by the adaptive beamformer if the signal were not present. With the signal present, however, minimizing the total output power causes the jammer to be admitted with just the right magnitude and phase to cancel the sinusoidal signal. Thus the signal sinusoid is admitted with a gain of 1, but just a "trickle" of the powerful jammer sinusoid is admitted to cancel the signal sinusoid perfectly and produce a net output of zero. The output power is minimized and the constraint is preserved, but the signal is lost in the process. Thus a smart jammer can kill the output signal by cancellation rather than by jamming with overwhelming interference.

If the input signal in the look direction is broadband (rather than sinusoidal) and the jammer is sinusoidal, the adaptive algorithm will (by varying the weights) modulate the sinusoidal jammer so that it will cancel some signal components at the jammer frequency and at neighboring frequencies. The signal spectrum will be notched at the jammer frequency. If the jammer signal contains a sum of sinusoids at spaced frequencies within the passband, the output signal spectrum will be notched at each of the jammer frequencies. This phenomenon could be troublesome even for spread-spectrum communications.

These and similar signal cancellation phenomena have already been observed and analyzed in Chapter 12 with adaptive noise-canceling systems which are much simpler than adaptive beamformers. In the simple canceler shown in Figure 12.1, for example, the primary input contains a useful signal, s, plus interference n_0. The reference input is separately obtained in practical systems and contains the interference n_1, which is related to the interference of the primary input. Generally, the relationship between the two interferences is unknown a priori, and the adaptive filter shapes the reference interference to replicate the primary interference (in the least-squared-error sense) so that subtraction will remove the interference from the primary input, providing a more useful output. We saw in Chapter 12 how an adaptive filter that minimizes output power in the system shown in Figure 12.1 causes the system output to be a best least-squares estimate of the useful signal s. Basically, the Howells–Applebaum sidelobe canceler in Figure 13.1 functions on the same principle, although in certain ways it is more complicated. Useful signals and jammer signals appear at both the primary and reference inputs, and spatial processing (i.e., array processing) is also involved.

If the reference input is a sinusoid as it is in Figure 12.6, the signal flow path from the primary input to the output behaves like a sharp, linear, time-invariant, notch filter, as indicated in Figures 12.8 and 12.9. This discovery came initially as a surprise because the adaptive filter itself is intrinsically nonlinear and varies with time. An analysis of the notch filter effect was presented in Chapter 12, and a more detailed analysis treating both single and multiple notch cases is contained in Glover [15, 16].

Let us reexamine Figure 12.6, which shows an adaptive noise canceler with two adaptive weights. The primary input is assumed to be an arbitrary signal which could be stochastic, deterministic, periodic, or transient. The reference input is assumed to be a pure cosine wave, $C\cos(k\omega_0 + \phi)$. The reference input is sampled directly with time step T, giving x_{1k}. After undergoing a 90° phase shift, it is sampled again, giving x_{2k}. The samplers are synchronous and are strobed at $t = 0, T, 2T, \ldots$.

A transfer function for the noise canceler shown in Figure 12.6 was derived using Figure 12.7. The signal flow path from the primary input to the noise-canceler output was seen to produce a notch filter having a pair of complex-conjugate zeros exactly on the unit circle at frequency ω_0. The bandwidth of the notch was seen in (12.60) to be proportional to μ, that is, proportional to the reciprocal of the time constant of the adaptive process.

In Figure 12.9 we saw the results of two experiments performed to demonstrate the adaptive system's notch-filter behavior. In the first experiment, the primary input was a unit cosine wave stepped at 512 discrete frequencies, the reference input was a unit cosine wave with frequency $\omega_0 = \pi/2$ rad, and the value of μ was 0.0125. The spectra of Figure 12.9 were computed using 512-point Fourier transforms, and the output power at each frequency is shown in Figure 12.9(a). As the primary frequency approaches the reference frequency, significant signal cancellation occurs. The weights do not converge to stable values but instead they "tumble" at the difference frequency, and the adaptive filter behaves like a modulator, converting the reference frequency into the primary frequency. The theoretical notch width for Figure 12.9(a) was given in (12.60).

In the second experiment of Figure 12.9, the primary input was composed of uncorrelated samples of white noise with unit power. The reference input and the processing parameters were the same as in the first experiment. The signal cancellation phenomenon is again evident in Figure 12.9(b), where an ensemble average of 4096 power spectra at the noise-canceler output is plotted.

Thus, in these experiments in Chapter 12, adaptive filtering of a reference cosine wave at a given frequency canceled the primary input components at adjacent frequencies. In particular, the result in Figure 12.9(b) indicates that under some circumstances, primary input components may be partially canceled and distorted even though they are not correlated with the reference input. This kind of non-Wiener cancellation is significant only when adaptation is rapid, that is, with large values of μ. When adaptation is slow, the weights converge to values that are nearly fixed, close to the Wiener least-squares solution, and though signal cancellation occurs, it is generally not significant since the notch is extremely narrow. In any event, whatever the value of μ, we can say that the primary input appears at the output after having gone through a notch filter. Fluctuation of the weight vector about the Wiener solution causes signal cancellation.

Other forms of signal cancellation phenomena also take place in adaptive beamformers. Suppose that the simple sidelobe canceler shown in Figure 13.1 is operating in an environment consisting of a signal plus a single jammer. The two array elements are omnidirectional, and both elements receive emanations from the signal source and the jammer. Assume that the jammer power is very much greater than the signal power, and that the number of degrees of freedom in the adaptive filter is sufficient to cancel the jammer but not sufficient to cancel both the jammer and the signal. Then, since the jammer is far more powerful than the signal, it will "grab" the degrees of freedom to effect its own cancellation. According to Wiener least-squares theory, the signal will have little influence on the adaptive weights and will appear at the system output together with certain small uncanceled jammer components.

The system shown in Figure 13.1 is similar to the adaptive noise canceler pictured in Figure 12.1, except that the reference input in Figure 13.1 contains a signal together with strong jammer inputs. After the weights have converged to the Wiener least-squares solution, the adaptive filter in Figure 13.1 will pass this signal and subtract the result from the primary signal, thereby introducing some signal distortion at the system output. We shall not concern ourselves with this type of distortion at this point because in many cases it will not be objectionable, and we are concerned here with non-Wiener phenomena.

As stated above, a least-squares solution is obtained only in the limit as the speed of adaptation is brought to zero (i.e., as μ is brought to zero). The weight dynamics inherent in adaptation give rise to modulation effects that cause signal cancellation, and it is these effects that we wish to examine further.

If, as we have assumed, the signal power in the reference input is low compared to the jammer power, we can neglect the effects of the signal on the adaptive filter. If we assume that the jammer is narrowband or sinusoidal, the input to the adaptive filter will cause a situation like the one represented in Figure 12.6. As discussed above, the signal flow path from the primary input to the system output will behave like a notch filter. Thus, both jammer components and signal components at and around the jammer frequency will be canceled at the system output.

An experiment was conducted to confirm this type of behavior. A Howells–Applebaum sidelobe canceler was configured with two omnidirectional elements placed one-quarter wavelength apart, and with four weights in the adaptive filter. A bandlimited signal was selected to arrive at $\theta = 90°$ with a center frequency at one-fourth of the sampling frequency and a bandwidth of 20 percent of the total bandwidth. A sinusoidal jammer was chosen to arrive at $\theta = 45°$ with the same frequency and with 100 times the signal power. Figure 14.16 shows the antenna pattern and the frequency response after convergence, and shows that the sidelobe canceler appears to be functioning in the manner described above. Figure 14.16(a) and (b) show that a 40-dB null has been formed in the jammer direction at the jammer frequency, and Figure 14.16(c) shows that the frequency response of the array in the signal direction is reasonably flat over the signal bandwidth. Overall, Figure 14.16 indicates that the performance of the Howells–Applebaum sidelobe canceler is nearly ideal.

However, steady-state observations of the antenna output spectra indicate otherwise. Figure 14.17 shows an ensemble average of the signal, jammer, and antenna output spectra with the sidelobe canceler operating with $\mu = 2.5 \times 10^{-4}$. The modulation effect and signal distortion inherent in this simple system are evident from Figure 14.17(c). The notch will always be present, and it can be narrowed only by slowing the rate of adaptation.

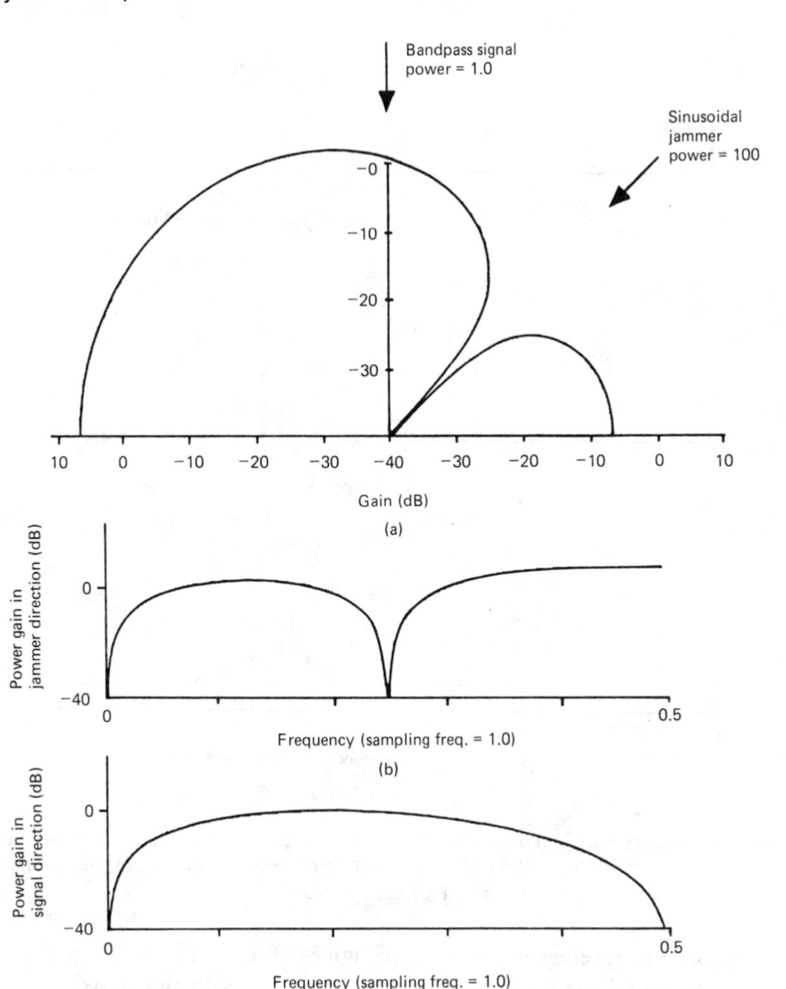

Figure 14.16 Behavior of a converged Howells–Applebaum sidelobe canceler: (a) directivity pattern plotted at jammer frequency; (b) frequency response in jammer direction; (c) frequency response in signal direction. From B. Widrow et al., IEEE Trans. on Ant. and Propagat., © May 1982.

The Frost adaptive beamformer exhibits the same signal cancellation effects that were observed with the Howells–Applebaum sidelobe canceler. To demonstrate signal cancellation in a Frost array, another experiment was conducted using a four-element array with four weights per element. The same signal used in the preceding experiment was again generated to be incident at $\theta = 90°$. The constraint in the look direction was set to unit gain and zero phase from zero frequency to half the sampling rate, that is, to a flat response at all frequencies. The jammer was sinusoidal at one-fourth the sampling frequency. In these experiments, ambient

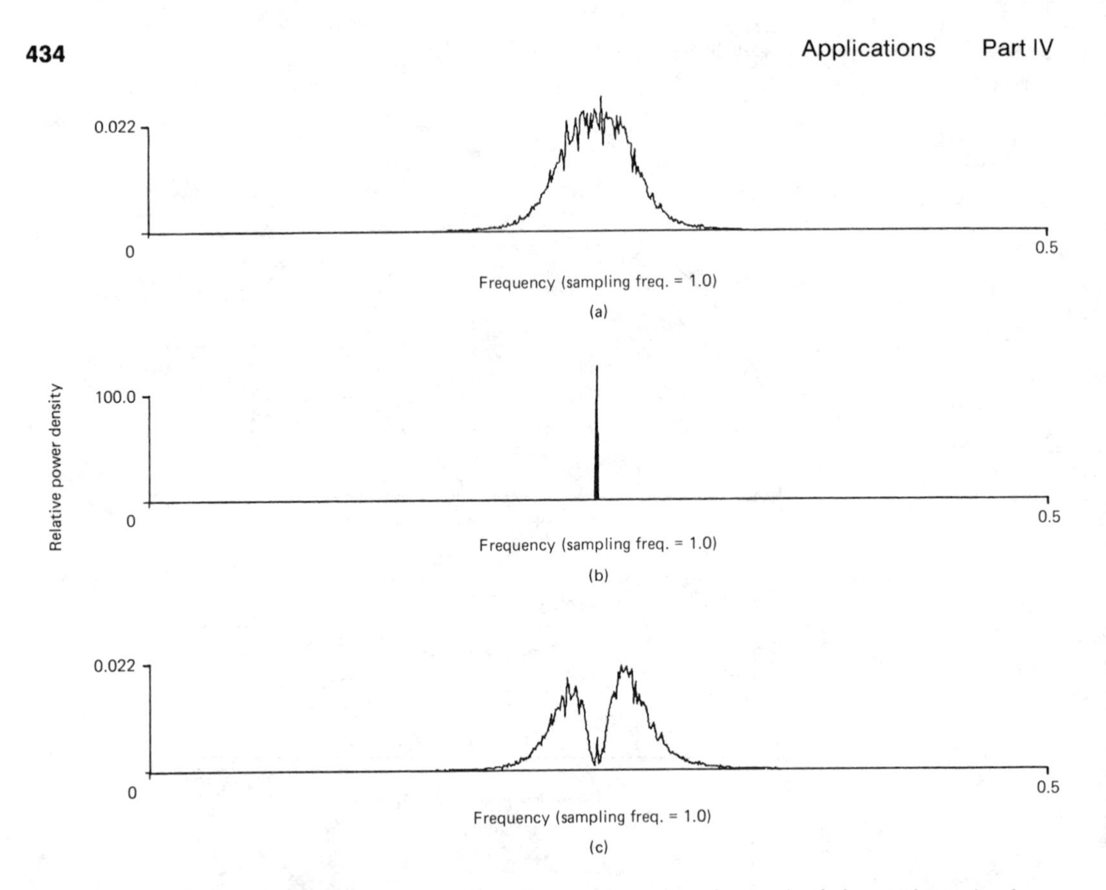

Figure 14.17 Power spectra of Howells–Applebaum beamformer simulation: (a) input signal spectrum; (b) jammer spectrum; (c) beamformer output spectrum. From B. Widrow et al., IEEE Trans. on Ant. and Propagat., © May 1982.

noise and receiver noise were negligible. Figure 14.18 shows the converged antenna pattern and the frequency response plots in both the signal and jammer directions. In the vicinity of the jammer frequency the look-direction gain is flat and the gain in the jammer direction is quite small; it was measured at 40 dB below the main beam gain. As in the Howells–Applebaum sidelobe canceler, the antenna pattern indicates that the adaptive beamformer is working perfectly.

However, Figure 14.19 shows again that all is not well. The bandpass signal whose spectrum is shown in Figure 14.19(a) was received by the Frost adaptive beamformer, and the spectrum of the sinusoidal jammer at $\theta = 45°$ is shown in Figure 14.19(b). Figure 14.19(c) shows the output power spectrum of the beamformer operating with $\mu = 10^{-3}$. The input signal appears at the output having gone through a notch filter. The notching effect is evident in the output signal spectrum and of course indicates a significant signal distortion at the beamformer output. The notch width is again given approximately by (12.60). The conditions for the derivation of the notch-width formula, that is, sinusoidal signals appearing with exact 90° separation at the input to the two weights, are not met with the 16-weight Frost

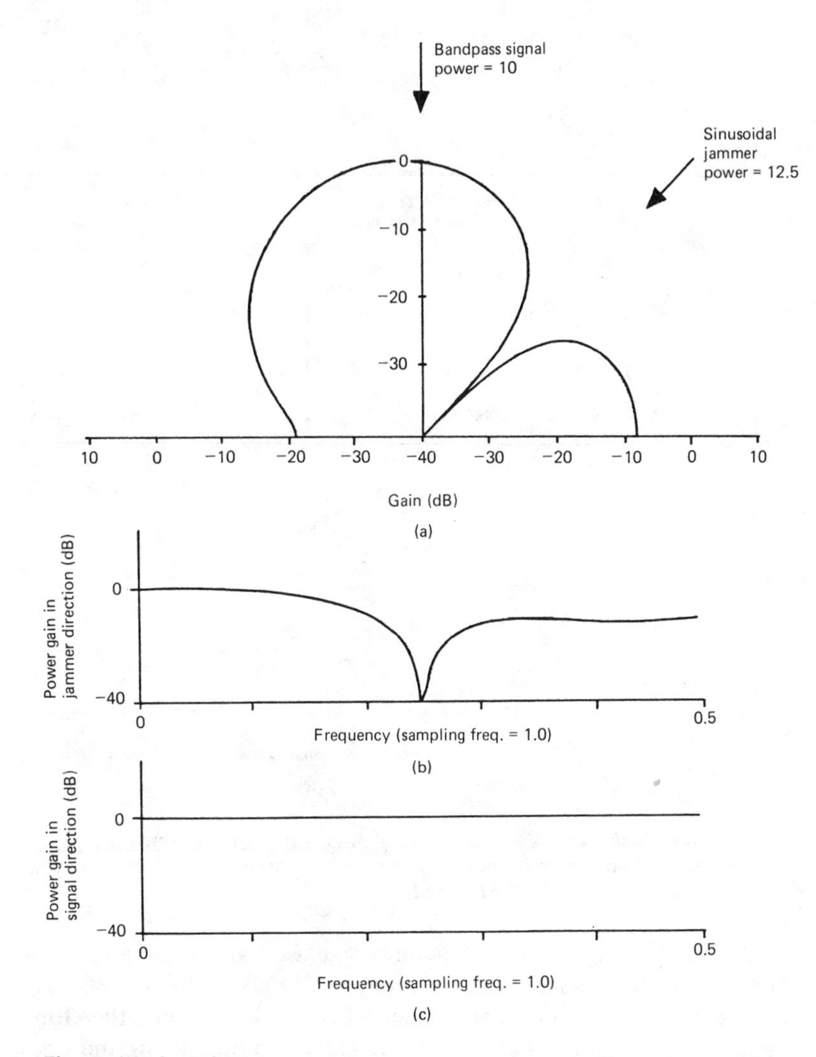

Figure 14.18 Behavior of a converged Frost beamformer: (a) directivity pattern plotted at jammer frequency; (b) frequency response in jammer direction; (c) frequency response in signal direction. From B. Widrow et al., IEEE Trans. on Ant. and Propagat., © May 1982.

processor described in this experiment; nevertheless, the simple formula in (12.60) gives at least an approximate prediction of notch width that is applicable at most jammer angles.

We can gain further understanding of the signal-canceling phenomenon by examining again the Griffiths-Jim configuration in Figure 14.4. For the sake of discussion, let the look direction constraint be unit gain and zero phase at all frequencies. With this constraint, the useful signal arriving from the look direction encounters unit gain from input to output, analogous to the direct primary signal

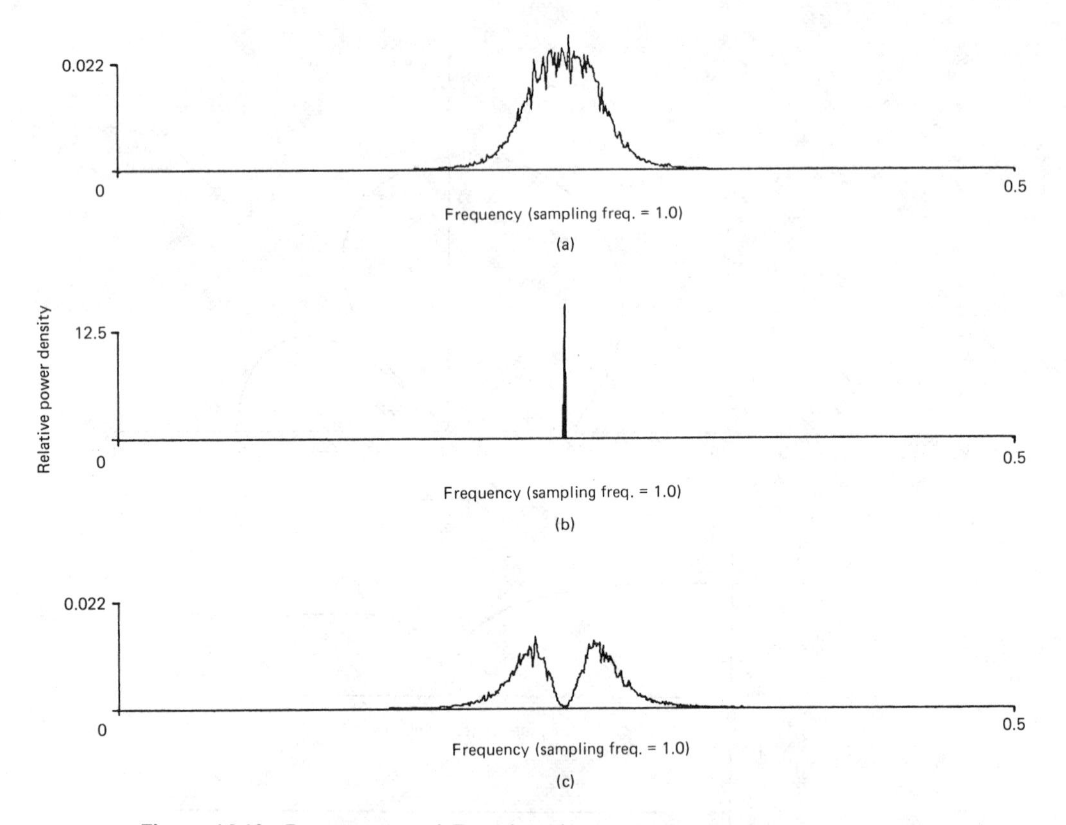

Figure 14.19 Power spectra of Frost beamformer simulation: (a) input signal spectrum; (b) jammer spectrum; (c) beamformer output spectrum. From B. Widrow et al., IEEE Trans. on Ant. and Propagat., © May 1982.

path shown in Figure 12.1. A jammer signal arriving at an angle other than the look direction encounters an adaptive filter, analogous to the reference signal path shown in Figure 12.1. A sinusoidal jammer off the look direction, therefore, causes fluctuations in the weights which effectively create a notch along the primary signal flow path to the output via the fixed-weight filter. Notching phenomena in this system are much like those of the adaptive canceler shown in Figure 12.1. Look-direction signals do not appear at the adaptive filter inputs, where only interference is present. Both signal and jammer are present in the primary signal flow path, and both signal and jammer experience notching at the jammer frequency. With high-speed adaptation, the notch could be very wide, incurring the risk of losing the signal in the jammer cancellation process, in effect "throwing the baby out with the bath water."

To explore the signal cancellation problem further, we now consider some additional experiments with the Frost adaptive beamformer. The jammer was again sinusoidal, while the target signal from the look direction was composed of white noise with unit power. Spectra of the beamformer outputs for different jammer power levels are shown in Figure 14.20. With the jammer power set at its lowest

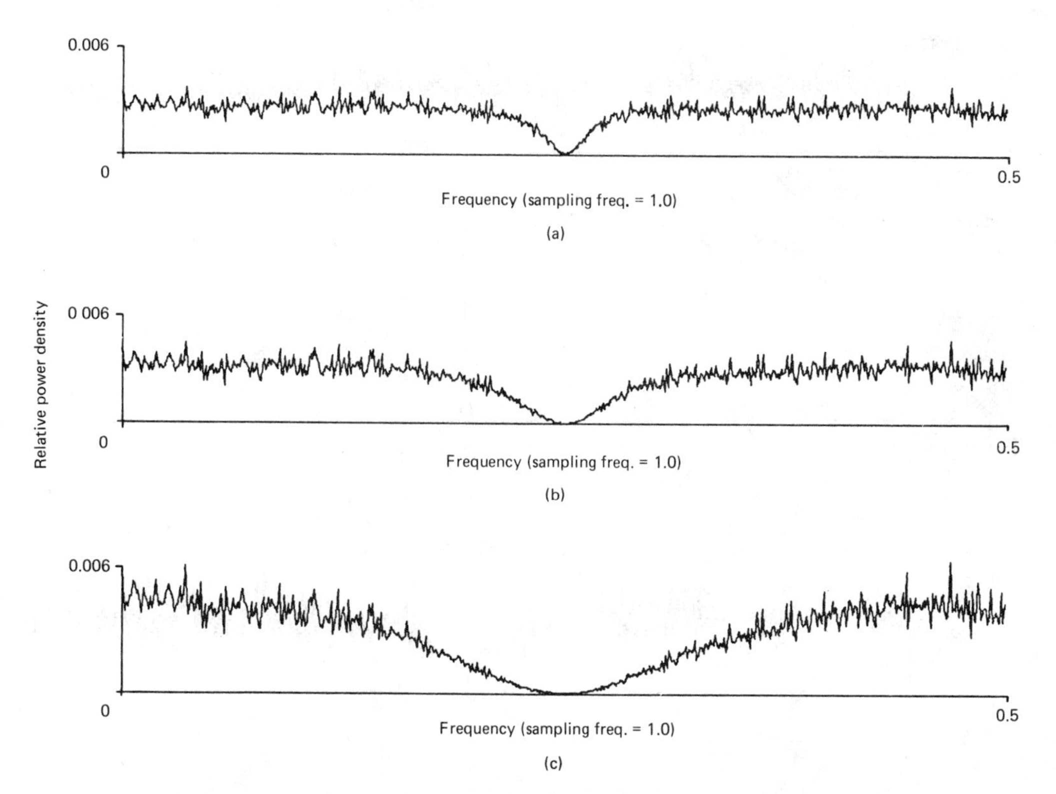

Figure 14.20 Frost beamformer output spectra for white-noise signal with power 1.0 and sinusoidal jammer: (a) jammer power = 12.5; (b) jammer power = 25.0; (c) jammer power = 50.0. From B. Widrow et al., IEEE Trans. on Ant. and Propagat., © May 1982.

level, the signal cancellation notch has the smallest bandwidth, as indicated in Figure 14.20(a). As the jammer power is increased with all other parameters constant, the notch width increases and Figure 14.20(c) shows the widest notch for the strongest jammer signal that was applied. In all three cases, (12.60) for the notch width is approximately correct.

Figure 14.21 shows the results of another experiment with the Frost beamformer. Here the target signal was white and the jammer was strong band-limited noise. Signal components were partially canceled over the jammer's spectral band and beyond, which of course implies extensive signal distortion. Results of this type occur only in cases of rapid adaptation. For the experiment of Figure 14.21, the time constant of the adaptive process was approximately equal to 20 time steps (i.e., five jammer cycles). The bandwidth of the jammer was approximately equal to 5 percent of its center frequency.

One way to prevent signal cancellation in adaptive beamformers is to isolate the target signal from the adaptive processor during adaptation. A method devised by K. Duvall [9, 18] accomplishes this type of isolation for the Frost beamformer. Another method, applicable to all of the adaptive beamformers described above, is

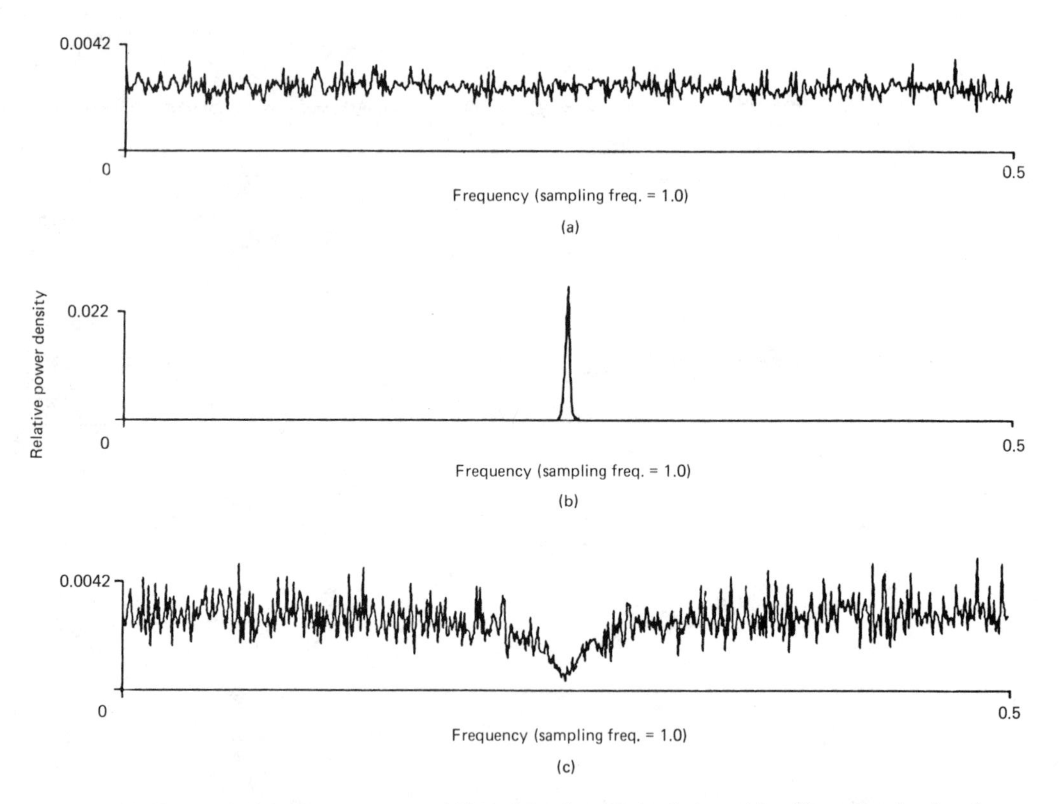

Figure 14.21 Power spectra of Frost beamformer simulation with white-noise signal and jammer: (a) input signal spectrum; (b) jammer spectrum; (c) beamformer output spectrum. From B. Widrow et al., IEEE Trans. on Ant. and Propagat., © May 1982.

based on frequency-hop spread-spectrum techniques and is described in the next section. Here we will describe Duvall's beamformer, which is based on the use of two signal-processing systems, one to perform the adaptation and the other to generate the system output signal. A block diagram is shown in Figure 14.22.

As we have seen, signal cancellation effects are due to an interaction between the signal and the jammer in the adaptive beamformer. Since this interaction is the root of the problem, it is useful to consider beamformer structures that somehow separate the signal from the jammer. Duvall's approach is based on this rationale. The system in Figure 14.22 has the usual linear array driving two beamformers. The upper beamformer is connected directly to the antenna elements and produces the useful array output signal. It is, however, a slaved beamformer rather than the adaptive beamformer that would usually be expected in this position. The lower beamformer is a Frost adaptive beamformer connected to the antenna elements through a subtractive preprocessor as in Figure 14.4. The preprocessor excludes the look-direction signal from the beamformer and admits off-axis jammer signals in a modified form. This adaptive beamformer generates a set of weights that provides a specified look-direction gain (determined by the Frost constraints) while minimizing

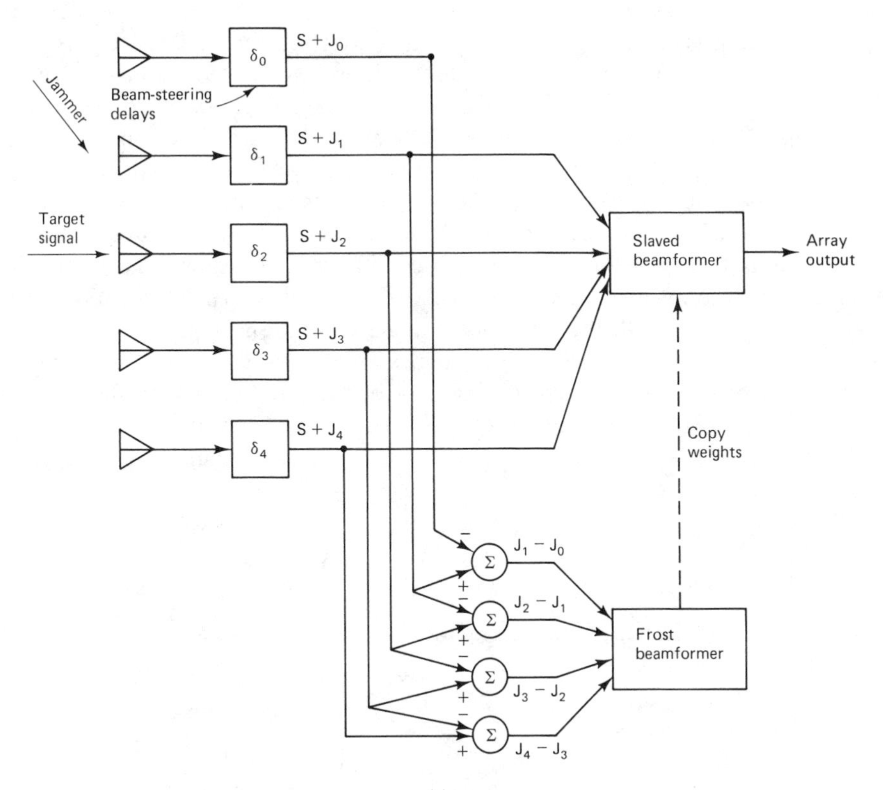

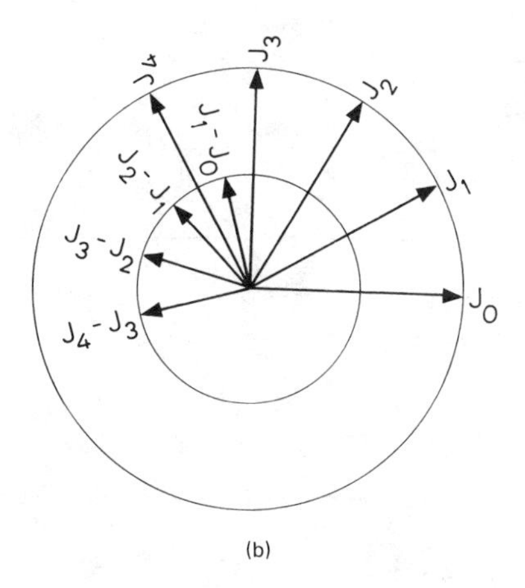

Figure 14.22 Duvall beamformer for eliminating signal cancellation: (a) structure; (b) phasor diagram showing jammer inputs to beamformers. From B. Widrow et al., IEEE Trans. on Ant. and Propagat., © May 1982.

(in the least-squares sense) the jammer contribution. These weights are copied into the slaved beamformer to provide the desired signal reception and jammer rejection.

The basis for the Duvall approach is the relationship between the jammer signals in the two beamformers. The phasor diagram shown in Figure 14.22(b) helps to clarify this relationship. For simplicity, assume just a single jammer. The jammer components received by the antenna elements are indicated by a set of equal-amplitude, uniformly spaced phasors J_0, J_1, J_2, J_3, and J_4. The uniform angular spacing results from the linear array with equally spaced omnis. The phasor inputs to the Frost beamformer are J_1–J_0, J_2–J_1, J_3–J_2, and J_4–J_3. These, too, are equal-amplitude phasors with the same phase-angle separations as the received jammer components. Since the relative phase angles are the same for the jammer components in both beamformers, correct alignment of the jammer null in the Frost beamformer assures correct alignment of the null in the slaved beamformer. Copying the weights will cause the slaved processor to have a main beam (resulting from the Frost constraints) in the direction of the target signal, and to have a null in the direction of

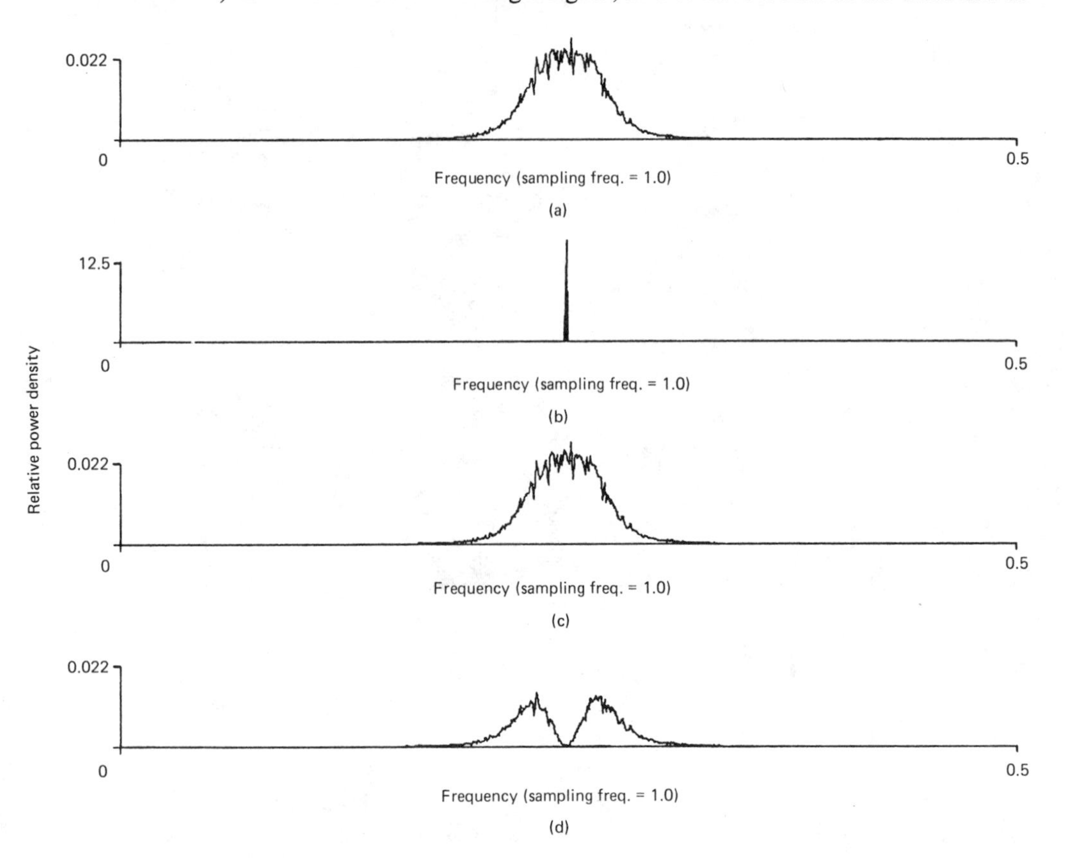

Figure 14.23 Comparison of Frost and Duvall beamformers: (a) input signal spectrum; (b) jammer spectrum; (c) Duvall beamformer output spectrum; (d) Frost beamformer output spectrum. From B. Widrow et al., IEEE Trans. on Ant. and Propagat., © May 1982.

the jammer. Although the phasor argument applies only to one jammer at one frequency, the principles of linearity and superposition can be used to show that the idea is applicable to multiple jammers and to broadband as well as narrowband jammers.

The beamformer block diagram shown in Figure 14.22(a) shows steering delays for broadband processes. Phase shifters could be used behind each array element for narrowband processes. The Frost algorithm can be applied as shown, or any other constrained adaptive algorithm could be used. The pilot-signal algorithm, for example, is easily applied in this system. Generalization of the subtractive preprocessor is also possible, and Jim has described a class of spatial filters [10] that offer greater flexibility than that shown in Figure 14.22(a).

Some experimental results with the Duvall beamformer are shown in Figures 14.23 and 14.24. Figure 14.23 compares the output spectrum of the Frost beamformer with that of a Frost-based Duvall beamformer, both adapting with a time constant

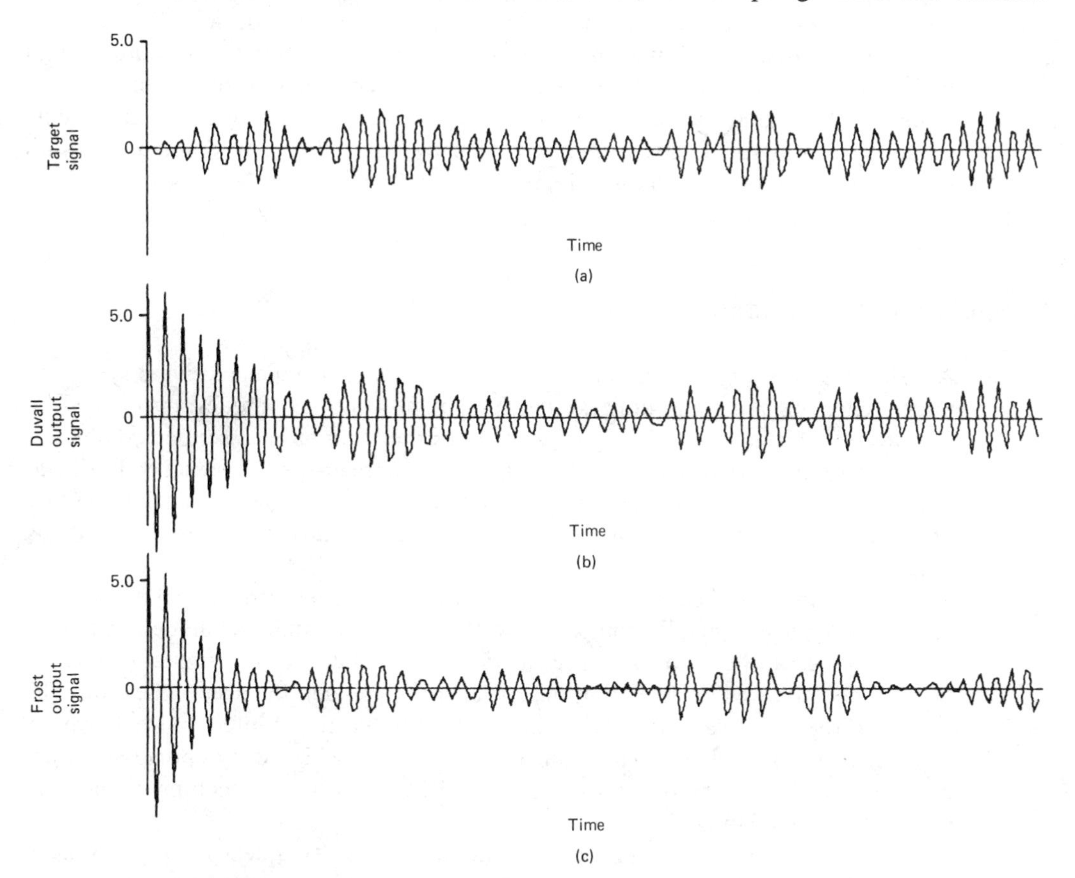

Figure 14.24 Comparison of time-domain outputs of the Frost and Duvall beamformers: (a) target signal input; (b) Duvall beamformer output; (c) Frost beamformer output. From B. Widrow et al., IEEE Trans. on Ant. and Propagat., © May 1982.

of approximately 20 samples. In both cases here, the array signal and jammer are the same as in the experiment of Figure 14.19. In Figure 14.23(d) the Frost beamformer shows evidence of strong signal cancellation, while the Duvall beamformer in Figure 14.23(c) shows no evidence of signal cancellation. In the time domain, Figure 14.24 compares the look-direction target signal with the output signals of the Frost and Duvall beamformers. In both cases the weights were initialized to zero, and adaptive transients are visible at the beginnings of the output tracings. Beyond the region where the transients exist, the signal distortion power in the Frost beamformer output was measured to be 6 dB below the input signal power. At the output of the Duvall beamformer, the distortion was measured to be 110 dB below the input signal power level.

The Duvall beamformer appears to be an important step toward the mitigation of the effects of signal cancellation. The concept is new, however, and the limitations on its performance have not yet been fully assessed. For example, the effects of component inaccuracies and array imperfections are not yet understood. Alternative techniques for steering nulls in the directivity pattern need to be studied, and performance measures need to be improved. Other methods for eliminating or reducing the signal cancellation effects of the adaptive process are also being pursued, such as spread-spectrum and frequency-hop techniques. Some results with these new techniques are discussed next.

FREQUENCY-HOP SPREAD-SPECTRUM TECHNIQUES

Another method for reducing the sensitivity of a radio or radar receiving system to interference or jamming utilizes the spread-spectrum techniques introduced in Chapter 9. This is a large family of techniques that cannot be described completely here, but we have seen that when data signals are spread-spectrum encoded prior to transmission, their bandwidth is increased greatly. The likelihood that interference or jamming will interrupt correct data decoding at the receiving end is thereby reduced.

Since adaptive beamforming can be done with wideband as well as narrowband signals, it is generally compatible with spread-spectrum techniques. Adaptive beamforming can be used to reduce the deleterious effects of directional interferers or jammers, and we have seen that the stronger the jammer, the deeper the null. The output of the adaptive beamformer may not be totally free of interference, however, and in such cases reliable data transmission can be achieved by spread-spectrum encoding of the data before transmission, and by appropriate decoding of the data received by the adaptive beamformer.

One of the major spread-spectrum methods is the "frequency-hop" approach, which is based on the following idea. Digital data can be encoded by dedicating a fixed period, called a time "chip," for the transmission of each data bit. Within this time chip, a sine wave at a specified carrier frequency is generated to represent a data bit, a "zero" corresponding to a specified phase and a "one" corresponding to a phase differing by 180°. This amounts to "phase-shift keying." To provide frequency hopping, the frequency of the sine wave carrier is changed from one chip to the next, hopping in a random fashion according to a random code known at both the transmitter and the receiver. The chip length is such that many cycles of the carrier are generated, with more cycles being generated for the higher frequencies. Correlation techniques are used at the receiving location to determine the phase of the received sine wave for each chip. With long chips (i.e., with a low binary data rate), the phases can be determined reliably in the presence of severe noise and in this way the original binary data can be recovered. Typical signals have carrier frequencies around 1 GHz, spread $\pm 5\%$ by frequency hopping. Typical hopping rates are from 100 to 1000 hops per second.

The frequency-hop spread-spectrum technique is of special interest because, using this technique, adaptive algorithms can be designed to greatly reduce or eliminate the signal cancellation phenomena that we have seen previously. The basic goal in these algorithms is to remove all target signal components from the adaptive process, yet make sure that the target signal appears in proper form at the array output.

Figure 14.25 shows a Frost adaptive beamformer equipped to work with frequency-hop signals. A system of this type having a four-element linear array was simulated. The antenna signals were fed to a bank of identical "frequency-hop notch filters" with flat frequency responses and linear phases except for frequency response notches at any specified single frequency. The frequency of these notches is changed with electronic switching and, at any instant, the notch frequency is chosen to correspond to the incoming target signal frequency. The target signal frequency varies from chip to chip in accord with a known random sequence. Synchronized by a real-time clock, a frequency code generator at the receiver reproduces this sequence and relays the signal frequency setting to the notch filters and to a local sine-wave generator. Much of the necessary detail has been omitted for clarity of presentation in Figure 14.25. The method is applicable to arrays of general geometry as well as linear arrays.

The frequency-hop notch filters keep signal components out of the adaptive process. Only interference can be present at the inputs to the Frost processor, in which the main beam constraints are sustained, as usual, and the jammers are nulled. The output signal of the Frost processor is not useful (aside from the implementation of the adaptive algorithm) since it does not contain the target signal. The weights of the adaptive processor are copied into a slaved processor having the same structure but no ability to adapt on its own. The inputs to the slaved processor come directly from the antenna elements without going through the frequency-hop notch filters, so the output of the slaved processor reproduces any target signal arriving from the look direction. In a computer simulation, the target signal from the look direction experienced unit gain and linear phase shift in accord with the Frost constraint that was set. When a bandpass jammer was simulated at 45° away from the look direction, it was deeply nulled.

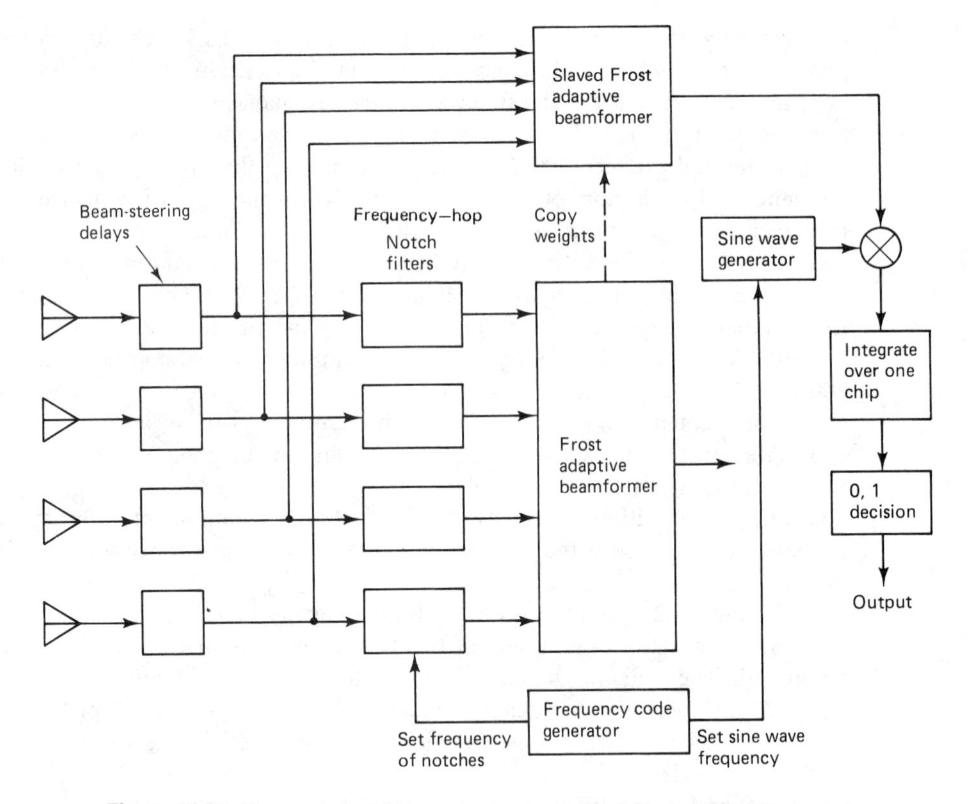

Figure 14.25 System designed to receive frequency-hop signals in the presence of noise.

The output of the slaved processor in Figure 14.25 is correlated with sine waves generated locally, with appropriate control of phase. The sine-wave frequency is selected by the frequency code generator as illustrated in the figure, and correlation is accomplished by multiplication followed by integration over each successive time chip. The integral of the product builds up during the chip interval as either a positive or a negative quantity, depending on the phase of the incoming signal, which in turn depends on whether the incoming signal corresponds to a zero or a 1. At the end of the chip interval, the integrator is initialized to zero in preparation for integration over the next chip interval.

Figure 14.26 shows results of two simulations of the integrator. Figure 14.26(a) shows the integrator output of a system with a slaved beamformer like that of Figure 14.25. One can see that the transmitted data sequence is 0, 1, 0, 0, 1, 1, then nothing. The correlations build up quite uniformly during the integration intervals, and almost no output results when no target signal is received. Figure 14.26(b) shows the output of a conventional Frost adaptive beamformer (no notch filters or slaved beamformer) with the same frequency-hop input signal. It is clear that signal cancellation in the Frost process is present, and that in some cases the signal

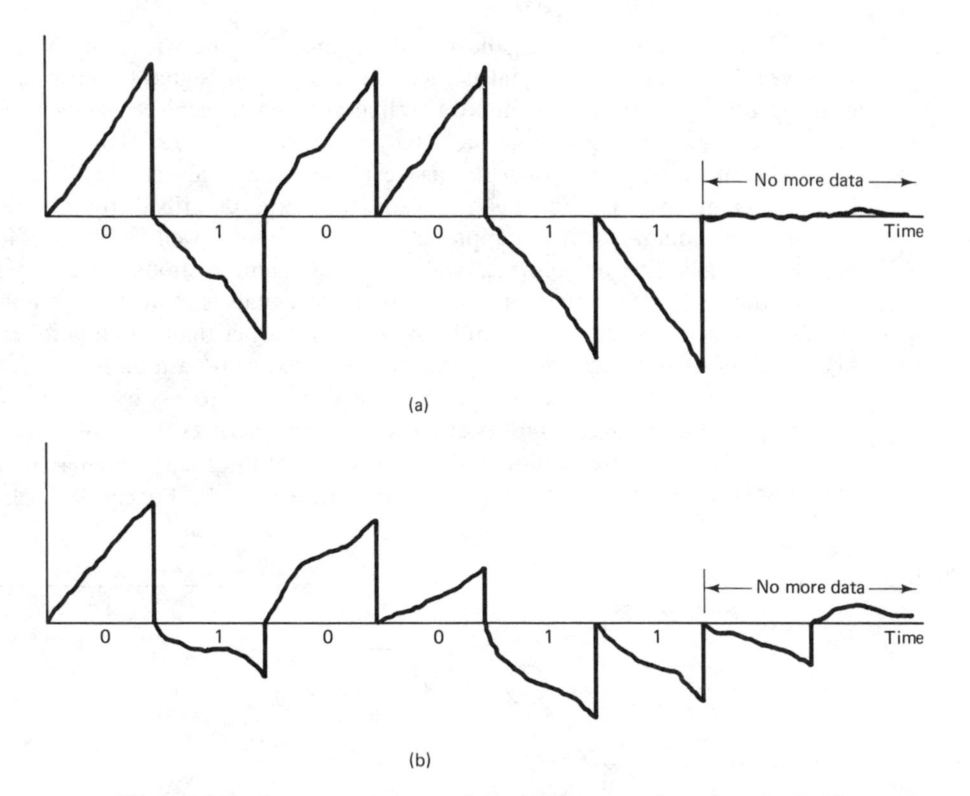

Figure 14.26 Integrator output waveforms: (a) frequency-hop adaptation in Figure 14.25; (b) with a conventional Frost beamformer.

distortion is severe enough to cause marginal data decoding. This example was developed and analyzed by Y. L. Su [19].

Thus the effectiveness of the scheme of Figure 14.25 is clearly evident. Signal cancellation is prevented through the use of frequency-hop signaling, by using frequency-hop notch filters to prevent the target signal components from entering the adaptive processor and influencing the weight settings. The weights form as if there is no target signal present, and are copied into a slave processor in order to obtain the target signal output. This same concept can be applied to all forms of adaptive receiving arrays.

BEAMFORMERS WITH SUPERRESOLUTION

The resolution of a conventional nonadaptive antenna or antenna array is limited by the well-known Rayleigh criterion [13]. An approximate formula for the 3-dB beamwidth of such an antenna is

$$\text{3-dB beamwidth} \approx \frac{\lambda}{d} \text{ radians} \tag{14.33}$$

where λ is the wavelength of the radiation and d is the width of the aperture. However, when the target signal is received with a high signal-to-noise ratio, it is possible for a maximum likelihood adaptive antenna to achieve a much narrower "adaptive beamwidth" giving a much sharper target bearing estimation.

The superresolution concept was developed and analyzed by W. F. Gabriel [17]. Different methods for bearing estimation are described by Gabriel. One is the "maximum-entropy" approach, which works well with a Howells–Applebaum adaptive beamformer which has an omnidirectional receiving pattern except where signals are present. The presence of signals is indicated by nulls in the receiving pattern. Since antenna nulls are always sharper than antenna lobes, signal bearings can be obtained more accurately from the beam pattern, and superresolution is the result. However, one can be fooled in this process by having spurious nulls appear with no signal inputs at corresponding bearings.

To analyze superresolution and its effects, we again discuss the maximum-likelihood adaptive beamformer using an analysis developed by Eugene Walach [1] for

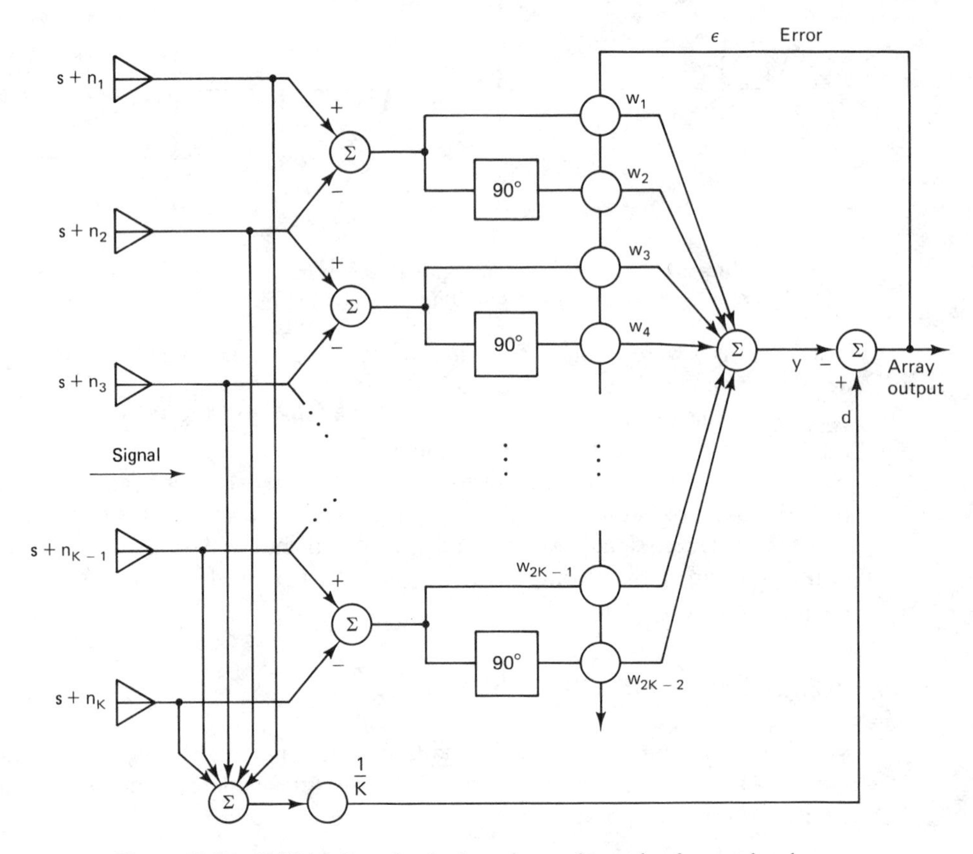

Figure 14.27 Griffiths–Jim adaptive beamformer for study of narrowband superresolution.

the Griffiths–Jim version in Figure 14.27. The receiving array is assumed to be linear and the signal is a single sinusoid given by

$$s = C \cos k\omega_0 \tag{14.34}$$

The receiver noises $n_1, n_2, \ldots, n_K$ are assumed to be white and mutually uncorrelated, each with power equal to σ_r^2. Steering delays have been omitted to simplify the beamwidth calculation, and the direction of arrival of the signal is indicated in the figure as the look direction (i.e., $\theta = 0°$). This is convenient for our discussion here, but of course the direction of arrival of the signal could vary. Maximum signal output results when the signal arrives from the look direction, because the main-lobe constraint gives the array unit gain in this direction. The array gain drops rapidly due to the adaptive cancellation process as the signal direction moves away from the look direction. An off-axis signal tends to be nulled by the adaptive beamformer. This is the basis of maximum-likelihood superresolution.

Let us now examine the 3-dB beamwidth of a maximum-likelihood adaptive beamformer as a function of the array parameters and the signal-to-noise ratio. As before, the spatial response of the beamformer is obtained by varying the direction of arrival of the signal. For each direction, the adaptive process is allowed to converge and the output signal power is measured, giving a point on the directivity pattern. At convergence, the optimal weight vector is given by the least-squares Wiener solution

$$\mathbf{W}^* = \mathbf{R}^{-1}\mathbf{P} \tag{14.35}$$

To determine this solution, we must first find $\mathbf{R}$ and $\mathbf{P}$.

Referring to Figure 14.27, with the signal arriving exactly from the look direction, the inputs to the weights contain no signal components, only receiver noise. Thus the correlations between the noises must be determined to obtain the R-matrix. The input to weight w_2 is the same as the input to w_1, but is delayed by an interval that causes a 90° lag at the signal frequency. Thus the noise components into w_1 and w_2 at the signal frequency are mutually uncorrelated. (We assume for this discussion that the noise power is concentrated at the signal frequency.) The same is not true of the inputs into w_1 and w_3, however. The receiver noise n_2 is present (with sign reversal into w_1) in both of these inputs. Since the power of each noise $n_1, n_2, \ldots, n_K$ is σ_r^2, the correlation of the noise inputs to w_1 and w_3 is σ_r^2. Since the noises are mutually uncorrelated, the power adds going into each weight, and the diagonal elements of the R-matrix are therefore all $2\sigma_r^2$. From these arguments, we can infer the R-matrix to be

$$\mathbf{R} = \sigma_r^2 \begin{bmatrix} 2 & 0 & -1 & 0 & 0 & \cdots \\ 0 & 2 & 0 & -1 & 0 & \\ -1 & 0 & 2 & 0 & -1 & \\ 0 & -1 & 0 & 2 & 0 & \\ 0 & 0 & -1 & 0 & 2 & \\ \vdots & & & & & \ddots \end{bmatrix} \tag{14.36}$$

This matrix pertains to the situation in which the signal arrives exactly from the look direction. It changes only slightly for small changes in the angle of arrival of the signal, because it is dominated by receiver noise. The signal components in the weight inputs remain small for small deviations of the signal arrival angle away from $\theta = 0°$. For our discussion here, we shall assume that **R** is fixed and given by (14.36).

The situation is different for the P-vector in the system of Figure 14.27. We can see that the desired response for the adaptive process is

$$d = s + \frac{n_1 + n_2 + \cdots + n_K}{K} \tag{14.37}$$

This desired response is precise when the signal arrives from the look direction, because then all the signal components received by the array elements are added in phase to produce the desired response. We again assume that the changes in these signal components will be negligible for small changes in the direction of signal arrival away from the look direction. With this assumption, we allow (14.37) to represent the desired response under all conditions of interest (i.e., near $\theta = 0°$).

To obtain the P-vector, we first consider the cross-correlation between the noise components of the weight inputs and the noise components of the desired response. The noise into weight w_1 is $n_1 - n_2$. The cross-correlation of interest is then

$$E\left[(n_1 - n_2)\frac{n_1 + n_2 + \cdots + n_K}{K}\right] = 0 \tag{14.38}$$

Similarly, the corresponding cross-correlations for the rest of the weights will be zero. We may therefore conclude that the receiver noise components have no effect on the P-vector, which must therefore be determined only by the signal components.

When the signal arrives from the look direction, the signal components of the weight inputs are all zero, and therefore the P-vector is zero. When the signal arrives close to the look direction, small signal components appear at the weight inputs and the P-vector is no longer zero. Suppose that the signal arrival direction is such that the signal into array element 1 leads the signal into element 2 by ψ radians. Referring to Figure 14.27, one can see that for small values of ψ the signal into w_1 is approximately

$$\begin{aligned} \text{signal to } w_1 &= C\cos\left(k\omega_0 + \frac{\psi}{2}\right) - C\cos\left(k\omega_0 - \frac{\psi}{2}\right) \\ &= 2C\left(\sin\frac{\psi}{2}\right)\cos(k\omega_0 + 90°) \\ &\approx C\psi\cos(k\omega_0 + 90°) \end{aligned} \tag{14.39}$$

This signal component is shifted 90° and thus uncorrelated with the signal component of the desired response which, from (14.37), is

$$s = C \cos k\omega_0 \tag{14.40}$$

Thus the signal components into all of the odd-numbered weights are uncorrelated with the desired response. On the other hand, because of the 90° lags, the signal components into the even-numbered weights will be correlated with the desired response in accord with

$$E\left[C\psi \cos(k\omega_0 + 90^\circ - 90^\circ) C \cos k\omega_0\right] = \frac{C^2\psi}{2} \tag{14.41}$$

Thus the P-vector is given by

$$\mathbf{P} = \frac{1}{2} C^2 \psi \begin{bmatrix} 0 \\ 1 \\ 0 \\ 1 \\ \vdots \\ 0 \\ 1 \end{bmatrix} \tag{14.42}$$

The least-squares solution can now be obtained by substituting (14.36) and (14.42) into (14.35).

We are concerned primarily with the signal component of the system output, ε in Figure 14.27, which appears as the difference between the signal component of the desired response and the sum of the signals through the weights. We have seen that the signal components at the inputs to the odd-numbered weights are all uncorrelated with the desired response. Thus, in the presence of receiver noise, each of the odd weights will converge to zero. On the other hand, the signal components into the even weights are all equal in amplitude and are all in phase with the signal component of the desired response. The output signal is therefore

$$\begin{aligned} \begin{pmatrix} \text{output} \\ \text{signal} \end{pmatrix} &= \begin{pmatrix} \text{signal component} \\ \text{of desired response} \end{pmatrix} - \begin{pmatrix} \text{sum of signals} \\ \text{through even weights} \end{pmatrix} \\ &= C \cos k\omega_0 - C\psi (\cos k\omega_0)\left(w_2^* + w_4^* + \cdots + w_{2K-2}^*\right) \\ &= C \cos k\omega_0 \left[1 - \psi\left(w_2^* + w_4^* + \cdots + w_{2K-2}^*\right)\right] \end{aligned} \tag{14.43}$$

When ψ is zero, the signal appears at the output with full power. Using (14.43), a half-power signal output corresponds to $\psi_{3\text{ dB}}$, given by

$$1 - \psi_{3\text{ dB}}\left(w_2^* + w_4^* + \cdots + w_{2K-2}^*\right) = \frac{1}{\sqrt{2}} = 0.707 \tag{14.44}$$

Using vector notation, the sum of the even weights can be expressed as

$$\begin{aligned}\left(w_2^* + w_4^* + \cdots + w_{2K-2}^*\right) &= [0 \;\; 1 \;\; 0 \;\; 1 \;\; \cdots \;\; 0 \;\; 1]\mathbf{W}^* \\ &= [0 \;\; 1 \;\; 0 \;\; 1 \;\; \cdots \;\; 0 \;\; 1]\mathbf{R}^{-1}\mathbf{P} \\ &= [0 \;\; 1 \;\; 0 \;\; 1 \;\; \cdots \;\; 0 \;\; 1]\mathbf{R}^{-1}\frac{C^2\psi_{3\text{ dB}}}{2}\begin{bmatrix}0\\1\\0\\1\\ \vdots \\0\\1\end{bmatrix}\end{aligned} \tag{14.45}$$

With the R-matrix taken from (14.36), Walach has determined that

$$[0 \;\; 1 \;\; 0 \;\; 1 \;\; \cdots \;\; 0 \;\; 1]\mathbf{R}^{-1}\begin{bmatrix}0\\1\\0\\1\\ \vdots \\0\\1\end{bmatrix} = \frac{K(K^2-1)}{12\sigma_r^2} \tag{14.46}$$

The derivation of this formula, which is exact, is lengthy and is therefore omitted here. Instead, we refer to Walach [1]. Combining (14.46) with (14.45) and (14.44), we obtain

$$\frac{C^2}{2}\psi_{3\text{ dB}}^2\frac{K(K^2-1)}{12\sigma_r^2} = 0.293 \tag{14.47}$$

The signal-to-noise ratio at each array element is defined as in (13.8):

$$\text{SNR} \triangleq \frac{C^2}{2\sigma_r^2} \tag{14.48}$$

With this definition, (14.47) becomes

$$\psi_{3\text{ dB}} = \sqrt{\frac{3.52}{(\text{SNR})K(K^2-1)}} \tag{14.49}$$

In this result we recall that $\psi_{3\text{ dB}}$ is the signal phase shift between two neighboring array elements required to obtain a 3-dB drop in output signal power, and K is the number of antenna elements in the array in Figure 14.27.

To find the beamwidth of the maximum-likelihood beamformer, we must now translate the signal phase shift, $\psi_{3\text{ dB}}$, into an arrival angle. The diagram in Figure 13.2 is helpful for this purpose. Again, the spacing between neighboring antenna elements is l, the angle of signal arrival in this case is $\theta_{3\text{ dB}}$, and the wavelength is λ.

Then, similar to (13.2), the signal phase shift is given by

$$\psi_{3\text{ dB}} = \frac{2\pi l \sin(\theta_{3\text{ dB}})}{\lambda} \tag{14.50}$$

We are assuming small angles where $\sin\theta \approx \theta$, so (14.50) therefore yields

$$\theta_{3\text{ dB}} \approx \frac{\lambda \psi_{3\text{ dB}}}{2\pi l} \tag{14.51}$$

Referring to Figure 14.27, the total aperture width d is seen to be

$$d = l(K-1) \tag{14.52}$$

Substituting this into (14.51) yields

$$\theta_{3\text{ dB}} \approx \frac{\lambda}{d}\frac{K-1}{2\pi}\psi_{3\text{ dB}} \tag{14.53}$$

The adaptive antenna beamwidth is twice $\theta_{3\text{ dB}}$, and thus our final result is

$$\begin{aligned}\begin{pmatrix}\text{adaptive}\\ \text{antenna}\\ \text{beamwidth}\end{pmatrix} &= \frac{\lambda}{d}\frac{K-1}{\pi}\psi_{3\text{ dB}} \\ &\approx \frac{\lambda}{d}\frac{K-1}{\pi}\sqrt{\frac{3.52}{(\text{SNR})K(K^2-1)}} \quad \text{radians}\end{aligned} \tag{14.54}$$

The basic assumptions in this derivation are that the phase difference at the signal frequency between neighboring antenna elements is a small angle, and that the beamwidth itself, given by (14.54), is a small angle.

As stated in connection with (14.36), the analysis leading to (14.54) also assumed that **R** is invariant for small angular deviations from the look direction. Taking into account second-order variations of **R** as a function of θ, Walach has determined the following more precise estimation of the 3-dB beamwidth:

$$\begin{pmatrix}\text{adaptive}\\ \text{antenna}\\ \text{beamwidth}\end{pmatrix} \approx \frac{\lambda}{d}\frac{K-1}{\pi}\sqrt{\frac{5}{(\text{SNR})K(K^2-1)}} \tag{14.55}$$

It is interesting to compare (14.55) with the corresponding formula for the conventional nonadaptive beamformer given by (14.33) (i.e., λ/d). They differ by the factor

$$\frac{K-1}{\pi}\sqrt{\frac{5}{(\text{SNR})K(K^2-1)}} \tag{14.56}$$

Whenever this factor is less than 1, superresolution is the result. Under the assumptions above, this factor is generally smaller than 1. As an example, let $K = 10$ antenna elements and let the SNR $= 1$. This factor then equals 0.204. The angular resolution of the adaptive beamformer is therefore close to five times sharper than that of the conventional time-delay-and-sum beamformer. If the SNR $= 10$ with

$K = 10$, the adaptive resolution is more than 15 times better than that of the conventional nonadaptive beamformer. If the SNR = 1 with $K = 20$, the adaptive resolution is almost seven times better than that of the nonadaptive beamformer. The adaptive and nonadaptive beampatterns are shown for this case in Figure 14.28 with a linear array of 20 elements spaced a half wavelength apart.

To apply superresolution in a practical situation, some experiments were performed using data recorded while a helicopter was flying within range of a sensor array. The sensors were low-frequency microphones at the earth's surface. A five-element circular array 30 m in diameter was used and acoustic emanations from the

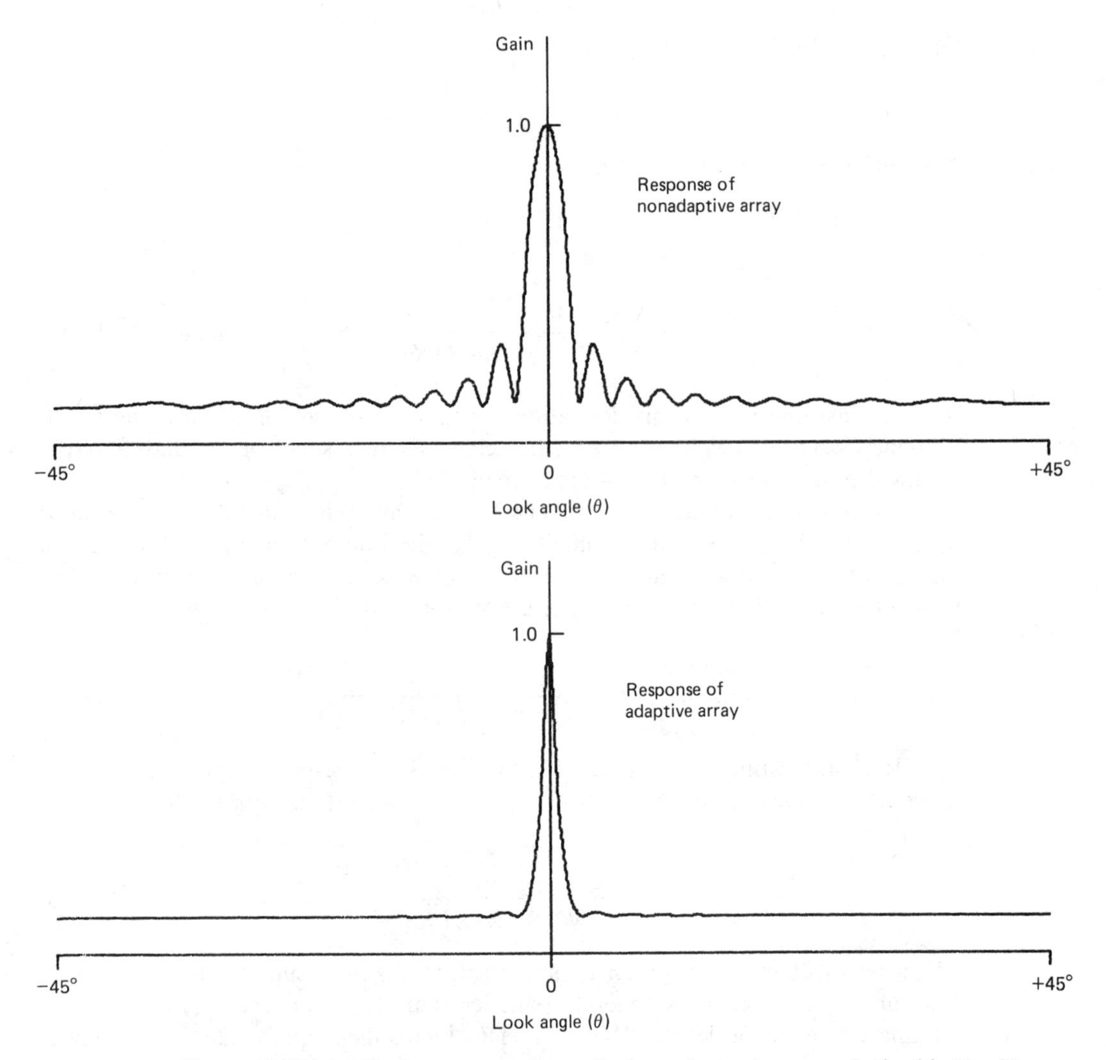

Figure 14.28 Amplitude responses of nonadaptive and adaptive arrays, both with $K = 20$ elements, SNR = 1, and spacing $l = \lambda/2$.

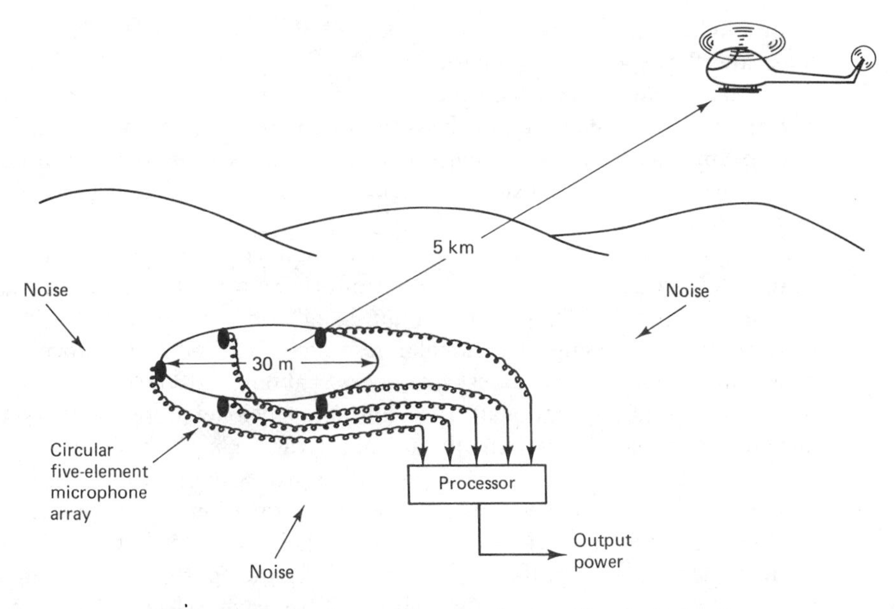

Figure 14.29 Passive receiving array for target bearing estimation.

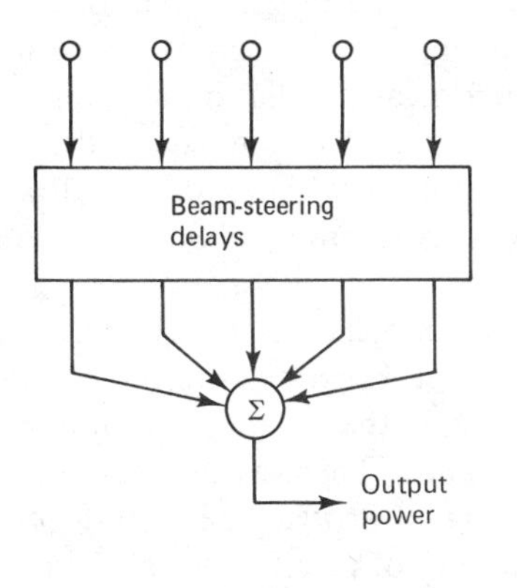

Steer beamformer through all angles.

Determine output power at each angle.

(a)

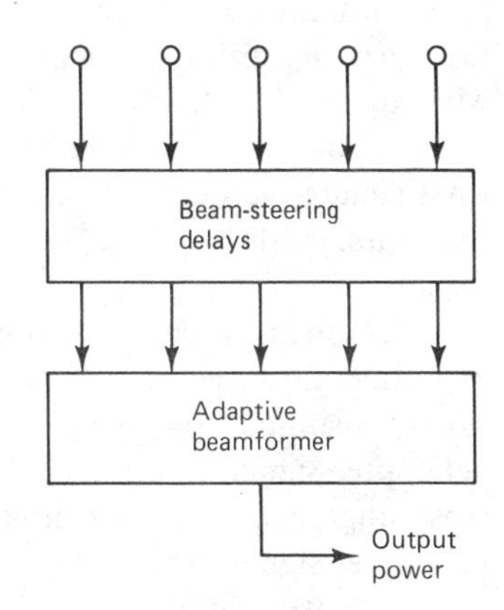

Steer beamformer through all angles.

Adapt beamformer to minimize output power at each angle subject to look-direction constraint.

Determine output power after convergence at each angle.

(b)

Figure 14.30 Conventional and adaptive beamformers for target bearing estimation: (a) conventional beamformer (resolution limited by array aperture); (b) adaptive beamformer (allows superresolution capability).

rotor blades were received with a good SNR at a range of 5 km. Figure 14.29 illustrates the experimental situation.

The conventional time-delay-and-sum beamformer and the Frost adaptive beamformer were compared in this situation, with the same steering delays used in both beamformers. With the helicopter in a fixed location, the output signal power was measured for selected steering angles covering 360°. Figure 14.30 illustrates the method of comparison.

Figure 14.31 compares the results obtained with conventional processing against those resulting with adaptive beamforming using the experimental data, and we can see that the adaptive maximum-likelihood scheme was clearly superior to conventional processing. The angular resolution with adaptive processing was approximately $\pm 2°$ and the largest sidelobe was about 25 dB below the response of the main lobe. With conventional processing, the sidelobes were so strong that it was almost impossible to determine the signal arrival angle. At best, the angular resolution was $\pm 20°$. The advantages of adaptive processing and the significance of the superresolution phenomenon in target bearing estimation are clear in this example.

To summarize superresolution, let us again consider how the Frost adaptive beamformer was used in this experiment. Using the appropriate steering delays, the look direction was scanned through 360°. In each selected look direction, the sensitivity of the receiving array in Figure 14.27 to incident radiation was constrained to have unit gain with linear phase shift. At the same time, the adaptive process minimized the total output power. The target signal, defined as any radiation arriving from the look direction, passed through the adaptive beamformer without distortion. Interference, defined as any radiation arriving from a direction other than the look direction, was eliminated or nulled as best possible in the least-squares sense. The beamformer output was the undistorted target signal with minimum additive noise, that is, the maximum-likelihood estimate of the target signal.

The helicopter signal originated from a fixed direction of about 137° in this experiment, as seen in Figure 14.31. As the Frost beamformer's look direction was slowly scanned toward this signal angle, the adaptive beamformer nulled the helicopter signal, treating it like interference. Then, as the look direction came into close alignment with the direction of signal arrival, the beamformer accepted the helicopter signal as the target signal instead of interference and reproduced it at the output. Further scanning of the look direction away from the helicopter signal angle again caused nulling. The result was small response to the helicopter signal unless the look direction was very close to the true angle of arrival. This type of performance is the basis of superresolution.

In other experiments, in the presence of multiple signals, the Frost beamformer tended to null all signals except for the signal arriving from the look direction. Not only did the adaptive beamformer provide sharp angular resolution, but it allowed one to "tune in" spatially on a signal from a selected look direction while adaptively "tuning out" other signals, interferences, jammers, and so on, from all other directions.

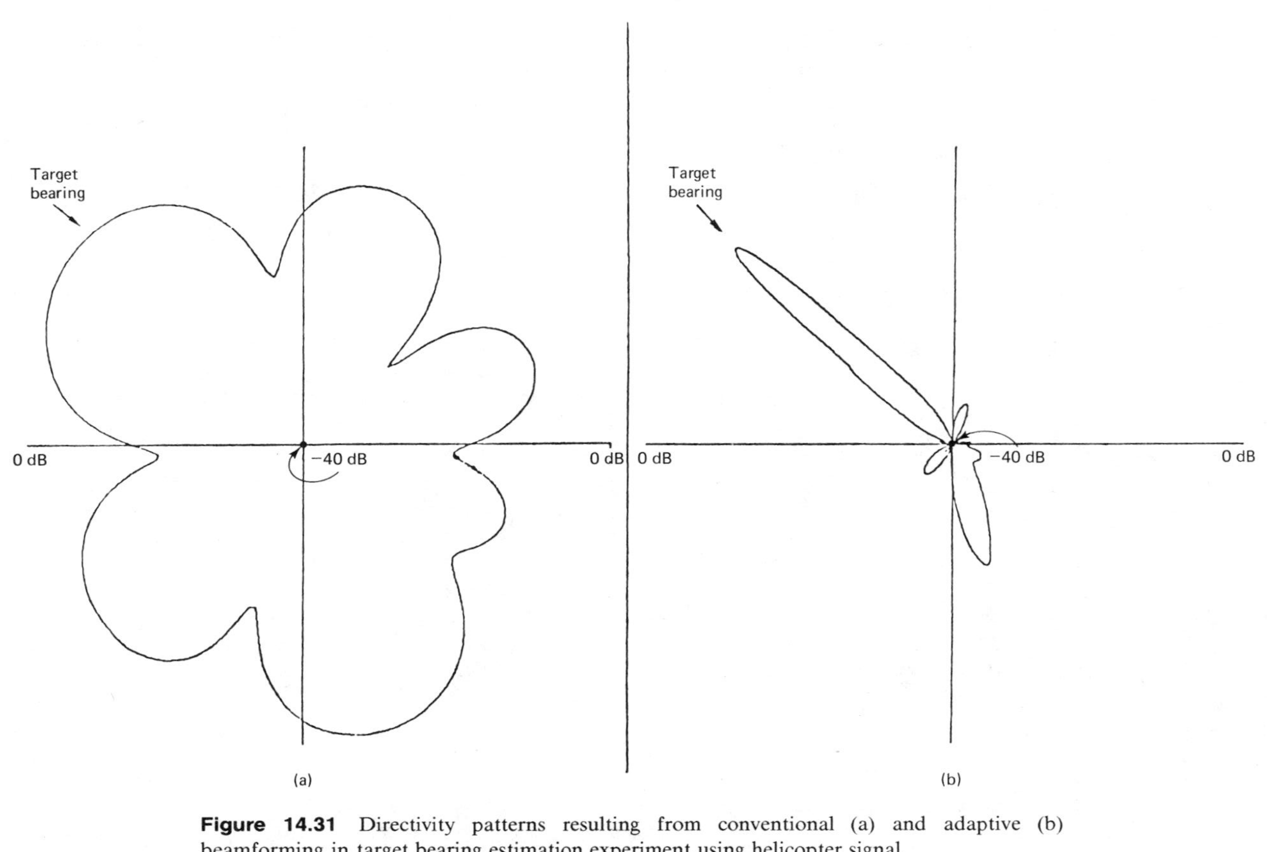

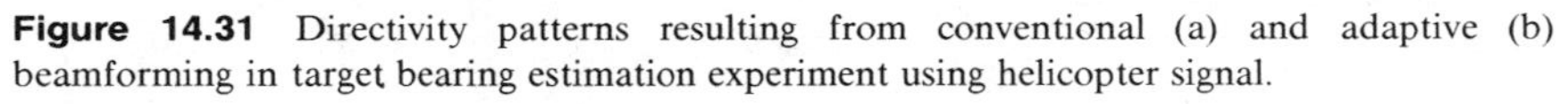

Figure 14.31 Directivity patterns resulting from conventional (a) and adaptive (b) beamforming in target bearing estimation experiment using helicopter signal.

EXERCISES

1. In the Griffiths algorithm (14.16), derive the expression given in Equation (14.20) for the acceptable range of μ.
2. Find the time constant of weight convergence for the Griffiths algorithm (14.16).
3. Find the minimum mean-square error at convergence for the Griffiths algorithm (14.16).
4. Devise a "leaky" version of the Griffiths algorithm, similar to Equation (13.27). What is its converged solution, $\mathbf{W}^*$? Is the introduction of leakage equivalent to the addition of input noise as in Chapter 13?
5. Write a version of the Griffiths algorithm (14.16) for the array shown in Figure 13.13, so that the look direction is:
   (a) At $\theta = 0°$.
   (b) At $\theta = 90°$.
   (c) At $\theta = 30°$.
6. Consider the adaptive beamformer shown below. The target signal is sinusoidal with unit power. The jammer is broadband, uncorrelated with the target signal, with unit power at the filter outputs.

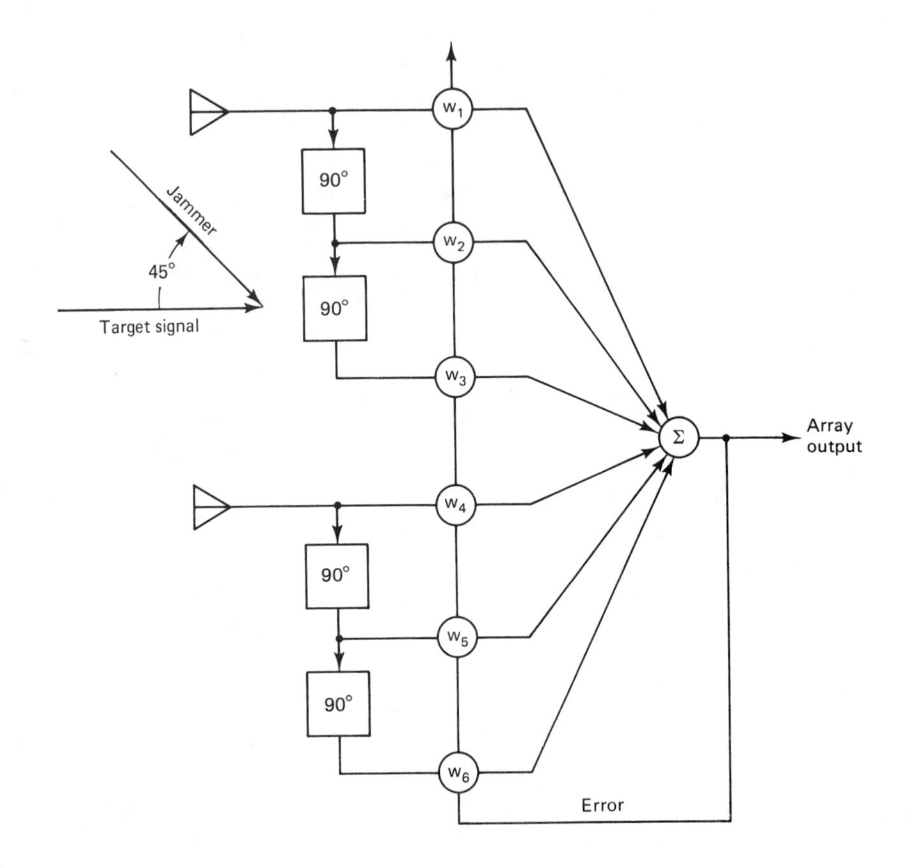

(a) For this beamformer, what is the P-vector needed for the Griffiths algorithm?

(b) Find the R-matrix and the optimum weight vector, $\mathbf{W}^* = \mathbf{R}^{-1}\mathbf{P}$, using the result in part (a).

(c) Find the output signal and noise power using $\mathbf{W}^*$. Compare the input and output SNRs.

7. Write a Frost algorithm, similar to Equation (14.30), for the array in Figure 13.13, with the look direction at $\theta = 0°$, and the gain constrained to one in that direction.

8. The "constrained modeling" situation shown below is representative of a class of systems like the Frost beamformer, in which there is a constraint on the weights. The constrained weights will adapt to model $H(z)$.

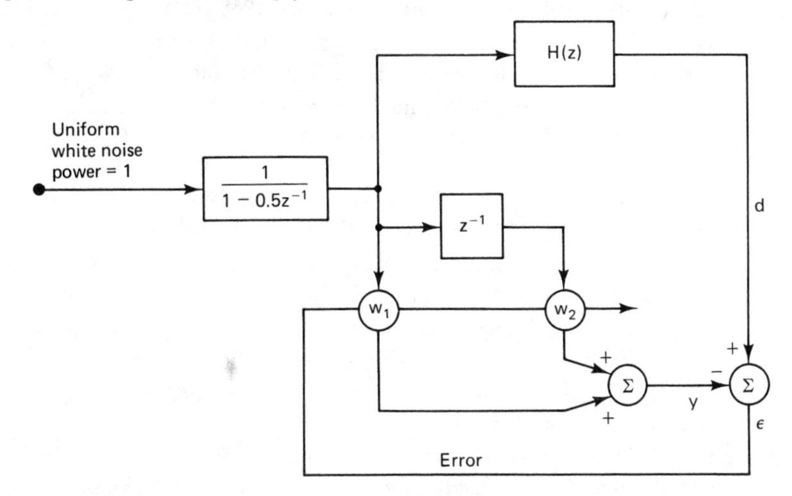

(a) Given $H(z) = 1/(1 - 0.3z^{-1})$, devise an algorithm to adapt w_1 and w_2 to minimize $E[\epsilon^2]$, subject to the constraint $w_1 + w_2 = 1$. Find the constrained optimal solution. Find the minimum MSE.

(b) Do part (a) using the constraint $w_1^2 + w_2^2 = 1$. Test the algorithm to see whether it will converge. Can you find the constrained optimal solution by analytical means?

9. Draw the diagram of a Griffiths–Jim adaptive beamformer similar to Figure 14.4, having four omnis in a circular array, with four tapped delay lines.

10. Draw the block diagram of a four-element pole–zero version of the Griffiths–Jim beamformer, similar to Figure 14.11.

11. Draw a block diagram showing how the Duvall beamforming concept is applied to the system in Exercise 10.

12. A conventional linear array with six equally spaced omnis has a 3-dB beamwidth of 0.1 for a particular target signal having 10 times the power of the receiver noise. What is the beamwidth when this array is connected to a maximum-likelihood adaptive processor constrained to have unit gain in the target direction?

13. Suppose that steering delays are included in Figure 14.26 to produce a look direction of $\theta = 45°$. Derive an expression similar to Equation (14.54) for the 3-dB beamwidth. State any assumptions that you made in the derivation.

REFERENCES AND ADDITIONAL READINGS

1. E. Walach, "On superresolution effects in maximum-likelihood adaptive antenna arrays," *IEEE Trans. Antennas Propag.*, vol. AP-32, no. 3, pp. 259–263, Mar. 1984.
2. L. J. Griffiths, "A comparison of multidimensional Wiener and maximum-likelihood filters for antenna arrays," *Proc. IEEE* (Letters), vol. 55, pp. 2045–2047, Nov. 1967.
3. L. J. Griffiths, "A simple adaptive algorithm for real-time processing in antenna arrays," *Proc. IEEE*, vol. 57, pp. 1696–1704, Oct. 1969.
4. L. J. Griffiths, "Signal extraction using real-time adaptation of a linear multichannel filter," Stanford Univ., Stanford, Calif., Jan. 1968 (Ph.D. dissertation).
5. E. J. Kelley and M. J. Levin, "Signal parameter estimation for seismometer arrays," MIT Lincoln Lab., Lexington, Mass., Tech. Rep. 339, Jan. 8, 1964.
6. J. Capon, R. J. Greenfield, and R. J. Kolker, "Multidimensional maximum-likelihood processing of a large aperture seismic array," *Proc. IEEE*, vol. 55, pp. 192–211, Feb. 1967.
7. O. L. Frost III, "An algorithm for linearly constrained adaptive array processing," *Proc. IEEE*, vol. 60, pp. 926–935, Aug. 1972.
8. O. L. Frost III, "Adaptive least-squares optimization subject to linear equality constraints," Stanford Univ., Stanford, Calif., June 1972 (Ph.D. thesis).
9. B. Widrow et al., "Signal cancellation phenomena in adaptive antennas: causes and cures," *IEEE Trans. Antennas Propag.*, vol. AP-30, p. 469, May 1982.
10. C. W. Jim, "A comparison of two LMS constrained optimal array structures," *Proc. IEEE*, vol. 65, pp. 1730–1731, Dec. 1977.
11. L. J. Griffiths and C. W. Jim, "An alternative approach to linearly constrained adaptive beamforming," *IEEE Trans. Antennas Propag.*, vol. AP-30, pp. 27–34, Jan. 1982.
12. R. P. Gooch, "Adaptive pole-zero array processing," *Proc. 16th Asilomar Conf. Circuits Syst. Comput.*, Nov. 1982.
13. Lord Rayleigh, "On the theory of optical images, with special reference to the microscope," *Phil. Mag.*, vol. 42, no. 5, p. 167, 1896.
14. R. P. Gooch, "Adaptive pole-zero filtering: the equation error approach," Stanford Univ., Stanford, Calif., June 1983 (Ph.D. thesis).
15. J. Glover, "Adaptive noise cancelling of sinusoidal interference," Stanford Electronics Lab., Stanford Univ., Stanford, Calif., Dec. 1975 (Ph.D. dissertation).
16. J. Glover, "Adaptive noise cancelling applied to sinusoidal interferences," *IEEE Trans. Acoust. Speech Signal Process.*, vol. ASSP-25, pp. 484–491, Dec. 1977.
17. W. F. Gabriel, "Spectral analysis and adaptive array superresolution techniques," *Proc. IEEE*, vol. 68, June 1980.
18. K. M. Duvall, "Signal cancellation in adaptive antennas: the phenomenon and a remedy," Stanford Univ., Stanford, Calif., Aug. 1983 (Ph.D. thesis).
19. Y. L. Su, "Cures for signal cancellation in adaptive arrays," Stanford Univ., Stanford, Calif., June 1984 (Ph.D. thesis).

Appendix A

A Portable Random Number Generator

We present here a simple, portable algorithm for generating a pseudorandom number sequence. The algorithm was chosen for use in engineering signal processing applications and has the following properties:

1. It works in any machine that allows integer words up to 32 bits in length, that is, positive integers up to $2^{31} - 1$.
2. It is easily initialized and can be started at any specified point in the sequence.
3. The samples in the sequence and in subsequences of reasonable length are distributed approximately uniformly between zero and one.
4. The samples are approximately independent and the spectrum is approximately flat.
5. The period of the sequence (10^6 samples) is long enough for most engineering applications.

THE ALGORITHM

The algorithm is a member of a class of remaindering algorithms that generate "linear congruential" sequences. The properties of these sequences are described by Knuth.[1] For our special case we generate integers recursively as follows:

$$I_{n+1} = (JI_n + 1) \bmod M \qquad n = 1, 2, \ldots, M - 1 \tag{A.1}$$

The integer sequence is then normalized to produce samples between 0 and 1. The initial value, I_0, is discussed below.

The period of the sequence in (A.1) obviously cannot exceed M. According to Knuth, one way to guarantee a period equal to M is to let

$$J = 4K + 1 \quad \text{and} \quad M = 2^L \tag{A.2}$$

where K and L are integers such that M is greater than J. The largest I_n in (A.1) is $M - 1$, and so in order for (A.1) to produce only factors less than 2^{31} in magnitude in accordance with property 1 above, we must have

$$J(M-1)+1<2^{31}, \quad \text{or}$$

$$\frac{4K+2^{31}}{4K+1}>2^{L} \tag{A.3}$$

Thus we have in (A.3) a relative bound on J and M in (A.1). We can establish a second constraint by noting that if J is small compared with M, then a subsequence beginning with a small value for I_0 in (A.1) will be monotonic until the quotient JI_n+1 exceeds M. For example, the trivial case is where $J=1$ and $I_0=0$. Then the sequence is $0,1,2,\ldots,M-1$. Thus, the sequence tends to be "more random" for larger values of J.

If we choose 10^6 as a minimum period for the sequence in accordance with property 5 above, we have $L=20$, and from (A.3), the largest possible K is 511. Thus, for our sequence in (A.1) we choose

$$M=2^{20}=1048576$$
$$J=4(511)+1=2045 \tag{A.4}$$

If in addition we choose (arbitrarily)

$$I_0=12357 \tag{A.5}$$

then the first 70 integers, I_1 through I_{70}, are as follows:

```
 104242  313963  325824  464321  575166  760775  746668
 209405  414218  874979  461400  893177  976150  786623
 128452  540341  846818  543835  652016  632625  823918
 899255  822748  603757  509114  950739  201352  723049
 143046 1024943  953588  784677  343186  318027  248096
 894113  793118  827815  480012  157405 1028970  800195
 620216  613337  177270  758431  147492  679829  887106
  95291  883536  134673  679374 1005207  439356  901965
  73242  882099  341736  498505  226854  445839  528212
 160261  578034  334379  133504  385921  679294  841607
```

We note that the integers are all between 0 and 1048576, and that at least this short sequence appears to have no undesirable patterns.

Since the usual random number routine produces samples between 0 and 1, we can create the nth random sample as

$$R_n=\frac{(I_n+1)}{(M+1)} \tag{A.6}$$

so that the random samples are all greater than zero and less than 1.

Using (A.1), (A.4), (A.5), and (A.6), we can write a subroutine to generate $I_1, I_2, \ldots, I_n$. With portability a requirement, automatic initialization becomes a problem. The initializing scheme is necessarily machine-dependent. The following version, however, will work in *most* computers:

```
      FUNCTION RANDOM(X)
      IF(INIT.EQ.12357) GO TO 1
      INIT=12357
      I=INIT
    1 I=2045*I+1
      I=I-(I/1048576)*1048576
      RANDOM=FLOAT(I+1)/1048577.0
      RETURN
      END
```

Notice in this routine that the internal variable INIT must not be equal to 12357 at load time and that both INIT and I must remain intact between calls. The initial setting of I = INIT guarantees that the random sequence will always begin as described above. The variable X is a dummy variable and serves no purpose in this version of the algorithm.

If one does not require the automatic initializing feature or wishes to provide variable initialization, the following version may be preferred:

```
      FUNCTION RANDOM(I)
      I=2045*I+1
      I=I-(I/1048576)*1048576
      RANDOM=FLOAT(I+1)/1048577.0
      RETURN
      END
```

In version 2, I is not changed except at initialization, where it is set to an initial value. Thus, for example, the following statements could be used with version 2 to initialize I to 12357 and generate the first two random numbers, say $R1$ and $R2$:

$$K = 12357$$

$$R1 = \text{RANDOM}(K)$$

$$R2 = \text{RANDOM}(K)$$

The argument (K) in the call to RANDOM in version 2 must, of course, not be changed elsewhere in the calling program. The advantages of version 2 over version 1 are (1) we do not need to assume that intermediate variables remain intact between calls and (2) the routines can be initialized arbitrarily, using any integer from 0 to 1048575 in the main program. The disadvantages are that the user must remember to initialize I (or whatever I is called in the main program) properly from 0 to 1048575, and that I must be left alone between subsequent calls.

An example of the use of version 2 is given in the following program, which prints the first 70 random samples corresponding with the above list of integers. Since I is initialized to 12357 in this example, the results are the same as if the first version above were used. The program and results are as follows:

```
      PROGRAM VERS2(INPUT,OUTPUT)
C-PRINT 70 RANDOM NUMBERS.
      DIMENSION R(70)
      K=12357
      DO 1 J=1,70
    1 R(J)=RANDOM(K)
      PRINT 2,R
    2 FORMAT(7F8.6)
      STOP
      END
C
      FUNCTION RANDOM(I)
      I=2045*I+1
      I=I-(I/1048576)*1048576
      RANDOM=FLOAT(I+1)/1048577.0
      RETURN
      END
/LGO
 .099414 .299419 .310731 .442812 .548521 .725532 .712078
 .199705 .395030 .834445 .440026 .851800 .930929 .750182
 .122502 .515310 .807589 .518642 .621811 .603319 .785750
 .857597 .784634 .575788 .485529 .906695 .192025 .689554
 .136420 .977462 .909412 .748327 .327288 .303295 .236604
 .852693 .756376 .789466 .457776 .150114 .981302 .763126
 .591484 .584924 .169059 .723296 .140660 .648336 .846010
 .090877 .842606 .128435 .647902 .958640 .419003 .860181
 .069850 .841235 .325905 .475412 .216346 .425186 .503743
 .152838 .551257 .318889 .127320 .368044 .647826 .802619
```

In the case of version 2, we note again that the period of the random sequence is 1,048,576 samples regardless of the initial value of I.

PROPERTIES OF THE RANDOM SEQUENCE

Some properties of the random sequence are shown in the figures below. First, Figure A.1 shows roughly how the amplitudes of subsequences of length 10,000 are distributed. For each of the 100 successive subsequences of length 10,000, a histogram was plotted. Each histogram has 10 cells in the range from 0.0 to 1.0, and all 100 histograms are plotted together in Figure A.1. The ordinate in Figure A.1 is relative frequency, so that 0.1 is the ideal ordinate value with 10 cells. We can see that there are no large excursions from the ideal value. Figure A.2 is similar except that the sequence length is 512 instead of 10,000. In Figure A.3, we have a plot of the "white noise" waveform obtained by plotting the first 1000 samples of

$$X(N) = \text{RANDOM}(\ ) - 0.5 \tag{A.7}$$

and we note in the figure the randomness of the sequence, and that the mean is approximately zero.

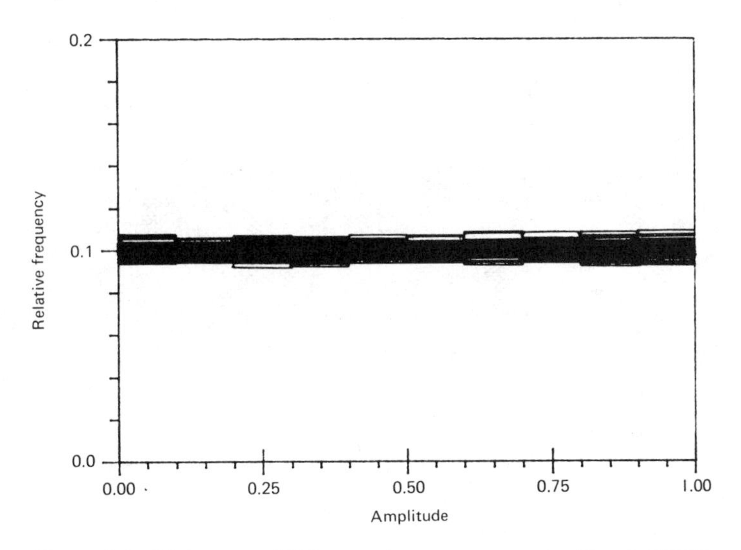

Figure A.1 Histograms of 100 successive sequences, each of length 10,000.

The autocorrelation function of the first 132,000 samples of $X(N)$ in (A.7), given by

$$E = \sum_{N=1}^{132,000} X^2(N)$$

$$R(K) = \frac{1}{E} \sum_{N=1}^{132,000} X(N)X(N+K) \qquad 0 \leq K \leq 100 \tag{A.8}$$

is plotted in Figure A.4. The normalizing factor, E, assures $R(0) = 1$, and we note

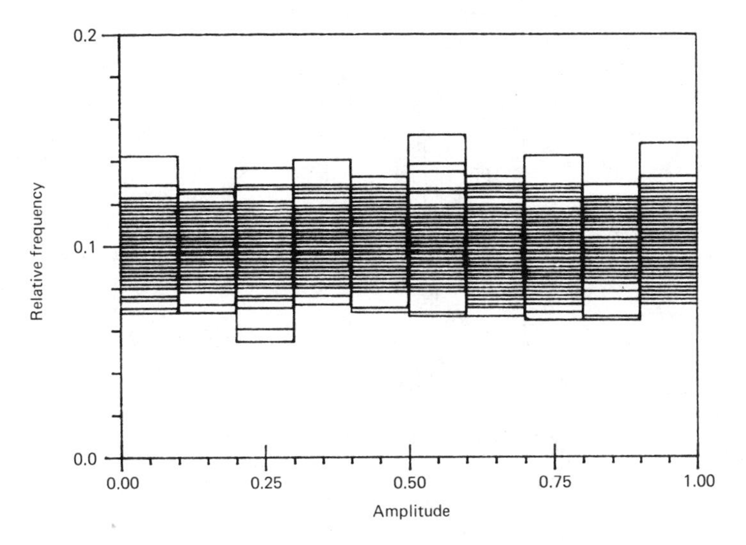

Figure A.2 Histograms of 100 successive sequences, each of length 512.

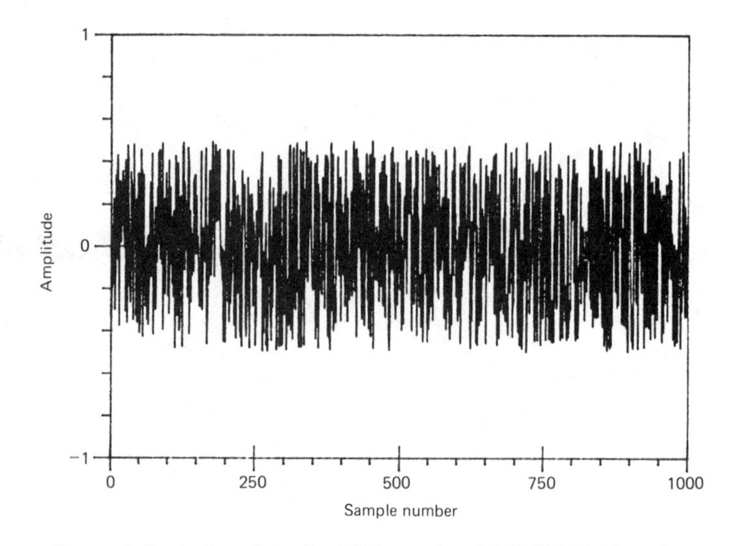

Figure A.3 A plot of the first 1000 samples of RANDOM ( ) − 0.5.

that $R(1)$ through $R(100)$ are close enough to zero to allow us to say that the samples, $[X(N)]$, are uncorrelated, that is, independent. A similar plot for the next 132,000 samples is presented in Figure A.5. Again, $R(K)$ is less than 1% of $R(0)$.

In Figure A.6 we have a power spectrum of the entire random sequence. The power spectrum is computed at 513 points as follows:

$$P(m) = \frac{1}{1024} \sum_{k=0}^{1023} |F_k(m)|^2 \qquad m = 0, 1, \ldots, 512 \tag{A.9}$$

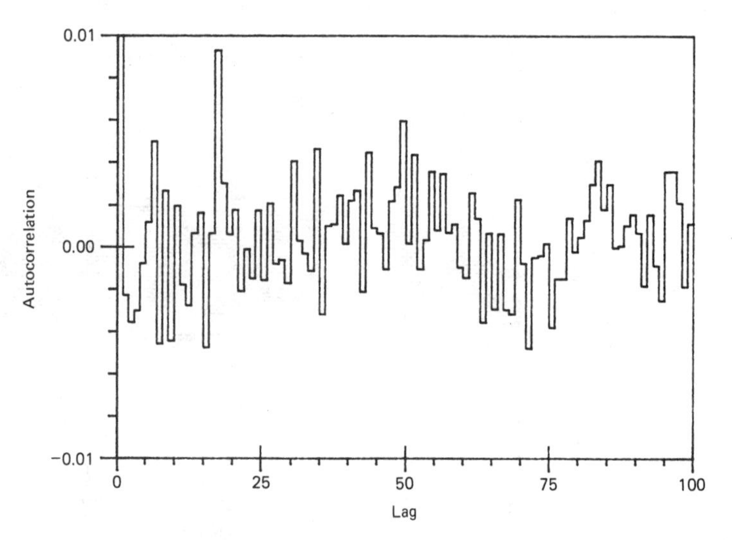

Figure A.4 Autocorrelation function of the first 132,000 samples of RANDOM ( ) − 0.5.

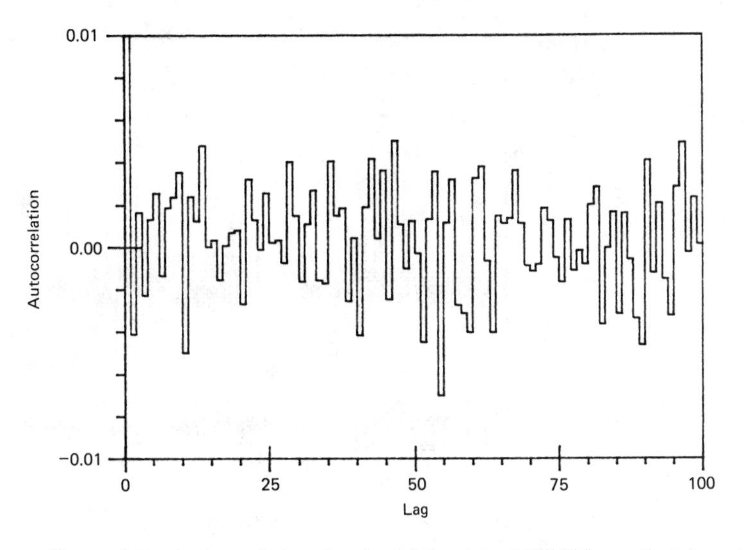

Figure A.5 Autocorrelation function of the second 132,000 samples of RANDOM ( ) − 0.5.

Each term in (A.9) is a periodogram component, that is, the squared magnitude of a discrete Fourier transform (DFT) component. The 1024 DFT's are computed from successive nonoverlapping blocks in the random number sequence in (A.7), $X(0)$ through $X(1048575)$, as follows:

$$F_k(m) = \sum_{n=0}^{1023} X(1024k + n)e^{-j2\pi mn/1024} \tag{A.10}$$

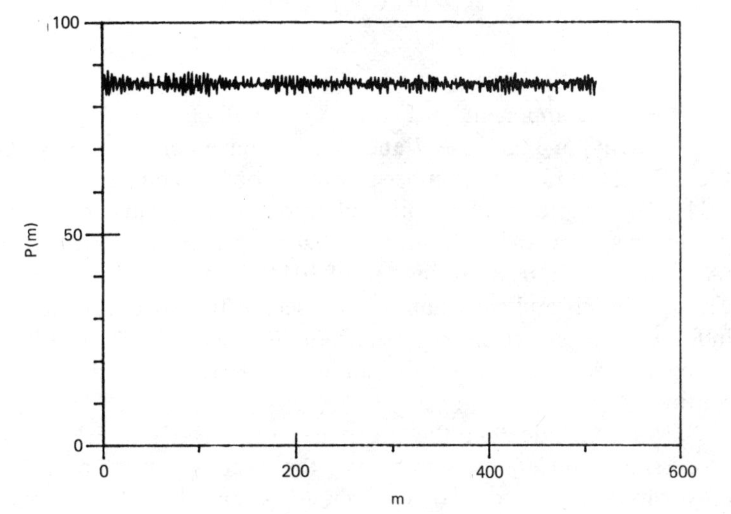

Figure A.6 Power spectrum of the random sequence obtained by averaging the periodograms of 1024 nonoverlapping segments, each of length 1024. The expected value is 85.3.

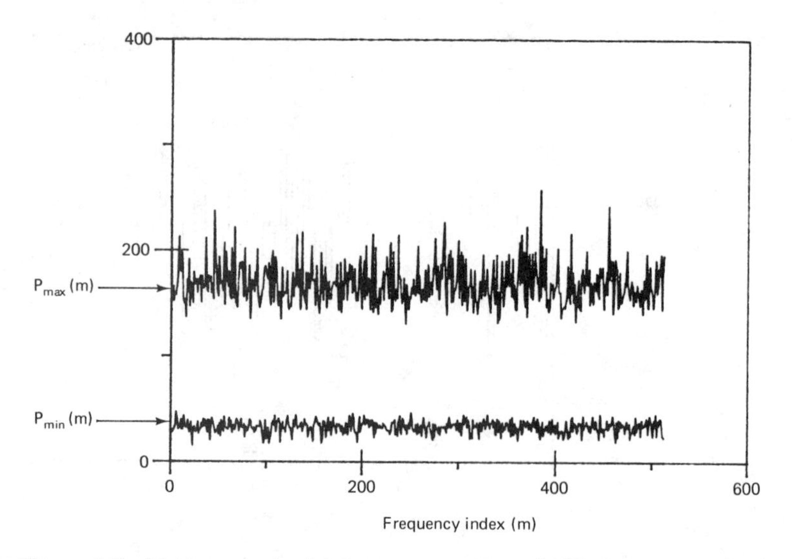

Figure A.7 Maximum and minimum power spectra of 102 sequences, each of length 10,240.

Thus, Figure A.6 is an average periodogram, and we note that the plot is reasonably flat over the complete range of frequencies. Since the variance of a uniform distribution is 1/12 of its range, we can also note that the theoretical spectral value in Figure A.6 is

$$\begin{aligned} E[F_k(m)] &= 1024E[X^2(n)] \\ &= 1024(1/12) = 85.3 \end{aligned} \tag{A.11}$$

The spectral components in Figure A.6 are all close to this expected value.

Finally, in Figures A.7 and A.8, we have compared RANDOM with RANF, the CDC-6600 random number routine. For this comparison a random sequence of 1,044,480 samples was partitioned into 102 segments, each of length 10,240. A power spectrum was computed for each segment as described in connection with (A.9) and (A.10), using 10 instead of 1024 nonoverlapping blocks for each sequence. The maximum and minimum of the set of 102 spectral values at each frequency index is then plotted in Figure A.7 for RANDOM, and in Figure A.8 for RANF. Using this comparison, we do not notice a significant difference between the two routines.

We conclude that the properties of RANDOM are satisfactory for signal processing applications. Other than the relatively short period (10^6 samples), the only known shortcoming of RANDOM is the relatively long execution time of the FORTRAN versions given above, which may be regarded as the price of portability. For example, a DO loop that generates 10^5 random samples requires an average of 2.55 s using version 2 of RANDOM on the CDC-6600, vs only 0.47 s for the library

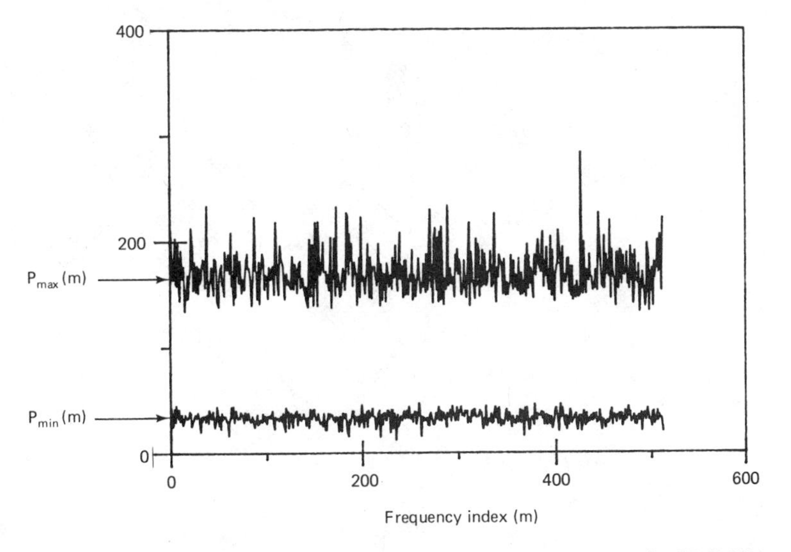

Figure A.8 Same as Figure A.7 with RANDOM replaced by RANF, the CDC-6600 random number routine.

routine, RANF, on the same machine. The longer time could, of course, be reduced by using machine language code.

REFERENCES AND ADDITIONAL READINGS

1. D. E. Knuth, *The Art of Computer Programming, Seminumerical Algorithms*, Vol. 2. Reading, Mass.: Addison–Wesley, 1969, Sec. 3.2.
2. L. Schrage, "A More Portable Fortran Random Number Generator," *ACM Trans. on Math Software*, pp. 132–138, June 1979.
3. S. D. Stearns, *A Portable Random Number Generator for Use in Signal Processing*, Sandia National Laboratories SAND 81–1933, October 1981.

Index

A

B

C

D

E

F

G

H

I

L

M

N

O

P

Q

R

S

T

U

V

W

Z